AIR WAR: VIETNAM PLANS AND OPERATIONS 1961 - 1968

By

Jacob Van Staaveren, Herman S. Wolk,
and
Stuart Slade

DEFENSE
LION
PUBLICATIONS

Air War – Vietnam

Acknowledgements

This study was prepared by Jacob Van Staaveren, historian on the staff of the Albert F. Simpson Historical Research Center, Maxwell Air Force Base, Alabama. Mr. Van Staaveren was born April 9, 1917, in Perrydale, the son of Siet and Christina Strikwerda Van Staaveren. He graduated from Perrydale High School in 1935 and Linfield in 1939. He received a master's degree in 1943 from the University of Chicago and pursued doctoral studies there in history and political science. From 1946 to 1948, Mr. Van Staaveren was a civil information, education and labor education officer with the Yamanashi Military Government Team in Japan. In 1949 he returned to Japan as part of a U.S. government team assigned to write the official history of the occupation.

From 1951-53, he worked in South Korea documenting the Korean conflict. He then worked until 1956 in the Military History Liaison Office of the United Nations Command and Far East Command in Tokyo. From 1958 to 1981, he was historian of the Office of Air Force History, later renamed the Center for Air Force History, in Washington, D.C. He retired in 1981, leaving behind a wealth of monographs and histories that set the highest of standards for those who followed him. He also had published or contributed to numerous other historical publications, including several on the war in Southeast Asia. He belonged to the Society for History in the Federal Government, American Historical Society, Japan-America Society and Association for Asian Studies.

Copyright Notice

Copyright © 2013 Lion Publications Inc, 22 Commerce Road, Newtown, Connecticut 06470 - USA. ISBN 978-0-9859730-8-7 No part of this compilation may be reproduced or transmitted in any form or by any means, electronic or mechanical including photocopying, recording or by any information and retrieval system without permission in writing from the publisher

Table of Contents

Acknowledgements ii
Copyright Notice ii
Table of Contents iii
   FOREWORD xiii
   CODE NAME DIRECTORY xvi
   ACRONYM DEFINITIONS xx
PART ONE PLANS AND POLICIES IN SOUTH VIETNAM 1952 - 1963 1
I. EARLY PLANNING 3
   Background 4
   The Counterinsurgency Plan. 5
   The Program of Action 6
   The Taylor Mission 7
II. STEPPING UP MILITARY ASSISTANCE 11
   Establishment of USMAC/V 11
   Establishment of 2d ADVON 12
   Deployment of USAF Aircraft 13
   Deployment of Support Equipment 15
III. PLANS AND OPERATIONS (December 1961-June 1962) 17
   Operational Planning 17
   USAF Operations and Augmentations 18
   The Interdiction Issue 20
IV. PLANS AND OPERATIONS (July 1962-December 1963) 23
   Planning For An Early Victory 23
   USAF Augmentation 24
   USAF/VNAF Operations 25
V. THE DISPUTE OVER AIRPOWER 29
   The JCS Review 29
   The Interdiction Issue Again 30

    The Problem of Army Aviation 31

    Problems of Command Relations 34

VI. TESTING CONCEPTS AND WEAPONS...**37**

    Supervision of Testing 37

    Test Results 39

    Defoliation..41

VII. USAF SUPPORT OF THE VIETNAMESE AIR FORCE **45**

    A Vietnamese Army Air Force? 45

    Build Up Of The VNAF...45

    The Problem of Jet Aircraft..47

VIII. THE OVERTHROW OF THE DIEM GOVERNMENT **49**

    Conflicting Evaluations of the War 49

    The Fall of the Diem Regime...51

    The "Number One" Problem 52

IX. SUMMARY, 1961 - 1963..**53**

APPENDIX 1 ....**57**

APPENDIX 2 ....**58**

APPENDIX 3 ....**58**

APPENDIX 4A **59**

APPENDIX 4B **59**

APPENDIX 5 ....**60**

APPENDIX 6 ....**61**

APPENDIX 7 ....**61**

APPENDIX 8 ....**62**

APPENDIX 9 ....**62**

APPENDIX 10 ....**63**

    Combat Casualties..63

PART TWO USAF PLANS & POLICIES IN SOUTH VIETNAM & LAOS 1964.....**65**

I. REVISED U.S.-SOUTH VIETNAMESE MILITARY PLANNING **67**

    General Khanh's Coup 68

    Plans to Revitalize Counterinsurgency Operations.................................69

    Plans to Increase Pressure On North Vietnam 71

New U.S. Policy Guidance 72
II. CONTINUED MILITARY AND POLITICAL DECLINE 75
The Search for Courses of Action 75
More U.S. Aid and Reorganization of MAC/V 76
More Planning for Operations in Laos and Vietnam 77
New U.S. Leadership and More Military Aid 80
III. THE GULF OF TONKIN INCIDENT AND AFTERMATH 83
U.S. Response in the Gulf of Tonkin 83
A New Round of Planning 86
New U.S. Guidance 89
The Low-Risk Policy 90
IV. THE BIEN HOA AIR BASE ATTACK AND AFTERMATH 93
The Bien Hoa Incident 93
The Problem of Base Security 95
Review of Future Courses of Action 96
Continuing Crisis and a New Incident 99
V. BUILDUP OF USAF FORCES IN SOUTHEAST ASIA 101
New Aircraft For the 1st Air Commando Squadron 102
Deployment of B-57s to the Philippines 103
Other USAF Augmentations Early in 1964 104
Buildup After The Gulf Of Tonkin Incident 105
More Transport and Reconnaissance Aircraft 105
Establishment of a Search and Rescue Unit 106
Retention of the 19th TASS 106
Airfield Expansion 107
VI. OTHER USAF ACTIVITIES AND PROBLEMS 109
USAF Support of the Vietnamese Air Force 109
Expansion of the VNAF 109
The Problem of 5th and 6th A-1H Squadrons 110
The Problem of Jet Aircraft 110

    Completion Of Helicopter Training 112
    VNAF Strength 112
    Air Force Representation in MAC/V 112
    Rules of Engagement 114
VII. BEGINNING OF AIR OPERATIONS IN LAOS 117
    Initial Lao and U.S. Air Activity 117
    Plans Against Infiltration 119
APPENDIX 1 121
APPENDIX 2 121
APPENDIX 3 121
APPENDIX 4 122
APPENDIX 5 123
APPENDIX 6 123
APPENDIX 7 124
APPENDIX 8 124
APPENDIX 9 125
APPENDIX 10 125
PART THREE USAF PLANS & OPERATIONS IN SOUTH EAST ASIA 1965 127
I. THE ALLIES STRIKE NORTH 129
    U.S. Restraint and Limited Pressure 129
    Attack Across The 17th Parallel 132
    Proposed Eight-Week Air Program 134
    Troop Deployments For Base Security 136
    Rolling Thunder Strikes Begin 137
II. DEBATE OVER STRATEGY 139
    USAF Opposition To Deploying Large Ground Forces 139
    New Assessments And The Army's 21-Point Program 140
    President Johnson's March Decisions 142
    Planning Allied Troop Deployments 143
    The Stepped-Up Air War 145
III. THE EXPANDING U.S. ROLE 147
    President Johnson's April Decisions 147

The Honolulu Meetings 148
Speeding Unit Deployments 150
The Air War In South Vietnam (April-June) 154
The Air War In North Vietnam And Laos (April-June) 156
New Command Arrangements 158
IV. PLANNING NEW DEPLOYMENTS 159
A Larger Force For Southeast Asia 159
Impact On The Air Force 161
New Agreements At Honolulu (27 September -7 October) 163
The Air War In South Vietnam (July-November) 164
Airfield Expansion And Security 167
The Air War In North Vietnam And Laos (July-November) 168
V. COMMUNIST GAINS AND U.S. RESPONSE 171
The Saigon Conference in November. 171
The Follow-Up 174
The Air War in December 175
At The End Of The Year: The Air Force View 178
APPENDIX 1 181
APPENDIX 2 181
APPENDIX 3 182
APPENDIX 4 182
APPENDIX 5 183
APPENDIX 6 183
APPENDIX 7 184
APPENDIX 8 185
APPENDIX 9 187
APPENDIX 10 188
APPENDIX 11 188
APPENDIX 12 189
APPENDIX 13 189
APPENDIX 14 190
PART FOUR AIR OPERATIONS 1966 191

Air War – Vietnam

I. OBJECTIVES OF THE AIR WAR AGAINST NORTH VIETNAM 193
    Background to Rolling Thunder 193
    The Air Force and JCS Urge Early Renewed Bombing 195
    Secretary McNamara's Views 196
    The Bombing Resumes and Further Air Planning 198

II. INCREASING THE AIR PRESSURE ON NORTH VIETNAM 201
    Air Operations and Analyses 201
    The Beginning Of Rolling Thunder Program 50 203
    The Rolling Thunder Study of 6 April 205
    Air Operations in May: Beginning of "Gate Guard" 207
    Highlights of June Operations 208

III. THE POL STRIKES AND ROLLING THUNDER PROGRAM 51 211
    Background of the POL Air Strikes 211
    The Strikes of 29 June 212
    The Mid-1966 Assessment 214
    The Beginning of Rolling Thunder Program 51 215
    The Tally-Ho Campaign. 218

IV. ANALYSES OF THE AIR CAMPAIGN 221
    Operational Studies 221
    The Effectiveness of Air Power 222
    Studies on Aircraft Attrition 225
    The Hise Report 226
    McNamara's Proposal to Reduce Aircraft Attrition 229
    Approval of Rolling Thunder Program 52 231
    The Furor Over Air Strikes "On Hanoi" 232
    Other Air Operations in November and December 233
    Assessment of Enemy Air Defenses 234
    Assessments of the Air War Against North Vietnam 236

APPENDIX 1 239
APPENDIX 2 239

APPENDIX 3 240
APPENDIX 4 241
APPENDIX 5 242
APPENDIX 6 243
APPENDIX 7 243
APPENDIX 8 243
APPENDIX 9 244
PART FIVE 245
THE SEARCH FOR MILITARY ALTERNATIVES 1967 245
I. THE SITUATION IN EARLY 1967 247
   The Joint Chiefs' and Air Force Views of the War 248
   U.S. and Allied Deployed Strength 248
   Adjustments in Deployment Planning 250
II. THE DEBATE OVER TROOP DEPLOYMENTS 253
   General Westmoreland's Proposals 254
   Air Staff-JCS Views of General Westmoreland's Requests 256
   OSD Request for Studies of Alternate Force Postures 257
   The Draft Memorandum to the President 260
   The U.S. Worldwide Military Posture 263
   Secretary Brown's Views on Deployments and Bombing 263
III. THE 525,000 U.S. TROOP CEILING FOR SOUTH VIETNAM 267
   The Saigon Conference of 7 - 8 July 267
   Approval of the 525,000 U.S. Troop Ceiling 269
   Refinements in the U.S. Troop List 271
   Plans to Increase South Vietnamese Forces 276
   Other Deployment Actions 277
IV. NEW STUDIES ON DEESCALATION & MILITARY ACTION 279
   The Threat in the Demilitarized Zone 279
   Deescalation Studies and Other Possible Actions 281
V. OTHER PROPOSALS TO SPEED UP PROGRESS IN THE WAR 287
   The Administration's Eight Programs for South Vietnam 287

The Westmoreland-Bunker Briefings 287
U.S. Strategy and Strength at the End of 1967 290
APPENDIX ONE U.S. MILITARY & AIRCRAFT STRENGTH 295
PART SIX TOWARDS A BOMBING HALT 1968 301
I. MILITARY & POLITICAL SITUATION, EARLY 1968 303
Evaluations of the War..303
Studies on a Bombing Halt and Negotiations To End The War 305
U.S. and Allied Strength in Southeast Asia 307
II. MILITARY CRISES LATE JANUARY - MARCH 309
Crisis in Korea 309
The Air Force at Khe Sanh...310
The 1968 Tet Offensive 316
III. WESTMORELAND SEEKS MORE TROOPS, AIRCRAFT & EQUIPMENT 321
Plans to Speed Up Deployment of American Troops. 322
General Westmoreland's Request for 206,000 More Troops.....324
IV. DEBATE OVER MORE DEPLOYMENTS & STRATEGY 327
Three Air Force Strategies 327
Response to the Westmoreland Troop Request 332
New Proposals for More U.S. and Allied Deployments 332
Air Staff Views of Other Proposals 335
V. R&D FOR SOUTHEAST ASIA, 1968...337
Debate Over Air Force R&D 337
Examining The SEAOR System 344
Countering The Enemy Defensive Threat................................347
Bombing, Interdiction, And Surveillance Operations...............352
Project Shed Light..360
VI. THE PARTIAL BOMBING HALT & REASSESSMENT OF RESOURCES....363
The President Decides to Halt the Bombing.............................364
Service Views on the Bombing Halt 365
Further Debate on Reserve Callups..366

Southeast Asia Deployment Program 6 368
Further Review of the B-52 Sortie Rate.. 369
Faster Buildup of South Vietnamese Forces.. 371
Air Staff / JCS Views on Negotiations 373
VII. FURTHER POLICY REVIEW & NEW PLANS...377
Review of the War in Saigon and Honolulu 377
Deployment Adjustments in the Remainder of 1968.......................... 381
Additional Planning to Build Up the RVN Forces 384
Post-Hostilities Planning.. 386
VIII. THE COMPLETE BOMBING HALT 389
Enemy Response and Revised Military Operations............................. 391
Decision to Lower the B-52 Sortie Rate 393
IX. SUMMARY 401
APPENDIX I...403
APPENDIX II..407
APPENDIX III...411
NOTES FOR PART ONE 415
Chapter I... 415
Chapter II.. 417
Chapter III... 418
Chapter IV 420
Chapter V 422
CHAPTER VI 424
CHAPTER VII.. 426
CHAPTER VIII... 428
NOTES TO PART TWO 431
Chapter I... 431
CHAPTER II... 432
CHAPTER III.. 434
CHAPTER IV 435

CHAPTER V	437
CHAPTER VI	439
CHAPTER VII	441
NOTES FOR PART THREE	**443**
Chapter 1	443
Chapter II	445
Chapter III	447
Chapter IV	450
Chapter V	453
NOTES FOR PART FOUR	**455**
Chapter II	456
Chapter III	459
Chapter IV	461
Chapter V	462
NOTES TO PART FIVE	**465**
Chapter I	465
Chapter II	466
Chapter III	468
Chapter IV	470
Chapter V	472
NOTES TO PART SIX	**475**
Chapter I	475
Chapter II	475
Chapter III	478
Chapter IV	479
Chapter V	481
Chapter VI	485
Chapter VIII	488

FOREWORD
By: Defense Lion Publications

During the middle and late 1970s, the United States Air Force Historical Research Center, Maxwell Air Force Base, Alabama produced a series of 17 monographs that detailed the history of the Vietnam War. These studies were classified as being Top Secret for many years and were only recently released to the public. The core of these monographs is the series that deal with the political, operational and technical development of the Air Force participation in the Vietnam War. These remarkable documents contain a wealth of historical data that explain the background and reasoning behind many controversial decisions.

This compilation has taken these monographs and assembled them into a single narrative. The documents have been painstakingly remastered and reset to the printed page but their editorial integrity has been scrupulously preserved. We have also added numerous photographs and other illustrations to the documents. Each monograph has been given a separate section within the compilation as a whole and has retained its own appendices as part of that section. However, the source notes have been moved to the end of the compilation and, in order to reduce unnecessary duplication, the glossaries for each section have been consolidated into a single listing.

The first part of this study outlines the role of Headquarters USAF in aiding the South Vietnamese effort to defeat the communist-led Viet Cong. The author begins by discussing general U.S. policy leading to increased military and economic assistance to South Vietnam. He then describes the principal USAF deployments and augmentations, Air Force efforts to obtain a larger military planning role, some facets of plans and operations, the Air Force-Army divergences over the use and control of air power, combat training and testing, defoliation activities and USAF support for the Vietnamese Air-Force. The study ends with an account of events leading to the overthrow of the Diem government in Saigon late in 1963. Because this study emphasizes plans and policies, no effort has been made to chronicle the hundreds of individual air actions in which USAF units participated.

The implications of this background starts to become apparent in the second section which emphasizes Headquarters USAF's plans and policies with respect to South Vietnam and Laos in 1964. In the first four chapters the author describes the progressive military and political decline of the Saigon regime, after two government coups, and the efforts by U.S. authorities to cope with this problem. He notes especially the view of the Air Force Chief of Staff, Gen. Curtis E. LeMay, frequently stated, that only air strikes on North Vietnam could end the insurgencies in South Vietnam and in Laos and bring stability to the Vietnamese government. This contrasted with administration efforts to devise an effective pacification program and, pending emergence of a stable government, its decision to adopt a "low risk" policy to avoid military escalation.

Air War – Vietnam

In the remaining chapters of the study, the author discusses briefly the major USAF augmentations, the expansion of the Vietnamese Air Force, the problem of service representation in Headquarters, Military Assistance Command, Vietnam, and the rules of engagement as they affected particularly air combat training. The section concludes with a brief review of the beginning of USAF special air warfare training for the Royal Laotian Air Force and the inauguration of limited USAF and Navy air operations over Laos to contain Communist expansion.

The third part of this study highlights USAF plans, policies, and operations in Southeast Asia during 1965, especially as they were significantly changed by the president's key decisions to bomb North Vietnam and transform the U.S. Advisory role in South Vietnam to one of active military support. The author focuses on U.S. participation in the development of policy for prosecuting the war, the build-up of U.S. military strength in the theater, and the gradually intensified air operations against enemy forces in South Vietnam, North Vietnam, and Laos.

The Air Campaign Against North Vietnam, 1966 reviews the political background and top level discussions leading to the renewed bombing campaign in early 1966, the restrictions still imposed on air operations, and the positions taken on them by the military chiefs. It discusses the various studies and events which led to the President's decision to strike at North Vietnam's oil storage facilities and the results of those mid-year attacks. It also examines the increasing effectiveness of enemy air defenses and the continuing assessments of the air campaign under way at year's end.

The fifth and final part of this volume deals with the growing realization during 1967 that the war situation was deteriorating at a frightening speed. While focusing on the Chief of Staff and Air Staff roles, the author necessarily has highlighted the plans and policies of higher authorities, the White House, the Office of the Secretary of Defense, the Joint Chiefs of Staff, and the recommendations of the Military Assistance Command, Vietnam. Topics covered include plans for the military buildup in Southeast Asia, political considerations associated with new force deployments, and the continuing debate on war strategy and the conduct of the air campaign in the North. In the final sections, it is possible to detect the growing realization that, somehow, the war had already been lost. A few voices still suggested that some sort of victory was in sight but, with fifty years of hindsight, their reasoning seems almost ingenuous in the light of the storm that was to break loose in the early months of 1968.

The final part of the first volume focuses on the roles of the Chief of Staff and the Air Staff and their proposals for the conduct of the air war. It examines the closely linked plans and policies of the White House, Secretary of Defense, and the Joint Chiefs of Staff, and the views of the Pacific Command and the Military Assistance Command, Vietnam. The siege of Khe Sanh and the 1968 Tet offensive had a major impact on the U.S. government's conduct of the war, particularly the President's decision to halt partially, and later completely, the bombing of North Vietnam in an effort to facilitate peace negotiations. This section also discusses U.S. efforts to hasten the modernization and self-sufficiency of South Vietnam's armed forces. The material on events in 1968 is supplemented by a chapter on the research and development activities for Southeast Asia written by Herman S. Wolk. This was actually a short, separate monograph but its content is so vital to properly understanding the context of the political and operational decisions taken in 1968 that it has been included here as Chapter V.

A second volume of this compilation will cover the process that led to America's retreat from Vietnam. The Vietnam War profoundly altered America's world view and its strategic outlook. Together these two volumes will provide a unique insight into how and why the decisions that led to those changes were made.

Air War – Vietnam

CODE NAME DIRECTORY

Listed below are the code names of certain air concepts, operations, programs, and aircraft cited in this study. The reader may find it helpful to refer to the list on occasion.

Arc Light — Use of B-52s to attack hostile ground targets

Banish Beach — Using C-130s to drop pallets of fuel oil drums that were then ignited by smoke grenades to achieve area denial.

Barrel Roll. — Initiated in December 1964, Barrel Roll missions were flown against troops, equipment and supplies provided by North Vietnam in support of the Communist-led Pathet Lao.

Blue Tree. — Reconnaissance north of 20th parallel in North Vietnam

Buffalo Hunter — SAC-conducted drone photographic reconnaissance in Southeast Asia.

Cobra Talon — Chinese Communist ICBM tracking radar

College Eye — EC-121D Airborne early warning & control aircraft

Combat Apple — SAC RC-135 Electronic Intelligence (ELINT) collection missions based at Kadena AB, Okinawa.

Combat Beaver. — An air concept developed by the Air Staff in conjunction with the other services during September-November 1966. It was designed to support a proposed electronic and ground barrier system between North and South Vietnam.

Combat Skyspot — MSQ-77 and SST-181 controlled bombing missions in Steel tiger, Route Package One and South Vietnam

Comfy Gator — Operational program with remote controlled equipment on C-130 aircraft.

Commando Fly — Urgent reinforcement plan for bolstering U.S. tactical air strength in SE Asia in the event of a North Vietnamese invasion of South Vietnam. Put into effect, April 1972

Commando Hunt — Interdiction program officially begun on 15 November 1968. Designed to destroy as many supplies as possible moving South.

Commando Scrimmage — A series of exercises held in Thailand during 1974 covering a potential renewed air attack on North Vietnam in the event of a major violation of the cease-fire

Credible Chase — Miniature gunship aircraft program

DAO — Defense Attaché Office, the post-1972 replacement for MACV

Air War – Vietnam

Duck Soup	A proposal considered during mid-1965 to use Air America to intercept North Vietnamese aircraft dropping supplies into northeastern Laos. Abandoned because of the risk that captured Air America pilots "would confirm to the communists the [the company's] paramilitary nature."
Duffel Bag	Exploitation of the sensor technology evolved in Igloo White for monitoring enemy movements and supply lines in South Vietnam
Flaming Dart	The initial Navy and Air Force retaliatory air strikes against North Vietnam on 7-8 and 11 February 1965.
Freedom Deal	Interdiction campaign in Cambodia
Freedom Train	May 1972 extended bombing campaign against North Vietnam
Frequent Wind	Helicopter evacuation of American civilians and at-risk Vietnamese from South Vietnam
Gate Guard	An air program designed to slow North Vietnamese infiltration toward the demilitarized zone. It began on 1 May 1966 in the northern part of Laos and then shifted into route package area I in North Vietnam.
Giant Scale	SAC-conducted aerial reconnaissance of Southeast Asia by SR-71 aircraft
Igloo White	Extension of Muscle Shoals program
Iron Hand	Operations begun in August 1965 to locate and destroy Soviet-provided SA-2 missile sites in North Vietnam.
Island Tree	A late 1971 program, recommended and approved from Washington for bombing suspected enemy troop concentrations along the Trail and dropping sensors to monitor effectiveness.
Lam Son 719	Cross-border operation into Laos, Feb 1971
Linebacker	May 1972 interdiction bombing campaign against North Vietnam. An expansion of Freedom Train
Linebacker II	Late 1972 interdiction campaign against North Vietnam. Included deployment of B-52s against targets in Hanoi and Haiphong.
Pave Sword	Laser target designation pod for F-4 aircraft
Paveway	Family of guided bombs using laser, electro-optical or infrared devices for guidance.
Port Bow	Deployment of additional B-52s in response to Pueblo Incident

Air War – Vietnam

Market Time	US Navy anti-infiltration blockade of SVN coast
Muscle Shoals	Use of acoustic and seismic sensors to locate enemy forces for air attack
Niagara	Overall plan for deployment of tactical and strategic air power to aid in the defense of Khe Sanh
Nimble Thrust	1973 airlift of military articles and services to the Cambodian Armed Forces
Rain Dance	Tactical air commitment used for friendly ground forces in Laos during 1969
Ranch Hand.	The use of defoliants sprayed by C-123 aircraft to strip away jungle cover from Viet Cong irregulars. The Ranch Hand campaign started in 1963.
Raven	USAF FACs in Laos (usually with a Lao observer aboard) under the direct control of the Air Attaché, Laos
Red Crown	Cruisers used to control air operations over North Vietnam,
Rolling Thunder	The major air campaign begun on 2 March 1965 which inaugurated regularly scheduled air strikes against North Vietnam.
Senior Book	Airborne communications intelligence activities
Shining Brass	Small scale ground and air attacks into Laos started in 1965
Snakeye	Bomb equipped with retarding fin structure
Steel Tiger	Initiated in April 1965, Steel Tiger strikes were made against infiltration routes south of the 17th parallel in Laos.
Surprise Package	Advanced AC-130A gunship provided with special equipment for improved offensive and survival capabilities
Tally-Ho	An air interdiction program started on 20 June 1966 in the southern part of North Vietnam, aimed at slowing the infiltration of North Vietnamese troops, equipment, and supplies through the demilitarized zone into South Vietnam.
Talon Vise	A 1974 contingency plan to assist the American embassy in Saigon to protect and evacuate American citizens and designated aliens.
Teapot	Joint USAF/USN system for controlling air operations over North Vietnam
Tennis Racket	A contingency plan for attacks against North Vietnam, published on January 29, 1974
Tiger Hound	Begun in December 1965, these strikes were aimed at infiltration targets in southern Laos. They featured for the

	first time in Laos the use of forward air controllers and airborne command and control for certain strikes.
Tight Jaw	Plans to "Vietnamize" the U.S. Army sensor program
Toan Thang	RVN cross-border operations into Cambodia
Wild Weasel	USAF aircraft, largely F-100Fs and F-105Fs, specially equipped with electronic and other devices to neutralize or destroy Soviet-provided SA-2 sites in North Vietnam.
Yankee Station	Carrier Force area off coast of North Vietnam
Yankee Team	USAF-Navy armed reconnaissance missions in the Laos panhandle. Yankee team strikes were allowed only under special circumstances.

ACRONYM DEFINITIONS

AA	Anti-Aircraft
AAA	Anti-Aircraft Artillery
AB	Air Base
A/C	Aircraft
ACTIV	Army Concepts Testing in Vietnam
AC&W	Aircraft Control and Warning
AD	Air Division
ADC	Air Defense Command
Addn	Addition
AFB	Air Force Base
AFCCS	Air Force Command And Control System
AFCHO	USAF Historical Division Liaison Office
AFCVD	Special Assistant for Sensor Exploitation (HQ, USAF)
AFLAG	Air Force Advisory Group
AFLC	Air Force Logistics Command
Aflds	Airfields
AFSC	Air Force Systems Command
AFTUV	Air Force Test Unit, Vietnam
AID	Agency for International Development
AGM	Air to Ground Missile
AIM	Air Intercept Missile
ALC	Air Logistics Command (Vietnam)
ALO	Air Liaison Officer
Altn	Alternate
AM	Amplitude Modulation
ANG	Air National Guard
ANZUS	Australia, New Zealand and United States
AOC	Air Operations Center
App	Appendix
Appns	Appropriations

ARDF	Airborne Radio Direction Finding
ARM	Anti-Radiation Missile
ARPA	Advanced Research Projects Agency
ARVN	Army of the Republic of Vietnam
ASD	Aeronautical Systems Division
ASOC	Air Support Operations Center
ASSS	Air Staff Summary Sheet
Asst	Assistant
Asst CS/I	Assistant Chief of Staff, Intelligence
Atchd	Attached
Auth	Authorized
AW	Automatic Weapons
BDA	Bomb Damage Assessment
BIAS	Battlefield Illumination Airborne System
BoB	Bureau of the Budget
BOBS	Beacon Only Bombing System
BPE	Best Preliminary Estimate
B/R	Barrel Roll
C/A	Course of Action
CAP	Combat Air Patrol
CAS	Close Air Support
CAS	Controlled American Source (CIA)
CBU	Cluster Bomb Unit
CDTC	Combat Test and Development Center
C-E	Communications – Electronics
CEP	Circular Error Probable
CFST	Control Field Service Team
CG	Commanding General
CG	Coast Guard
CHECO	Contemporary Historical Evaluation of Counterinsurgency
Chmn	Chairman

Air War – Vietnam

CIA	Central Intelligence Agency
CICC	Combined Interdiction Coordination Committee
CICP	Combined Interdiction Campaign Plan
CIDG	Civilian Irregular Defense Group
CINCPAC	Commander-in-Chief, Pacific
CINCPACAF	Commander-in-Chief, Pacific Air Forces
CINCSAC	Commander-in-Chief, Strategic Air Command
CJSC	Chairman, Joint Chiefs of Staff
CM	Chairman's Memo
CMC	Commandant Marine Corps
CMCM	Commandant Marine Corps Memo
CNO	Chief of Naval Operations
COIN	Counterinsurgency
Comd	Command
COMMAC/V	Commander, Military Assistance Command, Vietnam
COMUSMAC/V	Commander, U.S. Military Command, Vietnam
Conf	Conference
Const	Construction
CONUS	Continental United States
CRIMP	Consolidated Republic of Vietnam Armed Forces I&M Program
CRP	Combat Reporting Post
C/S	Chief of Staff
CSAF	Chief of Staff Air Force
CSAFM	Chief of Staff Air Force Memo
C/S USAF	Chief of Staff, USAF
CTZ	Corps Tactical Zone
CVA	Aircraft carrier
CVN	Nuclear-powered aircraft carrier
CY	Calendar Year
DA	Department of the Army
DAF	Department of the Air Force

DAO	Defense Attaché's Office
DART	Deployable Automatic Relay Terminal
DASC	Direct Air Support Center
DASK	Drift Angle Station Keeping
Dam	Damage
DCPG	Defense Communications Planning Group
DCS/P&O	Deputy Chief of Staff, Plans and Operations
DCS/P&R	Deputy Chief of Staff, Programs and Resources
DCS/S&L	Deputy Chief of Staff, Systems and Logistics
DDR&E	Director of Defense Research and Engineering
Def	Defense
Defol	Defoliation
Dep	Deputy
Des	Destroyed
DIA	Defense Intelligence Agency
Dir	Director, Directorate
DITT	Directorate of Targets, Tango Division (7th Air Force)
D/Ops	Directorate of Operations
D/Plans	Directorate of Plans
Dir/Ops	Directorate of Operations
Dir/Plans	Directorate of Plans
Div	Division
DJSM	Director, Joint Staff Memo
DMZ	Demilitarized Zone
Docs	Documents
DOD	Department of Defense
DSMG	Designated Systems Management Group
DSPG	Defense Special Projects Group
DRV	Democratic Republic of Vietnam (then, North Vietnam)
ECM	Electronic Countermeasure
Eff	Effectiveness

EOGB	Electro-optically guided bomb
EOS	Electronic Operational Support
Eval	Evaluation
EW	Early Warning also Electronic Warfare
FAC	Forward Air Controller
FAE	Fuel Air Explosive
FAG	Forward Air Guide
FANK	Khmer Armed Forces
FAR	Force Armee Royaume (Laotian Armed Forces)
FPJMC	Four-Party Joint Military Commission
FLIR	Forward Looking Infra Red
FM	Frequency Modulation
FMFPAC	Fleet Marine Force Pacific
FOL	Forward Operating Location
FSB	Fire Support base
Ftr	Fighter
FY	Fiscal Year
GCI	Ground Controlled Intercept
Gp	Group
GPO	Government Printing Office
GVN	Government of Vietnam
Hist	History
HONO	Honolulu
HSAS	Headquarters Support Activity, Saigon
ICBM	Intercontinental Ballistic Missile
ICC	International Control Commission
ICCS	International Commission of Control and Supervision
IDA	Institute of Defense Analysis
IFR	Instrument Flight Rules
I&L	Installation and Logistics
I&M	Improvement and Modernization

IFF	Identification Friend of Foe
Imp	Implement
Intvw	Interview
Infil	Infiltration
Invest	Investigation
IOC	Initial Operational Capability
IR	Infra Red
ISA	International Security Affairs
ISC	Infiltration Surveillance Center
JAOC	Joint Air Operations Center
JCS	Joint Chiefs of Staff
JCSM	Joint Chiefs of Staff Memo
JMA	Journal of Military Assistance
JMS	Journal of Mutual Security
JOEG/V	Joint Operational Evaluations Group, Vietnam
Jt	Joint
JTF	Joint Task Force
JUSMAAG/T	Joint United States Military Assistance Advisory Group Thailand
L&L	Legislative Liaison Office
LAPES	Low Altitude Parachute Extraction System
Lat	Latitude
LGB	Laser-Guided Bomb
LLLTV	Low Light Level Television
LOC	Lines of Communication
Long	Longitude
LORAN	Long Range Navigation
LST	Landing Ship Tank
Ltr	Letter
MAAG/V	Military Assistance Advisory Group, Vietnam
MAB	Marine Amphibious Brigade

Air War – Vietnam

MAP	Military Assistance Program
MACTHAI	Military Assistance Command Thailand
MASF	Military Assistance Service Funded
MAC/V	Military Assistance Command Vietnam
MC	Marine Corps
Mgt	Management
MiG	Mikoyan-Gurevich – producer of Russian fighter aircraft. Frequently used as a generic designation for Russian fighters in general
Mil	Military
MR	Memo for Record
Mtg	Meeting
MTI	Moving Target Indicator
MTT	Mobile Training Teams
Mun	Munitions
NATO	North Atlantic Treaty Organization
NCO	Non-commissioned Officer
NCP	National Campaign Plan
NLF	National Liberation Front
NMCB	Navy Mobile Construction Battalion
NOA	New Obligating Authority
NSA	National Security Action
NSAM	National Security Action Memorandum
NSC	National Security Council
NVA	North Vietnamese Army
NVN	North Vietnam
Ofc	Office
Off	Office(r)
Opl	Operational
Ops	Operations
ORI	Operational Readiness Inspection
OSD	Office, Secretary of Defense

OSD/ISA	Office of the Secretary of Defense, International Security Affairs
OSAF	Office, Secretary of the Air Force
OT&E	Operational Testing and Evaluation
Pac	Pacific
PACAF	Pacific Air Forces
PACOM	Pacific Command
PBD	Program Budget Decision
PCS	Permanent Change of Station
PDJ	Plaine des Jarres
Pers	Personnel
PIRAZ	Positive Identification and Radar Advisory Zone
Plcy	Policy
PMDL	Provisional Military Demarcation Line
PMS	Program Management System
POL	Petroleum Oil and Lubricants
Poltl	Political
Poss	Possible
POW	Prisoner of War
PPC	Photo Processing Cell
P.M.	Prime Minister
Prep	Prepared
Pres	President
Procur	Procurement
Prog	Program
Proj	Project
Prov	Province
PROVOST	Priority Research and Development Objectives for Vietnam Operational Support
PSAC	President's Scientific Advisory Council
QRC	Quick reaction capability
QRF	Quick Reaction Force

Qtr	Quarterly
R-Day	Beginning of withdrawals
RAAF	Royal Australian Air Force
R&D	Research and Development
Rcrd	Record
RDT&E	Research, Development, Test and Evaluation
Recon	Reconnaissance
Req	request
Res	reserve
RHAW	Radar Homing and Warning
RLAF	Royal Laotian Air Force
RLG	Royal Laotian Government
RLT	Regimental Landing Team
ROC	Required Operational Capability
ROK	Republic of Korea
ROKAF	Republic of Korea Air Force
RP	Route Package
Rprt	Report
Rqmts	Requirements
R/T	Rolling Thunder
Rqmts	Requirements
RTU	Replacement Training Unit
RVN	Republic of Vietnam
RVNAF	Republic of Vietnam Armed Forces
SA	Systems Analysis
SA	Secretary of the Army
SAB	Scientific Advisory Board
SAC	Strategic Air Command
SACSA	Special Assistant for Counterinsurgency and Special Activities
SAF	Secretary of the Air Force
SAFOS	Secretary of the Air Force

Air War – Vietnam

SAC	Strategic Air Command
SAF	Secretary of the Air Force
SAFOI	Secretary of the Air Force Office of Information
SAM	Surface to air missile
SAR	Search and Rescue
SAW	Special Air Warfare
SAWC	Special Air Warfare Center
SCNA	Self-Contained Night Attack
Scty	Security
SEA	Southeast Asia
SEADAB	SEA Database
SEAOR	Southeast Asia Operational Requirement
SEATO	Southeast Asia Treaty Organization
SECDEF	Secretary of Defense
Secy(s)	Secretary(s)
SIG	Special Interdepartmental Group
SIOP	Single Integrated Operational Plan
Sit	Situation
SLIC	Special Low-Intensity Conflict
SM	Secretary's Memo
SN	Secretary of the Navy
SNIE	Special National Intelligence Estimate
SOD	Secretary of Defense
SOF	Special operations Force
Spec	Special
SPO	Systems Program Office
SPOS	Strong Point Obstacle System
Stmt	Statement
STRAF	Strategic Army force
STRICOM	Strike Command
Strat	Strategic

Air War – Vietnam

SVN	South Vietnam
Sys	Systems
T-Day	End of Hostilities
Tac	Tactical
TAC	Tactical Air Command
TACAIR	Tactical Air
TACP	Tactical Air Control Party
TACS	Tactical Air Control System
TAPS	Tactical Air Positioning System
TDY	Temporary Duty
TEW	Tactical Electronic Warfare
TFR	Terrain Following Radar
TFS	Tactical Fighter Squadron
TOT	Time On Target
UE	Unit Equipment
USAF	United States Air Force
USAFE	United States Air Force, Europe
USAR	U.S. Army
USARPAC	U.S. Army, Pacific
USARSG/V	U.S. Army Special Group, Vietnam
USCG	U.S. Coast Guard
USIA	U.S. Information Agency
USIB	U.S. Intelligence Board
USMAAG/V	U.S. Military Assistance Advisory Group, Vietnam
USMAC/Thai	U.S. Military Assistance Command, Thailand
USMAC/V	U.S. Military Assistance Command, Vietnam
USN	U.S. Navy
VC	Viet Cong
VN	Vietnam
VNAF	Vietnamese Air Force
WESTPAC	Western Pacific
Wpn	Weapon
WSEG	Weapons Systems Evaluation Group

PART ONE

PLANS AND POLICIES IN SOUTH VIETNAM

1952 - 1963

Air War – Vietnam

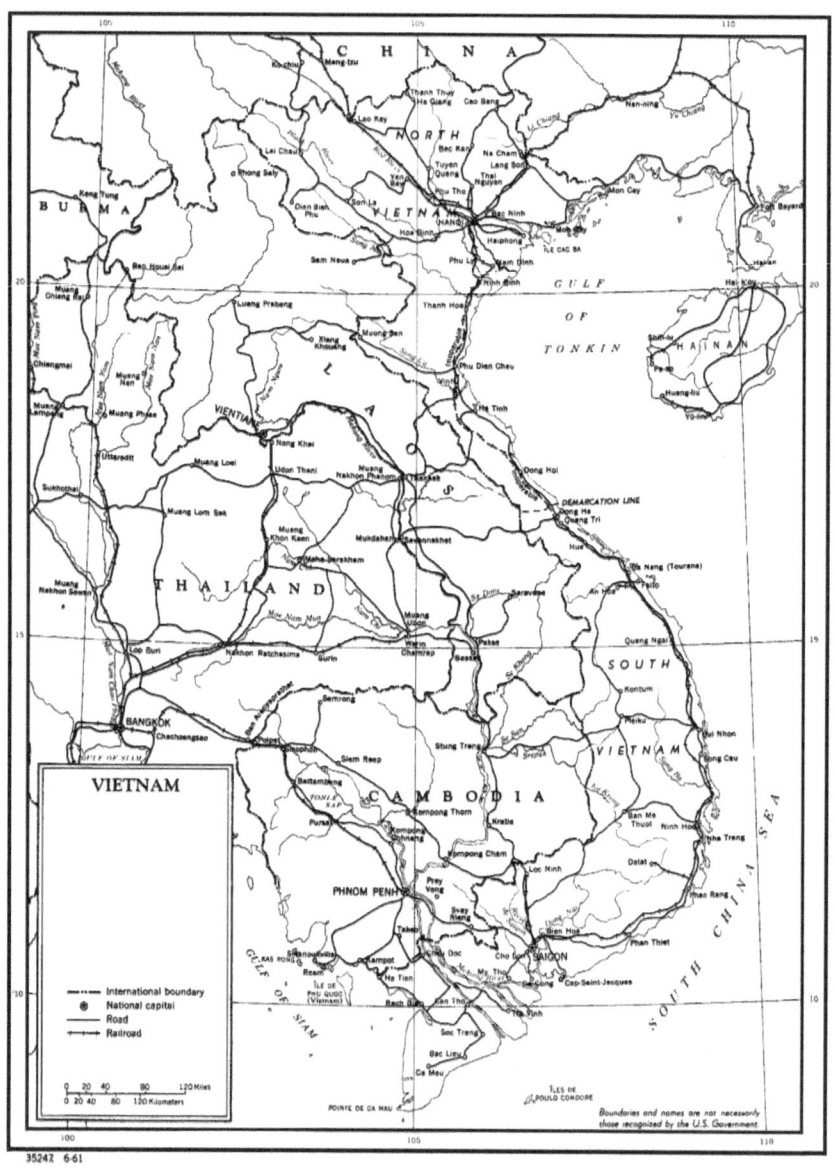

I. EARLY PLANNING

On 7 May 1954 the fortress at Dien Bien Phu surrendered to the Communist-dominated Viet Minh signaling the end of the rule of the French in Indochina that had begun in 1852. The Viet Minh (Vietnam Independence League), founded in May 1941 was a coalition of 15 revolutionary groups which had as a common objective: the abolition of French and Japanese rule in Vietnam. After World War II the Viet Minh gradually set up a Communist-controlled regime in North Vietnam which after the Geneva agreement became "The Democratic People's Republic of Vietnam."

At a conference held in Geneva, Switzerland, between 25 April and 21 July 1954, France agreed to the "full independence and sovereignty" of Vietnam, Laos, and Cambodia, new nations which evolved out of Indochina. Vietnam would be divided along the 17th parallel of latitude, with the French forces withdrawing south of that line, the Viet Minh north. Separate administrations on each side would consult in July 1955 on "free and general elections by secret ballot" in June 1956 to reunify the country. The newly created International Control Commission for Supervision and Control, made up of representatives of India, Canada, and Poland, would supervise the truce arrangement. [1]

Neither the government south of the 17th parallel nor the United States signed the Geneva agreement. Under Secretary of State Walter B. Smith asserted, however, that the United States would not use force to disturb the agreement, that it would view violation as a serious threat to international peace and security, and that it would continue to seek unity through free elections supervised by the United Nations.[2]

Meanwhile, South Vietnam prepared for nationhood. In July 1954 Ngo Dinh Diem became prime minister, and on 26 October 1955, following a referendum, president. On the same day he proclaimed the establishment of The Republic of Vietnam. In 1955, on the grounds that North Vietnam was violating the Geneva agreement and would not allow free elections and that his own country had not signed the agreement, Diem refused to undertake negotiations to reunify the country.[3]

President Ngo Dinh Diem
Source: Library of Congress

Air War – Vietnam

Background

The legacy of war found South Vietnam in political, economic, and social chaos. At the end of hostilities in 1954 its population of about 12.5 million (compared with 14 million in North Vietnam) increased by about 900,000 when refugees, largely Catholic, fled the Communist sector. Thousands of Communist guerrillas roamed the countryside, and private armies added to the disorder. And the lack of leadership, free of the taint of French or Viet Minh collaboration, exacerbated the nation's difficulties.[4]

To control unruly elements, the Diem government inherited from the French the Army of the Republic of Vietnam (ARVN), some 250,000 men. Since the French had occupied the high command positions, the army had virtually no qualified Vietnamese for staff officers. It was also woefully weak in artillery, heavy armor, engineering, and communications. Not until 1955 was the government able to assume effective administrative responsibility for the Army.[5]

The Republic of Vietnam Air Force (VNAF), also inherited from the French, had been organized in 1950 as air arm of the army to aid the French Air Force in the battle for Indochina. Until 1954, when it received its first combat aircraft, the VNAF flew only liaison and observation missions. Some of its aircraft were French, but most were obtained under the United States military assistance program.[6]

The outbreak of the Korean War prompted the U.S. government to send a military assistance advisory group (MAAG/V) to Saigon in July 1950, and on 23 December of that year the United States signed a mutual defense assistance agreement with France and Vietnam. In September 1954 the United States and six other nations signed the Southeast Asia Treaty Organization (SEATO) pact which included a pledge of military assistance, if requested, to South Vietnam. On 1 January1955 the United States agreed to send military assistance directly to South Vietnam and to assist in organizing and training its armed forces under the overall authority of the commander of the French forces remaining in the country.[7]

The United States briefly shared with the French the task of training and equipping the South Vietnamese military forces. At the request of the South Vietnamese government, the French withdrew their mission for the army in April 1956 and for the air force in May 1957. At this point, the United States became solely responsible for advising and supporting the Vietnamese armed forces.[8]

With U.S. financial support, South Vietnam reduced its armed forces to 150,000 men and stepped up it training program. South Vietnam also established a Civil Guard and a Self Defense Force to help control the groups that were spreading disorder. The Civil Guard, initially a paramilitary organization controlled by province chiefs, was later administered by the government's Department of Interior. The l0,000-man Self Defense Force, organized on a village basis with locally recruited personnel but headed by regular Army officers was attached to the government's Department of National Defense.

Plans And Polices 1952-63

In 1956 the air force became a separate arm of the Department of National Defense and in May 1957 it possessed four squadrons: one F8F, one C-47, and two L-19 for a total of 85 aircraft. None were combat ready. Authorized personnel strength was 4,000; the number assigned, 4,115. In fiscal year 1958, the VNAF was authorized 4,580, and shortly afterwards it had six squadrons.[9]

As the Diem government continued to manifest greater military political, and economic viability, the North Vietnamese decided in May 1959 to reunify the country by force. An insurgent group known as the Viet Cong that included about 3,000 armed guerrillas began a campaign of major subversion against South Vietnam. (Viet Cong is a derogatory abbreviation used in South Vietnam for "those who direct guerrilla warfare and also are subversive agents" that is, Vietnamese Communists. The term is not used in the north.) It drew its strength from former Viet Minh members who were ordered to remain underground in the south after the 1954 Geneva agreement, Viet Minh troops from the south who regrouped in the north, and elements of the southern population susceptible to Viet Cong recruitment. The insurgency was facilitated by the use of Laos as both corridor and sanctuary. Confronted with this Communist challenge, the United States in 1960 began to plan for and provide increased military and economic assistance to its embattled ally.[10]

Grumman F8F-2 Bearcats. In the late 1950s, these were the only combat aircraft in South Vietnam. These particular aircraft are Thai. Source: Royal Thai Air Force

The Counterinsurgency Plan.

During 1960 the Viet Cong became a dangerous threat to the established government in South Vietnam. The insurgents fought with arms left behind by the Viet Minh in 1954 or obtained from North Vietnam, and they also captured about 80 percent of the 3,700 weapons lost by the Vietnamese forces in 1960. During the year they not only conducted large, coordinated strikes but also 3,645 small ambushes, and they assassinated or kidnapped 2,647 village and hamlet officials. In the Mekong delta, the Viet Cong eliminated local government control and established a "liberated" area where they forcibly taxed the populace. Early in 1960, South Vietnamese intelligence estimated "hard core" Communist strength at 9,820, sympathizers at 2 million, and those "on the fence" at 2 million. According to this estimate, about one-third of the population either preferred Viet Cong rule or was indifferent to it.[11]

In April 1960, before the extensive growth in insurgency activities, Admiral Harry Felt, Commander-in-Chief, Pacific (CINCPAC) had prepared a plan aimed specifically at combating the Viet Cong. The JCS, after reviewing it, recommended, to Secretary of Defense Thomas S. Gates that all U.S. agencies concerned with South Vietnam develop a coordinated plan. After many revisions by American officials in Washington and Saigon, the coordinated plan was ready in January 1961 for final review by a next administration which had promised to give greater attention to all aspects of counterinsurgency.[12] For a discussion of this issue, see Charles H. Hildreth, "USAF Counter-insurgency Doctrines and Capabilities, 1961 - 1962 (AFCO, February 1964), pp 1 – 4.

The plan urged measures to remedy some political features of the Diem regime that created discontent. It stressed the need for personal security for the Vietnamese and for military, economic, and political reforms to achieve it. The plan also called for adding 20,000 men to the armed forces, raising their strength to 170,000, and improving the Civil Guard. On 30 January President John F. Kennedy and his Secretary of Defense, Robert S. McNamara, approved the plan and the outlay of $28.4 million for the armed forces and $12.7 million for the Civil Guard. The JCS approved implementation of the plan on 6 February.[13]

Although Headquarters USAF supported augmentation of Vietnamese armed forces, it thought the additional manpower allotted to the struggling VNAF was much too small. The VNAF would receive only 499 more men, 400 of these for AD-6 fighter and H-19 and H-34 helicopter units.[14]

In February 1961 the U.S. Ambassador to South Vietnam, Frederick E. Nolting, Jr., presented the counterinsurgency plan to President Diem. Because many provisions were unpalatable to him, Diem eventually issued only a few directives in support of it. He formed a committee to direct operations, transferred control of the Civil Guard from the Department of Interior to the Department of National Defense, developed plans to clarify authority for unified action under a single chain of command, and created corps and division tactical zones in place of military regions.[15]

The Program of Action

Increased Communist activity in South Vietnam and Laos prompted U.S. authorities to devise a program of action for the Diem government. Prepared by an interagency task force headed by Deputy Secretary of Defense Roswell L. Gilpatric, the new program incorporated much of the old one but was far broader. At a National Security Council meeting on 29 April 1961, President Kennedy approved numerous measures contained in the program: augmentation of the military assistance advisory group (MAAC/V) in Saigon to help train the expanding Vietnamese forces, shipment of radar surveillance equipment to detect Communist overflights and maintain aerial surveillance on the Laotian border, establishment of a combat development and test center, and expansion of the civic action and economic development program.[16]

On 11 May, the President approved a final draft of the program of action for South Vietnam. It was designed to prevent "communist domination, create a viable and increasingly democratic society, and institute mutually supporting actions of military, economic, psychological, and covert character." He asked for an assessment of the value and cost of further increasing the armed forces from 170,000 to 200,000 by creating "two new division equivalents" for the northwest border region. The President also directed the Department of Defense to continue its studies of the size and composition of the U.S. forces that might be needed for operations in South Vietnam should a meeting between Vice President Lyndon B. Johnson and President Diem scheduled for 11-13 May indicate such a need. On 13 May a Vietnam-U.S. communiqué stated, however, that both governments would build up existing programs of military and economic aid and that Vietnam's regular armed forces would be increased with U.S. assistance.[17]

Headquarters USAF strongly supported the program of action, suggesting only minor changes concerning personnel, equipment, and logistics. Previously, it had urged the preparation of this type of document for each area of the world where Communist encroachment existed or was expected. Secretary of the Air Force Eugene M. Zuckert called the program an "outstanding job" and a "realistic basis to [an] aggressive start in reversing [the] trend of events in Southeast Asia."[18]

In July and August President Kennedy made several other decisions relating to the program. After receiving JCS and OSD recommendations and the report of a U.S. financial survey group headed by the noted economist, Dr. Eugene Staley, he approved increasing the armed forces to 200,000 men. (In February 1962 they were raised to 205,000.) He made approval contingent on devising a satisfactory strategic plan to control the Viet Cong. The President deferred, however, a decision on Diem's request to raise military strength to 270,000 over a two-year period.[19]

With the Staley report as a guide, President Kennedy authorized more funds to carry out the program of action. He counseled U.S. officials to urge Diem to accept the program's reforms. And he directed that Diem be informed that the U.S. President agreed with the Staley Report's three basic tenets as they applied to the program of action: (1) security requirements should have first priority; (2) military operations could not achieve lasting results unless economic and social programs were continued and accelerated; and (3) it was in the interest of both countries to achieve a free society and a self-sustaining economy in South Vietnam.[20]

The Taylor Mission

These measures came too late. As the military situation worsened in South Vietnam and its neighbors, the JCS urged the deployment of SEATO troops to Laos to save that country and to protect the borders of South Vietnam and Thailand. (In May 1962 the United States sent combat troops to Thailand where they remained for several months.) But the President decided on alternate actions. On 11 October he authorized U.S. advisors to assist in counter-guerrilla operations against Techepone, Laos, a Viet Cong supply center. And, subject to Diem's

General Maxwell D Taylor.
Source: U.S. Army

concurrence, he authorized the dispatch of a detachment from USAF's Special Air Warfare Center to train the VNAF. Presaging additional U.S. involvement, he also ordered his Military Representative, General Maxwell D. Taylor, to Saigon to explore additional ways for more effective U.S. assistance. On the 24th, in a public letter to Diem, President Kennedy assured him of U.S. determination to help Vietnam preserve its independence.[21]

Composed of White House, State, Defense, and other officials, the Taylor Mission visited Southeast Asia from 15 October to 3 November 1961. In its report to the President, the mission warned that the Communists were pursuing

> "a clear and systematic strategy in Southeast Asia to by-pass U.S. nuclear strength rooted in the fact that international law and practice does not yet recognize the mounting guerrilla war across borders as aggression, justifying counter-attack at the source."

The mission noted that Viet Cong strength had risen from about 14,350 in July 1961 to 16,600 in November. But it also discerned Viet Cong weaknesses and the need to rely on terror and intimidation, reluctance to engage the ARVN openly, and fear of U.S. reaction. The Diem government estimated "positive" supporters of Communism within South Vietnam at 200,000, twice the number calculated by American sources.

The mission found that the Diem regime lacked confidence because of Viet Cong successes and uncertainty concerning U.S. policy in Laos. Because of inadequate intelligence, ground forces were engaged in static tasks. Command channels at both the provincial and national levels were unclear and unresponsive, and Diem's distrust of his military commanders exacerbated this feeling, But his government had certain assets, particularly the Army, Civil Guard, and Self Defense Force. The VNAF was ineffective because it lacked target intelligence and its command structure was incomplete; the Vietnamese Navy potential was not yet established.

The Taylor Mission recommended wide-ranging changes. It called for the U.S. military organization to change its relationship with the Diem government from advice-giving to partnership and to become something approximating an operational headquarters in a theater of war. The Diem regime should be brought closer to the people. There should be more emphasis on border control and additional covert operations in North and South Vietnam and in Laos. The United States should step up training and equipping of Vietnamese ground, air, naval, paramilitary, and special forces, and improve communication and intelligence organizations. It should build up MAAG/V to an 8,000 man force, place more emphasis on research and

Plans And Polices 1952-63

development, and give fast military and economic support to limited offensive operations. To provide more air support, the mission supported the dispatch of the USAF unit (Farmgate) and proposed the shipment of other aircraft and helicopters. Finally, it saw merit in the proposal of Admiral Felt and Ambassador Nolting that the United States should hasten this aid by immediately delivering units and equipment under the guise of helping the populace in recently flooded areas of the Mekong delta. [22]

The proposals were less forceful than those previously advocated by McNamara and the JCS. Observing that the fall of South Vietnam would lead to fairly rapid communization of neighboring nations, they desired deployment of a strong U.S. military force rather than a gradual entry of units. They proposed warning the North Vietnamese government of punitive action unless Viet Cong activities ceased. If North Vietnam and Communist China intervened, they believed that about 200,000 troops, including reserves, could contain the aggressor Although the United States faced a grave international situation over Berlin, McNamara and the JCS believed that this action in Vietnam would not seriously interfere with plans to defend the German city.[23]

After OSD consultations with State, in which the JCS did not participate, the two departments issued a milder memorandum in November. Warning of the military escalation that might result if U.S. troops were sent, the memorandum noted other possible dangers: failure because of Vietnamese apathy and hostility, political repercussions in the United States if only U.S. troops were used, and renewed Communist action in Laos that might prevent a political settlement in that country. The memorandum also pointed to advantages in obtaining third-country assistance for South Vietnam.[24]

President John F. Kennedy. Source Library of Congress

The President, after discussing the memorandum with the National Security Council, decided against the use of U.S. ground forces and adopted a policy of limited participation similar to that recommended by the Taylor Mission. On 22 November he directed that Diem be informed of our willingness to increase aid in a joint undertaking. the United States would provide more men and equipment, step up training, and help establish better communication and intelligence systems. Diem, in turn, would place South Vietnam on a war footing, mobilize its resources, give its government adequate authority, and overhaul the military establishment and command structure.[25]

On the basis of these instructions, Ambassador Nolting and Diem negotiated a bilateral agreement, and in December both governments announced its non-military features. In a White Paper, basically an appeal for world support, the

Department of State declared that North Vietnam had violated the Geneva agreement and that South Vietnam needed assistance. Other nations were asked to help.[26]

General Curtis LeMay's position on the Vietnam War is often misunderstood. His stated position was that the U.S. should decide what force level was needed to win the war quickly and decisively. If it was prepared to commit those forces, it should go straight to that level without any intermediate stages. If it was not prepared to commit that force level it should not get involved at all. Photograph source: U.S. Air Force.

Despite these measures, Gen. Curtis E. LeMay, USAF chief of staff, believed that the program for South Vietnam was still inadequate. On 5 December 1961 he obtained JCS support for another statement on the need for additional measures. The JCS asked McNamara on 13 January 1962 to inform President Kennedy and Secretary of State Dean Rusk of its belief that the United States should further bolster Dien and discourage factions seeking his overthrow. But Diem would need to cooperate by ending procrastination, authorizing his military commanders to carry out their plans, and providing an adequate basis for U.S. advice and assistance. If, on this basis, the Vietnamese could not control the Viet Cong, U.S. or allied forces should be introduced. In this eventuality, the JCS observed that the war would be peninsular and allow U.S. forces to utilize their experiences in World War II and Korea, the U.S. commitment would not seriously affect operations planned for Berlin and elsewhere, and that the Communists could sustain only limited forces because of logistic problems. McNamara sent these views to the President without endorsement, preferring to await the results of the current program.[27]

II. STEPPING UP MILITARY ASSISTANCE

The Kennedy Administration moved rapidly to help the embattled Diem government. On 27 November 1961 McNamara approved the establishment of a new military headquarters, headed by a four-star commander, to manage this country's limited participation in the war. U.S. military men would advise units of the Vietnamese armed forces while they engaged in combat. U.S. Army helicopters would be sent, plus USAF C-123 transports, T-28 fighters and a tactical air control system. McNamara also asked the JCS to prepare plans to use Vietnamese aircraft and helicopters in defoliant operations. Defoliants were chemicals which stripped the leaves off plants.[1]

This military aid raised an international legal issue, since the Geneva agreement prohibited the acquisition by South Vietnam of modern arms and restricted the size of foreign military advisory groups in that country. The Administration decided to abide by the agreement, but it believed that North Vietnam's violations gave South Vietnam legitimate grounds for self-defense, including accepting U.S. assistance, until these violations ceased. Therefore, the United States would not concede that this aid was a breach of the Geneva agreement.[2]

Establishment of USMAC/V

McNamara's plan to establish a new military headquarters in Saigon stirred considerable debate. The JCS strongly objected to a new headquarters in this area independent of CINCPAC, claiming that this would be incompatible with Admiral Felt's mission and responsibilities. The Joint Chiefs suggested instead the establishment of a subordinate unified command under Felt called "U.S. Forces, Vietnam" with the individual service component commands also in charge of the service sections of the Military Assistance Advisory Group, Vietnam (MAAG/V). As a precondition to altering the command structure, the JCS urged that the United States clearly define its objectives in South Vietnam and extract from the reluctant Diem government a commitment to a joint military program.[3]

The Department of State advocated arrangements less suggestive of major change. It proposed extending the authority of the Chief of MAAG/V over the additional U.S. forces and economic and intelligence activities. State also objected to a four-star Commander, believing this could be "an irrevocable and 100 percent commitment to saving South Vietnam."[4]

The conflicting views were reconciled. In mid-December McNamara and Rusk agreed to establish, in accordance with JCS views, a new subordinate unified command under CINCPAC and call it, as State later suggested, the U.S. Military Assistance Command, Vietnam (USMAC/V). the new command would be analogous to the U.S. commands in Taiwan, Korea, and Japan, Its chief would be a four-star commander, a rank McNamara considered "highly essential" to emphasize the positive impact of change in U.S. policy.[5]

Air War – Vietnam

After Presidential approval and the selection of Army Lt. Gen. Paul D. Harkins as Commander, MAC/V was established in Saigon on 1 February 1962. Responsible for carrying out U.S. military policy, Harkins was also authorized to discuss with the Vietnamese all facets of military operations. He reported to the Secretary of Defense through CINCPAC and the JCS. Coequal with the U.S. Ambassador to South Vietnam, Harkins could consult with him on all policy matters. Harkins also provided broad guidance to MAAG/V, now part of his command, on the military assistance program (MAP) for South Vietnam.[6]

USMAC/V was Army-oriented, and this quickly engendered a heated interservice conflict over the conduct of the war and especially over the use and control of airpower. The Air Force had good reason to be disappointed. In early planning, the services had agreed that the Air Force would hold the posts of chief of staff, J-2, and J-1. Harkins, however, selected a Marine lieutenant-general as his chief of staff. As a substitute, he proposed an Air Force officer for J-3 but under Army pressure he chose an Amy officer for this post. On 19 February, despite strong remonstrances by LeMay to McNamara and by the Pacific Air Force (PACAF) commander, Gen. Emmett O'Donnell, Jr., to Admiral Felt, the Secretary of Defense approved Harkins' selections.[7]

McNamara promised LeMay he would reconsider this decision if the circumstances warranted, but this prospect appeared dim. The service representation for Headquarters MAC/V was as follows: Army Commander (Gen.), J-3 (Brig. Gen.), J-4 (Brig. Gen.), and J-6 (Col.); Navy J-1 (Capt.); Marines Chief of Staff (Lt. Gen,); and Air Force J-2 (Col.) and J-5 (Brig. Gen.). Of the five general officers in key positions, the Air Force had only one. Numerically, it also felt underrepresented. Of the 105 officer spaces initially authorized, the Army had 54, the Navy and Marines 29, the Air Force only 22, Within Headquarters MAAG/V somewhat similar disparities existed.[8]

Establishment of 2d ADVON

The Air Force also had little voice in determining how its air units would function in South Vietnam. Without consultation, Admiral Felt determined that the Chief, Air Force Section, MAAG/V would be responsible for advising and training the VNAF and he would report to him (Felt) through the Chief, MAAG/V. The Chief, Air Force Section, MAAG/V would also command a special advanced echelon in South Vietnam to provide the VNAF with combat advisory training. He would also command through this echelon scattered PACAF detachments and elements in Southeast Asia. Wearing this second hat, he would report to Felt through O'Donnell, the PACAF Commander. Felt emphasized that the title of the advanced echelon should not imply a new command.[9]

On 15 November l961, Detachment 7, first unofficially and later officially designated 2d ADVON was established at Tan Son Nhut Airfield near Saigon as a provisional element of the 13th Air Force. The detachment was renamed 2d ADVON on 7 June 1962. In this study, it will be cited as 2d ADVON until its redesignation as 2d Air Division in October 1962. Subsequently, it became the only component command of MAC/V when that organization was established. On

Plans And Polices 1952-63

20 November Brig. Gen. Rollen H Anthis, Vice Commander of the 13th Air Force, was named commander of 2d ADVON and on 1 December, Chief, Air Force Section, MAAG/V.[10]

Deployment of USAF Aircraft

Well before MAC/V was established, U.S. military units were deploying to South Vietnam. On 11 October 1961 President Kennedy had authorized the dispatch of the first important USAF unit – Detachment 2 – an element of 4400th Combat Crew Training Squadron (Jungle Jim) stationed at Eglin AFB, FL.

On arrival at Bien Hoa Airfield in November, the detachment, nicknamed Farmgate, consisted of eight T-28s four SC-47s four B-26s (redesignated RB-26s since the Geneva agreement prohibited the entry of tactical bombers), and 151 officers and airmen. Operational control was vested in 2d ADVON, training in the Air Force Section MAAG/V, and as indicated, Gen. Anthis commanded both.[11]

The B-26 aka RB-26 aka A-26.

Source: U.S. Air Force

The primary mission of Farmgate was to train the Vietnamese in counter-guerrilla air tactics and techniques. There were restrictions on combat training operations. Under the rules of engagement approved by the President on 6 December, such operations were authorized only if the VNAF lacked the necessary training and equipment, combined USAF-VNAF crews were aboard, and the missions were confined to South Vietnam. Because of its special role, Farmgate aircraft bore Vietnamese markings.[12]

Air War – Vietnam

Since the Geneva agreement prohibited the entry of jets into South Vietnam, the Felt-Nolting proposal, which the Taylor Mission had supported, was adopted. On 20 October, the Air Force sent four RF-101s and a photo processing cell (PPC) to Tan Son Nhut, ostensibly to photograph areas in the Mekong delta in conjunction with flood relief. Nicknamed Pipestem, these aircraft in 31 days flew 67 reconnaissance sorties over South Vietnam and Laos to fulfill reconnaissance needs.[13]

RF-101 reconnaissance aircraft.
Source: U.S. Air Force.

On 29 October Felt directed PACAF to place four RF-101s and a PPC in Thailand. The aircraft and 45 men from the 45th Tactical Reconnaissance Squadron, 39th Air Division left Misawa, Japan, for Don Muang, Thailand. The unit (known as Able Mable) became operational on 8 November, overlapping briefly and then replacing the Pipestem flights. By the end of 1961, Able Mable had flown 130 sorties over South Vietnam and Laos. It made photos available to theater and national agencies within 24 hours. In February 1962 the unit had 55 men and a new PPC.[14]

In accordance with McNamara's decision of 27 November to accelerate military aid to South Vietnam, the Air Force in December dispatched 16 C-123 TAC transports and 123 men from Pope AFB, NC, to Clark AB, the Philippines. Nicknamed Mule Train, the squadron arrived at Tan Son Nhut in January 1962 to become the nucleus of an airlift buildup. It airlifted special forces for counter-guerrilla operations, airdropped supplies, and trained the Vietnamese.[15]

To conduct defoliation experiments, a group of six C-123s and 69 men (nicknamed: Ranch Hand) from TAC's special aerial spray flight at Langley AFB, Va. and Pope AFB, N.C., arrived at Clark in November 1961, then moved to Tan Son Nhut in January 1962. For psychological warfare activities, three USAF SC-47s, specially equipped for leaflet and loudspeaker flights, came to South Vietnam in December 1961 and were quickly operational. [16]

Deployment of Support Equipment

The United States sent support equipment to South Vietnam even before the visit of the Taylor Mission. Headquarters USAF, through the 13th Air Force, surveyed requirements for the radar surveillance equipment needed under the April 1961 program, but could not meet then immediately because all available USAF equipment was in use. On 11 September the JCS directed the Air Force to provide a combat reporting center (CRC), an essential element of radar surveillance. A CRC promptly left Shaw AFB, N.C. for Tan Son Nhut, where it went into round-the-clock operation on 5 October. The CRC came under the control of 2d ADVON after that unit was activated in November.[17]

To carry out Taylor Mission recommendations, McNamara on 27 November ordered a tactical air control system (TACS) deployed to South Vietnam. By joint agreement, the Vietnamese and U.S. Commanders retained operational control over their own aircraft with operations coordinated through a joint air operations center (JAOC). Activated at Tan Son Nhut on 2 January 1962, the JAOC was command post for 2d ADVON and VNAF and also liaison center with the Army and Navy. It was manned temporarily by 314 PACAF officers and men until regular-duty personnel arrived in February and March 1962.[18]

Established in accordance with a 13th Air Force operational plan (Barndoor), the TACS was assigned to 2d ADVON on 15 January and soon became operational, though with limited capability. In addition to the JAOC and the CRC, the TACS consisted of five forward air controllers (FACs); at Tan Son Nhut, two air support operations centers (ASOCS) - one in the north with the Vietnamese Army's I Corps at Da Nang, the other in the central highlands with the II Corps at Pleiku, and one combat reporting post (CRP) at Da Nang. when III and IV corps were established, two ASOCs were added at Can Tho in the south and in Saigon. The various elements of the TACS were interconnected by high-frequency voice and teletype radio circuits. [19]

The radars that controlled friendly aircraft also handled aircraft control and warning (AC&W). In accordance with the Barndoor plan, one USAF-operated AC&W radar was placed at Tan Son Nhut and another at Da Nang, while one VNAF-operated light radar was placed at Pleiku. These radars, plus one installed later at Ubon, Thailand (Barndoor II), provided radar air surveillance of South Vietnam and the surrounding territory.[20]

In January 1962 McNamara and the JCS also decided to establish a troposcatter communication system (Back Porch) under the operating responsibility of the Army. The Air Force installed the "backbone" equipment (AN/MRC-85) at Saigon, Nha Trang, Pleiku, and Da Nang in South Vietnam and at Ubon,

Air War – Vietnam

Thailand. This equipment, operated by Army and 150 USAF personnel, provided high-quality communications among U.S. military commanders, subordinate commanders, tactical field units, and as necessary, U.S. or SEATO forces. The Army installed the mobile equipment (AN/TRC-90) for 10 tributary links interconnecting the back-bone equipment and provided a signal battalion to operate it. The AN/MRC-85 equipment, installed by 1 September, provided 72 voice channels. The tributary lines added 24 channels. Several months later, under Back Porch II, the Air Force extended the troposcatter system to provide emergency communications between Saigon and Clark AB. [21].

III. PLANS AND OPERATIONS
(December 1961-June 1962)

Operational Planning

As the flow of men and materiel to South Vietnam increased, McNamara and his planners in December 1961 carefully studied short-and long-range operational plans. An early Outline Campaign Plan, drafted by CINCPAC for the Vietnamese, envisaged powerful strikes and the use of defoliants in Zone D of the III Corps area (a region near Saigon overrun by the Viet Cong). The plan also called for blows at guerrilla bases in I and II Corps and border areas and for mopping up and consolidation in central and northern areas.[1]

Since the Vietnamese could neither begin operations in Zone D immediately nor maintain their hold on areas already cleared, McNamara and military officials decided on a simpler plan to gain some initial successes. Known as Operation Sunrise, this plan called for securing and holding Binh Duong Province, where the government controlled only 10 of 46 villages. Based somewhat on successful British operations in Malaya, Operation Sunrise required three months for preparation, four months for military action, and two to three months for consolidation. It was slated to begin on 23 March 1962, and the Vietnamese would undertake shorter-range operations in the interim.[2]

Early in 1962 the Air Force proposed a quick reaction plan that would strengthen the government by demonstrating its concern for the safety of its people. Strongly supported by Zuckert and LeMay, this plan called for a quick reaction force composed of Vietnamese airborne troops and USAF-VNAF transport and strike aircraft deployed in nine areas of the country. Linked by a simple communication system to isolated villages, the force would respond within 10 to 30 minutes to a Viet Cong attack, LeMay thought that the plan would complement the strategic hamlet program then evolving, which in his opinion was too defensive.[3] The Vietnamese government [had] conceived the strategic hamlet program in 1961 and publicly announced support for it

Overhead view of a strategic hamlet
Source: U.S. Air Force

in February 1962, but it did not approve a national construction plan until August. Meanwhile, provincial governments built hamlets with little planning or coordination, and many were inadequately fortified and supported.

In March, the JCS approved the plan in principle and sent it to CINCPAC. The Army believed that the plan conflicted with the "clear and hold" concept of Operation Sunrise and asked for a Joint Staff study of a substitute plan. Despite strong USAF pressure, Felt believed that there should be only one master counter-insurgency plan for South Vietnam, and he adopted only certain features of the quick reaction plan.[4]

USAF Operations and Augmentations

Since USAF military units would be exposed to combat, Zuckert was concerned about the problem of public relations. On 4 December 1961, he asked OSD how to deal with possible Communist charges of bacteriological and chemical warfare. OSD responded that all U.S. activities should be explained as training or support for the Vietnamese even if incidental combat support operations were conducted, and that there should be no comment on reports to the contrary.[5]

U.S. air units began aiding Vietnamese ground troops against the Viet Cong in late 1961. The principal USAF unit, Farmgate, flew its initial combat training sorties on 19 December. Mule Train (C-123) flights began on 3 January 1962; Ranch Hand C-123s began defoliation operations on 13 January. U.S. Army helicopters inaugurated support flights on 23 December 1961, U.S. Marine helicopters in April 1962.[6]

The Fairchild C-123 transport was to become the archetypical aircraft of its type in Vietnam. Already considered obsolete by the early 1960s, it often seemed that it was used by everybody for everything. Source: U.S. Air Force

USAF activities fell into two categories: support and tactical. Support included airlift, liaison, observation, rescue, and evacuation; tactical consisted of combat training in close support and interdiction as well as combat airlift and reconnaissance missions. Close air support, provided primarily for the ARVN and Civil Guard, was directed by forward air controllers. Vietnamese requests for interdiction missions often were denied when jungle foliage made identification of friend and foe too difficult. In night operations, flare drops around a village or outpost under attack also deterred guerrillas who feared air strikes.7

USAF participation expanded during the first half of 1962 because Operation Sunrise, which began on 23 March, required all types of air support. Farmgate combat training sorties rose from 101 in January to 187 in June; transport and defoliation sorties from 296 to 1,102. Initial defoliation results were encouraging, but the Air Force suspended this type of operation from May to September for political reasons.[8]

There were occasional setbacks. On 11 February an SC-47 on a leaflet-dropping mission crashed, killing eight Americans (six Air Force and two Army) and one Vietnamese. The presence of so many Americans in the aircraft prompted public and Congressional inquiries. At McNamara's request, LeMay studied the psychological warfare mission and decided that the Vietnamese could perform it. The JCS then directed the transfer of the mission to the VNAF as soon as the Vietnamese were trained sufficiently. On 26 May, a Farmgate aircraft hit Da Ket, south of Da Nang, causing civilian casualties. Although the town was improperly marked on a map, military investigators attributed the accident to navigational error and relieved the crew of operational status. The mission was successful otherwise, since it caused an estimated 400 Viet Cong casualties.[9]

Under USAF tutelage, the VNAF increased its combat sorties in A-1Hs and T-28s from 150 in January 1962 to 389 in June. The VNAF flew its first T-28 sorties in March. And, in a 50-plane raid on 27 May against a Viet Cong headquarters in the central highlands, the VNAF destroyed warehouses and huts with 100 tons of fire bombs and explosives.[10]

A-1 Pulls Away After An Attack Run
Source: U.S. Air Force

Air War – Vietnam

Following reports of North Vietnamese aircraft operating south of the DMZ, a force of four F-102 Delta Daggers was sent to South Vietnam but found nothing. Ironically, fifty years later, the Vietnamese disclosed that An-2 biplane transport aircraft had been operating over the DMZ at that time. Since the maximum speed of the An-2 is less than the stalling speed of the F-102, exactly what the Delta Daggers could have done about them is unclear.
Photo Source: U.S. Air Force.

The possibility that enemy aircraft might contest Farmgate-VNAF air superiority led to a new augmentation of USAF strength. On 19-20 March surveillance radar at Pleiku and Man Iang detected unidentified aircraft. Conventional aircraft could not locate them, and PACAF quickly dispatched three F-102 and one TF-102 jet aircraft from Clark AB to Tan Son Nhut where they were placed on alert. Known as Operation Water Glass (redesignated Candy Machine in October 1963), these jets found no hostile aircraft, either at this time or at any time in 1962 and 1963, From April through July 1962 the F-102s deployed to South Vietnam at 10-day intervals, then alternated with a Navy detachment of three AD-5Q aircraft. In late 1962 the F-102s occasionally engaged in psychological warfare by creating sonic booms which disturbed Viet Cong siestas or nighttime sleep.[11]

In May the JCS authorized Sawbuck II, the deployment of a second C-123 transport squadron of 15 aircraft from Pope AFB, N. C., 12 going to Da Nang and 4 temporarily to Thailand. There were now 37 C-123s and 235 USAF personnel in South Vietnam under Mule Train and Sawbuck II. Concurrently, at the direction of the Chief of Staff, TAC established the Tactical Air Transport Squadron (Provisional 2), 464th Troop Carrier Wing, to bring Mule Train, Sawbuck II, and Ranch Hand C-123s under a single commander.[12]

Also in May, an upsurge of Communist attacks in Laos led to the dispatch of four additional night-photo RB-26s, two for Farmgate and two to Thailand. The latter joined Farmgate in December. [13]

The Interdiction Issue

The start of U.S. combat training activities almost immediately created political and military problems. Despite precautions, on 21 February 1962, a Farmgate aircraft erroneously bombed a Cambodian village in a poorly defined border area while participating in a four-day air and ground assault against the Viet Cong. Not only were President Kennedy, the Department of State, and OSD concerned with

Plans And Polices 1952-63

the ensuing diplomatic difficulties with Cambodia, but they feared that air strikes, if indiscriminate, could antagonize friendly Vietnamese.[14]

The Department of State questioned the wisdom of attacks on villages at all and doubted whether targets were being properly identified. It also alleged that the initial strikes alerted the insurgents, permitting them to escape. State recommended following the methods used successfully by the British in Malaya. the Air Force thought that the air attacks had not been failures because they had attained their objective of clearing the area of guerrillas. Moreover, since the insurgents had a sanctuary nearby, either in North Vietnam, Cambodia, or Laos, the British techniques were not necessarily valid in this instance. O'Donnell expressed his concern to LeMay that this initial reaction against the use of airpower might lead to additional restrictions on Farmgate training missions.[15]

General Anthis, Commander of 2d ADVON, conceded that complete target verification was not always possible since most tactical intelligence and requests for air strikes came from the Vietnamese. However, he defended Farmgate procedures as basically sound. In daytime no targets within five miles of the Laos-Cambodian borders could be attacked, and for night flights, only targets at least 10 miles from the borders. All targets were first marked by a forward air controller. Although McNamara warned against the consequences of harming innocents to kill a few guerrillas and suggested as a rule of thumb that pilots should weigh "risk against gains," he imposed no new rules of engagement on the Farmgate units.[16]

The O-1s flown by Forward Air Controllers were a vital link in the target process. After identifying a target, the FAC called for attack aircraft and marked the target. Strike aircraft could only attack after being "cleared hot" by the FAC. Source: U.S. Air Force

In March a U.S. Army team that had visited South Vietnam also concluded that indiscriminate bombing played into Viet Cong hands. Because the team failed to substantiate its allegations, no additional curbs were imposed on combat training. The team's additional observations that there were certain target identification problems and that the VNAF flew only daylight sorties were acknowledged by the Air Force which was trying to correct these deficiencies. The Air Force noted, however, that target identification was a problem that applied equally to ground attacks.[17]

PACAF thought that some of the Army charges were motivated by an Army plan to experiment with armed helicopters instead of relying on the VNAF and, when necessary, Farmgate aircraft for top cover and close support. In April LeMay visited South Vietnam and found no basis for "loose statements" which suggested

21

Air War – Vietnam

a careless attitude or incorrect procedures. He observed that while the Vietnamese selected the targets. the joint air operations center and air support operations centers carefully checked them, and forward air controllers in liaison aircraft marked them for attack.[18]

IV. PLANS AND OPERATIONS
(July 1962-December 1963)

In mid-1962 the conflict in South Vietnam appeared to many U.S. officials to have reached a turning point. In May, McNamara had visited South Vietnam and was "tremendously encouraged" for he found "nothing but progress and hope for the future" in the strategic hamlet and military training programs. Many U.S. military officers were also cautiously optimistic. Although the weekly average of terrorists incidents had declined only slightly from 414 between October and December 1962 to 394 between April and June 1962, Viet Cong casualties exceeded government casualties by a 5 to 1 ratio. And more guerrillas had surrendered or defected, while government troops had lost fewer weapons.[1]

Planning For An Early Victory

In July 1962 McNamara declared that the period of "crash" military assistance for South Vietnam was ending and that longer-range systematic planning was necessary. Assuming that the insurgency could be checked by the end of 1965, he directed the services to prepare a comprehensive three-year plan for training and equipping the Vietnamese and for removing most U.S. units from South Vietnam. As the Vietnamese assumed responsibility for their own defense, McNamara envisaged removing MAC/V entirely and leaving only a MAAG/V with about 1,600 personnel.[2]

In July McNamara also agreed to the transfer of responsibility for training the Vietnamese civilian irregular defense force (CIDG) from the Central Intelligence Agency to the Department of Defense, specifically to MAC/V. The CIDG was concerned with youth programs, commando units, civic action, and Viet Cong infiltration across the Laotian border.[3]

The services quickly prepared a plan to make the Vietnamese forces largely self-sufficient within three years, and McNamara approved it on 23 August. The plan later was revised extensive\y and integrated with a five-year U.S. military assistance program (MAP) for the Vietnamese and a national campaign plan (NCP). The Air Force portion of the plan called for accelerated training and equipping of the VNAF.[4]

MAC/V conceived the NCP in October 1962 to encourage the Dien regime to reorganize its military forces and to shorten the war by using its increased military resources in coordinated strikes against the Viet Cong. After the United States persuaded Diem to accept the plan, his government worked out the details aided by U.S. advisors. the NCP also was known as the "explosion" plan since military and paramilitary forces would "explode" into action on many fronts.[5]

The Department of State and the JCS became concerned that the NCP might prove overambitious and fail, undermining Vietnamese morale. MAC/V then scaled it down from a major "detonation" to a series of intense but highly coordinated small

Air War – Vietnam

operations that would extend the current effort. PACAF believed that the NCP could not fail completely because intensified action against the Viet Cong was bound to assure some success and any offensive would improve military morale and the will to fight.[6]

In accordance with NCP strategy, the Vietnamese would seek out and destroy enemy concentrations, clear and hold liberated areas, and establish fortified strategic hamlets in these areas. Working with plateau and mountain tribesmen, the government forces would achieve better border control. Aircraft would strafe Viet Cong zones, provide close fire support and reconnaissance, and transport men and equipment. The three phases of the NCP included preparation, execution, and consolidation.[7]

During the preparatory phase, Diem on 26 November realigned the military command structure and divided the country into four tactical zones and one military district. The second phase, requiring greatly stepped-up military and paramilitary operations with U.S. support, was scheduled to begin by 28 January 1963, the Vietnamese New Year's Day. But Diem procrastinated and decided not to launch the offensive until two-thirds of the population were in strategic hamlets, weakening the plan.[8]

On 18 June the Vietnamese forces finally received the order to launch the second phase on 1 July. The tempo of military activity then increased somewhat, but there were no spectacular victories. Harkins believed that the NCP had lost much of its usefulness. At the end of August, he informed Dien that government forces had failed to take full advantage of aerial reconnaissance, to pursue the Viet Cong, and to remain in conquered territory. They had fought too many one-day operations and not enough at night, and they had placed too little emphasis on psychological warfare, civic action, and the coordination of intelligence with operations. Responsibility for border surveillance had not been shifted from the special forces to the Corps commander, as proposed. And some Vietnamese Army commanders were reluctant to give their troops formal training.[9]

USAF Augmentation

Meanwhile, stepped-up military action and long-range planning required more USAF aircraft and personnel. In August 1962, with JCS approval, four USAF U-10B (L-28) aircraft arrived in South Vietnam to improve air-to-ground communications and target spotting and to provide faster air support. In October Harkins and O'Donnell proposed to augment Farmgate by five T-28s, ten B-26s, two C-47s, and 117 men. McNamara was cool to the proposal because it was contrary to his policy of shifting responsibility to the Vietnamese. But, after the JCS affirmed the Harkins-O'Donnell request, he approved it on 28 December and the President concurred shortly afterwards. This boosted Farmgate strength by February 1963 to 41 aircraft and 275 men.[10]

To help carry out the NCP, a second augmentation was approved in March 1963. The Farmgate sortie rate would be increased by 30 to 35 percent. This would be achieved, Felt decided, not by adding new T-28 and B-26 units but by doubling Farmgate personnel. The Army would deploy its own aircraft to support the

Vietnamese civilian irregular defense force rather than to rely on additional USAF aircraft, and this triggered a vigorous interservice debate. As a compromise, McNamara and the JCS authorized the Air Force to deploy an additional C-123 squadron (Sawbuck VII), one TO-1D squadron, and place one C-123 squadron on alert. The Sawbuck VII Squadron arrived in South Vietnam in April; the TO-ID squadron, consisting of 22 planes loaned from the Army, in August.[11]

Additional reconnaissance aircraft also were needed. In January 1953 two RF-101s (Patricia Lynn) joined Able Mable (the four RF-101s that had come in November 1961). In March, two RB-26Cs and two RB-26Ls (Sweet Sue) arrived, all capable of taking night photographs. The RB-25Ls also had an infrared capability. they were joined in June by two RB-57Es, both outfitted with night photo and infrared equipment. By mid-1963, 12 USAF aircraft and six U.S. Army Mohawks comprised the land-based reconnaissance strength in South Vietnam.[12]

The augmentations and expanded air activity led to personnel and organizational changes. At LeMay's request, the JCS on J2 April reassigned to PACAF for permanent duty the personnel in TAC units (Farmgate, C-123 units, and the new TO-1D squadron) who were on six-month temporary duty. This was done to stabilize manning, reduce training requirements, and make better use of experienced people.[13]

On 17 June Headquarters USAF disestablished Farmgate as a detachment of the Special Air Warfare Center and activated in its place the 1st Air Commando Squadron (Composite) at Bien Hoa Airfield, with Detachment 1 at Plei Ky airport and Detachment, 2 at Soc Trang airport. On 8 July, the squadron, with an approved strength of 41 aircraft and 474 men, was assigned to 34th Tactical Group, 2d Air Division. On 17 June Headquarters USAF also redesignated the 19th Liaison Squadron, equipped with TO-ID aircraft, as the 19th Tactical Air Support Squadron (light) and established it at Bien Hoa on I July. And on 4 November all USAF reconnaissance aircraft were brought together when PACAF established the 13th Reconnaissance Technical Squadron at Tan Son Nhut.[14]

USAF/VNAF Operations

Farmgate and VNAF units improved old tactics and devised new ones to cope with the Viet Cong. In August 1962 Farmgate crews began furnishing air support through a village air request net. They also discovered that napalm attacks were effective against guerrillas submerged in water, since burning napalm consumed air and forced the insurgents to surface. Farmgate crews also devised a better escort technique for helicopters ferrying Vietnamese troops. Two T-28s flew at different altitudes, permitting better observation and quick-firing passes against the enemy. By dropping colored smoke grenades to mark targets, pilots foiled Viet Cong attempts to confuse them with ordinary smoke grenades.[15]

Guerrilla ambushes of Vietnamese Army vehicle and train convoys had averaged two to three per week during the first half of 1962, but the VNAF significantly reduced this number. At Harkins' suggestion, Dien in August directed his Army commanders to use the VNAF to protect important convoys. Results were immediately gratifying. Between August and October 1962, the commanders made 506 requests for air convoys compared with only 32 for the first seven months of the year. An L-19 or several fighters in very dangerous territory provided escort and alerted ground troops accompanying the convoys. LeMay called this tactic a "big step forward", and Zuckert noted its success when he testified in February 1963 before a House committee.[16]

With USAF training and assistance, the VNAF improved its employment of aerial flares in night operations. Since these flares deterred the insurgents or forced then to break off attacks against villages and outposts, the VNAF began in August to place C-47 flare aircraft on airborne alert each night.[17]

To improve navigation of USAF and VNAF aircraft, in August the JCS approved installation by the Air Force of a Decca tactical air positioning system, and this British-made low-frequency system went into operation on 15 December. The Decca system, with three ground stations and 50 airborne receivers, provided over-the-horizon coverage and was more accurate than other available systems. A fourth ground station was added in 1963.[18]

The number of USAF sorties increased steadily during the year. Farmgate T-28s and B-26s averaging a total of only 15 aircraft for the 12-month period had flown 2,993 operational sorties, C-47s 843 (649 in support of the special forces), and C-123s 11,689. In addition, the transports carried more than 17,000 tons of cargo and airlanded or airdropped 45,000 Vietnamese. Exclusive of jet-aircraft missions, Farmgate, USAF transport, and other operational-type sorties at year's end totaled 15,867.[19]

USAF support constituted, of courser only a portion of all airpower employed. VNAF aircraft and helicopter strength totaled 180 by the close of December 1962, and its A-1Hs and T-28s had flown 4,496 sorties during the year. A Marine company with 20 rotary aircraft contributed to the air effort. Of major significance and considerable USAF concern was the expansion of U.S. Army aviation support in South Vietnam.[20]

Estimates of the damage inflicted by airpower varied. Headquarters USAF concluded that combined Farmgate-VNAF air strikes in 1962 accounted for 28 percent of the 25,100 Viet Cong casualties. (MAC/V estimated the casualties at 30,673 and later at 33,000). Of this total, Farmgate's T-28s and B-26s inflicted 3,200 and, in addition, destroyed about 4,000 structures and 275 boats. PACAF credited Farmgate aircraft with more than a third of officially recorded guerrilla casualties. The Defense Intelligence Agency attributed 56 percent to all U.S. aircraft employed.[21]

Although these statistics could not be verified easily, the Air Force believed that, by comparing the achievements of the 10,000 members of combined USAF/VNAF units with those of the 400,000 U.S. and Vietnamese Army, Navy, and paramilitary forces, air strikes accounted for a very high rate of enemy casualties in relation to the total effort. After visiting South Vietnam in December, Zuckert concluded that "the type of doctrine that is involved in our air commando operations is proving effective."[22]

In 1963 Farmgate crews trained the VNAF in night and instrument flying to develop an air close support capability during periods of darkness and inclement weather. the VNAF also assumed responsibility for most of the night flare drop missions. On reconnaissance missions, USAF aircraft also located sites for new strategic hamlets and roads. By May, six RF-101s and four RB-26s provided about 70 percent of all targeting information in South Vietnam.[23]

Airborne loudspeakers plus a "Chieu Hoi" or amnesty program, officially proclaimed by the Diem government on 19 April, reportedly encouraged Viet Cong defections. Since the VNAF was not carrying out this form of psychological warfare adequately, McNamara in May authorized USAF crews to participate more directly. At U.S. Army request, Farmgate loudspeaker sorties previously had been reported as "equipment test" missions.[24]

At mid-1963, there were nine loudspeaker aircraft, four USAF, four U.S. Army, and one VNAF. These planes broadcast information on resettlement,. amnesty, and strategic hamlets; warned civilians to leave dangerous areas; and carried the voices of defectors. Although results were difficult to measure, most U.S. officials considered the broadcasts useful and desired to increase them.[25]

In September 1963 the Viet Cong began taking advantage of political disorder in Saigon and stepped up the war. After the overthrow of the Diem regime on 1 November, the insurgents overran scores of inadequately defended strategic hamlets, and government casualties and losses mounted. During the week of the coup, the Air Force and the VNAF flew 380 combat and advisory sorties to aid 40 strategic hamlets.[26]

This high sortie rate was maintained through the end of the year. USAF non-jet operational sorties for 1963 totaled more than 42,000, a considerable jump from the nearly 16,000 in 1962. Of the 1963 total, B-26s and T-28s now averaging an inventory of 25 compared with 15 in 1962 - flew 8,522 sorties. Each USAF pilot flew 100 to 150 training sorties during his 12-month tour of duty. MAC/V estimated that USAF aircraft inflicted about 3,800 of the 28,000 insurgent casualties and destroyed about 5,700 structures and 2,600 boats. VNAF A-1H and T-28 sorties rose to 10,600 in 1963 from about 4,500 in 1962. U.S. Army aviation was employed at an even faster pace with 231,900 sorties claimed in 1963 as compared with 50,000 in 1962.[27]

Low-level air attacks became more hazardous as the accuracy of Viet Cong small arms fire improved. The insurgents scored 89 hits against Farmgate and other USAF planes during the last four months of 1962 but 257 in the first four months of 1963, a three-fold increase. About two-thirds of these were made when the aircraft was below an altitude of 1,000 feet, and none aircraft were lost. Of 24 November 1963 the enemy hit 24 U.S. and VNAF aircraft and helicopters, destroying five a one day high in the war. During the last three months of the year, 124 USAF and VNAF aircraft were hit, some with .50 caliber weapons. From November 1961 to March 1964, 114 U.S. aircraft were lost in South Vietnam: 34 USAF, 70 Army (including 54 helicopters), and 10 Marine (all helicopters).[28]

Air War – Vietnam

As antiaircraft fire, mechanical failure, and difficult terrain increased the aircraft attrition rate in 1963 and contributed to several B-26 and T-28 crashes, some Air Staff officers thought that the rules of engagement for U.S. aircraft should be changed to allow deployment of B-57 and F-100 jets. However, McNamara in March 1964 instead approved an Air Force proposal of September 1963 to replace the B-26s and T-28s with A-1Es.[29]

V. THE DISPUTE OVER AIRPOWER

As air support assumed a greater role in South Vietnam, Air Force-Army tension mounted over its use and control. Disagreements boiled to a head after a Vietnamese attack at Ap Bac, about 30 miles south of Saigon, on 2 January 1963. During the battle, Viet Cong ground fire hit 11 of 15 U.S. Army helicopters supporting the attack, downing five. The enemy inflicted severe losses, killing 65 Vietnamese and three Americans and wounding more than 100 Vietnamese and 10 Americans. For more than an hour, enemy fire pinned down 11 U.S. personnel.[1]

In reviewing the incident, Army officers accused the Vietnamese of lacking aggressiveness and refusing to heed advice. But the Air Force charged that the Army had failed to call on fixed-wing aircraft for cover because it was carrying out a close-support test of its armed helicopters. The two services could not agree on the reasons for the defeat.[2]

The JCS Review

Because of this disagreement, McNamara and the JCS decided on 7 January to send to South Vietnam a team of senior JCS and service representatives headed by the Army Chief of Staff, Gen. Earle G. Wheeler. Before the team left, service briefings laid bare doctrinal differences over the use of airpower in counter-insurgency operations. The Air Force believed that its system could meet any counter-insurgency requirements for reconnaissance, quick reaction, close support, air cover for helicopters or convoys, delivery of airborne troops and supplies, casualty evacuation, and communications. The Amy, conversely, maintained that it alone should be responsible for counter-insurgency since its organic air arm, weapons, and tactics were especially suited for land operations. It viewed the work of USAF's Special Air Warfare Center as trespassing on a mission traditionally assigned to the Army and Marines. The lessons learned about airpower in World War II and Korea, it argued, did not necessarily apply to South Vietnam where aircraft did not need to be as fast and where they needed to be based near the target. The Army demanded decentralized control of airpower in order to use its own support aircraft, whereas the Air Force wanted centralized control. Army and Air Force definitions of "close support" clearly differed.[3]

The JCS team went to South Vietnam, assessed military operations, and concluded in February that the United States should maintain its current level of aid for the Diem government and follow the three-year comprehensive plan for phasing out U.S. support. In commenting on the use of airpower, the team said that the Harkins-Anthis relationship was satisfactory but there were weaknesses in joint planning of air activities, reporting helicopter movements, and conducting logistic airlift. The team offered to furnish Harkins with experts to resolve airlift problems, but it thought that the joint planning and reporting difficulties could be ironed out at lower levels.[4]

Air War – Vietnam

Lt. Gen. David A. Burchinal, Deputy Chief of Staff for Plans and Programs
Source: U.S. Air Force

In a separate report, the USAF team representative, Lt. Gen. David A. Burchinal, Deputy Chief of Staff for Plans and Programs, Headquarters USAF, noted that the solution in South Vietnam depended on military, political, and economic factors, and he was less optimistic about an early victory. The Administration should cancel political restrictions and operations outside South Vietnam and on crop destruction. It should also give more authority to the American Ambassador in Saigon and to MAC/V. In the air war, Burchinal foresaw the need for jet aircraft, since conventional aircraft would become more vulnerable to Viet Cong automatic weapons. He recommended to Wheeler the return of test projects to the United States, removal of Howze Board issues, and a curb on the Army's generation of air requirements. *(The Army Tactical Mobility Board (known as the Howze Board after its chief, Lt. Gen. Hamilton H. Howze) recommended on 31 July 1952 that the U.S. Army assume part of the tactical close support mission. The board proposed that the Army obtain large numbers of fixed-wing aircraft, including transports and helicopters, and be responsible for their use and control. To the Air Force, this meant an encroachment upon a traditional USAF mission.)*

Burchinal believed that all aviation units should report to the JAOC, that armed helicopters should not be deployed until their usefulness had been determined, and that they then should operate under the same rules of engagement as Farmgate aircraft. He also urged assignment of a three-star USAF air deputy to the MAC/V staff, and the establishment of Army and Navy component commands similar to the 2d Air Division.[5]

As a result of the JCS team review, the Air Force won minor concessions, such as four more officer spaces on the MAC/V staff and Army support for an air deputy commander. But the limits and restraints on Farmgate operations remained in effect because the Administration was determined not to risk escalating the war and the Army largely controlled the U.S. military effort in South Vietnam.[6]

The Interdiction Issue Again

In March 1963 the Department of State again raised the subject of interdiction. Observing that Farmgate training aircraft flew numerous sorties of this type each month, W. Averill Harriman, Assistant Secretary of State for Far Eastern Affairs, solicited the views of Ambassador Nolting in Saigon. Harriman thought that air interdiction should be employed only against clearly defined enemy territory, He conceded that targeting procedures had improved and that no reliable information had indicated any undesirable effects. But, he stressed the political nature of the war, Vietnamese resentment against air strikes that might aid Viet Cong recruitment, the unsuccessful interdiction experience of the French, the political unawareness of provincial and district chiefs who supplied target information, and the restrictions of the 1954 Geneva agreement. To Harriman, the basic question was the political cost versus the military advantage of interdiction, whether by U.S. or Vietnamese pilots.[7]

Headquarters USAF considered the Harriman analysis as not wholly accurate and representing the views of only a small but influential minority in the State Department. The Air Staff especially disagreed that the war was only political or that occasional harm to innocents created a military problem. USAF planners thought that the State Department officials should study ground combat as well as air action when they assessed the effects of civilian casualties. The airmen noted that the small Farmgate-VNAF force had caused an important percentage of Viet Cong casualties. In April, Ambassador Nolting's reply to Harriman dispelled USAF concern. He recommended continuation, where necessary, of Farmgate interdiction-type sorties to restrict enemy movements, supplement VNAF efforts, and aid the national campaign plan.[8]

Because the interdiction issue again had been raised, Gen. Anthis in May explained again to U.S. officials the detailed and time-consuming method used to select and confirm targets. In interdiction sorties flown since January 1962, the targets selected were primarily enemy concentrations or buildings either used by the Viet Cong or abandoned by Vietnamese who had moved to strategic hamlets. By day, Farmgate crews hit targets only when marked by a VNAF forward air controller; by night, only targets illuminated by a C-47 flare ship in radio contact with Vietnamese ground forces. Military officials investigated all reports of targeting errors and, of 10 recent allegations, had verified only two.[9]

Although a State Department representative expressed concern about Farmgate combat training, McNamara made no comment. In May 1963, OSD and the JCS decided not to take any further action on this issue for the time being, but the Air Force expected that it would come up again.[10]

The Problem of Army Aviation

Despite the steadily-rising Farmgate sortie rate, the Air Force believed that the full potential of its air resources was not being employed. One reason was the rules of engagement that clearly limited USAF participation. Combat training sorties were permitted only if the VNAF lacked the necessary training and equipment and if combined USAF-VNAF crews were on board. There were also the time-consuming target identification procedures. In July 1962 PACAF urged that the provision requiring the presence of a Vietnamese crew member be rescinded, but Headquarters USAF could not overcome State and OSD objections.[11]

The major obstacle to the enlargement of the Air Force role in South Vietnam, however, was the U.S. Army. Its aviation arm, consisting of Mohawk, Caribou, and liaison aircraft and helicopters, grew by December 1962 to about 200 while the Air Force had only 63. In its support role, the Army frequently followed Howze Board concepts and used its aircraft outside the centralized tactical air control system (TACS) rather than call upon Farmgate and VNAF units. This practice brought the Army into a continuing, abrasive conflict with the Air

Force.[12]

After examining the TACS in operation, Lt. Gen. Gabriel P. Disosway, Deputy Chief of Staff for Operations, Headquarters USAF, concluded in December that its potential was high. He decried the Army practice of ignoring it because this led, in effect, to two separate tactical air control systems; one Air Force, the other Army. The Air Force thought that centralized control was a necessity. In a special forces attack on 10 August, for example, the Army had neither planned nor called upon the TACS for air cover, and the Viet Cong had escaped.[13] Another problem arose when USAF air liaison officers (ALOs) were assigned to ARVN divisions to advise them on air support. The Army insisted that these ALOs advise only the U.S. Army senior advisor to the ARVN commander. This dispute was fundamental, since it could determine whether Farmgate and VNAF or U.S. Army aviation would be employed for specific operations. Starting in mid-1962 USAF and Army leaders in South Vietnam tried to resolve this issue, but they had not succeeded by the end of 1963.[14]

In November 1962 Headquarters USAF acknowledged the lack of timely and accurate air intelligence and quick, reliable response to requests for air support. It ascribed this partly to inadequate delegation of authority within the Vietnamese forces, slow development of the VNAF and insufficient Vietnamese appreciation of and confidence in tactical airpower. But the Air Staff added that two contributing factors were the assignment of only Army intelligence advisors; 28 in all; to the single intelligence agency responsible for targeting and the requirement that forward air controllers report through an airborne air controller rather than directly to strike aircraft.[15] The Air Force also believed that the Army did not comply fully with the rules of engagement. Farmgate pilots, complying with combat training rules, flew in VNAF-marked aircraft, always carried a Vietnamese crew member, and received no official publicity. Army Mohawk and armed helicopter pilots seemed to interpret the rules more freely and engaged in close support missions, flew in U.S. marked aircraft, often did not carry a Vietnamese crew member and received official publicity.[16]

When U.S. forces began to support air-ground operations, USAF and VNAF ground communications for tactical air control were grossly incompatible with those of the Army. As a consequence, the services decided early in 1962 to retrofit AN/ARC-44 sets on all aircraft. But the Army, which administered the procurement contract, gave first priority to retrofitting its own aircraft rather than those of the Air Force and VNAF. After the OSD and JCS interceded, the Army agreed in June 1963 to meet the needs of the U.S. and Vietnamese Air Force.[17]

The two services also differed as to whether the Army's Caribou was preferable to the larger C-123 in counter-insurgency operations. The Army, using its own parameters, "proved" that the Caribou was more suitable because it could use 147 airfields in South Vietnam and the C-123 only 70. USAF analyses disproved this assertion.[18]

Despite USAF objections, the role of Army aviation in South Vietnam continued to expand. On 8 July 1963, MAC/V tightened Army control of air operations by establishing an aviation headquarters in each Vietnamese corps to plan and control Army and Marine aviation supporting it. In December the Army had 325 airplanes, or 47 percent of the 681 employed in South Vietnam. The Air Force had 117, the VNAF 228, and the Marines 20.[19]

Air War – Vietnam

The Army claimed that its Caribou transport aircraft were better suited to Vietnamese conditions. Source: U.S. Air Force.

Problems of Command Relations

*General Paul Donal Harkins
Source: U.S. Army*

The Air Force strongly believed that it could remove some of the restraints on USAF activities if it obtained a larger voice in the councils of the Army-dominated MAC/V. In April 1962, during a JCS meeting with McNamara, LeMay had charged that air planning often was omitted, that Anthis had difficulty seeing Harkins, and that neither Harkins nor his Chief of Staff, Marine Corps Maj. Gen. Richard G. Weede, properly understood air operations.[20]

Felt replied that Harkins and Weede were superior officers who were fully experienced in air-ground tactics and that Anthis could see Harkins at any time. He acknowledged inadequacies but noted that the VNAF was learning quickly and that the occasions when airpower was not used but

Plans And Polices 1952-63

should have been were exceptions rather than the rule.

Felt's detailed control also chafed the Air Force, since he assigned air units to MAC/V and fragmented USAF units among subordinate elements, limiting the responsibilities of both O'Donnell and Anthis. O'Dorrrell's primary authority consisted largely of providing logistic support or correcting problems reported by 13th Air Force or 2d ADVON. Gen. Disosway observed in December 1962 that the Air Staff did not always understand this.[22]

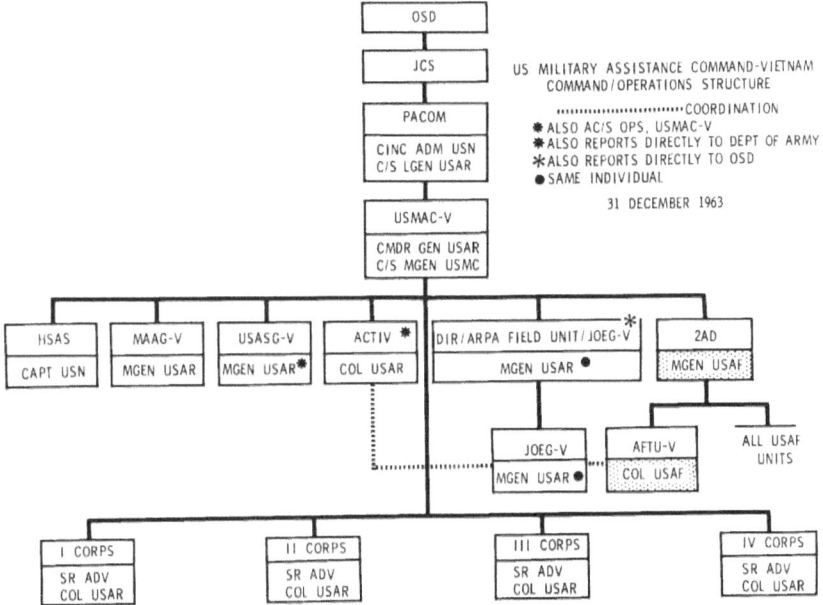

Since the Air Force had been denied the posts of chief of staff and chief of J-1, it urged the assignment of a three-star Air Force deputy commander to Harkins. Harkins and Felt agreed, and the JCS concurred on 22 August 1962, but McNamara decided in October that such a post was unnecessary. The Air Force then tried to secure the post of chief of staff when the Marine incumbent departed. But the Marine Corps adamantly opposed this, and the effort was abandoned. As noted earlier, the JCS team review early in 1963 resulted in four more officer spaces for the Air Force, two in J-3 and two J-4, but this was considerably less than it desired.[23]

In September Harkins and Felt agreed that the post of chief of staff should be filled by an Air Force general on 1 June 1964. They also agreed that five more administrative slots should go to USAF personnel. The JCS approved their decisions on 7 November. On 2 December, however, President Johnson directed the JCS to certify only "blue ribbon" men to MAC/V. After this injunction and another visit to South Vietnam, McNamara approved on 6 January 1964 the designation of Army Lt. Gen. William C. Westmoreland as deputy commander

Air War – Vietnam

and the transfers of J-1 from the Navy to the Army and J-2 from the Air Force to the Marine Corps. In the latter instance, the Air Force chief of J-2 was downgraded to deputy J-2 and, on orders of LeMay, reassigned.[24]

At the end of 1963, the Army held six of the nine top positions on the MAC/V staff (commander, deputy commander, J-1, J-3, J-4, and J-6), the Marine Corps two (chief of staff and J-2), and the Air Force one (J-5). Of 335 positions allocated in early 1964, the Army held 199, the Air Force 75, the Navy 42, and the Marine Corps 19. The Army was now in firmer control of planning and operations in South Vietnam than before. Reflecting this preeminent position, the Army had about 10,100 of the nearly 16,000 U.S. troops in the country at the end of 1963. The Air Force had 4,600, the Navy and Marine Corps 1,200.[25]

VI. TESTING CONCEPTS AND WEAPONS

As part of the program of action approved on 29 April 1961, President Kennedy authorized a combat development test center (CDTC) in South Vietnam. Composed of Americans and Vietnamese, CDTC was placed under the Vietnamese Joint General Staff in Saigon. In August it began experimenting with various projects, including the use of chemicals to destroy jungle foliage. The Americans in its field unit were members of OSD's Advanced Research Projects Agency (ARPA).[1]

On 5 September McNamara informed the services and other U.S. agencies that he wished South Vietnam to be a "laboratory for the development of organization and procedures for the conduct of sub-limited war." Some "laboratory" activities quickly became Army-Air Force combat test programs that engendered heated controversy over the use of tactical airpower.[2]

Supervision of Testing

To the Air Force, the Army desire to "verify" its Howze Board concepts by testing its aircraft in combat support in South Vietnam was an attempt to preempt certain traditional USAF roles and missions. In July 1962 LeMay proposed that a joint operational evaluation group (JOEG/V) in South Vietnam conduct meaningful tests to meet stated objectives. He hoped thereby to restrain the Army from introducing air units and equipment into Southeast Asia under the guise of testing. The JCS agreed, and on 21 July Felt established the group under the operational control of Harkins. Under its terms of reference, the JOEG/V would approve or disapprove test proposals by the JCS, the services, and other agencies. It would evaluate only combat tests having joint service implications.[3]

Since the ARPA Field Unit of CDTC was outside U.S. military channels, the JCS proposed that it too be placed under Harkins' operational control. McNamara decided instead to combine the administration of the unit and JOEG/V and create the post of director for both. The JCS and Harold Brown, OSD's Director of Defense Research and Engineering (DDR&E), selected Army Brig. Gen. Robert A York for the post, and McNamara approved his terms of reference on 31 October. York was responsible to Brown for CDTC activities and to Felt, through Harkins, for evaluating military operations and tests. All commands and services coordinated their tests both with York and the Vietnamese, included York's conclusions on test results, and made then available to the proper agencies.[4]

This centralized supervision of testing proved short-lived, On 11 September 1962 secretary of the Army Cyrus R. Vance proposed establishment of a separate Army test unit in South Vietnam. LeMay opposed this move vigorously in the JCS, arguing that it would duplicate JOEG/V functions, result in narrow conclusions, and permit the Army to transgress upon traditional USAF missions of close support, escort of airborne forces, and combat air cargo. The Navy and Marine Corps sympathized with the Army proposal, however. Felt also concurred with the

Army, provided that the test personnel and equipment remain in South Vietnam only for the duration of the project. In October McNamara formally approved the Army plan.⁵

Headquarters USAF then weighed various PACAF suggestions and decided that the Air Force also needed a special unit in South Vietnam to test concepts, tactics, aircraft, ordnance, and support equipment. These would complement but not duplicate special air warfare tests at Eglin AFB, Fla. Acting under OSD and JCS directives, LeMay in January 1963 ordered the establishment of a 12-man test unit as a special staff section within the 2d Air Division. ⁶

The Army Concepts Testing in Vietnam (ACTIV) was established in November 1962 as a permanent unit that would require initially 97 men. Since the Air Force would have a test unit also, Felt objected to this size. He approved the deployment of ACTIV on 7 January 1963 only after its roster had been trimmed to 50 and additional personnel assigned on a temporary basis. Sharing somewhat the Air Force view on this matter, Felt informed the JCS that the use of South Vietnam as a "test bed" was beclouding the primary U.S. objective of assisting the war effort.7

The JCS team that visited Vietnam early in L963 decided that there were too many test organizations and projects in that country and that their contributions should be appraised by MAC/V.⁸ In his separate report, Burchinal recommended, as had others in the Air Staff, that all-testing be withdrawn from Vietnam since it disrupted the task of defeating the Viet Cong. Subsequently, LeMay urged vigorously but unsuccessfully that U.S. Strike Command test divergent service concepts and doctrines. He decried interservice debates in the presence of an ally. He also pointed out that the Army did not withdraw its test units, thus adding to costs and logistic problems.⁹

Felt agreed with the JCS team that he and Harkins were in the best position to determine the validity of a test project. If they did not agree, the decision could go to the JCS; if ARPA desired a project despite JCS recommendations, the decision could rest with the Secretary of Defense. With the consent of DDR&E, the JCS submitted a similar recommendation to McNamara who approved it on 23 April.10

In May, the JCS asked Felt to prepare new terms of reference for consolidating combat development with research and development testing and engineering. The JCS then became deadlocked over an Air Force proposal to rotate the position of chief of this combined activity among the services and an Army proposal to delete a requirement that the JCS settle test problems affecting roles and missions. The Army objected to the first proposal because of its predominance in Vietnam, the Air Force to the second because only in the JCS did it possess a strong voice and possible veto on measures vital to its interests. And when Felt recommended that the combined activity be placed within military channels under Harkins, this was opposed by ARPA which favored a joint field agency with the commander responsible to both ARPA and Harkins.¹¹

Plans And Polices 1952-63

Reluctant to send a split paper to McNamara, the JCS finally asked its Chairman, General Maxwell Taylor, to decide upon the terms of reference. Taylor accepted some Air Force suggestions, but in the key decision he sided with the Army by deleting the requirement that projects with roles and mission implications be submitted to the JCS for approval. This gave CINCPAC rather than the JCS responsibility for settling such matters. In early January 1964 the terms of reference went to OSD.[12]

Test Results

The Army began unilateral testing in late 1962, the Air Force in early 1963. In conjunction with combat or special forces operations, the Army evaluated the Mohawk, armed helicopters, and the Caribou. In his February report, Burchinal declared that the Mohawk tests were designed to show how this aircraft could perform at less cost the USAF missions of artillery spotting, fire adjustment, reconnaissance, airborne command and control, and flank security. He believed that a test of this plane under combat conditions was unnecessary and added that USAF experience demonstrated that Army field maintenance for the Mohawk was unduly expensive and inefficient.[13]

After the JOEG/V-ARPA Field Unit evaluated the Mohawk tests, the JCS split over the conclusions. The Air Force disagreed that the Mohawk had "fully documented" its offensive capability and that Army direct, decentralized control showed better results than the centralized control exercised by the TACS. the Air Force also objected that the JOEG/V-ARPA Field Unit had violated its terms of reference by commenting on doctrinal issues.

Results of trials evaluating the OV-1 Mohawk were controversial.
Source: U.S. Army

Moreover, it stated that the unit's comparisons with other aircraft operating under different rules with different missions were invalid.[14]

Burchinal also had considered Army tests of armed helicopters to be of dubious value because no fixed-wing aircraft were employed for making comparisons. Army statistics on antiaircraft hits had omitted flying time and failed to differentiate between combat and combat-support sorties. LeMay pointed to the vulnerability of helicopters to ground fire, their weakness as "firing platforms" and the Marine Corps desire for fixed-wing aircraft as cover for its helicopters.[15]

The JOEG/V-ARPA Field Unit concluded, however, that armed helicopters were the most effective, single, aerial system for counter-insurgency and that they should provide the additional close support that fixed-wing aircraft could not give. Harkins thought the evidence insufficient to support the first conclusion, and Felt questioned the statistics indicating armed helicopters effectively suppressed ground fire. The Air Force questioned both conclusions. The JCS agreed with the

Air War – Vietnam

critics but split over whether the tests indicated a requirement for armed helicopters to protect transport helicopters. The Air Force believed, of course, that they did not.[16]

Armed helicopter evaluations were also disputed
Source: U.S. Army

In December, the JOEG/V-ARPA Field Unit concluded that the Army's Caribou tests demonstrated this transport's "extremely advantageous" characteristics for counterinsurgency, citing its short take-off and landing capabilities, light wheel pressure, and load adaptability. According to the testers, the Caribou could use air strips in the Mekong delta that heavier aircraft could not. They claimed that the Caribou was no more comparable to the C-123 than a two and a half-ton truck to a five-ton truck. On the merits of centralized versus decentralized control of the Caribou, they maintained that aircraft near a field commander were more responsible than those removed from his control. By the end of the year the JCS had not completed its study of this evaluation, but it was clear that the Air Force would not agree.[17]

Meanwhile, the Air Force unit had tested the YC-123H (a modified C-123B, capable of short-field take-off and landing), the U-10, and the Decca tactical air positioning system (TAPS).+ It concluded that the YC-123H could fulfill most airlift requirements in South Vietnam, operate from 88 percent of the airfields in that country, and almost satisfy the long-standing requirement for a 10-ton short take-off and landing aircraft with a 500 nautical-mile radius. The JOEG/V-ARPA Field Unit accepted this assessment but noted that Harkins believed this plane complementary to the Caribou, while the Air Force deemed it competitive.[18]

The U-10 had merits but was deficient in many areas. Source: U.S. Air Force

USAF testers decided that the U-10 was excellent for psychological warfare, support airlift, visual and manually controlled reconnaissance, and short take-off and landing. Forward air controllers had found it unsuitable, however, as it was also vulnerable to ground fire, had poor cockpit arrangements, and was not sufficiently maneuverable at high speeds. The JOEG/V-APRA Field Unit did not disagree.[19]

Tests of the TAPS indicated that it was promising but that its MK VII airborne equipment had experienced a major malfunction. As a consequence, the JOEG/V-ARPA Field Unit stated tentatively that the system was unreliable. Before it reached a definite conclusion, it awaited completion of ACTIV tests to determine whether TAPS was adaptable to helicopter operations.[20]

At LeMay's direction, the 13th Air Force used operational records to make tactical analyses of other USAF aircraft. The analysts assessed the T-28B as extremely effective and the B/RB-26 as effective also. But both planes were hindered by stringent target identification requirements, a shortage of VNAF crew members, and incompatible air-ground communication equipment. The analysts described the B/RB-26 as deficient in maneuverability, rate of climb, and dive angle capability, but they recommended its retention until the Air Force could replace it with a more suitable aircraft.[21]

To the analysts, the C-123B was a successful airplane and its replacement by the Caribou would be economically unsound and detrimental to counterinsurgency operations. They found that the TF-102 had demonstrated its identification capability in daylight.[22]

PACAF proposed a test of USAF tactical air support concepts, and the Air Staff in September 1963 requested that command to make the necessary preparations. This test would provide statistics on reaction times, responsiveness, and results of air strikes based on requests that used the USAF-operated TACS.[23]

Despite its interest in these tests, Headquarters USAF remained strongly convinced that testing in South Vietnam should cease because it interfered with the conduct of counter-insurgency operations. But OSD and the other services disagreed.[24]

Defoliation

The United States not only tested the effectiveness of defoliation as a counterinsurgency technique but also conducted defoliant operations against the Viet Cong. The spraying of jungle vegetation and crops had a twofold objective: reducing the danger of enemy ambushes and denying food to the Viet Cong. CDTC began testing in August 1961 but no large-scale operational plans were drawn up until after the Taylor Mission. On 21 November, Deputy Defense Secretary Roswell L. Gilpatric outlined for President Kennedy a carefully-controlled defoliation plan that was designed to support CINCPAC's outline campaign plan. To guard against ambushes, he proposed spraying a swath 200 yards wide on each side of the principal roads between Saigon and other key cities, roads peripheral to Zone D (the area near Saigon controlled by the Viet Cong), and Cambodian border areas through which guerrillas infiltrated. Gilpatric advocated spraying to deny food only after the friendly population had been resettled and fed. Six USAF C-123s would carry out tactical and border-control operations and specially-equipped Vietnamese helicopters, similar to those used by the British in Malaya in 1953, would destroy crops. He estimated that the program would cost $8 to $10 million dollars.[25]

Administration officials debated how the defoliation missions should be carried out. OSD and JCS favored open participation by aircraft and crews carrying USAF designations. The State Department, apprehensive about possible criticism by the International Control Commission, desired aircraft with Vietnamese markings and USAF crews in civilian attire. It was finally agreed that defoliation missions flown by USAF aircraft and crews should carry a Vietnamese crew member. Vietnamese markings were used only on a few special occasions. In Saigon, MAAG/V and Vietnamese officials worked out details of the Gilpatric plan. Harkins believed that defoliants would be effective in Zone D which had relatively few people, but Ambassador Nolting thought that their use might alert the Viet Cong. In December they agreed that defoliants could aid but not "win the battle" in that zone, an the Outline Campaign Plan was changed accordingly.[26]

Meanwhile, the Air Force deployed six C-123s and 69 men from TAC's Aerial Spray Flight at Langley AFB, Va., and Pope AFB, N.C. The aircraft, crews, and support personnel reached Clark AB on 6 November, and in January 1962 they proceeded to Tan Son Nhut Airfield. On 13 January three C-123s began spraying along 16 miles of a road between Bien Hoa and Vung Tau. They did not spray in Zone D since this was declared temporarily impractical.[27]

Defoliation remains probably the most controversial of all U.S. operations that formed part of the Vietnam War. The defoliant used, "Agent Orange" has become a by-word for counter-productive actions. Even fifty years later, the effects of the defoliation program remain a major problem for U.S. – Vietnamese relations. Picture source: U.S. Air Force.

As expected, Viet Cong propagandists attributed all dying plants to the spraying and warned that the chemicals had harmful effects. Certain Vietnamese claimed property damage from spraying, and a Vietnamese board evaluated the claims.

Some were valid, some were not. Ambassador Nolting feared that unsuccessful claimants night become antagonistic.[28]

In May1962, Harkins reported that in 21 areas sprayed, air-to-ground visibility had improved by 70 percent, ground visibility by 60. He thought that the C-123s could have achieved even better results with improved spraying gear and more herbicides. A subsequent evaluation indicated that defoliants were particularly useful in destroying mangrove but their effects had been overestimated in areas of mixed vegetation. Felt urged the JCS to authorize the spraying of grass, weeds, and brush around depots, airfields, and fields of fire. In the delta, 112 guerrillas had been frightened by defoliants and had surrendered, and Felt asked for an evaluation of psychological effects. He believed that in the future only three C-123s would be needed for defoliant operations.[29]

Communist propaganda and international negotiations on Laos prompted President Kennedy on 2 May to halt defoliation in South Vietnam temporarily and direct that testing continue in Thailand. The C-123s resumed spraying in South Vietnam from 1 September to 11 0ctober and achieved excellent results, according to the JCS, by using more herbicides and larger droplets. In six different areas, these sprayings using three gallons of defoliant per acre killed about 95 percent of the vegetation within 10 days. When one gallon per acre had been used in earlier operations, it took 20 to 60 days to obtain similar results.[30]

Because of these successful tests, the JCS recommended the following: (1) authority for Nolting and Harkins to order non-crop destruction projects; (2) defoliation around four communication routes and one power line; (3) additional testing of improved chemicals, dispersal equipment, and delivery techniques in the United States and the Panama Canal Zone; and (4) more attention to psychological aspects.[31]

On 13 October, Gilpatric agreed with the JCS that testing outside South Vietnam was necessary and that psychological aspects deserved more attention. He noted that DDR&E was stepping up research with herbicides. And on 27 November President Kennedy approved the other recommendations by authorizing Nolting and Harkins to order destruction of vegetation except crops, and by designating five new areas as defoliation targets.[32]

Meanwhile, crop destruction plans had been intensively reviewed. McNamara, Felt, Harkins, and Nolting favored a trial project, the spraying of 2,500 acres in Phu Yen province, but Secretary of State Rusk and Assistant Secretary Harriman were opposed. Rusk thought there was insufficient evidence that the crops belonged to the Viet Cong, feared adverse international reaction, and warned that a premature program could prompt the Viet Cong to step up attacks against strategic hamlets. Observing that the way to win a guerrilla war was to win the support of the people, Rusk argued that crop destruction ran counter to this rule. At best, he thought it should be attempted only in the latter stages of an anti-guerilla campaign.[33]

By the late summer of 1962 the maturity of crops and continued State Department opposition led to abandonment of the plan for spraying crops in large areas of Phu

Yen. Shortly afterwards, however, a limited program was approved for Phu Yen and Thau Thien provinces, which included spraying crops abandoned by Montagnard tribesmen to prevent their use by the Viet Cong. Thereafter, because of the delay in getting JCS approval and the advent of the dry season, there were no spraying projects until February 1963 when they were resumed until May. During this latter period, in accordance with Felt's recommendation, the number of USAF spray-equipped C-123s was cut to three and support personnel to seven officers and 12 enlisted men. [34]

In April, the JCS summarized defoliation operations since their inception. The aircraft had sprayed along 87 miles of roads and canals, around military installations, and on 104 acres of crops in two provinces. Herbicides had destroyed about 756,000 pounds of food without adverse effects on friendly Vietnamese. Conceding that it was difficult to measure military effectiveness precisely, the JCS thought that the benefits to reconnaissance from improved visibility and enhanced security made defoliation desirable and urged its continuation. The JCS believed that proper counter-propaganda actions would offset any adverse Communist charges.[35]

On 7 may, however, new State-OSD guidelines on defoliation contained so many restrictions that few operations were conducted afterwards. The Department of State basically opposed defoliation, especially crop destruction, because it might have adverse effects on friendly Vietnamese which the Communists could exploit. A small project was carried out in June, but a request to spray a 3,000-acre crop area was not approved at year's end. Ambassador Nolting and Felt again vouched for the usefulness of defoliation and recommended it as more efficient than the Vietnamese practice of burning, pulling, or cutting, but noted that the time-consuming procedures required for obtaining approval of defoliation missions negated their effectiveness. Because of the political restrictions and the limited period during the year that defoliation operations could be carried out, at the end of 1963, some military officials were seriously considering abandonment of the whole program.[36]

VII. USAF SUPPORT OF THE VIETNAMESE AIR FORCE

When the United States decided in late 1961 to step up its military assistance to South Vietnam, Headquarters USAF faced the task of enlarging an extremely small Republic of Vietnam Air Force. Some reasons for the VNAF's limited capability were inherent, such as the difficulty of quickly training poorly-educated Vietnamese. But the Air Force believed that another reason for VNAF weakness was the fact that the Army-dominated MAAG/V failed to appreciate the important role airpower could play in counter-insurgency. For example, the January 1961 agreement to increase the Vietnamese armed forces by 20,000 men included only about 500 spaces for the VNAF. Again, the border patrol proposed in the April program of action led to no immediate decision on VNAF employment. In mid-1961 the Air Force thought that VNAF's 4,765 men and 142 aircraft were much too small a part of a total Vietnamese military strength of about 170,000.[1]

A Vietnamese Army Air Force?

The U.S. Air Force was disturbed by U.S. Army efforts to encourage the Army of Vietnam to establish its own air force. In September 1961 U.S. and Vietnamese diplomatic and military representatives, including President Diem, agreed to four ARVN aviation units. U.S. Army officials then planned to transfer some VNAF aircraft to ARVN to carry out this agreement.[2]

When McNamara asked the JCS in 0ctober to review this proposal, that body could not reach an agreement. The proposal contravened long-established Air Force doctrine, and LeMay objected vigorously. He argued that the VNAF's administration, logistic, and maintenance responsibilities could not be separated from its operational activities. If divided, it could delay massing available airpower against a large opposing force. And, if the forces of the southeast Asia Treaty Organization entered the war, an air component would be needed to control all air power that might be used.[3]

In December, Felt asserted, and O'Donnell agreed, that "teamwork" rather than reorganization was necessary. McNamara then decided against an ARVN air corps, but he added that the VNAF needed to become more responsive to the requirements of ARVN corps commanders. Nevertheless, MAAG/V and then MAC/V continued to encourage the formation of an ARVN air corps, but without success.[4]

Build Up Of The VNAF

The Air Force provided aircraft, helicopters, and training personnel for the VNAF. Since USAF T-28Bs were not immediately available, the U.S. Navy in December 1951, sent the VNAF 16 T-28Cs and training personnel. The aircraft remained in the inventory. By April 1962, however, the Air Force had supplied the Vietnamese with 30 T-28Bs, a 52-man T-28 training unit, and 30 C-47 aircraft and pilots. Besides training VNAF personnel to fly C-47s, these pilots airlifted

Air War – Vietnam

livestock to Vietnamese outposts, quickly earning the sobriquet of "dirty thirty". They served until December 1963, logging about 20,000 flying hours.[5]

In April 1962 Gen. Anthis reported that VNAF training was proceeding satisfactorily although there were problems in training inadequately-educated Vietnamese to become pilots, mechanics, and radar specialists. Students had difficulty using the English language properly. It was also troublesome to obtain security clearances quickly for prospective pilots who were scheduled to train in the United States, especially after two dissident VNAF members bombed the government palace in February 1962. Another difficulty concerned some VNAF C-47 pilots who had been trained by the French and were reluctant to change their flying techniques.[6]

In 1962, the AD-6 (A-1H) was the most capable fighter in VNAF service. Source: U.S. Air Force

In April the VNAF possessed 63 fighters (19 AD-6s and 44 T-28s) and 117 support aircraft (C-47s, L-19/20s and H-34Cs). During the month, LeMay and an Air Staff group inspected the VNAF and found its fighters marginally adequate. The VNAF, the group decided, needed improved planes and more and better trained T-28 pilots. The VNAF commander, a colonel, had too low a rank compared to his ARVN counterpart. The group also supported the desire of the Diem government to obtain jet aircraft.[7]

The three-year comprehensive plan to train and equip the Vietnamese to defend themselves and to phase out major U.S. activities proposed by the JCS in July 1962, called for the Vietnamese regular and paramilitary strength to reach a peak of 575,000 in fiscal year 1964 and decline thereafter. (the JCS integrated this plan with the 1964-1969 military assistance program and the national campaign plan). The size of the VNAF would reach 15 squadrons (three fighter, four transport, one reconnaissance, four liaison, and four helicopter). To modernize the Vietnamese air arm, the United States would provide non-jet A-1H and jet F-5A/B fighters and non-jet RT-28 and jet RF-5B reconnaissance aircraft. These planes would be added to the six T/RT-33 jets programmed for delivery which the State Department had not yet approved. Two C-123 squadrons would strengthen VNAF transport capability. In the critical 1961- 1965 period, VNAF strength would rise to about 9,000 men.[8]

To enable the VNAF to absorb the new equipment and to reduce language and security problems, PACAF proposed that a larger portion of VNAF training be conducted in South Vietnam. The JCS approved the proposal on 25 April 1963, and Diem heartily endorsed it. (Earlier, the Vietnamese leader had informed Zuckert that 61 percent of VNAF training should be in South Vietnam and only 39 percent in the United States.[9]

Plans And Polices 1952-63

On 6 May, McNamara concluded that 1964 - 1969 MAP funds for South Vietnam would be insufficient to carry out the large contemplated program. Since an F-5 cost about $1 million, he vetoed the proposal to equip the VNAF with it on grounds of cost-effectiveness. A revised program for training more members of the VNAF in South Vietnam was quickly prepared and approved by McNamara on 27 May. It provided for the purchase and deployment of 25 U-17As plus a USAF detachment to train VNAF personnel in their use. It also augmented a USAF helicopter training detachment that had arrived in South Vietnam in January 1963. By December, when the VNAF had 228 aircraft, the stepped-up training program was well under way.[10]

South Vietnamese U-17A at Nha Trang. Source: U.S. Air Force

Meanwhile, on 1 July 1963, the government increased the VNAF personnel authorization from 7,651 to 8,897. In December it possessed 8,496 men: 805 officers, including 375 pilots, and 7,691 enlisted men. Although the Air Force trained most members of the VNAF either in South Vietnam or the United States, the U.S. Army and Navy also gave some assistance. Despite its efforts to make the VNAF operationally self-sufficient, the Air Force expected the shortage of aircraft control and warning, maintenance, and other technical personnel to continue until fiscal year 1965.[11]

The Problem of Jet Aircraft

From 1961 through 1963 Headquarters USAF strongly supported the assignment of jets to the VNAF for use in border surveillance. Assuming that these planes would eventually be authorized, the Air Staff programmed six T/RT-33s for the Vietnamese in the fiscal year 1961 USAF military assistance program.[12]

In October 1961 OSD and the JCS agreed that VNAF jet training was imperative because of the growing Viet Cong threat, the unstable situation in Laos, and the growing obsolescence of the AD-6s. On the 19th, the State Department instructed Ambassador Nolting to inform the Diem government that the United States would train Vietnamese to fly the six T/RT-33s. It asked the government not to publicize the offer until the two countries reached a decision concerning observance of the Geneva agreement which prohibited the use of jets in South Vietnam. After the Vietnamese completed their training, the United States would transfer the jets when it believed that they were needed and the pilots were able to fly them properly.[13]

In July 1962, after training had begun, McNamara questioned whether jets were needed in South Vietnam in place of conventional aircraft. He believed that the time had not yet come to violate the Geneva agreement. The Air Force, Felt, and Harkins urged the transfer of the six aircraft without delay, however.[14]

Air War – Vietnam

The T-33 was to have been the VNAF's first jet. Source: U.S. Air Force

In January 1963 the JCS also asked McNamara to authorize the transfer of jets, citing his statement of 8 October 1962 that called for a VNAF that could satisfy requirements. The JCS noted that better reconnaissance and other aircraft were needed for stepped up military operations and to counter heavier antiaircraft fire. In addition, there had been no significant political repercussions to the earlier entry of RF-101s and F-102s into South Vietnam. Zuckert endorsed this JCS position.[15]

OSD then decided to favor delivery of the jets, but State Department officials, led by Assistant Secretary Harriman, opposed the move. They argued that USAF pilots were not only better able to fly reconnaissance missions than the Vietnamese but were also subject to U.S. political control. If the VNAF flew jets, they claimed, the war would not be shortened but its terms, as understood by both sides, would change significantly. The International Control Commission and other nations in Southeast Asia would consider VNAF jet operations a violation of the Geneva agreement and a definite escalation of the war.[16]

When McNamara informed Taylor and Zuckert on 17 May of this opposition, he told them that the State Department might reconsider its stand at a later date if circumstances warranted, but he urged both men to take a "hard look" at plans for future jet deliveries to the VNAF, As mentioned earlier, he opposed any plans to equip the VNAF with F-5As. At the end of May, OSD informed CINCPAC that the T/RT-33s would not be transferred to the VNAF.[17]

Late in 1963, when the Viet Cong stepped up its antiaircraft attacks and inflicted heavy damage, the Air Force thought that the Administration might now permit use of high-performance jet aircraft (B-57s); with combined USAF-VNAF crews. A number of VNAF pilots had completed jet training in T-33s and could be ready to fly higher-performance jets in a relatively short time.[18]

VIII. THE OVERTHROW OF THE DIEM GOVERNMENT

In 1963 the "clear and hold" tactics adopted in the struggle against the Viet Cong appeared to be succeeding. At the end of 1962, MAC/V had reported that Vietnamese military units were reaching out from cleared areas and fragmenting enemy sources, and Viet Cong morale was low. According to one estimate, enemy casualties had mounted to an estimated 33,000 during 1962, more than double the 1961 figure, as against 13,000 for the government. Viet Cong desertions and weapon losses had increased while its attacks against the Vietnamese armed forces and populace had declined.[1]

Conflicting Evaluations of the War

Early in 1963, most U.S. officials were optimistic. Gen. Maxwell D. Taylor, who became JCS chairman on 1 October 1962, thought the Vietnamese forces were "on the road to victory." To a high State Department official, they were "beginning to win the war." McNamara observed that the Diem government now recontrolled an additional one-fourth of the population. This gave the government, according to Secretary Rusk, control of 951 villages or about half the total, compared with 8 percent held by the Viet Cong and the remainder uncommitted. Felt noted that Viet Cong attacks had dropped from about 100 weekly for the first half of 1962 to about 50 weekly in January 1963, and he pointed to the construction of about 4,000 strategic hamlets.[2]

The Air Force was less optimistic. Zuckert thought "real progress" had been made, but he saw a long struggle ahead. The Air Staff conceded that enemy casualties were high, but it observed that Viet Cong strength had risen from about 15,000 in January 1962 to 22,000 to 24,000 in December, with about 100,000 additional village and provincial forces and political and propaganda agents. In November 1962, the enemy had mounted battalion-size attacks, and the government had failed to seal the Laotian and Cambodian border against infiltration. Despite the increase in border control posts, the enemy continued to infiltrate into South Vietnam. Estimates of their number have varied greatly. A detailed MAC/V study in October 1964 arrived at the following figures: 1957-60, 4,500; 1961, 5,400; 1962, 13,000; 1963, 6,200 (including 580 civilian specialists). The infiltrators were believed to be largely retrained military personnel of South Vietnamese origin. The drop in numbers in 1963 appeared to indicate that the Hanoi government had used most of its South Vietnamese veterans of the French Indochina War and was relying on draftees of North Vietnamese origin. And the Diem regime was weak politically and needed to gain the support of the people.[3]

One Air Staff study stressed the political restrictions on USAF activities in South Vietnam which limited its participation largely to building up and training the VNAF. It noted that the U.S. Army efforts to "prove" by tests the Howze Board tactical concepts were preempting the traditional USAF role in close support. The study concluded that if the Army effort were successful, it might have an even greater adverse long-range effect on the future U.S. military posture than on the current war against the Viet Cong.[4]

A second study concluded that the Vietnamese forces were not winning. To improve U.S. military support, this country should dispose of the Army-Air Force doctrinal battle and eliminate all but essential testing. An air deputy commander in USMAC/V should improve air-ground operations. The United States should deploy more USAF aircraft, step up VNAF training, remove political restrictions against defoliation, and encourage third-country aid, particularly by the Chinese Nationalist Air Force. Finally, there should be more overt and covert strikes against North Vietnam despite the increased risk of military escalation.[5]

U.S. newsmen frequently criticized the war effort also, contrasting the pessimistic reports from lower U.S. echelons with those of top officials. These newsmen believed that the Vietnamese lacked sufficient offensive spirit and that Diem lacked public support and interfered with the military to prevent the rise of a rival leader. So severe were some of these criticisms that Felt, in November 1962, informed OSD that there might be a well-planned "whispering campaign" against military activities in South Vietnam that merited investigation.[6]

A Senate foreign relations subcommittee, headed by Senator Mike Mansfield, visited South Vietnam and, in its report early in 1963, doubted that optimism was justified, It warned that U.S. involvement in lives and resources night reach "a scale which would bear little relationship to the interests of the United States or, indeed, to the interests of the people of South Vietnam."[7]

Attempts to prevent infiltration by using air strikes proved unsuccessful.
Source: U.S. Air Force

Notwithstanding the critics, the counsels of optimism continued to prevail. In May 1963 U.S. officials again concluded that most "indicators" of progress were more Viet Cong casualties, defections, and fewer attacks were favorable. The strategic hamlet program showed rapid progress, and the Diem forces would begin to carry out the much-delayed national campaign plan on 1 July. But important problems remained, especially the infiltration of insurgents and the concealed delivery of supplies from Laos and Cambodia. To reduce the flow, U.S. and South Vietnamese officials agreed on 1 May to conduct air and ground operations closer to the border areas than had previously been allowed. In addition, the JCS considered proposals to expand covert military operations against North Vietnam to convince the government of that country that it must stop aiding the Viet Cong or suffer more serious reprisals. Both the Army and CINCPAC prepared specific plans for such operations.

LeMay believed that the Army plan of "hit and run" airborne and amphibious raids near the coast line was too restrictive. On 22 May, the JCS approved a concept for expanding such covert activities.[8]

Reflecting the general U.S. confidence at the time, McNamara in May asked for a plan to train enough Vietnamese so that about 1,000 U.S. military personnel could return to the United States by the end of 1965. Suggested by the British Advisory Mission to South Vietnam, this action would demonstrate the U.S. intention to withdraw, indicate that Vietnamese forces were winning, and blunt the growing opposition to the Diem government. Headquarters USAF hoped that the withdrawal would reduce the spiraling testing activity in South Vietnam which, it believed, was interfering with the war effort.[9]

The Fall of the Diem Regime

Although optimistic, U.S. officials were aware of the dangers that might result from the political and religious conflicts in South Vietnam. Ambassador Nolting observed on 5 May 1963 that U.S. relations with the Diem regime had deteriorated because Diem considered our Laos policies equivocal, resented our alleged intrusion in Vietnamese affairs, and believed the Mansfield Report a criticism of his regime. Two days later, Diem's security forces fired into a Buddhist demonstration, killing several people. Subsequently his regime faced more demonstrations, dramatic protests by self-immolation, and talk of a military coup. To defend itself, it arrested many Vietnamese and in late summer temporarily declared a state of martial law.[10]

Weighing the possibility of a debacle, the services drew up plans for evacuating by air and sea about 41,600 noncombatants. For this eventuality, PACAF placed 46 aircraft, mostly C-130s, on alert in Okinawa in August. The United States continued to back Diem, but President Kennedy on 2 September warned that without public support the Vietnamese government could lose the war. The United States renewed its efforts to persuade Diem to stop oppressing his people, but without success.[11]

Despite the political and religious disorders, U.S. officials up to 1 November 1963 were still optimistic. On 2 October McNamara and Taylor, after visiting South Vietnam, still hoped to withdraw 1,000 U.S. troops by the end of the year and complete most of America's military task by the end of 1965. JCS optimism was based on Vietnamese achievements. About 8,300 strategic hamlets had been built for 9.7 million Vietnamese, and 5,200 village and hamlet radio sets had been installed. Overall Viet Cong strength had decreased from 123,000 in November 1962 to 93,000 a year later, and about 14,000 insurgents had defected since April 1963. Except for the swampy Mekong delta, the Vietnamese appeared to have made good progress in clearing northern and central areas and in opening roads and rail lines.[12]

On 1 November, the political and military situation changed drastically. A military junta, headed by Maj. Gen. Duong van Minh, overthrew the Diem government and shot both Diem and his brother, Ngo Dinh Nhu, the following morning. On the 5th, the junta formed a civilian provincial government. Military

Air War – Vietnam

leaders stated that one reason for the coup was their belief that Nhu, Diem's chief political advisor, was negotiating an unacceptable compromise with North Vietnam to settle the war. For political and other reasons, more than 400 Vietnamese officers were soon discharged and others placed on leave without pay.[13]

The "Number One" Problem

Although the political and military situation deteriorated after the coup, the United States announced on 14 November its intention to withdraw as planned about 1,000 troops engaged in engineering, ordnance, medicine, and similar tasks. Beginning 3 December, these troops, which included 274 USAF personnel, departed from South Vietnam.[14]

The political and military setback following the coup did not change basic U.S. policy toward South Vietnam. After conferring with the National Security Council, President Johnson on 26 November asserted that the principal U.S. objective would still be to assist the new government to consolidate itself, win public support, and defeat the communists. To implement this policy, the United States would attempt to persuade the new government to concentrate its efforts within the Mekong delta. U.S. military planners would consider the possibility of more action against North Vietnam and the Communists in Laos. This country would make a greater effort to improve relations with Cambodia (a Viet Cong sanctuary) and also show the world how the insurgents were controlled and supported by nations outside South Vietnam.[15]

Declaring Vietnam to be the "number one" problem of the United States, President Johnson on 2 December directed the JCS to send only the best U.S. military personnel to that country. By year's end, U.S. and Vietnamese military leaders were preparing a new pacification plan which, they hoped, would reverse the recent tide of defeat.[16]

Meanwhile, as the insurgents continued their offensive, the Administration directed more attention to controlling the flow of men and supplies from Laos and Cambodia. In view of the political obstacles to "hot pursuit" and inspection, especially in Cambodia (where in November 1963 U.S.-Cambodia relations reached a new low when the Cambodian government terminated U.S. economic and military assistance), McNamara in January 1964 urged more high and low reconnaissance missions. The JCS desired a still bolder program, recommending that the United States temporarily assume tactical direction of the war and deploy more U.S. forces, including combat units, if necessary. They also suggested that MAC/V be responsible for all U.S. programs in South Vietnam, U.S. pilots overfly Cambodia and Laos, and the South Vietnamese conduct operations against North Vietnam and Laos. Whether any of these recommendations would be adopted in 1964 remained to be seen.[17]

Plans And Polices 1952-63

IX. SUMMARY, 1961 - 1963

The growing Communist menace to South Vietnam in 1959-1960 found the U.S. government responding gradually. In late 1961 an initial program of action stressing military training and economic projects was deemed insufficient. As a result, President Kennedy sent his Military Representative, General Taylor, and other U.S. officials to South Vietnam to assess the threat. The Taylor Mission recommended more military and economic aid and greater, although limited, U.S. participation in training, advisory, and support activities. McNamara and the JCS thought that the situation in both South Vietnam and Laos merited the use of SEATO or U.S. combat forces. But fearing military escalation, the Administration generally accepted the Taylor Mission's program.

By late 1961, U.S. military units and advisory and training personnel were deploying to South Vietnam. The Air Force deployed a small special air warfare unit eventually nicknamed Farmgate, one C-123 transport squadron, and other support aircraft and equipment including a tactical air control system. The basic mission of Farmgate was to advise and train the Vietnamese Air Force. Combat training missions with combined USAF-VNAF crews were authorized only when the VNAF was unable to fulfill all air support needs. In February 1962 a U.S. Military Assistance Command, Vietnam, was established in Saigon to coordinate all U.S. activities in support of the Vietnamese.

In mid-1962, initial evaluations of limited "clear and hold" and other support operations were optimistic. In the limited air war, USAF combat training and transport sorties increased, defoliation tests were promising and USAF strength had been augmented. But the Air Force chafed under the restraints imposed by the Department of State, OSD, and the Army. These restraints limited air strikes for fear they would harm friendly Vietnamese, create undesirable political repercussions, and escalate the war. Equally disturbing to the Air Force was its subordinate military planning role under both CINCPAC and MAC/V, especially the latter. This contributed to Air Force failure to win approval, of some of its own concepts for defeating the Viet Cong, such as the quick reaction plan of early 1962. There was also a growing Air Force-Army dispute over tactical air control.

Although Farmgate sorties increased, new air tactics evolved, and Farmgate-VNAF air strikes accounted for a high percent of Viet Cong casualties, the political restrictions on Farmgate activities remained. Air Force-Army differences over the use of airpower in counter-insurgency were intensified as the Army began testing "Howze Board" tactical air concepts that, the Air Force believed, pee-empted its own long-established tactical roles and missions. The conflict reached the highest OSD level when a strike against the Viet Cong on 2 January 1963 resulted in high losses, allegedly because of inadequate use of air support. A JCS team reviewed the incident, the war's conduct, and Air Force grievances, but the Air Force won only minor concessions. Because of Army or OSD opposition it also failed to obtain the post of chief of staff or to create the post of air deputy commander in MAC/V.

Meanwhile, in late 1962 and early 1963, most top officials remained hopeful about the war's progress on the basis of enemy casualties, defections, reduced terror strikes, and the progress of the Diem government's strategic hamlet program designed to isolate the populace from the Viet Cong. But much of the U.S. news media, pointing to the ineffectiveness of the Diem government and the Vietnamese forces, thought that the optimism was unjustified. A Senate foreign relations subcommittee questioned the wisdom of growing U.S. involvement.

Some Air Force officers who took a somber view thought that the war was being lost. Observing the increasing value of VNAF and Farmgate missions in stopping or deterring Viet Cong attacks against villages, outposts, strategic hamlets, and rail and road convoys, and for inflicting casualties, destroying equipment and supplies, and inhibiting enemy movement, they urged greater use of air-power. They also recommended jet aircraft for both USAF and VNAF units to conduct air strikes more effectively and to counter the effects of increased antiaircraft fire. They urged the removal of political restrictions against border flights, defoliation, and other activities.

In early 1963 U.S. authorities, in the light of growing military requirements, authorized the Air Force and the U.S. Army to augment partially their air strength in Vietnam. This would enhance the mobility of the Vietnamese Army and paramilitary forces, provide additional air support for a national campaign plan designed to shorten the war, and permit the withdrawal of most U.S. units, except training, by the end of 1965. A decision to accelerate the training and equipping of the VNAF added to the Air Force's commitment.

This picture echoed around the world and brought about the collapse of the Diem government inside South Vietnam.
Source: National Public Radio

In the spring of 1963, rising religious and political unrest against the Diem regime was highlighted by Buddhist and student demonstrations. As political deterioration continued, U.S. efforts to persuade the regime to be less oppressive were unsuccessful. Most U.S. authorities continued to believe that U.S.-South Vietnamese military operations still presaged success. But the government's unpopularity and the belief that it harbored secret neutralization plans led on I

Plans And Polices 1952-63

November to a military coup d'etat. In subsequent weeks Viet Cong attacks increased to take advantage of the political disorder. The 1st Air Commando Squadron (previously Farmgate) and the VNAF flew large numbers of sorties to aid strategic hamlets overrun or threatened by the Communists.

The Air Force believed that the use of Vietnam as a testing ground for new concepts in warfare was detrimental to the primary objective of actually winning the war. Air mobility was one of the concepts that was evaluated in Vietnam and the extent to which its development diluted the effects of tactical air power is still debated. Photo source: U.S. Army.

The immediate post-coup period vitiated much of the previous two-year's military and economic gains and demonstrated the persistent, growing Viet Cong strength. Although programs and tactics were reviewed, there were few indications that U.S. Government policies limiting direct USAF participation, permitting the use of Army tactical air concepts, and encouraging Army aviation testing, would be greatly modified. In fact, personnel changes in MAC/V placed the day-to-day conduct of the war even more firmly in Army hands. In air support the Army's domination was dramatized by the greater number of aircraft on hand and sorties flown compared with the Air Force. However, heavier aircraft attrition from ground fire, McNamara's request for more air reconnaissance of borders, and the slow progress of the VNAF suggested the possible use of more USAF aircraft, including jets.

Air War – Vietnam

At the end of 1963 President Johnson asserted that the United States would help the new South Vietnamese government consolidate itself and win the support of the people. Observing that the war was America's "number one" problem, he directed the use of only "blue-ribbon" U.S. military personnel. As a gesture of confidence, 1,000 U.S. officers and men, including 274 from the Air Force, were returned to the United States in December. But as 1964 began the JCS was increasingly apprehensive of Viet Cong strength and advocated stronger U.S. action against border areas and North Vietnam. They urged temporary overall U.S. direction of the war. Whether the political rules of the war would be significantly relaxed as the JCS counseled and as the Air Force had recommended, remained to be seen.

APPENDIX 1

Farmgate Combat Training Sorties							
	1962*			1963			Grand Total
Operational	B-26	T-28	Total	B-26	T-28	Total	
Close air support	150	446	596	660	1,077	1,737	**2,333**
Interdiction	334	346	680	1,432	1,383	2,815	**3,495**
Escort Helicopter	21	359	380	98	450	548	**928**
Escort Aircraft	21	69	90	137	307	444	**534**
Escort Convoy	30	16	46	91	48	139	**185**
Escort Train		14	14		35	35	**49**
Air Cover	67	129	196	410	501	911	**1,107**
Armed Recon	31	282	313	52	724	776	**1,089**
Photo Recon	429	121	550	523	11	534	**1,084**
Visual Recon	9	20	29	62	272	334	**363**
Defensive	9	10	19	164		164	**183**
Other	39	41	80	45	40	85	**165**
Total	1,140	1,853	2,993	3,674	4,848	8,522	**11,515**
Non-Operational							
Administrative	479	967	1,446	299	573	872	2,318
Flying time (Hours)	**3,953**	**4,505**	**8,458**	**9,464**	**8,554**	**18,048**	**26,506**

*Includes Dec 1961 in these appendices where applicable.

Source: Data Control Br. Sys Dic, Dir of Ops, DCS/ P&O

APPENDIX 2

Results of Farmgate Missions			
	1962	1963	Total
Enemy Killed	3,200	3,256	6,456
Enemy Wounded		556*	556
Structures Destroyed	4,000	5,750	9,750
Structures Damaged		6,253*	6,253
Boats Destroyed	275	2,643	2,918
Boats Damaged		302*	302
* Includes figures for 1962			

APPENDIX 3

USAF U-10 and TO-1D Sorties			
Type Aircraft	1962	1963	Total
U-10	351*	2,404	2,755
TO-1D		3,957 **	3,957
* Began operational flights in Sep 1962			
** Began operational flights in Jul 1963			
SOURCE: Memo, M/G R.F. Worden to C/S USAF, 23 Jan 1964, subj: JCS Briefing by Gen. Anthis.			

APPENDIX 4A

USAF C-123 Sorties and Logistic Activities			
	1962	1963	Total
Sorties	11,689	24,429	36,118
Passengers	54,734	142,124	196,858
Troops Airlanded	32,906	1,349	34,255
Training Troops Dropped	8,952	2,072	11,024
Combat Landing Team Troops Dropped		47	47
Cargo Airborne Resupply (Tons)	1,973.1	613.5	2,586.6
Cargo Airlifted (Tons)	15,346.5	32,396	47,742.5
Flying Time (Hours	17,842	29,255	47,097

APPENDIX 4B

USAF SC-47 Sorties and Logistic Activities			
Operational	1962	1963	Total
Reconnaissance	12		12
Flare Drop	21	51	72
Airborne Alert		5	5
Paradrops	1	293	294
Special Forces Support	649	2,578	3,227
Radio Relay	4	5	9
Other	147	42	189
Non-Operational/Administrative	1,376	1,428	2,804
Flying Time (Hours	836	5,289	8,125

Source: Data Control Br. Sys Dic, Dir of Ops, DCS/ P&O

APPENDIX 5

VNAF A-1H and T-28 Sorties							
	1962			1963			Grand Total
Operational	A-1H	T-28	Total	A-1H	T-28	Total	
Interdiction	969	1,379	2,348	1,605	3,331	4,936	**7,284**
Air Support	0	234	234	500	493	993	**1,227**
Escort Helicopter	80	407	487	116	374	490	**977**
Escort Convoy	93	36	129	74	106	180	**309**
Escort Aircraft	26	51	77	116	387	503	**580**
Escort Train	27	52	79	520	278	798	**877**
Air Cover	384	211	595	790	443	1,233	**1,828**
Armed Recon	26	144	170	154	1,099	1,253	**1,423**
Visual Recon	0	0	0	0	24	24	**24**
Air Defense	12	0	12	6	48	54	**66**
Other	257	108	365	49	148	197	**562**
Total	**1,874**	**2,622**	**4,496**	**3,930**	**6,731**	**10,661**	**15,157**
Non-Operational							
Administrative	1,204	3,730	4,934	1,263	3,717	4,980	9.914
Flying time (Hours)	7,179	7,778	14,957	9,914	12,757	22,671	37,628

Source: Data Control Br. Sys Dic, Dir of Ops, DCS/ P&O

APPENDIX 6

U.S. and VNAF Military Aircraft			
	1961*	1962*	1963*
USAF	35	63	117
VNAF	152	180	228
U.S. Army	40 (approx)	200	325
U.S. Marine Corps	0	20	20
Total	227	463	690
*As of December each year			
SOURCE: Project CHECO Southeast Asia Report, Oct 61 – Dec 63, pt 1 chart 1-2; Office, Asst for Mutual Security, DCS/S&L			

APPENDIX 7

U.S. Aircraft Lost, 1 Jan 1962 – 31 Mar 1964			
	Fixed Wing	Rotary	Total
USAF	34	*	34
U.S. Army	16	54	70
U.S. Marine Corps	*	10	10
Total	40	64	114
*No USAF or Marine aircraft of these types			
SOURCE: Report of Air Force Study Gp on VN, May 64, in OSAF			

APPENDIX 8

USAF Aircraft Destroyed and Damaged						
Type	1962			1963		
	Destroyed by Enemy	Destroyed Other Causes	Damaged*	Destroyed by Enemy	Destroyed Other Causes	Damaged
B-26	1	0		3	3	60
C-123	1	3		0	3	66
C-47	1	1		0	0	10
T-28	2	0		2	2	72
TO-1D	0	0		1	1	13
U-3	0	0		0	1	0
U-10	1	0		0	0	8
Total	6	4		6	10	229

* No records available for 1962

SOURCE: Memo, M/G R.F. Worden to C/S USAF, 16 Apr 1964, subj: Addit A/C (A-1Es) for RVN, in Plans RL (64) 38 - 9

APPENDIX 9

U.S. Military Personnel						
	Dec 61	Jul 62	Dec 62	Mar 63	Sep 63	Dec 63
Army		6,155	7,885	8,718	10,795	10,119
Air Force	421*	1,699	2,422	3,256	4,444	4,630
Navy		320	447	585	668	757
Marines		648	535	584	551	483
Total		8,822	11,289	13,143	16,458	15,989

* Excluding Air Force Section MAAG/V

SOURCE: State Rpt. Trends in Counter-Insurgency, 21 Sep 63; msg 271045, 2d AD to PACAF, 27 Apr 64

APPENDIX 10

Combat Casualties

U.S. Vietnamese and Viet Cong Battle Casualties

	1961	1962	1963	Total
South Vietnam	9,000	13,000	19,000	41,000
Viet Cong	13,000	33,000	28,000	74,000
United States	0	101	491	592
Total	22,000	46,101	47,491	115,592

U.S. Casualties By Type*

	1962**	1963	Total
Killed in Action	21	72	93
Wounded in Action	80	406	486
Missing in Action	0	13	13
Non-battle Deaths	34	37	71
Non-battle Injuries	45	73	118
Total	180	601	781

U.S. Combat Casualties, Dec 1961 – Dec 1963 ***

Killed in Action	27
Wounded in Action	22
Missing in Action	4
Total	53

* SOURCE: Hist of 13 AF, Jul – Dec 63, p 53

** Includes Dec 1961

*** SOURCE: Rpt of AF Study Gp in VN, May 1964, in OSAF

Air War – Vietnam

PART TWO
USAF PLANS & POLICIES
IN SOUTH VIETNAM & LAOS 1964

Air War – Vietnam

Vietnam and Laos 1964

I. REVISED U.S.-SOUTH VIETNAMESE MILITARY PLANNING

At the beginning of 1964 the South Vietnamese government, now headed by Maj. Gen. Duong Van Minh, had not recovered from the overthrow of former President Ngo Dien Diem on 1 November 1963. The breakdown in authority enabled the Viet Cong (Vietnamese Communists) to overrun many strategic hamlets and military outposts and achieve other successes. Buoyed by victories, improved organization, and increasing North Vietnamese and other Communist bloc aid, their momentum continued into the new year. U.S. estimates placed hard-core Viet Cong strength at 22,000 to 25,000, and irregular forces at 60,000 to 80,000. Compared with January 1963 estimates, hard-core cadres had increased modestly and irregular forces had declined slightly despite losses of about 1,000 monthly from deaths, wounds, capture, and defections.[1]

Viet Cong units showed increasing tactical expertise during 1964. Guerilla warfare is ruthlessly Darwinian; only the most competent survive meaning that those who do are dangerous opponents indeed. Source: Vietnamese People's Liberation Army.

Despite setbacks, South Vietnamese forces engaged the Viet Cong in scores of actions, mostly in the southern part of the country. In the first five weeks of 1964 they averaged 56 battalion-size or larger operations per week, but smaller actions, while less frequent, were more effective, accounting for one half of reported enemy killed. Ground action was accompanied by a rising level of air support by USAF's 1st Air Commando Squadron (previously Farmgate) and the Vietnamese Air Force (VNAF). Summarizing the military situation for the JCS, Adm. Harry

D. Felt, Commander-in-Chief, Pacific (CINCPAC) and Gen. Paul D. Harkins, Commander, U.S. Military Assistance Command, Vietnam (COMUSMAC/V) said that the most suitable Vietnamese tactics required good intelligence, communication security, and large and small actions to "clear and hold" former enemy territory.[2]

After the fall of Diem, top U.S. military and diplomatic officials reviewed their Vietnam planning. Headquarters MAC/V prepared a new pacification plan to replace the poorly executed and moribund national campaign plan of 1963. The U.S. Ambassador in Saigon, Henry Cabot Lodge, advocated a broader civic action program as he perceived a Viet Cong shift from military to political tactics. Lodge stressed the need for trained political teams to acquaint the rural populace with the Saigon government's objectives in education, land reform, health, and other areas. He urged a beginning in Long An Province where Viet Cong control was virtually complete.[3]

The JCS pressed for stronger measures. On 22 January, it recommended to Secretary of Defense Robert S. McNamara that the United States should deploy more forces, assume temporary tactical control of the war, and make MAC/V responsible for the entire U.S. effort in South Vietnam. It favored air and ground actions to halt the flow of personnel and supplies from Laos and Cambodia, and air and sea strikes against North Vietnam.[4]

The U-2 started reconnaissance missions over NVN in 1964. Source: U.S. Air Force

McNamara expressed special interest in employing more reconnaissance to detect Communist infiltration. In response to a query, Gen. Curtis E. LeMay, USAF Chief of Staff, prepared a list of Air Force and VNAF aircraft in the theater available for this purpose and said that more were scheduled to arrive. One decision reached was to begin high altitude U-2 flights in February over North and South Vietnam, Laos, and Cambodia.[5]

General Khanh's Coup

Meanwhile, a power struggle within the Minh government led, on 6 January 1964, to the establishment of a military triumvirate. Twenty-four days later Maj. Gen. Nguyen Khanh, Commander of the Vietnamese Army's I Corps, organized a bloodless coup d'etat against the triumvirate. Khanh emerged as Chairman of the Military Revolutionary Council and, on 8 February 1 took over as Premier of the country with General Minh elected to the ceremonial post of head of state. In justifying his actions, Khanh charged that the three-month old Minh regime had failed to make progress in effecting political, social, and economic reforms and was susceptible to the influence of a neutralist officer faction. He also accused

Vietnam and Laos 1964

President Charles De Gaulle, of France, of attempting to interfere in Vietnamese affairs.[6]

In his coup, Khanh enjoyed the strong support of Col. Nguyen Cao Ky Commander of the Vietnamese Air Force (VNAF) since 16 December 1963. On 5 March 1964 Colonel Ky was promoted to Brigadier General. U.S. officials subsequently expressed hope that the new government would, as it promised, step up operations against the Viet Cong. On 17 February McNamara told a House committee that the Khanh government appeared to have considerably more popular support than its predecessor and was pursuing more effective strategic hamlet and "clear and hold" programs. The Defense Secretary reaffirmed plans to withdraw most U.S. troops by the end of 1965.[7]

Maj. Gen. Nguyen Khanh
Source: *Vietnamese Govt*

To improve U.S. assistance to the new government, President Johnson established an interdepartmental committee (known as the Sullivan Committee, it was headed by William H. Sullivan, Assistant to Undersecretary for Political Affairs, W. Averill Harriman) to manage U.S. policy and operations in South Vietnam, ordered the prompt fulfillment of all aid requests from Ambassador Lodge, asked that U.S. dependents be encouraged to return voluntarily, and directed a speed-up in shaping a "credible deterrent" against North Vietnam. The President also announced that McNamara would again visit Saigon to review the military situation.[8]

Plans to Revitalize Counterinsurgency Operations

As a result of Premier Khanh's promising leadership, the Chairman of the Joint Chiefs of Staff (JCS), Gen. Maxwell D. Taylor, asked the JCS for a new plan to revitalize counterinsurgency and recommendations to stabilize the government and prevent new coups. The JCS quickly recommended stepped up intelligence and operations in border areas, financial relief for areas taxed by both the government and the Viet Cong, more U.S. military and civilian advisors at all government levels, better civilian programs to gain popular support, more effective crop destruction in Viet Cong areas, and increased effort to win the support of U.S. news media. It studied the possibility of combining the Military Assistance Advisory Group, Vietnam (MAAG/V) with MAC/V, endorsed the latest Vietnamese national pacification plan, and urged the preparation of a civilian plan wherein new "Life Hamlets" would replace strategic hamlets. The JCS cautioned that only Vietnamese civilian administrators, in the long run, could stabilize an area cleared by military forces.[10]

The new national pacification plan was scheduled to begin on 3 February but the Khanh coup caused a delay. After he approved it on the 17th, government

Air War – Vietnam

ministers changed the name to the Chien Thang or "victory" national pacification plan. Based on a "spreading oil drop" concept, it consisted of two phases. First, military operations would destroy or expel the Viet Cong. Secondly, the Viet Cong "infrastructure" or cells would be liquidated and replaced by new and "friendly" organizations. There would be expanded civic action programs designed to improve police, education, health, welfare, economic, and other activities to win the confidence of the people. A national pacification council, headed by Premier Khanh, was created to oversee the plan.[11]

An air plan subsequently prepared by the Pacific Air Forces (PACAF) to aid pacification called for enlarged and better coordinated close support and interdiction programs with more aircraft placed on continuous alert to provide faster reaction. As the "oil drop" spread and liberated areas widened, pockets of Viet Cong would be rooted out by heavier day and night air attacks. Because of VNAF limitations, USAF aircraft and personnel would be needed for combat training strikes and to provide reconnaissance for aiding border control. PACAF believed that the expanded use of airpower was essential to weaken enemy morale, increase his casualties and defections, win support of fence-sitting Vietnamese, and demonstrate Vietnamese and U.S. determination.[12]

Eugene M. Zuckert
Source: U.S. Govt

The JCS endorsed Ambassador Lodge's proposal (supported by the State Department) to recapture Long An Province from the Viet Cong. The Air Force especially believed that air support would be vital to the operation. Secretary of the Air Force Eugene M. Zuckert informed McNamara that USAF and VNAF units could transport medical and other supplies, and provide aerial loudspeakers for broadcasting to the Vietnamese. Political teams, if attacked, could quickly radio for air support and airborne troops.[13]

Some U.S. officials considered the Lodge plan impractical. The U.S. Minister-Counselor in Saigon (and sometimes Acting Ambassador), David G. Nes, thought that the JCS directive to implement the plan revealed "an almost total lack of comprehension" of the Vietnam problem. General Harkins and Admiral Felt agreed that an immediate offensive in Long An Province was not possible. Harkins pointed to inadequate Vietnamese civic action planning, conflicting provincial military priorities, and a "bizarre" command structure that permitted pacification troops to be transferred. As a consequence, Ambassador Lodge's proposal was soon abandoned.[14]

Although the Air Force Chief of Staff concurred with JCS proposals to revitalize the counterinsurgency program, he urged still bolder U.S. measures. A 12 February intelligence report, General LeMay observed, warned that without a marked improvement in efficiency, the Vietnamese government and armed forces "at best had an even chance" of withstanding the Viet Cong in the coming weeks

Vietnam and Laos 1964

and months. Regardless of the threat of escalation, LeMay thought that the time for a military showdown had arrived, and that the U.S. government should explain to the American people the extent of Communist subversion in South Vietnam and announce its determination to defeat it.[15]

Plans to Increase Pressure On North Vietnam

With its hopes raised by the seemingly strong Khanh government, the administration was not ready to follow LeMay's counsel. However, on 21 February, McNamara asked the JCS to assess ways to apply more pressure on North Vietnam to persuade it to end support of the insurgents in the South and in Laos. They were to include actions such as special air and sea non-nuclear attacks which would be least likely to escalate the conflict and cause adverse third country reaction. In addition, he asked them to suggest how best to deter Hanoi and Peking from dispatching troops throughout Southeast Asia.[16]

In a partial reply on 2 March the JCS recommended selected air attacks immediately on North Vietnam for "shock" effect as part of a coordinated diplomatic, psychological, and military program. These attacks could be followed by additional air and amphibious attacks, sabotage, and harassment of the North's fishing and shipping in ascending severity. Some of these activities would be under the aegis of special Plan 34 that provided for limited operations such as mining of waters, bombardment of selected installations, sabotage, radio broadcasts, and leaflet drops.

GIỜ CHƯA PHẢI LÚC TRỞ VỀ VỚI GIA ĐÌNH SAO ?
ANH CHỌN CẢNH NÀO TRÊN NÀY ?

U.S. aircraft dropped leaflets in Southeast Asia. This one says, "Is now the time to return home to your family? Which scenery would you choose?" Source: U.S. Air Force

For the air and sea assault program, VNAF's effort could be augmented by 1st Air Commando Squadron and B-57 aircraft. Additionally, there should be preparations for armed reconnaissance of military supply lines between North Vietnam and Laos and China, air strikes of industrial targets in the Hanoi-Haiphong area, mining of waters, and a maritime blockade of the North. The Joint Chiefs also foresaw the need for limited Vietnamese incursions, with U.S. support, into Laos and Cambodia to reduce Viet Cong infiltration from and escape into these sanctuaries. They prepared a special memorandum for McNamara on this subject.[17]

The JCS considered it unlikely that the proposed graduated attacks would result in any large-scale Chinese intervention. In the dry season, it thought, the Chinese could support logistically 13 infantry divisions, less artillery and armor, and North Vietnam 9 divisions. Estimated air strength in South China, Hainan Island, and North Vietnam was placed at 400 jet fighters and 125 light bombers. Chinese sea power was limited. Although China could order land, sea, and air attacks simultaneously against South Korea, Taiwan, and other areas, it could not sustain a major assault in more than one region at a time. [18]

McNamara's 21 February request also prompted the JCS to ask CINCPAC to prepare an air and naval plan against North Vietnam and China. Previously, the Air Force excepted, the services had opposed the concept behind such a plan: the Army and Marine Corps because it was "unthinkable" not to provide for sizeable ground forces; the Navy because of concern lest an Air Force commander exercise control over Navy air. In response, CINCPAC on 1 June issued Operational Plan 38-64. The JCS approved it in July. While basically concerned with air and naval actions, Plan 38-64 also required the use of sizeable ground forces. [19]

New U.S. Policy Guidance

Meanwhile, Washington's review of the U.S. role in South Vietnam and the possibility of air strikes on the North receive4 much publicity. Apparently, the administration hoped that hints of more forceful action would have a deterrent effect on Hanoi. As part of the reassessment, McNamara departed for Saigon. [20]

Accompanied by General Taylor and other officials, the Defense Secretary reached South Vietnam early in March. He toured the countryside with Khanh to build up the Premier's image and dramatize U.S. support. However, he found the situation had deteriorated. There was virtually no "clear and hold" program and few directives were flowing from the new government. Nevertheless, McNamara and Taylor remained "guardedly optimistic," if Khanh stayed alive and in power. They still believed most U.S. personnel could be withdrawn by the end of 1965. For example, McNamara thought that the aircraft of the USAF 0-1 squadron could soon be transferred to the expanding VNAF, and that its personnel, as well as a U.S. Marine helicopter squadron, could depart by mid-1964. [21]

For the immediate future, more U.S. assistance was needed. McNamara authorized additional manpower for MAC/V, continuation of special operations under Plan 34A the integration of the Vietnamese civilian irregular defense group (CIDG) into the regular armed forces, and aerial mining training for the VNAF. He refused, however, to approve any relaxation in the rules of engagement for the 1st Air Commando Squadron, and held in abeyance a decision on the recent JCS proposal to replace B-26s with jet B-57s. He said restrictions on defoliation activities would remain in effect and believed that the United States should "stay out of this business." [22]

McNamara's report to President Johnson contained 12 major recommendations. Although the JCS considered them insufficient and again urged air attacks on North Vietnam, the President approved them on 17 March after conferring with the National Security Council. Generally they expanded or accelerated programs

Vietnam and Laos 1964

already in effect: support for the government's mobilization plans, a 50,000 increase in Vietnamese regular and paramilitary strength, more compensation for the military, improved organization, establishment of a truly Vietnamese offensive guerrilla force, more equipment for the Vietnamese Army and Navy, addition of a third VNAF fighter squadron and the replacement of all T-28s with A-1Hs, continued high-level reconnaissance flights over South Vietnamese borders, and support for more rural reform and a civil administration corps to work at the province, district, and hamlet level. The President also restated U.S. support for the Khanh government and opposition to more coups.

Most importantly, the President approved, for the first time, planning to permit on 72-hour notice retaliatory air strikes and on 30-days notice graduated strikes against North Vietnam and Vietnamese "hot pursuit" of Viet Cong units crossing into Laos. (Pursuit approval followed a South Vietnamese-Laotian agreement on resuming diplomatic relations and military planning. Vietnamese units over battalion size would require the approval of Laotian Premier Souvanna Phouma.) But any U.S. support of pursuit into Cambodia would be contingent on U.S. Cambodian relations. In 1963 Cambodia rejected further U.S. aid and broke diplomatic relations with South Vietnam. Throughout 1964 U.S.-Cambodian relations grew worse. A poorly defined border resulted in several erroneous bombings of villages by the Vietnamese and, on 24 October, in the downing of a USAF C-123 by Cambodian gunners, killing eight U.S. personnel. During the year Cambodia strengthened its ties with Hanoi, Peking, and Moscow. At year's end diplomatic talks in New Delhi, India, to resolve differences proved fruitless.

In separate decisions in March, the administration approved the transfer of three B-57 squadrons from Japan to the Philippines and the beginning of USAF special air warfare (SAW) training of Lao and Thai pilots in Thailand because of the Communist danger in Laos.[23]

T-28s used for training Thai, Vietnamese and Laotian pilots in Thailand. Source: U.S. Air Force.

Air War – Vietnam

Meanwhile, at JCS request, Felt and Harkins quickly developed plans in accordance with Presidential decisions. On 30 March, Felt sent Operational Plan 37-64 to the JCS. A three-part plan, it provided for limited U.S. air and ground support for Vietnamese operations for border control and retaliatory and graduated strikes, using VNAF, USAF, and Navy aircraft, against North Vietnam. The JCS approved it, with amendments, in July. Thereafter it evolved into one of CINCPAC's most comprehensive plans for stabilizing the military situation in South Vietnam and Laos, and three other CINCPAC plans eventually were incorporated into it. In June Harkins completed MAC/V Operational Plans 98-64 and 98A-64 for limited U.S. support of cross-border operations into Laos. [24]

Vietnam and Laos 1964

II. CONTINUED MILITARY AND POLITICAL DECLINE

Although the President's 17 March decisions showed U.S. readiness to bring military pressure against the Communists in Laos and North Vietnam as well as in the South, the military and political situation in South Vietnam continued to deteriorate. The Army's low morale and irresolute leadership was increasingly manifest and not easily overcome by the infusion of more U.S. advice and military and economic aid. Some advisors on the scene credited many Viet Cong victories to Vietnamese apathy rather than to Viet Cong skill. [1]

The Search for Courses of Action

Alarmed over Communist gains, the JCS launched into another review of the military situation and in mid-April completed a new study for McNamara. The chiefs split in their recommendations. General LeMay and the Commandant of the Marine Corps strongly advocated immediate Vietnamese expansion of operations against North Vietnam backed by U.S. low-level reconnaissance and other forms of assistance. But the Army and Navy chiefs demurred, apparently feeling that momentarily the Saigon government was in no position to shoulder more military responsibility and risks. In subsequent months the Air Force and the Marine Corps again would be aligned on the side of more forceful action while the other two services recommended a more cautious approach. [2]

In April Secretary of State Dean Rusk flew to Europe and Southeast Asia seeking "more flags" in South Vietnam from America's NATO and SEATO allies. After his return to Washington, Rusk proposed additional political and financial measures to strengthen internally the Saigon regime. To "signal" Hanoi, he recommended establishing a U.S. naval presence at Touraine or Cam Rhan Bay, more visible air training flights over Vietnam, and a diplomatic effort to impress upon Hanoi's leaders the benefits from "leaving its neighbors alone." He opposed another Geneva conference until the military situation improved. [3]

The JCS agreed that Rusk's proposals would improve the situation in the South but were insufficient to "turn the tide" to victory. Only greatly intensified counterinsurgency operations and a "Positive" program of military pressure against the North could do this. [4]

Gen. Earle G. Wheeler, Army Chief of Staff, after visiting South Vietnam, recommended that USAF air commando strength be increased to three squadrons, all equipped with A-1Es. He also recommended a "Hardnose" operation in Laos to disrupt Communist infiltration, and continuance of Plan 34A activities to help siphon off North Vietnam's resources. [5]

More Viet Cong successes and a lagging Vietnamese pacification program prompted President Johnson, in May, again to send McNamara and General Taylor to Saigon. Premier Khanh confessed he was unable to cope with the political problems. About 8,000,000 Vietnamese, he thought, were under Saigon's

Air War – Vietnam

control but 6,000,000 were not, although all of the latter were not necessarily under the Viet Cong. But the Communists had the initiative as demonstrated by the loss of 200 of 2,500 villages since September 1963, the rise of "incidents" to 1,800 per month, and fewer casualties. Vietnamese forces, in turn, were suffering greater losses in casualties, weapons, and from desertions. Their morale was low and recruiting was difficult. [6]

More U.S. Aid and Reorganization of MAC/V

After his conferences, McNamara announced plans to enlarge the Vietnamese regular and paramilitary forces and provide other aid. The VNAF would receive more aircraft and a 100-percent increase in pilots. Observing the frequent changes in Vietnamese government and military leaders, the Defense Secretary conceded it would be a "long war," thus finally abandoning hope for withdrawing most U.S. forces by the end of 1965. On 19 May, President Johnson asked for and Congress shortly approved $125 million to finance the additional military and economic aid. [7]

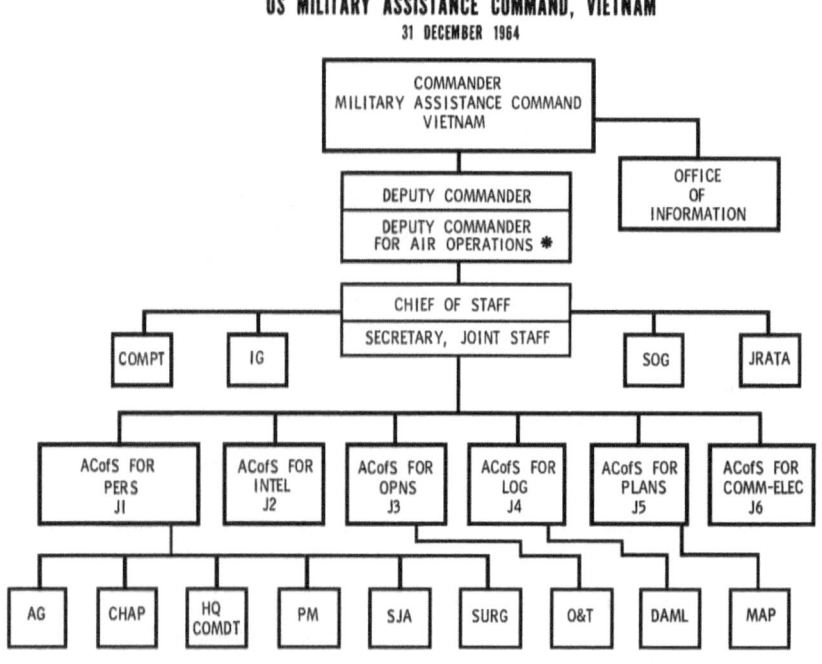

* PROPOSED IN LATE 1964 BUT NOT FULLY APPROVED BY JCS AND SOD UNTIL 10 MAY 1965

In conjunction with these decisions, the administration streamlined its activities in Saigon by combining the Military Assistance Advisory Group, Vietnam (MAAG/V) with MAC/V. Initially studied in February as a possible way to help revitalize counter-insurgency operations, the consolidation was opposed vigorously by General LeMay and the Navy and Marine Corps chiefs. They feared

Vietnam and Laos 1964

it might lead to the establishment of an Army specified command and would produce insignificant personnel and financial savings. Generals Taylor and Wheeler thought otherwise, however, and McNamara on 8 April concurred. The consolidation became effective on 15 May.[9]

As a result of the change, the Air Force Section MAAG/V was redesignated the Air Force Advisory Group, MAC/V and placed under the operational control of the 2d Air Division. But military assistance program (MAP) responsibilities remained with the enlarged MAC/V.[10]

Still under JCS and Defense Department consideration were Sullivan Committee proposals to increase drastically the number of U.S. advisors in South Vietnam to improve government efficiency, pacification, and paramilitary training.[11]

More Planning for Operations in Laos and Vietnam

In addition to devising measures to strengthen South Vietnam, administration planning addressed itself increasingly to neighboring Laos and North Vietnam.

In Laos, the Communists had long enjoyed a sanctuary for infiltrating men and arms to the Viet Cong. In April, Communist-led Pathet Lao forces attacked Laotian neutralist and right-wing forces, jeopardizing the 14-nation agreement of 23 July 1962 on the neutrality of Laos. Cautiously responding to both threats, U.S. authorities on 5 May instructed General Harkins to begin limited U.S.-Vietnamese planning for small ground patrols, aided by unmarked aircraft and helicopters. And on 19 May, USAF and Navy aircraft began "Yankee Team" reconnaissance over Laos to aid friendly Laotian air and ground forces and observe infiltration routes. The administration desired to obtain a cease-fire and restore the military status quo ante.[12]

Laos and surrounding countries. Note, this is a 2012 map. Source: CIA

The administration also reviewed more plans and the risks involved in striking North Vietnam. At McNamara's request, the JCS studied additional "telegraphing" actions along with specific military pressure against Hanoi. It warned that certain types of actions, like deploying more U.S. forces to Southeast Asia and the Western Pacific, could lead to international demands for another Geneva-type conference before Hanoi altered its policy. Telegraphing actions in themselves, the JCS thought, would have little effect: only "positive" offensive measures could convince Hanoi that its support of the Viet Cong and the Pathet Lao no longer would be tolerated.[13]

LeMay believed that the war was being lost. Administration authorities had directed the JCS on 20 May to tighten its rules of engagement for U.S. air support within South Vietnam to lessen U.S. involvement. With respect to strategy against the North, LeMay pointed to two years of unsuccessful efforts to compel Hanoi to decide to end its subversion by examples of U.S. determination. The objective, he said, should be to destroy the North's capability, and to achieve this he proposed conveying the "message" by attacking sharply two important targets supporting the Viet Cong and Pathet Lao: Vinh and Dien Bien Phu. [14]

The Commander of this Pathet Lao unit is reading a letter from "higher authority". Source: Lao Patriotic Front

In this instance the Army and Navy chiefs agreed with LeMay but General Taylor considered the risk too great as both were huge targets. Air strikes would require hundreds of sorties for several days, be unnecessarily destructive, retard eventual "cooperation" with Hanoi, challenge the Communist bloc, and escalate the war. Of three JCS proposals considered, a massive air attack on all significant targets, a series of lesser attacks, and limited attacks to show U.S. will, Taylor favored the. last although he asked Felt to prepare for all three. McNamara agreed with Taylor's conclusion. PACOM's commander submitted the plans to the JCS early in July.[15]

JCS advocacy of air strikes against North Vietnam had strong support in the State Department. The chairman of its Policy Planning Council, Walt W. Rostow, although opposed to a large-scale U.S. ground commitment in Southeast Asia, agreed that the United States should demonstrate its willingness to use air and naval power to stop the insurgencies in South Vietnam and Laos. Warning of possible defeat, he said this would mean preparing for war to gain a political

An intriguing picture, courtesy of the Lao Patriotic Front. It is labeled as a Pathet Lao BTR-152 but is clearly a U.S.-built half-track. It was probably captured from the French in the early 1950s.

Vietnam and Laos 1964

objective as in Cuba in 1962. [16]

Early in June, Rusk, McNamara, Taylor, and top field officials met in Honolulu to review the political and military situation. Rusk indicated that Premier Khanh's position was shaky and McNamara was pessimistic about the success of internal reform measures. In the war there was danger that the Viet Cong might push from Laos to the sea through Quong Ngai Province, cutting South Vietnam in half, and this was forcing Khanh to concentrate military forces in the north rather than in the south.

The conferees agreed that air strikes against North Vietnam should be authorized by Congress and preceded by an augmentation and redistribution of U.S. forces in the western Pacific and Thailand. Taylor postulated three levels of strikes against the North: using only the VNAF to demonstrate U.S. will; using USAF's 1st Air Commando Squadron and the VNAF to destroy Hanoi's will; and using the 1st Air Commando Squadron, the VNAF, and other U.S. air units to destroy Hanoi's ability to support the Viet Cong. In the event the Chinese Communists intervened, McNamara thought air attacks could reduce the Chinese effort by 50 percent if enough conventional bombs were available, but this would not resolve the problem of coping with 5 to 18 Chinese divisions. Felt believed that the United States would run out of aircraft before enough conventional bombs were dropped to defeat the Chinese. On the other hand, to resort to nuclear weapons, said Rusk, was "a most serious" matter and he foresaw the possibility of Soviet counteraction elsewhere to U.S. strikes on the North. [17]

The conferees further agreed to provide more U.S. military and economic aid for the Khanh government. Another decision required the services to review their available shipping, manpower, reconnaissance, airlift, ordnance, and command post resources, and future requirements to sustain the "escalation" phases of CINCPAC's Operational Plans 32-64 and 37-64. McNamara directed the Army to prepare for the dispatch of an infantry brigade and asked the JCS to submit a joint U.S.-Thai military plan for defense of the Mekong delta and for punitive action against Communist forces in northern Laos. [18]

There was more planning against the threat in Laos. Limited U.S.-Vietnamese planning was authorized on 5 and in late June the JCS sent McNamara MAC/V's plans for Vietnamese cross-border operations. Fuller consultation with Saigon was now required but the State Department would not allow this until

Damage caused by a communist ground attack on Luang Prabang airfield.
Source: U.S. Air Force

political objections raised by the U.S. Ambassador in Laos were resolved. The delay greatly troubled the Air Staff. [19]

Laotian planning also figured in a JCS reply to the National Security Council (NSC) request for guidance. Deeply concerned over the growing U.S. commitment in Southeast Asia, the NSC in July asked for a restrictive program that would aid the counterinsurgency effort in South Vietnam and reduce the defeatism of South Vietnam and its leaders, but minimize U.S. participation and the risk of military escalation.

Even as early as 1964, North Vietnamese infiltration into the South was including equipment such as this armored car. Source: U.S. Air Force

The Joint Chiefs offered three courses or action: ground cross-border operations into Laos against infiltration targets, air strikes on Laotian infiltration routes, and selected air attacks on North Vietnam with unmarked aircraft. The JCS warned, however, that while its proposals would have some military and psychological value provided the effort did not absorb counterinsurgency resources, they would not significantly affect Communist support for the Viet Cong. And they might aggravate the political situation in Laos. [20]

Overall planning trends were now strongly weighted toward expanded use of airpower. In late July, the JCS directed CINCPAC to plot 94 key North Vietnam targets, a list subsequently included in CINCPAC's 37-64 plan. [21]

New U.S. Leadership and More Military Aid

Coincident with planning operations against Laos and North Vietnam were changes in U.S. military and diplomatic leadership in Saigon. On 20 June Gen. William C. Westmoreland, deputy to General Harkins, became the commander of MAC/V. On the 23d President Johnson announced that General Taylor would succeed Ambassador Lodge (Gen. Taylor officially succeeded Lodge on 2 July) and that Alexis Johnson would become Deputy Ambassador, a newly created post. General Wheeler, the Army's Chief of Staff, succeeded Taylor as JCS chairman. On 30 June Adm. U.S. Grant Sharp succeeded Admiral Felt as CINCPAC. The changes were accompanied by a new warning to the Communists on the 28th by President Johnson. He said that the United States was prepared to "risk war" to preserve peace in Southeast Asia and would continue to stand firm to help South Vietnam maintain its freedom. [22]

Almost simultaneously MAC/V asked for more U.S. military advisors, units, and equipment. For expanded air operations the Army would provide 27 more CH-1B helicopters and 16 CV-2B Caribou transports (and a few supporting aircraft), while the Air Force would deploy a fourth C-123 squadron (16 aircraft), 25 A-1Es (for the second combat training squadron approved on 5 May), and six HH-43B helicopters for a search-and-rescue (SAR) unit. There would be more air liaison officer and forward air controller (ALO/FAC) teams for stepped up combat training and close air support operations. [23]

MAC/V's request was followed by more South Vietnamese set-backs in July. The Viet Cong stepped up its attacks in the Mekong delta, Vietnamese forces suffered a major defeat in Chuang Province, and on the 20th there was another coup attempt in Saigon. U.S. officials now estimated Viet Cong strength at 34,000 with about 30 percent of the infiltrators coming from the North, and irregular forces at 68,000. Concluding that counterinsurgency activities were insufficient and that only direct pressure on the North could defeat the Viet Cong, Premier Khanh's government agreed to U.S.-Vietnamese planning for such action without a firm U.S. commitment. [24]

Firefighters at Phan Rang AB, South Vietnam use an HH-43 to battle a simulated aircraft fire. Source: U.S. Air Force

Meeting with McNamara on 20 July, the JCS generally supported MAC/V's proposals except for additional Army helicopters and Caribous. LeMay and the Commandant, Marine Corps, strongly believed that the Army aviation units required more justification in view of available USAF and VNAF aircraft for close support and airlift. They were subsequently overruled by the Defense Secretary. [25]

After assessing MAC/V's ability to absorb quickly the additional personnel, aircraft, and equipment, the administration announced on 27 July that about 5,000 more U.S. military personnel would go to South Vietnam, raising the total there to 21,950. After adjustments, the [increase] was reduced to 4,800 personnel. Most of the manpower and equipment would arrive by 30 September as MAC/V wished, but some units could not be absorbed or sent until November and December. These were the fourth C-123 squadron, the SAR unit, five A-1Es, 20 (of 40 requested) ALO/FAC teams, and 336 jeeps. More civilian technical advisors also would be sent. For certain units, final approval to deploy was still pending. [26]

A badly-damaged A-1E forced down during operations against the Viet Cong. Source: U.S. Air Force.

Vietnam and Laos 1964

III. THE GULF OF TONKIN INCIDENT AND AFTERMATH

In March, May, and July the administration was forced to provide more aid for South Vietnam. Counter-insurgency operations were proving ineffectual in the face of demoralized Vietnamese leadership and rising Viet Cong strength and aggressive tactics. As a consequence, planning focused increasingly on airpower as a means to reverse defeats. Early in August, the Communists supplied the provocation needed to launch an air attack on North Vietnam.

U.S. Response in the Gulf of Tonkin

On 2 August the U.S. Navy destroyer *Maddox*, part of a patrol in the Gulf of Tonkin, detected three hostile patrol boats closing in at high speed. After three warning shots failed to halt them, the destroyer opened fire with its 5-inch batteries. One boat was disabled but succeeded in firing two torpedoes that missed the *Maddox* by 200 yards; a second boat lost power and retired, and a third, also struck, passed 1,700 yards astern the *Maddox*, firing a machine gun. In response the United States reinforced the patrol by adding a destroyer (the *C. Turner Joy*) and an aircraft carrier (*Ticonderoga*). On the night of 3 August enemy boats again attacked the patrol. In return fire, one was presumed sunk.

North Vietnamese motor torpedo boat making its run against USS Maddox on August 2, 1964. Photograph taken from USS Maddox, source U.S. Navy

North Vietnamese torpedo boat under fire on August 2, 1964. Photograph taken from USS Maddox, source U.S. Navy

On 4 August, immediately after the second attack, Admiral Sharp proposed and the JCS and the President agreed to conduct punitive air strikes against North Vietnam. These were launched on 5 August when Navy A-1 Skyraiders, A-4 Skyhawks, and F-8 Crusaders from the *Ticonderoga* and the *Constellation* flew 64 sorties, attacking four torpedo bases at

Air War – Vietnam

Hon Gay, Loc Chao, Phuc Loi, and Quang Khe and an oil storage facility at Vinh. The code name for the air strike was "Pierce Arrow." Eight boats were destroyed and 21 damaged and the Vinh oil facility, representing about 10 percent of North Vietnam's oil storage capacity, was 90 percent destroyed. Two aircraft, an A-1 and an A-4, were shot down by antiaircraft fire over Hon Gay killing one pilot. The other was taken prisoner. Two other aircraft were hit but returned safely. No USAF aircraft participated in these strikes. [1]

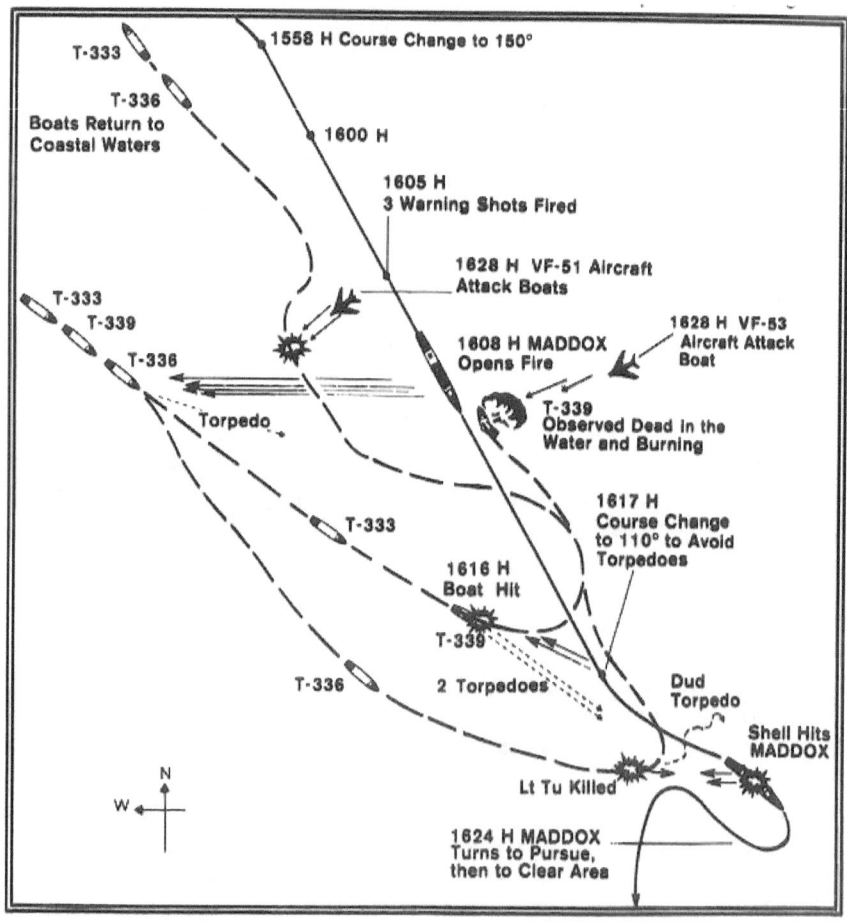

Track Chart of Incident of August 2, 1964. Note that the "shell hit" on Maddox was actually a 14.5mm bullet that did trivial damage. Source: U.S. Navy.

Simultaneously, the President publicly warned the Communist world not to support or widen aggression in Southeast Asia and McNamara, with the President's approval, announced the dispatch of more U.S. reinforcements to the area. The initial deployment of air units was called "One Buck," and subsequent deployments were "Two Buck," "Three Buck," etc. Pacific theater shifts brought

Vietnam and Laos 1964

50 additional USAF aircraft (B-57s, F-102s, RF-101s) to South Vietnam and 26 (F-105s, F-100s, KB-50s) to Thailand. Other aircraft (F-105s, C-130s) from the United States went to U.S. bases in Japan, Okinawa, and the Philippines. From its First Fleet on the Pacific Coast the U.S. Navy sent the supercarrier *Ranger*, 12 destroyers, an antisubmarine task force, and selected Marine units. The Army sent additional aviation and ground units. Tours of duty for tactical units deployed in support of CINCPAC Plan 37-64 were extended indefinitely. Total U.S. force authorization for South Vietnam was raised to 23,308. [2]

On 7 August, at the request of President Johnson, the Congress approved overwhelmingly a resolution assuring the Chief Executive of support: [3]

> "• • • the Congress approves and supports the determination of the President, as Commander in Chief, to take all necessary measures to repel any armed attack against the forces of the United States and to prevent further aggression.
>
> "• • • Consonant with the Constitution of the United States and the Charter of the United Nations and in accordance with its obligations under the Southeast Asia Collective Defense Treaty, the United States is, therefore, prepared, as the President determines, to take all necessary steps, including the use of armed force, to assist any member or protocol state of the Southeast Asia Collective Defense Treaty requesting assistance in the defense of its freedom."

Track Chart of Incident of August 4 1964. Many of the observations reflected in this chart were later determined to be inaccurate, being the product of radar ghosting and anomalous propagation. Over the years much has been made of these errors but the historical record shows that mistakes of this kind are not uncommon and are a feature of many naval actions. During the Second World War, there were may reports by all sides of sinkings and engagements that turned out to be erroneous.
Source: U.S. Navy

Signed by the President on 10 August, the resolution was similar to those approved by Congress during the crises in the Formosa Strait in 1955, in the Middle East in 1958, (See *Air Operations 1958: Lebanon and Taiwan*, Defense Lion Publications 2012) and in Cuba in 1962.

Air War – Vietnam

Chinese MiG-19s flying out of Hainan started to appear over the Gulf of Tonkin shortly after the alleged second incident, adding to the impression that the 'second attack' had been an orchestrated event. This illustrates an old truth of the intelligence business; "if one seeks confirmation of something, one will find it regardless of whether it is there or not". Photo source: U.S. Air Force

Hanoi, Peking, and Moscow accused the United States of "provocative" action and pledged continued support for the insurgents. Some neutralist nations and U.S. allies were concerned about the reprisal strikes on North Vietnam but others, such as Thailand, were heartened. Tension increased as Chinese MiGs on Hainan Island were observed flying periodically toward South Vietnam. There were "scrambles" of USAF F-102s and Navy F-4s and F-8s to meet them. On 8 August, one such operation involved 30 U.S. jets. Meanwhile, on the 7th, reconnaissance showed 36 MiG-15s and -17s on Phuc Yen Airfield in North Vietnam, flown in presumably by Chinese- or Soviet-trained Vietnamese pilots. [4]

The buildup of combat aircraft in Southeast Asia and in other parts of the Pacific and the possibility of air action focused attention on the problem of command and control. Admiral Sharp concluded that his Operational Plan 99-64 (to cover military operations against North Vietnam and to stabilize the situation in Laos) now was more relevant than Operational Plan 37-64 (to stabilize the military situation in South Vietnam). Therefore, he proposed to control land-based air forces through his component commanders. PACAF, as the Air Force component command, would control 13th Air Force and 2d Air Division aircraft. Sharp believed this would allow MAC/V, which was inadequately manned for jet combat operations, to concentrate on counter-insurgency actions and only monitor 2d Air Division activities. [5]

A New Round of Planning

To the dismay of the JCS, the confrontation in the Gulf of Tonkin did not result in follow-up strikes. Instead, the administration pursued a "holding action" to await Communist response and place upon Hanoi the onus for escalating the war. Over strong JCS objections, the administration halted temporarily the Navy's patrol in the gulf, some special operations under Plan 34A and slackened support for T-28 strikes in Laos.[6]

*USS Maddox in March 1964.
Source: U.S. Navy*

State Department and other agency proposals were reviewed intensively. To the extent these proposals provided additional (if limited) pressure on the North, and for U.S.-Vietnamese planning, VNAF training, cross-border activities, and similar measures, the JCS agreed with them. But it considered such actions insufficient. Administration leaders, conversely, believed that in view of a weakening Saigon government the situation demanded U.S. prudence and, for the moment, no further escalation.[7]

Premier Khanh's regime, meanwhile, was given only a 50-50 chance to remain in power. Apprehensions about the stability of his government arose when the Military Revolutionary Council on 16 August ousted General Minh as president, elected Khanh to that post, and promulgated a new constitution giving him near dictatorial powers. These changes set off more Buddhist rioting and other civil disturbances, culminating in late August in a one-week "resignation" by Khanh.

Later, Ambassador Taylor observed ruefully that there was "no George Washington in sight" in Saigon. However, he said that there was no alternative to continued U.S. support because of the dire effects an American defeat in Southeast Asia would have in Asia, Africa, and South America. He averred publicly that Viet Cong insurgency could not be defeated by military means in the foreseeable future. A U.S. intelligence report stated that the odds were against the emergence of a stable government in Saigon but suggested one might be created after the release of pent-up pressures and the sobering effects of instability were realized fully by the Vietnamese.[8]

Air War – Vietnam

The USS Turner Joy. The original text incorrectly refers to this ship as the C. Turner Joy. Source U.S. Navy

The JCS continued to review and comment on many proposals. On 24 August it sent McNamara another list of North Vietnam air targets, which, if bombed, would possibly end Hanoi's support of the Viet Cong and Pathet Lao. The targets were divided into five categories: airfields, lines of communication, military installations, industrial sites, and certain others suitable for armed reconnaissance missions. [9]

On the 26th the JCS recommended a number of priority actions that should be taken without delay. They included: resumption of patrols in the Gulf of Tonkin and in support of Plan 34A operations; retaliatory air strikes in response to large-scale Viet Cong or Pathet Lao actions; attacks against the Viet Cong leadership; Vietnam-Thai-Lao air operations with U.S. support on communication lines in the Laotian corridor; "hot pursuit" into Cambodia; stricter patrols of the Mekong and Bassac Rivers; more pacification projects with the emphasis on the Hop Tac program around Saigon (The Hop Tac program, concentrating on seven provinces around Saigon, began in September. Initial results were meager but by the end of 1964 it was one of the few areas where pacification efforts showed some success) and buildup of U.S. combat units.

As the JCS were doubtful if these proposals would deter Hanoi, it asked additionally for more U.S. forces to support CINCPAC's 37-65 plan and the inauguration of air strikes on North Vietnam. The JCS believed that only stepped up and forceful action could prevent a complete collapse of the U.S. position in Southeast Asia. [10]

Despite much unanimity on what should be done, the JCS was divided over the timing and severity of the proposed strikes on the North. General Wheeler and the Army and Navy chiefs agreed with Ambassador Taylor that the United States

Vietnam and Laos 1964

should not create an incident by an immediate attack but respond appropriately to the next Viet Cong strike on a U.S. unit. General LeMay and the Marine Corps chief argued, however, that time was running out and that air strikes were imperative. They advocated a retaliatory U.S.-Vietnamese air attack after the next "significant" Viet Cong incident, if only a battalion-size operation, in accordance with the 94-target plan, and more public statements on U.S. determination to defend South Vietnam.[11]

Laotian T-28 operations depended largely on semi-covert U.S. support.
Source: U.S. Air Force

LeMay was greatly distressed over U.S. policy. He believed that the "message" delivered to the Communists on 5 August in response to their attacks in the Gulf of Tonkin had been nullified by other U.S. actions. There was the apparent leak to the press, for example, of a CIA study indicating U.S. desire to negotiate, and the reduction of Laotian Air Force T-28 strikes in Laos. He perceived undue concern over escalation and the desire to strengthen Saigon politically before striking North, whereas air strikes, in his view, would strengthen Saigon's political base.

Believing that U.S. restraint was being practiced to the point of inadequacy, LeMay urged unsuccessfully for the immediate implementation of the JCS' recommendations of 26 August and the deployment of more ground forces to Thailand.[12]

New U.S. Guidance

Out of the interminable high level conferences and policy reviews, the President's chief advisors emerged in early September with new proposals. Concluding that the internal political turmoil would leave the Khanh government in the next two or three months too weak to allow the United States to risk military escalation, they drew up a "low risk" program. The objective was to improve Vietnamese morale but also to show that the United States "meant business."[13]

On 10 September President Johnson approved part of the program: resumption of U.S. Navy patrols, with air cover, beyond the 12-mile limit in the Tonkin Gulf; resumption of Plan 34A air, leaflet, and maritime operations; U.S.-Laos discussions on allowing limited air and ground action in Laos by the Vietnamese supported by Lao pilots and possibly U.S. armed reconnaissance; preparations to retaliate against the North for the next important Viet Cong attack on a U.S. or Vietnamese unit; and specific aid measures, regardless of cost, such as pay raises for Vietnamese civilians or for special projects that would help the Khanh

Air War – Vietnam

government. The President emphasized that the "first order of business" was to strengthen the political fabric of the country.[14]

The Low-Risk Policy

Cautiously, the administration pursued its "low risk" policy against North Vietnam. On 15 September, the JCS authorized resumption of a patrol in the Gulf of Tonkin. But on the 18th, there occurred another incident between the patrol and Communist craft. U.S. ships fired on them in the darkness and the JCS ordered Sharp to prepare for reprisal strikes. But a search of the waters disclosed no positive evidence of an attack, although the Navy was convinced one was made. As a consequence, the administration refused to sanction an air strike. And to avoid another incident, it suspended, despite CINCPAC and JCS recommendations to the contrary, further patrols until early December.[15]

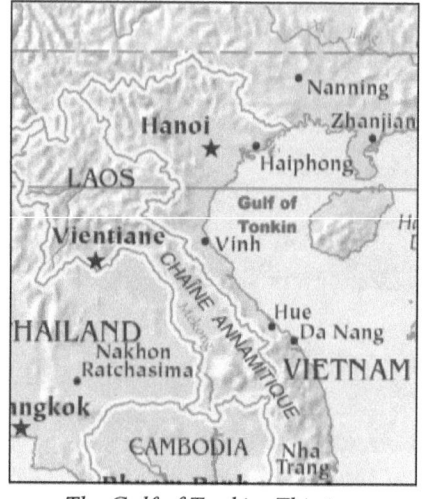

The Gulf of Tonkin. This is a modern map but it has a strange relevance to the Vietnam War. Today territorial limits in the Gulf are still disputed only now it is Chinese warships patrolling the waters where the Maddox was attacked almost 50 years ago.
Map source: CIA.

The continuing concern over escalation prompted more preparations to use airpower. On 21 September the JCS approved CINCPAC's Operational Plan 39-65. It was designed to counter a Chinese attack alone or in league with North Vietnam and North Korea against South Vietnam, South Korea, or other parts of Asia. To the Air Force, the plan was a milestone in that it provided for the destruction by air of the enemy's primary military, economic, and logistic targets-"where it would hurt the most." Heretofore, the. Army and Marine Corps had opposed an air plan on the premise that airpower alone was no substitute for ground forces.[16]

The JCS also revised its plans for air strikes against North Vietnam. At the suggestion of LeMay, who pointed to the danger of air opposition (especially after 36 MiGs arrived at Phuc Yen Airfield in August), the service chiefs approved a change in the 94- target objectives. Air strikes, if conducted, would inflict maximum damage on selected targets. This contrasted with the initial strategy of diffusing strikes among the targets and causing less damage on individual ones. When completed on 17 December, the revision required an increase in USAF's force structure in Asia.[17]

The President's approval of U.S.-Laotian discussion on Vietnamese cross-border operations to reduce the infiltration of men and materiel through Laos into South

Vietnam and Laos 1964

Vietnam again spurred preparations on this long-delayed project. Headquarters MAC/V estimated that from January to August 1964, 4,700 Communists had entered South Vietnam from 1959 to August 1964, the total was 34,000 with 31,500 of them military personnel. In July the JCS had sent MAC/V's plans to McNamara. Now there was more discussion on the type and extent of U.S. support. The Air Force and Army debated the relative value of air and ground action with the Army asserting that airpower would be restricted by the jungle canopy and the weather. [18]

On 30 September the JCS agreed to an air-ground plan to support the Vietnamese. It provided for coordination with the Yankee Team-Laotian Air Force operations already under way in Laos. USAF aircraft would help to suppress antiaircraft fire and strike difficult targets, such as bridges. Ground forces, with attached U.S. advisors, beginning 1 November would penetrate into three areas up to 20 kilometers. [19]

But political turbulence in Saigon [was illustrated by] another coup attempt against Premier Khanh on 13 September. On 26 September a High National Council was established, charged with setting up, if possible, a civilian government. Frequent personnel changes in the Vietnamese high command and difficulties with Montagnard tribesmen (some of whom had begun to revolt in September) prompted the administration to limit and finally to postpone the venture. On 7 October a State-Defense directive forbade for the time being any U.S. strike participation and permitted only combat air patrol. On the 21st McNamara ordered the JCS to limit the project to planning only. A few days later General Westmoreland reported that Saigon's political weakness would preclude any cross-border undertaking until 1 January 1965. [20]

Meanwhile, General LeMay pointed to a disturbing intelligence report showing, he thought, that Saigon's political problems were virtually beyond resolution. He again urged the JCS to agree to an immediate air response to the next "significant" move such as a battalion-size or a terrorist attack. He recommended a strike by VNAF A-1Hs with USAF F-100s and F-102s and Navy aircraft providing cover. As U.S. intelligence indicated that the Communists had every reason to regard favorably present trends, LeMay thought it unlikely that they would provoke the United States, even if U.S.-Vietnamese forces struck North. His assessment that the Communists probably would not attempt another provocative act (as in the Gulf of Tonkin) was shared by the Army. [21]

But the JCS agreed only to somewhat less precipitous courses of action, mostly old, a few new, inside and outside of South Vietnam, all in a new order of ascending severity. Sent to McNamara on 22 October, the JCS paper observed, however, that the USAF and Marine Corps chiefs believed that "time was running out," and that there was no alternative to a prompt air strike on North Vietnam. McNamara promised to convey their views to the White House but advised that Ambassador Taylor was reluctant to increase pressure on Hanoi while Saigon was without a responsible government. [22]

Air War – Vietnam

So critical was the situation that preparations began for a possible collapse of the South Vietnamese regime and the emergence of an unfriendly one that might ask for the withdrawal of U.S. forces. For this eventuality, the Air Force considered steps to protect major U.S. airfields and redeploy U.S. and friendly Vietnamese air and ground units to Thailand, the Philippines, and elsewhere. To prepare for any contingency, LeMay directed his commanders to assess their ability to support PACAF's plans and to report any inadequacies or the need for more guidance.[23]

On 30 October U.S. pessimism about Saigon's political future was tempered slightly. General Rhanh voluntarily resigned as Premier to allow Saigon's former mayor, Tran Van Huong, the new Premier, to install South Vietnam's first civilian government since the overthrow of President Diem a year earlier.[24]

Vietnam and Laos 1964

IV. THE BIEN HOA AIR BASE ATTACK AND AFTERMATH

In addition to a new Vietnamese civilian government, the end of October also witnessed a new policy crisis. An impression that the Viet Cong, seeing only auguries of success, might refrain from another dramatic strike against the United States was dispelled quickly.

B-57 crippled in the mortar attack on Bien Hoa AB. Source U.S. Air Force

The Bien Hoa Incident

On the night of 31 October to 1 November, Viet Cong troops eluded successfully Vietnamese army security guards around Bien Hoa Air Base, creeping within 1,500 meters of the control tower. They fired about 80 rounds of mortars for 30 minutes against the tower, the packed flight line, and the bivouac area. The attack was costly. The Air Force suffered 7 aircraft destroyed (6 B-57s and 1 H-43 helicopter) and 16 damaged (13 B-57s and 3 H-43s). VNAF losses were 3 aircraft destroyed (all A-1Hs) and 5 damaged (3A-lHs and 2 C-47s). In addition, three houses, a mess hall, vehicles, and fuel tanks were destroyed or badly damaged. U.S. casualties were 4 personnel killed and 30 badly wounded plus 42 personnel with lesser wounds. Vietnamese casualties were 2 killed and 5 wounded. [1]

Within 5 minutes after the attack began, base defense teams and aircraft sprang into action, but the enemy escaped. The next day 800 Vietnamese troops,

Air War – Vietnam

supported by helicopters, likewise could find no trace of the guerrillas. Momentarily, the losses were a blow to PACAF. And coming on the eve or a national holiday to celebrate the first anniversary of the fall of the Diem government on 1 November 1963 in South Vietnam and an American presidential election, the incident, according to news media, was a blow to U.S. prestige.[2]

Top U.S. officials: Admiral Sharp, General Westmoreland, Ambassador Taylor, and JCS expected the administration to order immediate reprisal air strikes. The JCS, having suddenly resolved the major differences over the timing and severity of military reprisal, orally gave unanimous support on 1 November. But the administration again demurred. Compared with previous Viet Cong incidents, it believed that the attack on Bien Hoa differed mainly in degree and damage done and was not, necessarily an act of major escalation. There was reluctance to retaliate simply because the attack was directed primarily at the United States, and deep concern lest a strike against the North would trigger, in turn, air and ground action by Hanoi and Peking. And there was the overriding need to establish political stability in Saigon.[3]

Smoke billowing from the POL dump at Bien Hoa. Source U.S. Air Force

The administration's initial response was to order the immediate replacement of the destroyed B-57 aircraft, warn Hanoi and Peking not to expect a change in U.S. policy in Asia after the American elections (on 3 November), and express encouragement about the latest complexion or the Saigon government and a few recent military successes. Publicly, Washington officials differentiated between the Bien Hoa and Gulf of Tonkin attacks, asserting that there would have to be "broader reasons" for making a retaliatory strike against North Vietnam.[4]

On 4 November, still convinced that a U.S. riposte was in order, the JCS reaffirmed its views and urged McNamara to approve immediately armed reconnaissance of infiltration targets in North Vietnam up to 19 degrees latitude, and strikes against the Techepone and Ben They areas and two bridges in Laos. Within 60 to 72 hours, the JCS said, there should be night strikes against Phuc Yen Airfield in the North by 30 B-52's, and VNAF and U.S. strikes on some of the other "94 targets." It further recommended instant deployment of Marine or Army units to provide more security for the Bien Hoa and Da Nang air bases, and the evacuation of U.S. dependents from Saigon.

The JCS warned that the Communists and America's Southeast Asia allies might misconstrue U.S. restraint. In response to another query from McNamara, the Joint Chiefs assured him that U.S. forces could deal with any military "response" by Hanoi or Peking, and expressed confidence in the stability of the new Huong

Vietnam and Laos 1964

government to permit "positive" U.S. action. They objected to Ambassador Taylor's proposal for a "tit for tat" strike policy henceforth against the North. [5]

Again JCS counsel was not accepted. Subsequently, McNamara informed the Joint Chiefs that their views were being considered in interdepartmental deliberations on future U.S. action in Southeast Asia.[6]

The Problem of Base Security

If retaliatory strikes against North Vietnam were not warranted, a review of U.S. base security was. Its weaknesses now underwent thorough scrutiny.

Since late 1961, primary responsibility for base security rested with the Vietnamese armed forces. Periodically the Air Force had asked for more protection, especially for Tan Son Nhut, Bien Hoa and Da Nang. Air Force concern rose after the Gulf of Tonkin incident in August and the deployment of B-57s from Clark AB, the Philippines, to Bien Hoa. Some improvements were made, enabling the JCS, on 1 September, to agree that security was adequate. [7]

But security was largely in the hands of the Vietnamese and was effective only to the extent they accepted the responsibility. From mid-1964 on, the progressively weakening Saigon government reduced, in turn, Vietnamese concern and protection. As a consequence, General LeMay on 28 September ordered another review of base defenses. Oversaturation at Bien Hoa was quite apparent and this resulted, fortuitously, in a decision to redeploy on 31 October, only hours before the Viet Cong attack on the air base, 20 B-57's from Bien Hoa to Clark AB. This saved many bombers from destruction or damage. [8]

On the eve of the attack, defense measures at the three main airfields consisted of joint USAF-VNAF manning of the inner and Vietnamese Army manning of the outer perimeter. There were also special command posts, and helicopters and flare aircraft on alert. [9]

The provision of guard towers and perimeter control was an early response to the Bien Hoa attack. However, guard towers are not much use without reliable people within them. Photo Source: U.S. Air Force

Air War – Vietnam

As a result of losses at Bien Hoa, a board of inquiry was convened by USAF Maj. Gen. Milton. D. Adams of MAC/V's staff. The joint research and test agency (JRATA) unit was directed to examine tactical air base needs. Other studies were undertaken. Top Air Force leaders urged changes in the U.S.-Vietnamese agreement to allow U.S. combat troops, Army or Marine, to secure and control an 8,000 meter area around each airfield. [10]

But Sharp, Westmoreland, and Taylor opposed the use of combat troops, asserting they would be ineffectual. The troops would lack language and area knowledge and authority to search private dwellings, cause political and psychological difficulties, and encourage the Vietnamese to relax still more their security efforts. Sharp recommended to the JCS only 502 more police-type personnel for base defense: 292 Air Force, 52 Army, and 153 Marine personnel. For backup, there was afloat offshore a marine brigade and a special landing force. [11]

LeMay thought differently. Pointing to the lack of surveillance, the ease of infiltration, and the prospect of more damage to U.S. property, he wanted Sharp to reassess the ability of the Vietnamese to provide base security. If they were unable to do so, U.S. combat troops, he reiterated, should be used. For the interim, he and the other service chiefs accepted Sharp's proposal to augment base defense strength by 502 personnel and, on 23 December, sent this recommendation to McNamara. No decision had been made by the end of the year. [12]

Review of Future Courses of Action

Having again elected not to respond to a "provocation," the administration launched into another review of U.S. policy.

On 1 November, immediately after the Bien Hoa attack, the State Department proposed three "options": continue existing policies and take no reprisal action except to Viet Cong "spectaculars" like Bien Hoa; apply immediately more military pressure to show firm U.S. determination but also willingness to negotiate; apply graduated and carefully controlled military pressure in concert with political action to end Hanoi's support of the South Vietnam and Laos insurgencies. They formed the basis of a report by the NSC Working Group, now headed by William F. Bundy, Assistant Secretary of State for Far Eastern Affairs. The group favored the third option and its pursuit for six to eight months while the door to negotiations was left open. In subsequent days the three alternatives were refined extensively.[13]

Chinese propaganda leaflet. This poster predates the AK-47 becoming the iconic image of the Viet Cong, Source: U.S. Air Force

The consequences of North Vietnam strikes were thoroughly reviewed. At White House request, the JCS on 14 November sent an analysis of possible Hanoi-Peking reaction. The Joint Chiefs believed that the fear of massive retaliation would prompt the Communists to rely on propaganda and diplomacy rather than on enlarging the war. If the Chinese Communists felt compelled "to do something," they might enter Laos, perhaps at the invitation of the Pathet Lao, but not North Vietnam unless Vietnamese or U.S. forces occupied territory in either Laos or in the North, or attacked Chinese soil. Admittedly, the Chinese might intervene for "irrational" reasons or through miscalculation. But on balance, the risks inherent in striking North Vietnam were preferable to continuing the current policy or withdrawing from Southeast Asia. As a precaution, the JCS favored the deployment of two additional USAF fighter squadrons, more USAF reconnaissance and tanker aircraft, and another Navy carrier to Southeast Asia. Except for the latter phases of CINCPAC's 32-64 and 39-65 plans, there would be no logistic difficulties in carrying out the 94-target attack. [14]

Sending another Navy carrier was a preferred JCS option.
Source: U.S. Navy

The Air Force especially did not think air strikes on the North would trigger a major air and land war nor lead to an untenable U.S. negotiating position, two objections raised by the working group. [15]

In reply to another McNamara request, the JCS sent him proposed U.S. objectives if the policy of graduated military pressure was adopted. [16]

On 23 November, in another paper, the JCS informed McNamara that there were five rather than three courses of action that should be considered:

(1) withdrawal from South Vietnam and Laos (and abandonment of U.S. objectives);

(2) continue current policy with improvements where possible (with no likelihood of attaining U.S. objectives);

(3) graduated military and political pressures as proposed by the NSC Working Group (with inconclusive objectives and high risk as the uncertain pace could encourage enemy buildup);

(4) graduated military pressure to reduce North Vietnamese capability to support the insurgencies in South Vietnam and Laos (probably achieving U.S. objectives);

(5) rapid and forceful military pressure (involving the least risk, casualties, and costs, insuring less possibility of enemy miscalculation and intervention, and most likely to achieve U.S. objectives).

The JCS recommended adoption of the fifth course of action. [17]

Air War – Vietnam

Having examined JCS and other agency viewpoints, President Johnson on 2 December issued another policy guide for South Vietnam. It followed most closely a sixth view submitted by the Office of International Security Affairs in OSD. The President concluded that South Vietnam's problems were two; government instability and Viet Cong insurgency as aided by the North. But the two problems were of unequal importance. Viet Cong actions were only contributory whereas a stable government in Saigon, in accordance with recent policy, was of paramount importance. Thus the United States could not risk preventing its establishment. This was the antithesis of the longheld Air Force and lately JCS' position that gave top priority to ending North Vietnam's support for the insurgency. [18]

After this decision, the President instructed Ambassador Taylor to "consult urgently" with South Vietnam's leaders to improve the internal situation in their country. Taylor foresaw no immediate need for more U.S. military personnel, now numbering about 22,000, nor for major changes in prosecuting the war except in tactics. [19]

T-28s over Laos. Source: U.S. Air Force

The President approved limited but graduated military pressure, largely by air. A two-phase program required heavier Laotian T-28 strikes and U.S. armed reconnaissance (Barrel Roll) missions along infiltration routes in the Laos corridor and special Plan 34A maritime operations against the North. The air attacks would be primarily psychological, warning Hanoi of U.S. strength. There would also be initial steps to end the flow of U.S. dependents to Saigon. [20]

After a transition period or unspecified duration between. the first and second phases, additional military pressure for two to six months would be exerted. There would be more high- and low-level reconnaissance and maritime operations against the North, and heavier strikes against infiltration routes near the South Vietnam-Laos border. This stage would require some augmentation of U.S. strength and include the deployment of 150 or more U.S. aircraft and the alerting of ground forces for Southeast Asia. [21]

As the program of graduated military pressure began, Taylor, on returning to Saigon, plunged into a series of conferences with Premier Huong and other Vietnamese and U.S. officials. They discussed the use of $60 to $70 million in U.S. aid to speed up economic and rural development, more effective measures against Communist infiltration, expansion of the Vietnamese military and police forces, and other topics. A joint communiqué on 11 December on the meetings reaffirmed U.S. support for the Huong government. [22]

Vietnam and Laos 1964

With respect to increasing Vietnamese military strength, the JCS on 17 December approved a MAC/V proposal to add 30,309 men to the regular forces (for an authorized total of 273,908),and 110,941 to the non-regular forces. The VNAF would gain 342 spaces. The augmentation would also require 446 more U.S. military advisors. The new U.S. authorized manpower ceiling in South Vietnam was 22,755 (revised from 23,308.). [23]

On 13 January 1965, McNamara approved the JCS recommendations subject to final approval by the State Department. [24]

Continuing Crisis and a New Incident

The administration's latest attempt to create political stability in Saigon while simultaneously applying low-key military pressure on the Communists was disrupted in mid-December by another political upheaval. Buddhists began a new drive to unseat Premier Huong and bitterly attacked Ambassador Taylor. There were more military setbacks. On 20 December a group of "Young Turks" led by Air Commodore Ky (during 1964 the rank of VNAF's commander changed from brigadier general to air commodore) and Brig. Gen. Nguyen Chan Thi, Commander of the Army I Corps, overthrew the civilian-oriented High National Council and arrested some of its members. This partial coup, which left U.S. officials close to despair, put the military through the Armed Forces Council again in the ascendancy and left the tenure of Premier Huong in doubt. [25]

Premier Tran Van Huong. South Vietnam's crippling political instability just kept getting worse. Source: Democratic People's Republic of Vietnam

The U.S. government tried to be firm. Ambassador Taylor in Saigon and Secretary Rusk in Washington warned that unless civilian rule was restored, the United States might have to review its aid and other commitments to South Vietnam. On the 26th, administration officials directed all U.S. military advisors to withdraw from advance planning of non-routine military and civilian operations until the future of U.S. aid was clarified. This strong stand drew a sharp blast from General Khanh, now siding with the Young Turks, who severely criticized Taylor for interfering in Vietnamese affairs. In the closing days of 1964, the political crisis eased and Huong was still Premier although the High National Council had not been reconstituted. [26]

In the midst of the political turmoil, the administration's restraint was again challenged on 24 December when the Viet Cong bombed the U.S.-occupied Brink Hotel in Saigon. The blast killed two Americans and wounded 64. Forty-three Vietnamese were wounded. The JCS recommended an immediate reprisal air attack on Army barracks at Vit Thu Lin in North Vietnam. CINCPAC alerted

Air War – Vietnam

Navy air, rather than PACAF, for the reprisal, if authorized. Again the administration chose not to respond. [27] Between 3 February and 27 December 1964, the Viet Cong engaged in 61 attacks against U.S. personnel, exclusive of the Gulf of Tonkin incident. The attacks included grenades thrown at vehicles and into bars, sniper fire, entry into U.S. compounds and bombing of hotels.

As 1965 began, administration policy of seeking a political solution in Saigon first rather than a military victory against the Viet Cong was in question. Observing that the coups were getting worse and that current U.S. strategy was not working, General LeMay reiterated his view that the only alternative was to strike North Vietnam, although he said the hour was so late this might not stop the aggression. He foresaw danger lest rioting spread to the Vietnamese armed forces, the only cohesive element in the country, and the possible loss of everything in South Vietnam including American lives. He recognized the fact that the Chinese Communist might intervene and believed that the United States should be prepared to take care of them, by air. Using only conventional ordnance, this would be a major task. In a big war, he thought, a few nuclear weapons on carefully selected targets would be a more efficient way "to do the job." [28]

Vietnam and Laos 1964

V. BUILDUP OF USAF FORCES IN SOUTHEAST ASIA

While the administration sought desperately in 1964 to halt the political and military decline in South Vietnam, the demand for more aircraft rose.

At the end of 1963 U.S. and Vietnamese fixed wing and rotary aircraft in South Vietnam totaled about 690. The Air Force possessed approximately 120, all controlled by Headquarters, 2d Air Division at Tan Son Nhut Airfield near Saigon. Its major units were the 33d and 34th Tactical Groups, the 315th Troop Carrier Group, and the 23d Air Base Group. Also under the 2d's control was the 35th Tactical Group in Thailand. On 31 January the 2d's commander, Maj. Gen. Rollen H. Anthis, was replaced by Maj. Gen. Joseph H. Moore, Jr. [1]

The 2d's aircraft consisted of 22 O-1s, 49 C-123s, 6 RF-101s, 2 RB-57s, 6 F-100s, 4 F-102s, 13 T-28s, and 18 B-26s. The F-102s were stationed at Don Muang Airport, Thailand. The B-26s and T-28s were assigned to the 34th Group's 1st Air Commando Squadron (previously Farmgate), a combat training unit. To limit U.S. combat training participation, the 1st operated under rules of engagement that severely circumscribed its activities. USAF efforts in 1962 and 1963 to change the rules were unsuccessful. [2]

In the spring of 1964 two circumstances led to a critical shortage of aircraft for the 1st Air Commando Squadron. In one instance, investigation of a B-26 crash at Hurlburt Field, Fla., in February showed that the aircraft had experienced structural failure. As a consequence, the B-26s in South Vietnam were grounded temporarily, then permitted to fly on restricted basis and, in March,

The crash of a B-26 led to doubts over the structural integrity of these aircraft.
Source: U.S. Air Force

withdrawn from combat-type activities. Meanwhile there were T-28 operational losses including one that killed Capt. Edwin C. Shank, Jr., on 24 March. These losses further reduced the 1st Air Commando's inventory to the detriment of its combat training mission. To meet the many requests for air support, nine T-28s were borrowed from the VNAF, currently in the process of exchanging these aircraft for single-seat A-1Hs. They would be used until two-seat A-1Es, also previously scheduled for the 1st Air Commando Squadron, arrived. [3]

Shortly after these events, certain letters written by Captain Shank, published posthumously; and news articles alleged that U.S. pilots were poorly equipped and flying obsolete aircraft. This triggered Congressional investigations of U.S. air activities in South Vietnam. Secretary Zuckert testified that both the B-26 and T-28 had been drastically changed and carefully tested before being sent overseas

Air War – Vietnam

and had performed outstandingly. He conceded that in one or two instances of non-combat accidents, structural failure may have been a factor. He defended combat training activities and said that more efficient A-1 Skyraiders were replacing the B-26 and T-28 aircraft used by the 1st Air Commando Squadron and the VNAF. [4]

New Aircraft For the 1st Air Commando Squadron

The A-1E provided close air support to ground forces, attacked enemy supply lines, and protected helicopters rescuing airmen downed in enemy territory. This particular aircraft flown by Maj. Bernard Fisher on March 10, 1966, when he rescued a fellow pilot shot down over South Vietnam. For this deed, Fisher received the Medal of Honor. Source: U.S. Air Force

In September 1963 the Air Force had recommended replacing 1st Air Commando aircraft with two-seater A-1Es. Later it had suggested replacing the B-26s with B-26Ks, a radically modified plane. But deliveries could not begin until mid-1964 and 1965, respectively. This circumstance, plus its desire for faster reacting fighter-bombers and mounting concern over anti-aircraft fire and VNAF operational inadequacies prompted the Air Force to press for the interim use of jets. The JCS agreed and asked McNamara's approval to employ B-57s then in Japan. These aircraft were scheduled for redeployment to the United States in June 1964 and transfer to the Air National Guard. [5]

As administration policy still prohibited jets for combat training in South Vietnam, McNamara turned down the Joint Chiefs' request and said all 1st Air Commando and VNAF fighter aircraft would be replaced by A-1s. On 16 March

Vietnam and Laos 1964

the JCS ordered the Air Force to carry out his instruction. To assure quick replacement, the A-1 modification program was immediately accelerated. [6]

Meanwhile, there was also pressure to increase the number of combat training aircraft because of Communist gains and rising military and political deterioration in South Vietnam. Statistics on aircraft attrition and casualties were disturbing. They showed that from 1 January 1960 to 4 February 1964 antiaircraft fire accounted for 70 of 113 U.S. personnel killed. [7]

Backed by reports from Harkins and Felt, the JCS on 29 April asked McNamara to raise the authorized combat training strength from 31 to 50 aircraft and the manpower ceiling to 280 men. Two squadrons of A-1Es, each with 25 aircraft, would permit traditional four-plane flight tactics against ground fire: two for flak suppression and two for combat training strikes on targets while flying escort for helicopters, trains, and vehicles. [8]

Although McNamara during the March meetings in Saigon and Honolulu had expected that a rapid VNAF buildup would permit an early phase-out of the 1st Air Commando Squadron, on 5 May he approved the JCS request. Simultaneously he approved re-equipping USAF's SAW unit at Eglin AFB, Fla., with the same type of aircraft. As a consequence, 85 A-1Es shortly were designated for modification. [9]

The A-1Es started arriving in mid-1964.
Source: U.S. Air Force

The first six Skyraiders arrived at Bien Hoa AB on 30 May and began operations on 1 June. Air Force officers in the field praised highly the performance of these aircraft. Fifteen Skyraiders had arrived by the end of July. As more were sent to South Vietnam a second combat training unit, the 602d Fighter Squadron (Commando), was established. Authorized 66 personnel, it transferred on 1 October from TAC to PACAF and on the 18th from PACAF to the 2d Air Division. [10]

At year's end the 1st and 602d squadrons possessed 48 Skyraiders. The delivery of nine more early in 1965 would make the 602nd fully operational. [11]

Deployment of B-57s to the Philippines

In justifying the interim use of jets for combat training, PACAF's commander, Gen. Jacob E. Smart, argued that the presence of RF-101s and F-102s in Southeast Asia had not provoked the Communists to escalate the war. Despite the 1954 Geneva Agreement, which prohibited the introduction of new military armament

Air War – Vietnam

into Vietnam, the administration had approved the use of Army jet-powered helicopters. Smart also observed that the United States had not signed the agreement.[12]

As Harkins and Felt were in general accord with these views, LeMay on 21 February asked JCS concurrence to transfer three squadrons of B-57 light bombers from Yokota AB, Japan, one to South Vietnam and two to Clark AB, the Philippines. On the 29th the Joint Chiefs agreed and shortly afterwards sent their recommendation to McNamara. They expected quick approval as U.S. officials were seeking new ways to force Hanoi to halt its support of the Viet Cong and Pathet Lao.[13]

But during the March conferences in Saigon and Honolulu, the Defense Secretary rejected the Joint Chiefs' counsel. He said lack of airpower was not a major problem, the jets would have no impact on winning the war, and the issue would only cause difficulties with the State Department. As has been noted, McNamara directed the replacement of 1st Air Commando B-26s and T-28s by A-1Es.[14]

Deployment of B-57s was delayed but finally took place. Source: U.S. Air Force

Although denying the use of B-57s in South Vietnam, McNamara desired their withdrawal from Yokota to make room for other U.S. units. Their departure would also help ease the U.S. balance of payments problem with Japan. As a consequence, the JCS on 30 March again urged their redeployment, but only to Clark AB. Their presence would strengthen the U.S. military position in Southeast Asia.[15]

Still confronted with a critical military situation, McNamara the next day authorized the transfer of 48 B-57s and 1,081 personnel to the Philippines until 30 June 1964. After the State Department worked out the arrangements with the Tokyo and Manila governments, PACAF on 7 May began flying the aircraft to Clark AB.[16]

After another trip to Saigon and Honolulu, McNamara in mid-May extended authority to maintain the B-57s at Clark AB until 1 January 1965, but the prohibition against their use for combat training in South Vietnam was still in effect at the end of the year.[17]

Other USAF Augmentations Early in 1964

A rising Communist threat in Laos also brought more USAF aircraft to South Vietnam and Thailand.

In March, a special air warfare (SAW) detachment arrived at Udorn, Thailand. Using 4 T-28s and, later, three C-47s, the detachment trained Lao and Thai pilots. To support Yankee Team missions over Laos, the JCS on 8 June directed the movement of eight F-100s from Clark AB to Da Nang Airport from where they began operations the next day. The administration's decision to use jets in Laos

Vietnam and Laos 1964

was due to the different military situation in that country. In July, four RF-101s transferred from Okinawa to Tan Son Nhut, raising to 10 the number of these aircraft at that base.[18]

By July, USAF had in Thailand a SAW unit at Udorn, 6 F-100's at Takhli, 4 F-102s at Don Muang, 4 KB-50s at Don Muang and Korat, and 2 H-43Bs for search and rescue at Nakhom Phanom: near the Laotian border.[19]

Including. auxiliary and allied aircraft, the 2d Air Division controlled about 155 aircraft in South Vietnam and Thailand on the eve of the Tonkin attack.[20]

Buildup After The Gulf Of Tonkin Incident

On 4 August, immediately after the Communist attack, McNamara announced the dispatch of reinforcements to Southeast Asia. USAF deployments included three fighter-bomber squadrons from the United States to the Philippines and Japan,. and two squadrons of the much-debated B-57s from the Philippines to South Vietnam. The major movements to and within the Pacific area were as follows:

Type of Aircraft	Number	From	To
KB-50s	4	Yokota AB, Japan	Takhli AB, Thailand
B-57s	36	Clark AB, PI	Bien Hoa AB, SVN
F-100s	4	Clark AB, PI	Takhli AB, Thailand
F-100s	36	CONUS	Clark AB, PI
RF-101s	2	Misawa AB, Japan	Tan Son Nhut AFLD, SVN
RF-101s	6	CONUS	Kadena AB, Okinawa
F-102s	6	Clark AB, PI	Da Nang ARPT, SVN
F-102s	6	Clark AB, PI	Tan Son Nhut AFLD, SVN
F-105s	18	Yokota AB, Japan	Korat AB, Thailand
F-105s	18	CONUS	Yokota AB, Japan
C-130s	18	CONUS	Clark AB, PI
C-130s	18	CONUS	Naha AB, Okinawa

In subsequent weeks additional aircraft arrived or were retained in South Vietnam.

More Transport and Reconnaissance Aircraft

The Gulf of Tonkin incident hastened a final decision to add a fourth C-123 squadron to the 315th Troop Carrier Squadron. The JCS, on 4 August, recommended and McNamara on the 7th approved its deployment. On 8 October the unit was activated at Tan Son Nhut and the aircraft arrived shortly afterward.

Air War – Vietnam

This raised to 64 the number of C-123s in South Vietnam. By December, augmentations brought the total to 72. [22]

To improve night reconnaissance the JCS on 4 September recommended and McNamara approved the dispatch of two more RB-57E's with improved infrared, sensor, and navigation systems. This would provide a total of four "Patricia Lynn" special reconnaissance aircraft for the 13th Technical Reconnaissance Squadron. The third aircraft arrived in December. [23]

Establishment of a Search and Rescue Unit

The July decision to dispatch a professionally trained USAF search and rescue (SAR) unit followed several Army and Marine helicopter personnel losses in rescuing downed USAF and VNAF pilots in South Vietnam. Previous search and rescue operations in South Vietnam had centered in Pacific Air Rescue Center's Detachment 3. But rescue missions were largely carried out by the U.S. Army and Marine Corps or by the VNAF, often with inadequately equipped helicopters and poorly trained crews. After approval by the JCS, three H-43F helicopters and crews on temporary duty (TDY) reached Bien Hoa on 14 August. A permanent unit, Detachment 4, Pacific Air Rescue Center, was activated on 20 October. After receiving six HH-43B helicopters and 86 personnel, Detachment 4 became fully operational on 5 November. Three helicopters and crews were placed at Bien Hoa and Da Nang, respectively. Also stationed at Da Nang were three HU-16 flying boats for sea rescue of downed pilots. The H-43Fs were sent to rescue units in Thailand. [24]

HU-16s tasked for air-sea rescue.
Source: U.S. Air Force

Retention of the 19th TASS

A decision also was made to retain 22 O-1s of the 19th Tactical Air Support Squadron (TASS). Used primarily for visual reconnaissance and forward air control (FAC), the 19th was organized at Bien Hoa in July 1963 and scheduled to transfer to the VNAF by 30 June 1964. [25]

As both the Air Force and the Army used O-1s, the question of whose aircraft should be transferred was debated vigorously. In March 1964, General Harkins reaffirmed the decision to transfer the 19th TASS's O-1s. But the shortage of FAC aircraft prompted the Air Force in April to suggest keeping the 19th's personnel and employing T-28s scheduled for phase-out from both the 1st Air Commando

Vietnam and Laos 1964

Squadron and the VNAF. The need for more FAC aircraft appeared essential after McNamara, in May, ordered a further buildup of the VNAF. [26]

Air Force appeals to retain the 19th were rejected. On 8 August the squadron was deactivated and personnel began to depart. Meanwhile, the Air Force attempted to keep the 19th operating pending receipt of a JCS fact-finding team report. The team subsequently affirmed the shortage of O-1s to meet growing air support needs. With Westmoreland and Sharp now in agreement, the JCS on 15 September informed McNamara that the squadron not only should be retained but its authorized strength increased by 49 officers and 131 enlisted men. Also, more MAP U-17's should be procured for the VNAF in lieu of the USAF 0-l's that had been scheduled for transfer. [27]

On 28 September, McNamara agreed with the Joint Chiefs' recommendation but the 19th was not reactivated and reassigned to the 34th Tactical Group until 16 October. In the preceding weeks it had lost many of its personnel and much of its effectiveness. The necessity for USAF 0-ls was further supported early in December when the JCS agreed that the 19th should have 30 aircraft and 215 men. It also desired reduced crew-aircraft ratios to pet the assignment of more qualified VNAF O-1 pilots as forward air controllers and air liaison officers. By 31 December McNamara had not rendered a decision on these two proposals.[28]

Thus successive augmentations during 1964 raised the total of USAF aircraft in South Vietnam by year's end to 221 compared with 117 at the end of 1963. In addition, USAF's overall posture was strengthened measurably by new deployments to Thailand, the Philippines, Japan, and Okinawa. The USAF buildup, especially after the Gulf of Tonkin incident in August, presaged a new phase in the war that would begin in February. [29]

Airfield Expansion

The USAF buildup was not without problems. There was aircraft overcongestion on airfields in both South Vietnam and Thailand, aggravating the security problem as demonstrated vividly on 1 November when the Viet Cong attacked Bien Hoa. To lessen the danger, PACAF, on 24 November, ordered the repositioning of several units in South Vietnam to other bases. [30]

In addition, airfield expansion was accelerated in both countries, especially at the six primary jet airfields of Tan Son Nhut, Bien Hoa, and Da Nang in South Vietnam and Takhli, Korat, and Don Muang in Thailand. On 29 December OSD approved expenditures for architectural-engineering services for two of the biggest projects: a second runway at Da Nang and a new airfield at Chu Lai on the coast. Work on Can Tho Airport in the Mekong Delta, begun in February 1964, produced a usable runway by October; the project was nearing completion at year's end. Important expansion was programmed or begun at numerous smaller airfields. [31]

Air War – Vietnam

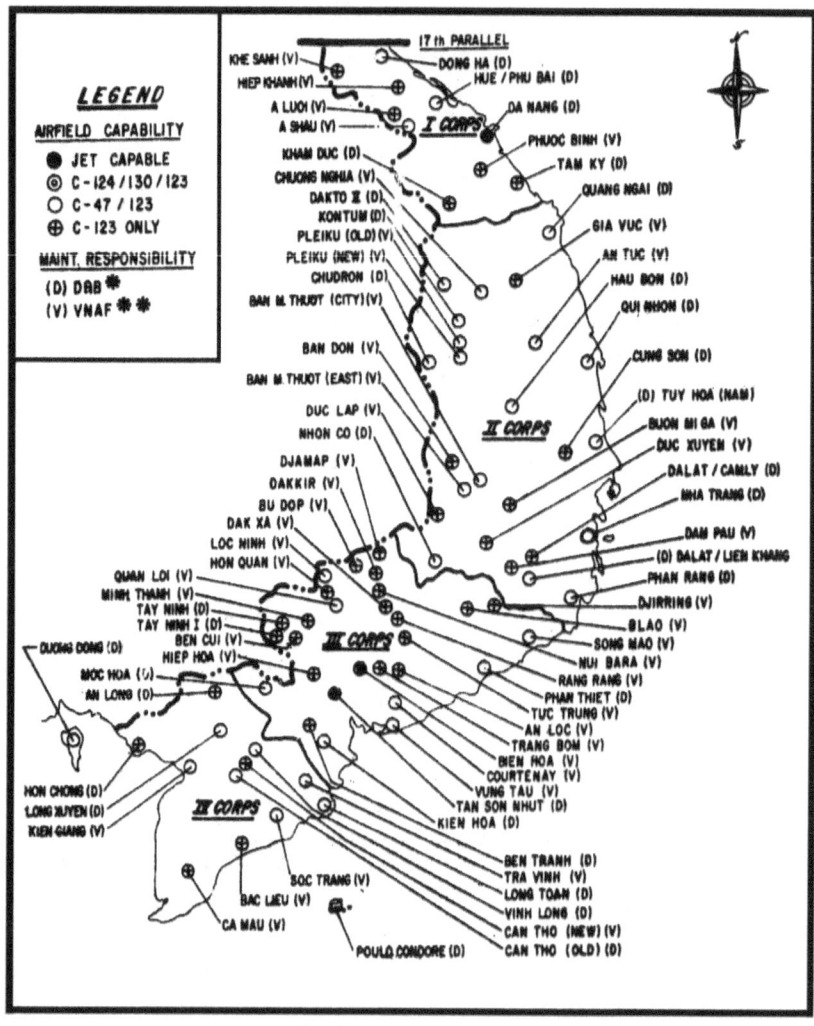

* DEPARTMENT OF AIR BASES, A CIVILIAN COMPONENT OF THE VIETNAMESE ARMY.
** VIETNAMESE AIR FORCE

AIRFIELDS NOT LISTED ABOVE ARE NORMALLY THE RESPONSIBILITY OF THE PROVINCE CHIEF.

Vietnam and Laos 1964

VI. OTHER USAF ACTIVITIES AND PROBLEMS

USAF Support of the Vietnamese Air Force

Throughout 1964 the Air Force continued its training program for the Vietnamese Air Force. It was also concerned with the problems of service representation in MAC/V and rules of engagement for combat training operations.

Expansion of the VNAF

At the end of 1963 the Vietnamese Air Force (VNAF) possessed 228 aircraft in nine squadrons: 2 fighter (A-1Hs and T-28s), 1 tactical reconnaissance (RT-28s and RC-47s), 2 helicopter (CH-34's), 3 liaison (O-1s and U-6s) and 1 transport (C-47's). On 16 December of that year, the VNAF acquired a new commander, Colonel Ky, who quickly won a reputation as a highly motivated and popular leader. [1]

As a result of previous decisions, more aircraft arrived early in 1964. A second A-1H Skyraider squadron was activated in the VNAF during January and flew its first operational mission on 18 March. RT-28D's reached the VNAF in February and aircraft for a third A-1H squadron at the end of April. The Skyraiders came from U.S. Navy resources. Thus Navy personnel performed the operational and maintenance training function. [2]

After his visit to Saigon and Honolulu in March, McNamara submitted new recommendations to the President to enlarge the Vietnamese armed forces. Approved on the 17th, they called for a 50,000 increase in Vietnamese regular and paramilitary forces and other forms of assistance. [3]

As part of the VNAF fighter aircraft buildup, McNamara directed the replacement of all T-28's (many of which were subsequently were made available for the use of the Thai and Laotian air forces) by A-1H's, and an increase in A-1H strength from three to four squadrons to enable the South Vietnamese to carry out their own combat support activities. But General Smart asserted that USAF forces would still be needed to "fill the gap," as the VNAF still showed some reluctance to fly at night and on weekends and were often slow in making air strikes. McNamara replied, however, that it would be cheaper to build up the VNAF than to give the USAF more aircraft. [4]

The Defense Secretary continued to pursue this policy in May When he again visited Saigon and Honolulu. He directed MAC/V to develop a plan for additional expansion of the VNAF and the eventual phase-out of the 1st Air Commando Squadron. His decisions would give the VNAF 339 aircraft by 1 June 1965. These would include 150 A-1Hs (six squadrons) and 300 A-1H pilots by February of that year. This goal was attainable, McNamara thought, if the VNAF's pilot-aircraft ratio were raised from 1 to 1 to 2 to 1 to compensate for poor motivation and a low combat sortie rate, and if the incoming RT-28's were exchanged for more Skyraiders. In addition, 0-1 squadrons would increase from 2 to 4 (40 to 80 aircraft), and C-47 squadrons from 2 to 3 (32 to 48 aircraft). In subsequent weeks McNamara approved UE increases that would boost total liaison aircraft to 120 and helicopters (with a fourth squadron added) to 80.[5]

The Gulf of Tonkin incident in August and the continued military and political decline in South Vietnam showed, however, that the VNAF would not be able to carry the main air burden for counterinsurgency activities in the foreseeable future. More, not less, aircraft were needed despite the VNAF buildup. As noted, in September McNamara agreed with a JCS recommendation to retain the USAF 19th Tactical Air Support Squadron whose O-l's had been scheduled for turnover to the VNAF. In lieu of the transfer, 20 more U-17's were programmed for the Vietnamese.[6]

The Problem of 5th and 6th A-1H Squadrons

There was one exception to the trend in late 1964 toward enlarging both the USAF and VNAF forces. Virtually until the end of the year, administration authorities hoped to phase out the 1st Air Commando Squadron after the VNAF's 5th and 6th A-1H squadrons were activated. After the decision in May to add the latter, Saigon and Pentagon planners wrestled with the problem of establishing realistic activation schedules.

A 2d Air Division plan, staffed through MAC/V and PACOM, initially proposed activating the 5th and 6th squadrons in November 1964 and January 1965, but the Air Staff considered these dates too optimistic. The JCS agreed and, on 24 July, proposed January and March 1965, but McNamara took no action. On 15 October the JCS proposed July and December 1965 but urged retention of USAF's two combat training squadrons until all six VNAF A-1H squadrons were fully operational. Thereafter USAF would keep only a residual training capability in South Vietnam. The Joint Chiefs pointed to the greater Viet Cong activity, aircraft losses by ground fire, and a general insufficiency of aircraft for close support, as justifying extended retention of the USAF capability.[7]

On 6 November McNamara approved the JCS-proposed A-1H activation schedule only. He deferred a decision on retaining the 1st Air Commando Squadron until the fifth VNAF A-1H squadron was operational.[8]

Because of the worsening military situation, Ambassador Taylor, in December, proposed an additional stretch-out for the last two Skyraider squadrons in order to allow 1st Air Commando and VNAF pilots to use B-57's. This was rejected by the JCS.[9]

The Problem of Jet Aircraft

As in 1962 and 1963, the possible use of jet aircraft by the VNAF was periodically reviewed. In May 1964, after McNamara had approved the movement of B-57's from Japan to Clark AB, Admiral Felt informed the JCS that rising air needs might require the use of the bombers by either the USAF or VNAF while both were changing to A-1s.[10]

Administration policy not to assign jets to the Vietnamese Air Force was unchanged. It authorized, however, six VNAF pilots to take 15 hours each of B-57 familiarization training. By 23 July all six had completed flying and received excellent performance ratings. [11]

A handful of Vietnamese pilots trained to fly the B-57 but the type never entered Vietnamese service. Source: U.S. Air Force

Although the JCS had agreed to the familiarization program, it believed that the B-57's should remain in USAF hands. VNAF jets, if and when approved, should consist of other types. After the Gulf of Tonkin incident and the sighting of MIG-15s and -17s on an airfield near Hanoi, the JCS proposed to McNamara the development of a VNAF air defense capability. It suggested sending 15 pilots to the United States for jet training in 1965, and the assignment of 10 F-5s to the VNAF in 1966. [12]

McNamara disagreed. On 25 September he informed the JCS that the United States rather than the VNAF should provide air defense in the foreseeable future. He also said jets would not contribute to the VNAF's counterinsurgency effort and would compete with other air support resources. In November the JCS resubmitted its recommendation but McNamara again turned it down. The VNAF had not yet attained full capability with four A-1H squadrons, he observed, and accelerated aircraft deliveries for the 5th and 6th A-1H squadrons promised to create more problems. [13]

The JCS made no further effort during the remainder of the year to introduce the jets. General LeMay had favored giving the VNAF a few B-57s but he agreed that none should be assigned until all six A-1H squadrons were operational. [14]

Completion Of Helicopter Training

In July Air Training Command's 917th Field Training Detachment stationed at Tan Son Nhut completed the training of its last class or VNAF helicopter pilots and mechanics. Begun in January 1963, this helicopter training program was the first the Air Force had conducted outside of the United States. Despite a formidable language problem and the hazards of climate and antiaircraft fire, the detachment trained 98 pilots and 102 mechanics for the VNAF.[15]

CH-34 Source: U.S. Army

VNAF Strength

At the end of 1964 the VNAF possessed 280 aircraft, a net increase of 52 for the year. There were now four fighter squadrons (A-1Hs and a few T-28s), four helicopter squadrons (CH-.34's), four liaison squadrons (0-ls, U-6s, and U-17s), and one support wing (C-47s and RC-47s), but some authorized aircraft had not yet been received by the units. By 15 January 1965 the VNAF's authorized strength was 11,276 of which 10,849 were assigned. Students in training totaled 1,775 in Vietnam and 345 in the United States.[16]

Air Force Representation in MAC/V

Army domination of MAC/V, the top U.S. command structure in South Vietnam, continued to trouble the Air Force during 1964. Of the nine key positions in MAC/V at the beginning of the year, only one (J-5)was held by a USAF officer. Previously, the summer and fall of 1963 when impending vacancies arose in the posts of chief of staff and deputy commander, the Air Force had urged assignment of one of its general officers. Harkins and Felt agreed that at least the chief of staff position should be filled by the Air Force.[17]

However, when McNamara withheld his approval, Harkins in March 1964 asked for Army Maj. Gen. Richard G. Stilwell to replace the outgoing chief of staff, Marine Brig. Gen. Richard G. Weede. The JCS split over the issue. Taylor and the Army Chief, General Wheeler, concurred. The Navy and Marine Corps chiefs agreed conditionally, asserting that as a matter of principle all three top MAC/V positions should not be held by the same service. LeMay was opposed. But McNamara, on 10 April, supported the majority opinion.[18]

On 12 June the Joint Chiefs split again over filling the post of deputy commander being vacated by General Westmoreland who replaced Harkins as commander on the 20th.(Westmoreland had asked for an Army officer and suggested that a senior Air Force officer, if needed, would be more effective in Bangkok as deputy commander to MAC/Thai.) LeMay and the Navy and Marine Corps chiefs backed an Air Force designee for the post but Taylor and Wheeler supported Westmoreland's request.

Vietnam and Laos 1964

Taylor informed McNamara that in view of the nature of counter-insurgency, it was "hardly conceivable" that the post could be filled from a service other than the Army. On 18 July, McNamara again sided with the Army, allowing that service to hold the three top posts in MAC/V. [19]

In conjunction with actions on consolidating MAAG/V with MAC/V, the JCS at the end of July asked newly arrived Admiral Sharp, PACOM's commander, to survey the command structure of MAC/V and report on manning and service representation. The survey, however, was delayed due to the heavy U.S. augmentations that followed the administration decisions in July and the Gulf of Tonkin incident on 4 August. [20]

The U.S. buildup, especially of USAF units, slightly improved the Air Force's command position in Southeast Asia. On 7 August the post of deputy commander, 2d Air Division was established at Udorn, Thailand. There was some initial uncertainty about its function, but it was finally determined that the deputy commander would "conduct, control, and coordinate all USAF matters pertaining to assigned and attached Air Force units, activities, and personnel in support of U.S. and Allied air operations in Laos." This made him responsible to the 2d Air Division rather than to MAC/V. The basic service makeup of MAC/V was unchanged. [21]

Pressing for JCS support to have Sharp prepare as soon as possible a manpower report on MAC/V, General LeMay in late August pointed to the trend from a joint to unilateral service (Army) U.S. command structure. This was evidenced not only by the fact that there was only one senior Air Force officer in MAC/V, but also by the subordinate role of USAF advisors and air liaison officers at Vietnamese corps and division level compared to Army advisors, and by the absence of a senior VNAF representative or senior USAF advisor at the Vietnamese Joint General Staff level. Until there were USAF advisors of appropriate rank to advise Vietnamese Army commanders, LeMay said, he could not be assured that USAF and VNAF units were being utilized fully in the war effort. On 2 September he again voiced concern to the JCS, citing the need to improve air-ground coordination in the war against the Viet Cong. [22]

Shortly afterward, Admiral Sharp, in conferences with the JCS, indicated that he would abide by McNamara's decisions on filling the top MAC/V posts, although he (Sharp) personally favored appointing an Air Force deputy air commander to MAC/V. In the event the war escalated, he said he would "fight the war" through his component commanders since MAC/V did not have enough skilled Air Force specialists. In deference to Army views, Sharp also indicated that he would not support an Air Force proposal to place USAF full colonels at Vietnamese corps level. [23]

On 29 September Westmoreland made a partial concession to the Air Force. He informed Sharp that he would appoint General Moore, the 2d Air Division commander, deputy commander for air operations, a new post that would be an additional duty for Moore. Sharp supported the recommendation but the Air Staff objected to creating such a lesser position. It would add to Moore's workload and

Air War – Vietnam

fail to give Headquarters, MAC/V the balanced service representation it needed. The Air Force reiterated its desire for a deputy commander within the Headquarters MAC/V staff structure and hoped Sharp would reconsider his position and support the Air Force's view. Prospects were not encouraging. In November Sharp sent the JCS a new joint table of distribution proposed by MAC/V for additional U.S. manpower that provided for a deputy commander for air operations. At year's end the JCS had not acted on it nor on new proposed MAC/V terms of reference.[24]

Thus, despite the rapid USAF buildup in Southeast Asia, MAC/V at the end of 1964 remained an Army-dominated command. Its top positions now numbered 10, of which they occupied all but two: commander, deputy commander, chief of staff, J-1, J-3, J-4, J-6, and commander of the joint research and test agency (JRATA) that had been established on 11 February 1964 to bring together all test agencies in South Vietnam. The Marine Corps held the J-2 slot and the Air Force the J-5. The incumbent of J-5, Maj. Gen. Milton D. Adams, had held this post since 7 December 25, 1962.[25]

Rules of Engagement

A major Air Force objective was to obtain administration approval to relax the rules of engagement for the 1st Air Commando Squadron. Adopted in late 1961, these rules authorized operations when the VNAF lacked the necessary training and equipment, combined USAF-VNAF crews were aboard, and the missions were confined to South Vietnam. In addition, the aircraft carried VNAF rather than USAF markings and there were strict target verification procedures. Previous USAF efforts to modify the rules were unsuccessful.[26]

A-1Es were useful because they were two-seaters (unlike the single-seat A-1H) and could carry the Vietnamese crew member demanded by the RoE.
Source: U.S. Air Force

Vietnam and Laos 1964

Because of the rising need for air support and the slow growth of the VNAF, the 1st Air Commando sortie rate increased. It felt that more effective air support would be possible if the rules were relaxed, but administration officials retained them for political reasons. Meanwhile, U.S. Army aviation appeared to be interpreting the rules more freely, their armed helicopters carried U.S. markings, and their pilots received more public recognition, a circumstance that greatly troubled the Air Force. [27]

In March and May 1964, after visits to Saigon and Honolulu, McNamara reaffirmed the rules for the 1st Air Commando Squadron. The official view was that, despite U.S. assistance, the war was primarily Vietnamese and that there was Presidential understanding that the 1st Commando's activities were temporary until the VNAF "could do the job." [28]

In April and May the role of the 1st Air Commando became a public issue after the publication in the press and Life magazine of the letters of Captain Shank, who died on 24 March in the crash of a T-28. As noted earlier, he complained about inadequate aircraft and equipment. But Shank's letters also indicated that the Commando pilots often engaged more in combat than in training. Former Commando pilots and top U.S. officials were called to testify before special Senate and House investigating subcommittees. [29]

General LeMay took the occasion to urge the JCS to persuade McNamara to change the rules of engagement, as the United States had more to lose than gain by denying a fact of USAF activity in the war. [30]

LeMay was unsuccessful. Indeed, on 20 May the JCS tightened the rules of engagement: 1st Air Commando pilots could fly only bona fide combat training missions against hostile targets with VNAF pilots in training and not with Vietnamese "observers" (the intent being to eventually eliminate the squadron and leave combat support to the VNAF); no armed helicopters should be used as a substitute for close air support strikes; and U.S. advisors should be exposed to combat only to the extent that U.S. advisory duties required this. [31]

General Smart, PACAF's commander, believed that the latest JCS guidance left unclear whether 1st Air Commando pilots should "fight or not." Nor was the Air Force's disenchantment with the rules dispelled by MAC/V's continued freer interpretation of them for armed helicopters, despite the injunction against combat-type missions except to protect vehicles and passengers. [32]

Four months later military deterioration in South Vietnam again forced a change in the rules. With Westmoreland's and Sharp's support, the JCS recommended that the 1st Air Commando be authorized to fly with either VNAF observers or student pilots, to fly with USAF pilots alone for immediate air support if requests were beyond the VNAF's capability or if no VNAF crew member was available (PACAF believed that this change alone would increase the 1st Air Commando's average monthly sortie rate from 497 to 960) and to assign a dual training and combat support mission to the 1st Air Commando. On 25 September McNamara agreed to only one change: either a VNAF observer or a student pilot could be

Air War – Vietnam

used, thus reverting to a practice in effect prior to 20 May. The JCS sent an implementing directive on 14 October. [33]

Meanwhile, the possibility of Communist air activity after the Gulf of Tonkin incident resulted in a general relaxation of the rules of engagement for other USAF and Navy air activities. Decisions in August and September gave General Westmoreland or Admiral Sharp greater authority to engage enemy aircraft over South Vietnam, Thailand, and Laos and in international airspace, and to attack hostile vessels in international waters. [34]

Vietnam and Laos 1964

VII. BEGINNING OF AIR OPERATIONS IN LAOS

As increased Communist activity in Laos also threatened South Vietnam, the administration in 1964 took new measures to bolster the tenuous leftist-neutralist-rightist coalition government of Premier Souvanna Phouma. Laotian neutrality, first guaranteed by the 1954 Geneva Agreement and later by the 14-nation declaration of 23 July 1962, was in constant jeopardy because of repeated Communist-led Pathet Lao violations and North Vietnam's use of Laos for infiltrating men and arms to the Viet Cong. [1]

Initial Lao and U.S. Air Activity

Although the Royal Laotian Air Force (RLAF) received limited aid under the U.S. military assistance program (MAP), the 1954 and 1962 accords restricted training in that country. To improve the tiny RLAF, in December 1963 PACAF proposed deployment of a USAF special air warfare unit to Thailand. Its presence would permit training of Lao, and perhaps Thai pilots in counter-insurgency tactics and techniques. In January and February 1964, after coordinating with U.S. Ambassadors in Vientiane and Bangkok and the two governments concerned, OSD and the State Department concurred. On 5 March the JCS directed the Air Force to send a SAW unit to Udorn, Thailand, for six months. General LeMay promptly instructed Headquarters, TAC to dispatch Detachment 6, 1st Air Commando Wing with four T-28s and 41 personnel. Nicknamed Water Pump, the detachment arrived at Udorn on 1 April.[2]

In addition to providing counterinsurgency training, the detachment was to provide logistic support, sponsor Lao-Thai cooperation, and augment, if necessary, the RLAF if the Pathet Lao and North Vietnamese forces should resume an offensive. Despite objections of the Chief, Joint U.S. Military Assistance Advisory Group, Thailand (JUSMAAG/T), CINCPAC assigned operational control to the Commander, 2d Air Division because of the similarity of the detachment's mission with that of the 1st Air Commando Squadron in South Vietnam. JUSMAAG/T was the ranking U.S. military officer in Thailand under General Harkins who also served as COMMAC/Thai.

In April a right-wing coup attempt upset the shaky coalition government. It triggered a resurgence of Pathet Lao attacks on neutralist and right-wing forces in the Plaines des Jarres. When Premier Phouma asked for help, the United States responded by stepping up its aid to the RLAF. It also released ordnance, enabling the RLAF to begin air attacks on Communist positions on 18 May.[4]

On the same day the JCS directed CINCPAC to use USAF and Navy aircraft for medium and low-level "Yankee Team" missions over the embattled area. Previous USAF reconnaissance missions over Laos with century-series aircraft began in 1961 under the Pipestem and Able Mable programs. Following the signing of the Laotian neutrality agreement on 23 July 1962, the missions were discontinued on 1 November of that year. On the 19th May, RF-101's stationed at Tan Son Nhut made the first flight. On the 21st Seventh Fleet RF-8A's and RA-3B's were used to inaugurate the Navy's participation in the program. The 2d Air Division was assigned coordinating responsibility for the Lao-U.S. air operations. Only search and rescue flights were permitted from Thai bases. Air attacks above 20 degrees North latitude were prohibited. [5]

Publicly acknowledging the U.S. operations, the State Department said they were requested by the Laos government because of the inability of the International Control Commission to obtain information on recent attacks on neutralist and right-wing forces. The administration also considered dispatching combat troops to Thailand, as in 1962, in a "show of force." [6]

Since only the RLAF performed air strikes, more T-28s were urgently needed. At the request of the U.S. Ambassador to Laos, T-28s of Detachment 6, after re-marking, were loaned temporarily to the Laotians giving them a total of seven. On 20 May, 10 more T/RT-28s from South Vietnam (where the 1st Air Commando Squadron and the VNAF were replacing them with A-1s) were loaned to the RLAF. Together with subsequent augmentations, about 33 were available by late June.

Because of the pilot shortage, Thai Air Force personnel, with their government's approval, were trained and joined the Laotians in flying operational missions. Some pilots of Air America, a small U.S. contract airline, also received combat training. [7]

Ambassador to Laos had asked for MAP-financed C-47s for the RLAF. Admiral Felt and General LeMay immediately endorsed the request. Subsequently concurring, the JCS on 30 June directed the Air Force to provide the necessary training. Three C-47s and 21 personnel were sent to join Detachment 6 in Thailand, arriving there on 24 July. The unit began immediately to give air and ground crew training to the Laotians. [8]

U.S. Yankee Team missions, begun originally on a temporary basis, were extended by the JCS on 25 May for an indefinite time period. These flights had a fourfold mission: to provide intelligence for friendly Laotian forces including assessment of RLAF bombings, determine the extent of Communist infiltration and aid to the Viet Cong, encourage allies, and demonstrate U.S. resolve to check communism in Southeast Asia. [9]

Early in June two Navy aircraft were downed in Laos by antiaircraft fire. As a consequence, on the 6th the JCS authorized Yankee Team pilots to engage, with restrictions, in retaliatory fire. For this purpose, USAF deployed eight F-100's from Clark AB, the Philippines, to Da Nang Airfield. On the 9th, supported by SAC KC-135 tankers, several of these aircraft made the first USAF jet strikes of the war against antiaircraft sites and selected military targets. After the Gulf of Tonkin incident, newly arrived USAF F-105's, at Korat AB, Thailand, were employed in conjunction with search and rescue missions only. The changing circumstances led to frequent revisions in the rules of engagement. In July seven new or revised rules were issued with respect to reconnaissance, altitude, and retaliatory strikes. [10]

Vietnam and Laos 1964

By late June and July Lao-Thai-Yankee Team reconnaissance, interdiction, and airlift operations had been a major factor in stabilizing the military situation in Laos. The defense of Muang Soui, a vital area near the Plaines des Jarres, was bolstered and later an "Operation Triangle" further improved the position of noncommunist forces. Clearly the rapid USAF training of inexperienced Lao and Thai pilots had "paid off" and LeMay commended highly the work of Detachment 6. In September the JCS extended the detachment's tour for 120 days and in December until September 1965. Also in December LeMay assigned one U-10B and four more men to the detachment to begin a limited medical civic action program for Thai people. At the end of the year the detachment possessed eight aircraft and 66 personnel. In addition to providing valuable information on Communist activity in Laos and infiltration into South Vietnam, Yankee Team and Water Pump missions had raised Laotian morale.[11]

In July the JCS approved LeMay's proposal to delegate to CINCPAC more responsibility for air activity in Laos. It desired faster mission approval, relaxation of the rules of engagement, night strikes on Communist convoys on "Route 7," and more direct participation by U.S. and Thai pilots. But Secretary McNamara did not endorse these proposals. High administration policy required the approval of each mission and as available air resources seemed sufficient, there would be no deeper U.S. involvement for the time being in Laos.[12]

To improve command and control of U.S.-Lao-Thai air operations, the post of deputy commander, 2d Air Division was established at Udorn, Thailand, on 7 August.[13]

Plans Against Infiltration

The more stable military situation in Laos after mid-1964 contrasted with the political and military deterioration in South Vietnam. After the President approved additional planning for air and ground operations in Laos, U.S. diplomatic representatives in Bangkok, Vientiane, and Saigon met with PACOM and MAC/V officials to examine ways to reduce infiltration of men and arms through the Laos corridor. Reaching initial agreement on about 22 targets, (after subsequent OSD-JCS-State Department coordination, the JCS on 10 November approved a list of 28 targets) PACOM and MAC/V developed an air-ground plan requiring Yankee Team and RLAF air strikes and U.S.- aided Vietnamese ground attacks a short distance into Laos. The JCS approved the plan on 30 September.[14]

As political disarray in Saigon increased and infiltration appeared more menacing, the JCS in October repeatedly urged McNamara to adopt the 30 September plan that would require, in addition to RLAF operations, considerable Yankee Team participation in striking "hard" targets, suppressing flak, and providing high cover in case North Vietnamese MiGs tried to intervene.[15]

The plea for more U.S. air support also received the unanimous endorsement of the recently-formed Southeast Asia Coordinating Committee (SEACOORD). In August, General Taylor [had] proposed establishing SEACOORD and a military component, SEAMIL, to improve coordination of U.S. policy in Laos, South

Air War – Vietnam

Vietnam, and Thailand. Washington authorities approved SEACOORD in September but as SEAMIL threatened to bypass CINCPAC, it was strongly opposed by the Air Force, Navy, and Marine Corps chiefs. On 9 December McNamara agreed not to alter the military command structure. The committee desired approval of RLAF strikes on Mia Gia pass, a vital transit point on the Laotian-North Vietnam border. Citing latest intelligence, the committee said that stronger action was needed outside of South Vietnam to produce the desirable psychological and military impact on the Communists. Without U.S. air there might be unacceptable RLAF losses and a doubt as to U.S. resolve in South Vietnam and Laos. [16]

But, as noted earlier, the continued political turmoil in Saigon precluded any modification of State-OSD directives and allowed planning only for the proposed air-ground operations in the Laos corridor. General Westmoreland, in late October, foresaw no likelihood of beginning cross-border activity until after 1 January 1965. [17]

On 18 and 21 November two USAF Yankee Team aircraft, an F-100 and an RF-101, were lost to ground fire. Whereupon LeMay proposed and the JCS approved a recommendation to conduct retaliatory flak suppression strikes along two infiltration routes. Again, the administration took no action pending another searching reappraisal of U.S. policy in Southeast Asia. One proposed course of action was to employ U.S. ground forces in the Laos panhandle. The Joint Chiefs had not officially considered such a deployment1 and they advised McNamara that it appeared prudent to implement previous JCS recommendations before undertaking ground operations. [18]

On 2 December after Ambassador Taylor had conferred with NSC and other top U.S. officials, the administration approved very limited and highly controlled measures for exerting more pressure on North Vietnam. They included U.S. strikes on infiltration routes and facilities in the Laotian corridor, armed reconnaissance missions every three days with flights of four aircraft each, but no over-flights of North Vietnam. Nicknamed Barrel Roll, the missions had a primarily psychological purpose: to "signal" Hanoi of the danger of deeper U.S. involvement in Southeast Asia. The JCS quickly sent implementing instructions to Admiral Sharp. [19]

After the Laotian government approved the initial targets and routes, Barrel Roll missions began on 14 December. USAF F-100s from Da Nang and F-105s from Thailand flew the first mission. Navy F-4Cs and A-1Hs began on the 17th. Like Yankee Team, Barrel Roll missions were tightly controlled by Washington. [20]

Thus 1964 witnessed the initial employment of limited U.S., Lao and Thai airpower in Laos. Events in Laos figured increasingly in U.S. planning to thwart a Communist takeover in that country and in defending South Vietnam. By the end of the year Yankee Team aircraft of the Air Force and Navy had flown 1,257 photo, escort, and weather sorties. One hundred and fifteen aircraft received ground hits on 56 missions and each service lost two aircraft. By 2 January 1965 six Barrel Roll missions had been flown with no aircraft lost. [21]

Vietnam and Laos 1964

APPENDIX 1

U.S. Military Personnel In South East Asia. 31 Dec 64				
	Vietnam	Thailand	Total
Army	14,679	3,120	17,799
Navy	1,109	99	1,208
Marine Corps	900	37	937
Air Force	6,604	1,027	7,631
TOTAL	23,292	4,283	27,525
SOURCE: Hist, CINCPAC, 1964, Chart I-6			

APPENDIX 2

U.S. Aircraft In South East Asia. 31 Dec 64			
	Vietnam	Thailand	Total
Army	509*	2***	511
Navy	0	0	0
Marine Corps	29**	0	29
Air Force	221	75	296****
TOTAL	759	77	836
* consisted of 182 fixed wing and 327 rotary			
** consisted of 25 rotary and 4 fixed wing. Total as of 27 Jan 65.			
*** consisted of one fixed wing and one rotary.			
**** Included 13 SAR rotary variously stationed in South Vietnam and Thailand.			
SOURCE: Hist of 2d AD, Jul - Dec 64, Vol I, pp 69 - 70 & Vol II, pp 22, 116; USAF Mgt Survey, 1 Feb 65; MAC/V Command Hist, 1964, pp 59 and 128.			

APPENDIX 3

U.S. Aircraft In South East Asia. 31 Dec 64				
	Vietnam	Laos	Thailand	Total
United States	759	0	77	836
South Vietnam	280	0	0	280
Laos	0	67**	0	67
Australia	6*	0	8	14
New Zealand	0	0	2	2
TOTAL	1,045	67	87	1,199
* Six Caribous arrived in Aug 64.				
** Includes 18 T-28's and 12 RT-28's received from Vietnam.				
SOURCE: Hist of 2d AD, Jan - Jun 64, Ch 1, p 98, Jul - Dec 64, Vol I, pp 22, 25,				

and 116; USAF Mgt Survey, 1 Feb 64; MAC/V Command Hist, 1964, pp 59 & 128; Journal of Mil Asst, Dec 64, p 167.

APPENDIX 4

USAF Flying and Sorties in South Vietnam 31 Dec 64		
Type Aircraft	Flying Hours	Sorties
T-28*	4,073	2,328
B-26**	2,009	622
C-47	5,073	3,659
C-123	37,537	25,327
O-1F	20,020	11,213
RF-101	4,936	2,081
RB-57C	1,328	638
U-3	1,411	161
U-10	2,914	2,015
A-1E***	9.149	2,698
TOTAL	88,450	50,742

* Ended operations in Jun 64.
** Phased out in Mar 64.
**** Began operations in Jun 64
SOURCE: USAF Mgt Survey, 1 Feb 65

Vietnam and Laos 1964

APPENDIX 5

USAF Aircraft Losses In Southeast Asia 1964			
Type Aircraft	Combat Losses	Operational Losses	Total
T-28*	7	2	9*
B-26**	1	1	2
C-47	0	1	1
C-123	1	1	2
O-1F	3	0	3
RF-101	1	0	1
RB-57C	0	0	0
B-57	6**	1	7
U-3	0	0	0
U-10	0	1	1
A-1E	7	1	8
F-100	2	0	2
F-105	1	0	1
F-102	0	1	1
KB-50	0	1	1
HH-43	1	1	1
TOTAL	30	10	40

* Includes T-28s loaned to the Royal Laotian Air Force but accountable to the 2d AD.
** Destroyed by Viet Cong Attack on Bien Hoa AB, 31 Oct - 1 Nov 64.
SOURCE: Data Control Br, Sys Div, Dir of Ops, DCS/P&O

APPENDIX 6

U.S. Casualties From Hostile Action in Vietnam			
Fatalities	Dec 61-Dec 64	USAF Casualties 1964	
Army	181	Killed in action	24
Navy	4	Wounded in action	94
Marine Corps	11		
Air Force	51*		
TOTAL	247		118

*2d Air Division source shows 56 fatalities.
SOURCE: Hist, CINCPAC,1964, Chart IV-6; Hist of 2d AD, Jul - Dec 64, Vol II, p 29.

Air War – Vietnam

APPENDIX 7

Type Aircraft	VNAF Aircraft Losses 1962-1964						
	1962		1963		1964		
	Hostile	Accdt	Hostile	Accdt	Hostile	Accdt	Total
T-28	2	1	4	3	1	3	14
A-1H	5	1	3	2	12	12	35
U-17	0	0	0	0	0	3	3
H-34	0	0	1	0	5	1	7
O-1	0	0	3	1	2	10	16
C-47	0	0	0	1	0	3	4
U-6	0	0	0	0	0	0	0
RT-28	0	0	0	0	1	1	2
UH-19	0	0	0	0	0	3	3
L-19	0	0	0	0	0	0	0
TOTAL	7	2	11	7	21	36	84
SOURCE: USAF Mgt Survey, 1 Jan and 1 Feb 65.							

APPENDIX 8

VNAF Sorties Flown	
Type Aircraft	Sorties Flown
T-28	2,958
A-1H	9,456
C-47	3,561
U-17	984
U-6A and O-1A	21,697
UH-19 and CH-34	14,059
TOTAL	52,715
SOURCE: Data Control Br, Sys Div, Dir of Ops, DCS/P&O	

APPENDIX 9

South Vietnam and Viet Cong Military Strength 31 Dec 64

South Vietnam

Regular and paramilitary forces	535,851
Desertions (in 1964)	(73,379)

Viet Cong

Regular forces	32,500
Irregular forces	60,000-80,000

SOURCE: Hist of 2d AD, Jul - Dec 64, Vol II, pp 24 - 26.

APPENDIX 10

Vietnam and Viet Cong Deaths and Weapon Losses 1961 - 1964				
	Combat Deaths		**Weapon Losses**	
	S Vietnam	Viet Cong	S Vietnam	Viet Cong
1961	4,000	12,000	5,900	2,750
1962	4,450	21,000	5,200	4,050
1963	5,650	20,600	8,250	5,400
1964	7,450	16,800	14,100	5,900

SOURCE: Testimony of Secy McNamara, 4 Aug 65, before Senate Subcmte on DOD Appropriations for 1966, 89th Cong, 1st Sess, pt 2, pp 765-66.

Air War – Vietnam

PART THREE
USAF PLANS & OPERATIONS IN
SOUTH EAST ASIA 1965

Air War – Vietnam

Plans and Operations 1965

I. THE ALLIES STRIKE NORTH

At the beginning of 1965 the Republic of South Vietnam was in a state of military and political decline. Its regular, regional, and Popular forces, numbering about 510,650, had been seriously weakened during previous months by defeat and desertions. A most severe setback had occurred from 26 December 1964 to 2 January 1965 at Binh Gia where the Viet Cong virtually destroyed two Vietnamese Marine battalions.[1]

Augmented by combat forces infiltrating from North Vietnam, the Viet Cong was becoming stronger. January estimates placed Viet Cong strength at 29,000 to 35,000 "hard core" guerrillas and 60,000 to 80,000 irregular forces. The communists generally avoided large engagements and directed their "hit and run" attacks and terrorism against Vietnamese irregular forces, the police, and the civilian population. These tactics were increasingly successful.[2]

A Viet Cong terrorist bombing killed 2 Americans and injured 107 people at the Brinks Hotel, Saigon, on Christmas Eve 1964. Source: U.S. Air Force.

Political instability exacerbated military difficulties. Demonstrations and strikes by Buddhists and other groups in the larger cities against the civilian-led government of Premier Tran van Huong, who had been installed in office on 4 November 1964 occurred with greater frequency. Huong's rule came to an abrupt end on 27 January 1965 when the Vietnamese Armed Forces council ousted him leaving only a facade of civilian government. Meanwhile, the power struggle impeded military operations since elements of the Vietnamese Air Force (VNAF), for example, had to be on constant "coup alert." Top U.S. officials were deep\y concerned by this internal conflict. Gen. Curtis E LeMay, USAF Chief of Staff, expressed fear that the disorders could infect and destroy the Vietnamese armed forces, the only cohesive group in the country.[3]

U.S. Restraint and Limited Pressure

The interminable military and political crises had forced the United States to send increasing amounts of military and economic aid in an effort to avert a collapse. At the beginning of 1965, 23,292 U.S. military personnel were serving in South Vietnam. The Air Force had about 6,604 men and 222 aircraft assigned to 2d Air Division headed by Lt. Gen. Joseph H. Memore, Jr. (General Memore had been promoted to Lieutenant General on 25 June 1965.) The Air Force contingent included two air commando squadrons (The lst and 602d) with about 48 non-jet

A-lEs for "combat advisory" support of Vietnamese ground forces. in addition, there were in South Vietnam, 72 C-123Bs, 10 B-57s, 3 RB-57s, 30 F-100s, 6 F-102s, 12 RF-101s, 22 O-1Fs and auxiliary aircraft.

An additional 4,283 American military personnel including 1,027 Air Force in Thailand backstopped U.S. activities in South Vietnam, flew limited air missions over North Vietnam and Laos, and aided Thai and Lao forces. The USAF units in Thailand, also assigned to the 2d Air Division, possessed 83 USAF aircraft. Aircraft in Thailand consisted largely of 18 F-105s, 15 F-100s, 4 F-102s, 10 T-28s, 10 RT-28s, and 8 search-and-rescue helicopters. The use of these aircraft for "out of country" missions was restricted, however, because the Thai government feared becoming too deeply involved in the conflict in Southeast Asia.[4]

In accordance with decisions made late in 1964 by President Lyndon B. Johnson, stronger U.S. military action with its attendant risks was withheld pending emergence of a more stable Saigon regime. As a consequence the 2d Air Division continued patiently to train, support, and work with a Vietnamese Air Force that, partly oriented toward political affairs, was distracted from the war effort. USAF combat advisory missions remained encumbered with numerous "rules of engagement," including a prohibition against the use of jet aircraft for air strikes and against A-IE sorties without the presence of a Vietnamese "observer" or "student pilot" on board. The latter injunction, a long-standing handicap, became an increasing hindrance because of a shortage of "trainees."[5]

A Viet Cong mortar attack on Bien Hoa Air Base in November 1964 marked a major escalation in hostilities. Four Americans died and 72 were wounded. Source: U.S. Air Force

Air base security was precarious. The Johnson administration was reluctant to dispatch combat troops to guard air bases as requested by the Air Force, and lesser security measures were adopted. After the costly Viet Cong attack on Bien Hoa AB on 1 November 1961, the 2d Air Division initiated "crash" measures to improve the defenses of the three major bases of Bien Hoa, Da Nang, and Tan Son Nhut. Much remained to be done in 1965, such as completing revetment construction for safer aircraft dispersal, making more thorough air base patrols, adding more Vietnamese security forces and counter-mortar and ground-surveillance radar, obtaining better intelligence and improving population control. The joint chiefs of staff (JCS) had recommended deployment of a Marine Hawk battalion from Okinawa to Da Nang, but this still awaited final approval.[6]

In January 1965 Adm. U.S. Grant Sharp, commander-in-chief, Pacific (CINCPAC), warned that the air bases remained vulnerable. Considering the limited resources at hand, the Air Staff thought he had taken all "practical steps" possible. Gen. John P. McConnell, who became USAF chief of staff on 1 February, asked the Joint Staff to monitor base security actions and to keep the JCS fully apprised of them.[7]

The U.S. restraint in South Vietnam was matched by limited action against North Vietnam and its infiltration of men and supplies through Laos. A draft national security council (NSC) memo, dated 29 November and revised on 1 December 1964, had outlined a two-phase program beginning on 14 December that called for very selective use of military power against the North.

In Phase I, begun on schedule and lasting about 10 days, more high-level reconnaissance missions were flown over the North and maritime operations, with VNAF cover, were stepped up south of the 18th parallel in accordance with the special covert operation plan 34A.* No air strikes against the North were permitted. in Laos there was a measured increase of Royal Laotian Air Force strikes against communist Pathet Lao-North Vietnam forces, USAF-Navy "Barrel Roll" armed reconnaissance in Northern Laos against infiltrating personnel and supplies supporting these communist units, and USAF-Navy "Yankee Team" reconnaissance in the panhandle against specified infiltration routes. The main objective was to "signal" Hanoi that the United States was determined not to permit a Communist take-over of South Vietnam.

Air Operations Still Had To Comply With Severe Restrictions
Source: U.S. Air Force

Beyond Phase I, the draft NSC memo provided for either a continuation of these actions without change or a transition to other very limited measures. The latter would include withdrawal of U.S. dependents from South Vietnam, more air deployments, low-level reconnaissance over the North, and then air strikes on infiltration routes near the border. The NSC desired to give an impression of steady, deliberate action. U.S.-South Vietnamese forces would begin Phase II with more air strikes and other military activities against the north. Both Gen. William C. Westmoreland, Commander of the Military Assistance Command, Vietnam (MAC/V), and Maxwell D. Taylor, U.S. ambassador to South Vietnam, agreed that there was little chance of finding a successful solution to the war without advancing to Phase II.[8]

When Phase I ended in mid-January the administration was still reluctant to apply increased pressure on Hanoi. The JCS urged more frequent and extensive armed reconnaissance in Laos, less restraint in selecting targets, and less Thai government restrictions on flying USAF strike missions from Thai bases. On the 29th the JCS recommended reprisal air strikes on Northern targets within 24 hours after the next communist act of terrorism in the south. The Joint Chiefs observed that Ambassador Taylor now agreed this might deter further acts of this type.[9]

Secretary of Defense Robert S. McNamara did not act immediately on these recommendations. Vietnamese military reverses continued, however, and the U.S. government moved to provide more assistance including seeking allied aid. in accordance with Office of the Secretary of Defense (OSD) guidelines, the JCS since late 1964 had been planning an international force for South Vietnam composed of as many as 22 nations that would require U.S. logistic support (it soon became apparent that not many nations could participate in such a force). On 27 January, after conferring with McNamara, the JCS Chairman, Gen. Earle G. Wheeler, asked the JCS to consider the dispatch of 80,000 to l00,000 more U.S. ground troops to the embattled country.[10]

Other events presaged the use of more air power. At Binh Gia at year's end, Vietnamese marines had suffered heavy losses, despite assistance provided mostly by armed Army helicopters. After a MAC/V investigation, Westmoreland issued new directives requiring more use of fixed-wing aircraft for close air support. Since the Viet Cong might step up its activities during the annual Vietnamese lunar holiday or "Tet" from 2 through 6 February, he also requested and the President on 27 January approved the use of USAF jet combat aircraft in an emergency.[11]

Attack Across The 17th Parallel

The Viet Cong precipitated the next major U.S. decision. During the annual Vietnamese celebrations early in February, virtually all large-scale military activity ceased. However, in the early hours of the 7th, as Tet ended, an insurgent unit, using recoilless rifles, rifle grenades, and 81 mm mortars struck the air base at the Vietnamese II Corps headquarters at Pleiku and an air strip at Camp Holloway, about six kilometers distant. The 10-minute attack at Pleiku destroyed 5 helicopters and damaged 11 others and 6 fixed-wing aircraft. American losses at both sites were 7 dead and 109 wounded.12

President Johnson immediately authorized a reprisal air strike against the north, ordered the withdrawal of U.S. dependents, and directed the deployment of a Hawk air defense battalion from Okinawa to Da Nang AB. Indicating other

Plans and Operations 1965

measures might soon follow, he declared the United States had no choice but to "clear the decks" to show America's determination to help South Vietnam fight to maintain its independence.[13]

The reprisal strike, also carried out on 7 February, opened a new phase of the war. Under the code name Flaming Dart I, 49 aircraft of the Seventh Fleet bombed and strafed barracks and staging areas used for infiltration near Dong Hoi, slightly north of the demilitarized zone. One A-4 aircraft and its pilot were lost and seven A-4s and one F-8 damaged by anti-aircraft fire. The presence in Hanoi at the time of Soviet Premier Alexei S. Kosygin led the administration to assure the Russians that the air attack was not related in any way to the Premier's visit. Other planned missions, canceled because of poor weather, were carried out on the 8th. Led by Air Vice Marshal Nguyen Cao Ky, 24 VNAF A-1Hs attacked Vinh Linh, another transportation and military installation above the demilitarized zone. They were supported by 6 USAF A-1Es, 20 F-100s, and 3 RF-101s. The USAF aircraft were used for flak suppression, as patrols for rescue and to counter enemy aircraft, and as escort for bomb damage assessment. The Navy separately hit the Dong Hoi area again.[14]

Meeting with the JCS on the 8th, McNamara asked for and the Joint Chiefs sent him recommendations for an eight-week program of air attacks on the north as a reply to any further "provocations." On the 10th, preparing for more action, Pacific Air Forces (PACAF) moved an F-100 and an F-105 squadron to Da Nang AB and two similar squadrons to Thailand. (For a discussion of USAF activities in Thailand see Project CHECO rprt, USAF Operations from Thailand, 1961 - 1965, in AFCHO.) [15]

The next day Viet Cong terrorists blasted a U.S. enlisted man's barracks at Qui Nhon, killing 21 and wounding 22 Americans and killing 14 Vietnamese. This act, coupled with Viet Cong ambushes, capture of a district town, attacks on the railway system, and assassinations of Vietnamese civil and military officials during a 72 hour period, triggered the largest retaliatory strike of the war thus far. Named Flaming Dart II, 28 VNAF A-1Hs and 20 USAF F-100s, 3 RF-101s

Air Force F-100Ds Struck Back At North Vietnam. Source: U.S. Air Force

and one F-100 weather reconnaissance aircraft hit Chap Lee. Simultaneously, 111 Navy aircraft struck Chahn Hoa not far from Dong Hoi.[16]

The administration again announced that the bombings were in response to Hanoi's provocations. Subsequently, McNamara stated that the attacks on North Vietnam had three main purposes: to raise South Vietnamese morale, to reduce the flow of infiltrating men and material and increase its cost, and to force Hanoi

Air War – Vietnam

at some point toward negotiations. Meanwhile, looking to possible future operations, the administration approved the dispatch, from 11 to 13 February, of 30 B-52s to Guam and 30 KC-135s to Okinawa. Designated Arc Light, these bombers and tankers of the Strategic Air Command (SAC) initially were earmarked for high-altitude, all-weather bombing of important targets in the North.[17]

Proposed Eight-Week Air Program

The joint chiefs sent their eight-week air assault program to McNamara on 11 February. It called for two to four U.S.- VNAF strikes per week, contained a list of Viet Cong actions requiring reprisal, recommended U.S. military deployments, and suggested measures to improve base security and steps to guard against intervention by Hanoi and Peking.

Enter The Gray Lady. B-52Ds deploy to Anderson Air Force Base in Guam.
Source: U.S. Air Force

The JCS recommended initial attacks against North Vietnam targets along "Route 7" south of the 19th parallel and near the Laos border. The JCS proposed sparing enemy airfields unless Communist aircraft intervened. There would be closer coordination of all air action in North and South Vietnam and Laos. Supplementary actions against the North would consist of Vietnamese sea harassment, more U.S. bombardment of targets, resumption of special navy patrols offshore and continuation of Plan 34A activities. in Laos there would be intensified air-ground attacks on selected infiltration points. To carry out this program, the JCS wished to deploy about 325 more aircraft to the western pacific to deter or cope with any escalation that might result. This would include, besides the dispatch of 30 B-52s to Guam, deployment of 9 more USAF tactical fighter squadrons and a fourth aircraft carrier. Some Marine and army units would go to Thailand and other units would be alerted.

As for the risks, the JCS believed that only Hanoi might intervene directly. The Chinese and the Soviet Union would react primarily with propaganda attacks and diplomatic efforts, although the Chinese communists might send "volunteers" into North Vietnam or northern Laos as a threat to escalate the war and as a challenge to the Soviet Union. The United States could resist intervention by Hanoi and Peking by putting into effect either CINCPAC's Operation Plan 32-64 (for the defense of mainland Southeast Asia), or Operation Plan 39-65 (an offensive air and naval plan for Southeast Asia and Mainland China). Only in the latter stages of plan 32-64 and to an undetermined extent in 39-65 would there be significant logistic, transportation, and personnel problems. These views were reaffirmed on 4 March.[18]

Plans and Operations 1965

The service chiefs agreed on the foregoing measures but, for different reasons, considered them inadequate. General McConnell thought the JCS recommendations of late 1964 spelling out heavier air strikes on the North remained valid. General Wheeler backed deployment of more USAF and other air units but pressed for an integrated air program against the north's transportation system, especially railroads. He also believed, along with Gen. Harold K. Johnson, Army Chief of Staff, that three U.S. ground divisions might have to be sent to Southeast Asia. The JCS chairman directed the Joint Staff to examine the possibility of placing one or two of these divisions in Northeast Thailand and a third, Augmented by allied personnel south of the Demilitarized Zone in South Vietnam.[19]

All of the eight-week air program was not approved immediately but some recommendations, such as the deployment of B-52s to Guam were quickly accepted. Meanwhile, the Viet Cong shifted their main effort from terrorist acts to the I and II corps area in the central highlands of the South where battalion-sized units inflicted heavier casualties on the Vietnamese forces and threatened to split the country at the corps boundary line.[20]

To thwart such a plan, the Vietnamese and the Americans moved more ground and air units to that region. VNAF A-1Hs and USAF combat advisory A-1Es and AC-47s struck hard at the insurgents, causing substantial casualties. AC-47s equipped with gatling guns had been used successfully for the first time on 15 December 1964. in accordance with the President's Authorization of 27 January 1965, the JCS approved Westmoreland's request to employ jet combat aircraft in an emergency.

The three 7.62mm gatling guns on the AC-47 put out a spectacular display of firepower. This AC-47 is actually firing over Saigon in 1968.
Source: U.S. Air Force

The first jet combat strike of the war was flown on 19 February when 24 B-57s hit a target area in Phuoc Tuy Province. The "emergency" stricture remained until 10 March when the JCS permitted the MAC/V and 2d Air Division Commanders to use South Vietnamese-based U.S. jet or non-jet aircraft for missions in or out of the country when the Vietnamese Air Force could not perform them. The JCS also rescinded requirements for carrying VNAF observers or student pilots and for placing VNAF markings on USAF's two A-1E squadrons. Some high U.S. Embassy officials expressed concern that these decisions might result in the killing of friendly civilians and create more enemies.[21]

While the tempo of military operations rose in February, new political upheavals occurred in Saigon. On the 16th Phan Huy Quat emerged as the new Premier. On

Air War – Vietnam

the 19th, another coup attempt was smashed, largely by the intervention of the VNAF led by Marshal Ky and by the negotiations conducted by Brig. Gen. Robert E. Rowland, chief of the Air Force Advisory Group in headquarters MAC/V. Then on the 22d, The Vietnamese Armed Forces Council deposed its chief of staff, Lt. Gen, Nguyen Khahn, replacing him with Maj. Gen. Tran Van Minh. Again U.S. officials in Washington and Saigon were dismayed by political turbulence that diverted attention and effort against the Viet Cong.[22]

Troop Deployments For Base Security

The bold communist strikes in February posed a new crisis in air base security. Within the JCS, McConnell and Gen. Wallace M. Greene, Jr., the Marine Corps commandant, stressed the urgent need for more U.S. forces to guard the bases regardless of cost. They noted that MAC/V expected more attacks on these sites and that a security analysis indicated the need for the equivalent of one U.S. division plus additional engineers. On 20 February, the joint chiefs warned McNamara that the security problem was compounded by the questionable integrity of some Vietnamese troops who had recently demonstrated against their government and the United States. They doubted that the Vietnamese alone could repel an all-out Viet Cong attack on Da Nang AB, the "number one" communist target since it was the springboard for reprisal strikes on North Vietnam, air operations in Laos, and certain Plan 34A operations. Other insecure places were the Saigon-Bien Hoa-Vung Tau area, Nha Trang, and Cam Ranh Bay. As a first step, the JCS recommended the dispatch of the 8,500-man 9th Marine Expeditionary brigade to Da Nang and Marine reinforcements to the western Pacific. Simultaneously it reaffirmed the need for the eight-week air assault program against the Hanoi regime.[23]

Air Base Security. The M-16 with grenade launcher suggests that this photograph dates from later in the Vietnam War.
Source: U.S. Air Force

Ambassador Taylor opposed the placement of large numbers of U.S. marine forces around Da Nang AB and on 24 February the JCS reduced their requirements. However, as new MAC/V and Central Intelligence Agency (CIA) reports on the 25th underlined the gravity of the military and political situation in South Vietnam,* U.S. officials announced that day that more American troops would be sent. The first elements of a 3,500-man Marine unit arrived at Da Nang on 8 March and the entire unit, including its own air arm, was shortly in place.

Plans and Operations 1965

Secretary Of State Dean Rusk said that the Marines would provide "close in security" and would not engage in "pacification operations" although they would "shoot back" if attacked.[24]

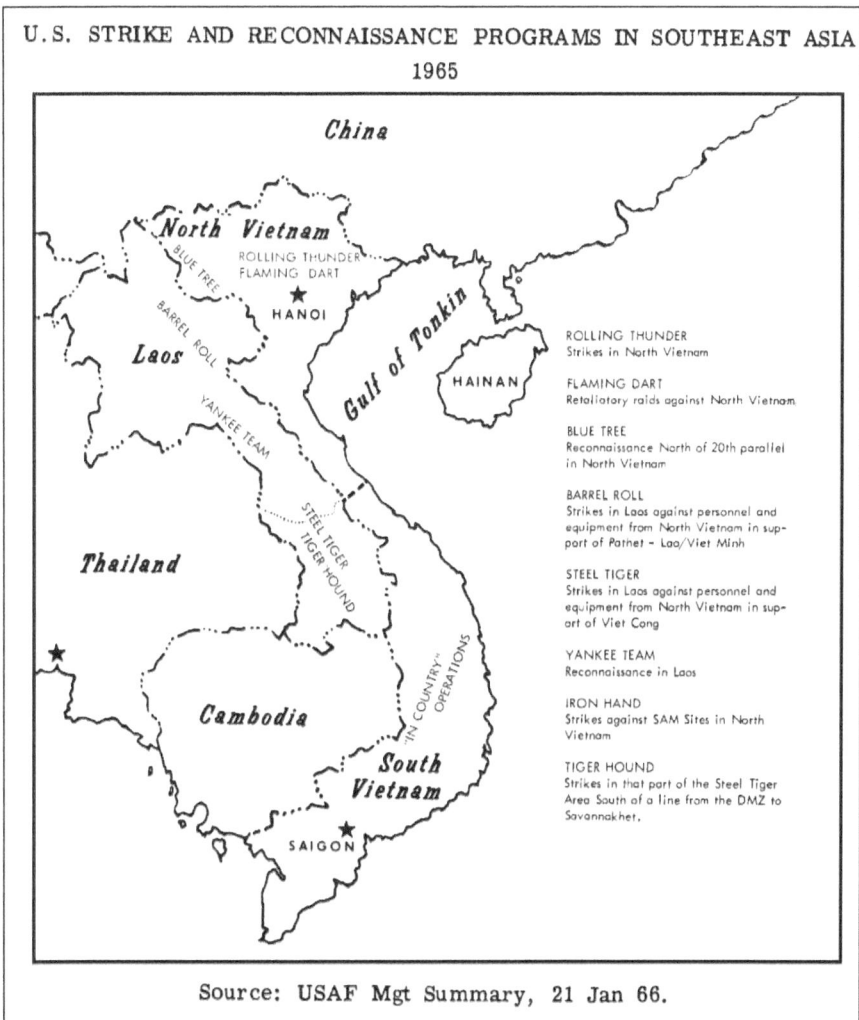

Rolling Thunder Strikes Begin

Meanwhile, extensive planning of .new air strikes against the North neared completion. Several were scheduled for late February but were postponed because of poor weather and the political turmoil which affected the Vietnamese Air Force. But on 2 March "Rolling Thunder" began when 104 USAF aircraft (B-57s, F-100s, F-105s, and refueling KC-135s) plus 19 VNAF A-1Hs hit Quong Khe and Xom Bang. B-52s on Guam were alerted but not used. This was the first strike on the North in which USAF aircraft played the dominant role. It was also the first time that the U.S. government abandoned its policy of purely retaliatory response for official spokesmen asserted that the strike was part of a continuing effort to resist aggression.25

Although the attack was considered "very successful" the loss of four USAF aircraft, three to anti-aircraft fire, caused concern. Deputy Secretary of Defense Cyrus R. Vance convened a meeting attended by Secretary of the Air Force Eugene M. Zuckert and other USAF officials to consider using the high-flying B-52s for pattern bombing in either North or South Vietnam to avoid communist ground fire. The Air Staff and SAC recommended reserving B-52s for use against major targets in the north. The idea of B-52 pattern bombing was not seriously considered again until April. [26]

The A-1H was the VNAF's most advanced and effective combat aircraft and was used by them for strikes on targets inside North Vietnam.
Photo source: U.S. Air Force

On 14 March, 24 VNAF A-1Hs supported by U.S. jets, in the second Rolling Thunder operation, struck weapon installations, depots, and barracks on Tiger Island, 20 miles off the North Vietnamese coast. The next day in the third Rolling Thunder strike more than 100 U.S. aircraft (two-thirds Navy, one-third USAF) hit an ammunition depot near Phu Qui, only 100 miles southwest of Hanoi. The earlier hesitancy about bombing the North had disappeared. in addition to Yankee Team and Barrel roll activities in Laos and the open U.S. air participation in South Vietnam after 10 March, Rolling Thunder was a third separate air campaign aimed at bringing the communists to the negotiating table. The attacks, tightly controlled by top U.S. officials in Washington, were carefully planned by the JCS without formal service participation.[27]

Plans and Operations 1965

II. DEBATE OVER STRATEGY

The United States had met the growing Viet Cong challenge by unleashing more air power in the South, dispatching marine combat troops to secure Da Nang AB, and beginning air strikes against the North. The administration now engaged in an intense debate over future strategy that would determine the type and extent of further U.S. and JCS participation in these discussions revealed major differences of opinion as to what should be done.

USAF Opposition To Deploying Large Ground Forces

Early in the year, a JCS plan to dispatch a large international force to South Vietnam had fallen through because of a lack of allied support. The Air Staff had opposed this plan, declaring it contradicted prior JCS views on the proper U.S. course of action. If such an international force were possible, the Air Staff thought it should be limited to air, naval, and marine units under the aegis of the ANZUS or SEATO alliances.[1] The ANZUS treaty (signed by Australia, New Zealand, and the United States) came into force on 29 April 1952. The SEATO treaty (signed by Great Britain, France, Pakistan, Thailand, Philippines, New Zealand, Australia and the United States) came into force on 19 February 1955.

In fact, the Air Staff opposed placing any sizeable U.S. ground forces in South Vietnam for combat. It questioned the wisdom of sending 80,000 to 100,000 U.S. troops to that country, as proposed by Wheeler on 27 January. It believed this would require partial U.S. mobilization, create tremendous logistic requirements, take months to accomplish, prove very costly, invite rather than deter Chinese intervention, and adversely affect America's world-wide military posture. The Air Staff favored invoking, if necessary, CINCPAC's air and naval 39-65 plan to deter, or failing that, to defeat the Chinese.[2]

The Army disagreed. It argued that the United States should be prepared for Chinese communist intervention after either limited or massive air and naval attacks on North Vietnam and Laos. Destruction of the North, the army claimed, would certainly lead Hanoi to ask for and Peking to provide large-scale assistance. Adequate U.S. ground forces would be needed to secure essential U.S. bases and facilities and deter such intervention because otherwise, according to CINCPACs estimates, Chinese and North Vietnamese forces could seize Saigon in D plus 60 days and Bangkok in D plus 65 days.[3]

The deep cleavage between USAF and Army strategic thinking was further demonstrated during

Gen. John P McConnell.
Source: U.S. Air Force

Air War – Vietnam

JCS discussions over the relevancy of CINCPAC's 32-64 and 39-65 plans. The Air Force disliked the first plan which called for limited operations in Southeast Asia, selected air strikes, extensive logistic support, and the use of reserve and National Guard units. General McConnell strongly argued for adoption of the second plan which would permit employment of superior U.S. air and naval strength against Asian manpower. He said it would require fewer reserve forces, promised to deter the Chinese more effective\y, and, if they entered the war, would bring then to terms.[4]

Concerned over this interservice debate and confusion about respective requirements, Secretary Zuckert wrote to Deputy Secretary of Defense Vance and expressed the view that the basic issue should be whether the plans were feasible logistically, politically acceptable, and credible to the communists. Vance, in turn, asked the JCS to review all of CINCPAC's contingency plans and U.S. ability to reinforce NATO and meet its other military commitments.[5]

Gen. Earle G. Wheeler.
Source: U.S. Army

On 11 March, Wheeler informed the JCS that neither plan 32-64 nor plan 39-65 was feasible. The first could not be carried out within a stipulated time and had been overtaken by events (the dispatch of Marine forces to Da Nang). The second was impractical because it was unlikely that the United States would make a quick political decision to use it. He directed the Joint Staff to prepare new recommendations for air, ground, and Naval deployments to the Pacific to insure holding Southeast Asia, Taiwan, and Korea and to permit, if necessary, air and naval operations against China.[6]

McConnell did not oppose further study of U.S. strategic requirements, but he disagreed with the concept inherent in Wheeler's request to the Joint Staff. in view of U.S. worldwide commitments, he warned of excessive logistic requirements and possible imbalance of the military force structure. He reaffirmed his confidence that the air and naval 39-65 plan could check intervention by Hanoi and Peking.[7]

New Assessments And The Army's 21-Point Program

Meanwhile, reports from South Vietnam pointed to a larger U.S. involvement. On 25 February a MAC/V analysis of the military situation in all four Vietnamese corps areas agreed with a grave CIA appraisal issued the same day. Observing that the pacification effort had virtually halted, Westmorelend foresaw in six months a Saigon government holding only islands of strength around provincial and district capitals that were clogged with refugees and beset with "end the war" groups asking for a negotiated settlement. The current trend presaged a Viet Cong takeover in 12 months, although major towns and bases, with U.S. help, could hold out for years. To "buy time," permit pressure on North Vietnam to take effect, and reverse the decline, he proposed adding three Army helicopter companies, flying

Plans and Operations 1965

more close support and reconnaissance missions, opening a "land line" from Pleiku in the highlands to the coast, and changing U.S. policy on the use of combat troops.[8]

Sharp generally concurred with these recommendations but advised the JCS that the full use of air power in North and South Vietnam was the most important measure that could be taken to improve the military situation quickly. He also advocated obtaining better intelligence and naval bombardment of the north's coastal installations. And he warned that a coup by Lt. Gen. Nguyen Chanh Thi, the Vietnamese I corps commander was possible, and this would be an "undesirable" change.9 There was now fear at the highest administration level that the entire Vietnamese military effort night collapse. This led to another visit to South Vietnam from 5 to 12 March of a high-ranking military and civilian mission headed by General Johnson, the Army's Chief of Staff.[10]

In Saigon, the mission was briefed by Ambassador Taylor who stressed the historical, racial, and religious factors that prevented establishment of a unified country. He said these were the chief causes of the U.S. failure thus far, and he saw no quick results regardless of massive American Aid. On 14 March General Johnson sent the JCS and McNamara a 21-point program. It included but went beyond Westmoreland's prescription.

Despite conducting the first bombing missions carried out by jet aircraft, the operations of these B-57Bs was not having the desired effects on the Viet Cong. Source: U.S. Air Force.

For South Vietnam, Johnson proposed more U.S. and, if possible, allied troops, more helicopters and O-1 aircraft, possibly more USAF fighter-bombers (after further MAC/V evaluation), better targeting, accelerated airfield expansion, more special operations, and additional logistic, construction, advisory, civic action, and financial measures. He proposed that the additional troops secure the bases of Bien Hoa, Tan Son Nhut, Nha Trang, Qui Nhon, and Pleiku or defend Kontum, Pleiku, and Darlac provinces in the II corps area of the highlands. Either deployment could free many Vietnamese battalions for combat. Johnson preferred the second alternative but recognized, as did Westmoreland, that this could require a "clarification" if not a "change" in U.S. combat policy.

To step up pressure on North Vietnam, Johnson asked for the rescission of many restraints on air strikes. For Laos he favored reorienting Barrel Roll operations to allow, air strikes on infiltration routes separate from those directed against the communist-led Pathet Lao and North Vietnamese units.[11]

Carl Rowan, director of the United States Information Agency (USIA) also accompanied the Johnson mission. He prepared a 16-point program which included recommendations for an increase in psychological warfare operations including leaflet-dropping and broadcasting. To carry out these activities he asked for 20 more U-10 aircraft or helicopters.[12]

President Johnson's March Decisions

After reviewing these recommendations, President Johnson on 15 March authorized new military measures to reverse the trend in South Vietnam, increase Viet Cong casualties, and "make them leave their neighbors alone." Approving most of General Johnson's program, he directed: (1) deployment of three more helicopter companies within 30 days and three more Army O-1 companies and three more USAF 0-1 squadrons totaling 185 aircraft within 120 days; (2) establishment of a joint U.S.-Vietnamese target and analysis center; (3) use of the Seventh Fleet for more air and surface patrol and air strikes; and (4) accelerated construction of airfields, including emergency work at Da Nang and Chu Lai. He also directed that additional advisory support be provided Vietnamese regional and popular forces and that agreements be sought with Australia and New Zealand to provide more assistance.

The Forward Air Controllers were vital to the conduct of air operations. Maj. James Harding had the remarkable and unique achievement of flying 101 missions over North Vietnam in this O-1, light, single-engine forward air control aircraft. He flew a total of 596 combat missions in Southeast Asia, and was awarded the Air Force Cross while flying an A-1 Skyraider in Laos in 1972.
Source: U.S. Air Force

To increase the pressure on North Vietnam, the President rescinded orders that the Air Force fly air

Plans and Operations 1965

strikes only with the VNAF and hit only primary prescheduled targets. He gave field commanders more flexibility in timing air strikes because of weather or other delays, allowed low-level reconnaissance south of the 20th parallel and authorized air and naval harassing operations against coastal staging areas, including the use of special "De Soto" sea patrols and Plan 34A operations. He deferred action on several of General Johnson's recommendations, including dispatch of more U.S. combat troops, until he received more data from the State Department, USIA, and other agencies.[13]

Planning Allied Troop Deployments

Until 1 April when President Johnson made additional decisions, the dominant issue was the proposed large-scale deployment and possible combat use of more American and allied troops. The JCS, CINCPAC, MAC/V, and the U.S. Embassy in Saigon, examined at least 10 separate proposals. Four principal recommendations emerged. They called for deploying: (1) one U.S. Army division in the central highlands around either Pleiku or Saigon to prevent infiltration and to permit the release of Vietnamese security units for combat; (2) a U.S. or multinational (SEATO) force south of the 17th parallel; (3) one South Korean division in the Saigon area; and (4) undetermined forces in enclaves along the coast.[14]

The army, in accordance with General Johnson's views, favored stationing one division initially in the II corps area near Pleiku. The Marine Corps initially favored the "enclave" concept with units stationed at strategic locations along the coast from the demilitarized zone to the Mekong plus others in Thailand to secure bases in that country and act as a deterrent. It also advocated "direct military involvement by U.S. troops" because of the political instability of South Vietnam and the unreliability and opportunism of its military leaders.[15]

Reconciled to the deployment of ground troops, McConnell supported the "enclave" concept and thought that two divisions in South Vietnam and one in Thailand would suffice. But, feeling that the Army and Marine Corps proposals were oriented too much on South Vietnam, he presented another option - a 28-day air program against North Vietnam to destroy all targets on the 94-target list. He proposed beginning the air strikes in the Southern part of North Vietnam and continuing at two-to-six-day intervals until Hanoi itself was attacked. "While I support appropriate deployment of ground forces in South Vietnam" McConnell wrote, "it must be done in concert with [an] overall plan to eliminate the source of [the] insurgency." Simultaneously, other forces would support Vietnamese operations. McConnell believed that this proposal was consistent with previous JCS views on action against the North and would be a strong deterrent against open Chinese intervention.[16]

Later, after the JCS adopted a 12-week air strike schedule against the North that was acceptable to the Air Force, McConnell withdrew the 28-day program. Meanwhile, on 20 March, he joined the other service chiefs in warning McNamara that direct U.S. military action was imperative and recommending that the Marines at Da Nang AB conduct counter-insurgency operations. The Joint Chiefs

Air War – Vietnam

also urged the following deployments: (1) the remainder of the Marine brigade to Da Nang; (2) a U.S. combat division and supporting forces to the Pleiku area "as soon as logistic support was assured"; (3) a Korean Army division, if available, for counter-insurgency and base security; and (4) four of the nine USAF squadrons recommended on 11 February and 4 March.[17]

McConnell informed the JCS that the Air Force could resupply an Army division at Pleiku by flying 16 C-130s from Saigon to four nearby small airports where 10 USAF CH-3C helicopters would complete delivery of items to units not served directly by the C-130s. Admiral Sharp advised the Joint Chiefs that the proposed forces would require 18,000 to 20,000 more U.S. logistic personnel, including 4,500 previously requested for a logistic command in the theater.[18]

Plans were for C-130s to do resupply: Source: U.S. Air Force

On 25 March the JCS submitted another proposal on Marine deployments. Then on the 29th, the Army, in a surprise move informed the other services that it planned to send the 1st Cavalry Division (Air Mobile) to the Pleiku area to assist the Vietnamese in its defense and to secure the communication lines from Pleiku to Qui Nhon on the coast.[19]

The Air Staff believed, however, there was no need for a Division-size force near Pleiku and endorsed the enclave concept being supported in varying degrees by Ambassador Taylor, Westmoreland, Sharp, and others. Taylor also was opposed to deploying too many ground troops to South Vietnam. Furthermore, the JCS had not decided on the requirement, organization, or mission of an air-mobile division. The Air Staff agreed that resupply of the division would be risky and that the concept was strategically questionable. Defense of the highlands could best be achieved from coastal enclaves after logistic support was assured.[20]

The JCS recognized the seriousness of the military situation. As March ended, it asked for immediate increases in funds, a separate military assistance program for Southeast Asia, improved communication systems, faster response to Admiral Sharp's requests, exemption of Southeast Asia from the balance of payments goals, authority to extend military terms of service and to consult with Congress on the use of reserve forces, relaxation of military and civilian manpower ceilings, and a substantial increase in military air transport in and out of South Vietnam.[21] McNamara did not reply formally until 14 May when he observed that many of these recommended actions would not be carried out unless one of Admiral Sharp's major contingency plans was put into effect.[22]

Plans and Operations 1965

The Stepped-Up Air War

While administration officials weighed the cost and risk of a larger U.S. commitment, the pace of the war quickened.[23] To blunt Viet Cong attacks on South Vietnamese forces, both the Air Force and the VNAF had increased the number of their combat sorties. A high communist casualty rate was expected from the first authorized employment of USAF combat jet aircraft on 19 February and the rescission on 10 March of the major restrictions on all air operations in the south. The arrival in March of 45 more O-1s for VNAF visual reconnaissance and forward air control duties further enhanced the air effort.[24]

On 31 March, in a major attack on a Viet Cong stronghold, USAF aircraft set fire to Boi Loi woods in Binh Duong Province. Called Operation Sherwood Forest, C-123s first defoliated the area and then dropped fuel drums which were ignited by attached flares. A-1Es and B-57s fed the flames with napalm, but a rain storm extinguished the blaze. This attack, coupled with previous bombings and a psychological warfare leaflet-loudspeaker effort, induced several thousand civilians to leave the area.[25]

Against North Vietnam, the initial Rolling Thunder strikes on 2, 14, and 15 March were followed by more frequent USAF, VNAF, and Navy attacks. Beginning on 21 March, they struck targets four days in a row. On the 30th, another USAF F-105 squadron arrived at Korat AB, Thailand, from Okinawa to bolster USAF fighter-bomber strength. Ambassador Taylor affirmed that the air program had produced a "very clear lift in morale" in South Vietnam.[26]

The F-4C was an impressive multi-role aircraft. Source: U.S. Air Force

Although communist aircraft did not interfere with the Rolling Thunder attacks, enemy aircraft trails were sighted on 15 March about 50 miles from a target area. The presence of 34 MiG-15s and -17s on Phuc Yen airfield near Hanoi and additional MiGs and Il-28 bombers on the nearby Chinese Communist island of Hainan also disturbed the Air Force. On 17 March, McConnell proposed and the JCS three days later recommended the immediate dispatch from the United States to Thailand of a USAF F-4C squadron, one of the nine proposed by the JCS for Asia on 11 February and 4 March. The multipurpose F-4C could be used for air defense, "cover" for reconnaissance, and strikes in North Vietnam and Laos. The State Department quickly obtained the concurrence of the Thai government.[27]

As noted earlier, McConnell withdrew from JCS consideration his proposed 28-day air strike program against the North in light of a new 12-week program, drawn up by the JCS in accordance with guidelines from McNamara. The Joint Chiefs informed the Defense Secretary, however, that Rolling Thunder strikes could be made more effective by: (1) relaxing the rules of engagement; (2) giving field commanders more discretion to conduct medium and low-altitude reconnaissance flights and to determine tactics, escort, areas of operation, and exceptions to the rules of engagement; and (3) listing targets south of the 20th parallel to be hit in the ensuing weeks.[28]

The objectives of the USAF-Navy air program in Laos did not change during March. Yankee Team reconnaissance aircraft were moderately successful in surveillance of known targets, intelligence-gathering on Communist Pathet Lao movements, and assessment of bomb damage of targets struck by the Laotian Air Force. The number of Barrel Roll sorties increased from 67 in February to 211 in March. However, these efforts were hampered by the continuing restraints placed on operations by the Thai and Lao governments and Washington. McConnell and the other service chiefs could do little about the policies of Bangkok and Vientiane, but they agreed that the air activities in Laos would be more effective if Admiral Sharp was given freedom of action and they urged that Washington relax its control.[29]

Plans and Operations 1965

III. THE EXPANDING U.S. ROLE

At the end of March 1965, about 31,000 U.S. military personnel, 7,500 of them Air Force, were in South Vietnam. At least 15 key JCS recommendations aimed at arresting the military decline in that beleaguered country still awaited action. The United States was now openly participating with air strikes in the South and had begun air attacks against the north and stepped up air activity in Laos. As administration leaders considered new, major decisions, the services were poised to send more forces. Four USAF fighter squadrons were on alert in the United States for immediate deployment with five more prepared to move shortly afterward.[1]

President Johnson's April Decisions

On 1 and 2 April the President again made several major decisions. He approved the dispatch of two more Marine battalions, one F-4B Marine air squadron, and support elements. Most important, he authorized their "more active use" in South Vietnam under conditions to be established and approved by the Secretary of Defense in consultation with the Secretary of State. The President further approved sending 18,000 to 20,000 additional U.S. troops for logistic duties and to fill out existing units. He reaffirmed support of General Johnson's 21-point program and the effort to obtain "significant" combat elements from Australia, New Zealand, and Korea. He also stressed the need for faster movement of aircraft and helicopter units to Southeast Asia.

Marine Corps F-4Bs supported the arrival of the Marines. Source: U.S. Navy..

For North Vietnam, the President directed a slowly rising Tempo of Rolling Thunder operations, more leaflet missions, and more measures to counter the threat of enemy fighters. He said that aerial mining and blockade proposals against the north required more study, and that 12 CIA suggestions for additional covert and other activities should be explored quickly. For Laos he asked for stepped up air attacks on infiltration routes in the panhandle. The president also

Air War – Vietnam

approved a 41-point nonmilitary program prepared by Ambassador Taylor and directed him to seek the concurrence of the Saigon Government for these moves.²

In a restrained public statement, the administration announced that it would send several thousand more advisors and security forces to South Vietnam to protect installations, provide more economic aid, possibly increase the intensity of the war against the north, and help the Saigon Government increase its regular military, paramilitary, and police forces by 16,000 men. Taylor described the decisions as neither "a fundamental change in strategy" nor "sensational."³ Meanwhile, the president prepared for, and on 7 April launched, a major peace offensive. He asserted that the United States was willing to engage in "unconditional discussions" with the communists. He also proposed a billion-dollar development program for Southeast Asia.

The RB-66s were to prove their worth. Source: U.S. Air Force

Immediately after the President's military decisions, McNamara ordered more Marine and Army units to the Da Nang-Hue-Phu Bai areas along the coast. Army engineering and fuel units were to move to Thailand when that government consented. USAF units, some long on alert, also proceeded to Asian bases. During the first six days of April, McNamara directed the deployment of one F-4C squadron to Ubon and Udorn AB's and one F-105 squadron to Takhli AB in Thailand, one RB-66 squadron to Tan Son Nhut AB, South Vietnam, two fighter squadrons and one C-130 squadron to Okinawa, one fighter squadron to Taiwan, and two C-130 squadrons to the Philippines.⁵

To comply with the President's injunction of 15 March to increase Viet Cong casualties, each service submitted suggestions. McConnell proposed continuous O-1 aerial surveillance, a better air and ground alert in each corps area, an airborne command post to facilitate communication between forward air controllers (FACs) and strike aircraft, and simpler procedures for requesting air strikes to eliminate delays.⁶

On 9 April the President also approved the USIA's 15-point program for stepped-up psychological warfare.⁷ On the 13th he authorized the dispatch of more Marine forces and the Army's 173d Airborne Brigade. The Marines began arriving the next day, bringing to 8,000 the number of Leathernecks guarding the Da Nang AB and nearby facilities. Advance units of the 173d did not arrive until 3 May.⁸

The Honolulu Meetings

Other major ground deployments were under consideration. On 1April McNamara asked the JCS for the plan proposed by Wheeler in February to send two or three

Plans and Operations 1965

more divisions to Southeast Asia. After a meeting at Honolulu from the 8th to the 10th and a JCS meeting on the 12th, McNamara and the JCS agreed to adopt the enclave concept proposed earlier by the Marine Corps to introduce and support more U.S. and allied forces. The plan called for the United States to secure installations and enclaves along the coast, conduct operations from them, secure inland bases, and then conduct operations with Vietnamese units from these bases. The plan went to McNamara on 17 April.[9]

On 20 April key U.S. officials from Washington and Saigon again convened at Honolulu to continue deliberations on the U.S. build-up. As they met, Vietnamese units were still plagued by defeats and desertions; the increased U.S. application of air power had scarcely begun to rectify the situation. The conferees did not expect the Viet Cong and North Vietnamese forces to capitulate immediately and thought a favorable settlement of the war possible only in six months to two years. It might come as much from Viet Cong failure in the south as from the punishment inflicted by air attacks on the north. The communists had to be denied victory before a political solution could be reached.[10]

On the Air War, McNamara advised MAC/V to concentrate on South rather than North Vietnam and to "slip" Rolling Thunder operations if necessary. He said that close air support strikes should have priority over other types of air action. The Marine corps would provide its own close air support,. He thought better air organization was needed in using A-ls, B-57s F-100s, F-4s, and A-4s. Navy aircraft, he said, would not be required in the south except for large saturation strikes similar to the Black Virgin operation of 15 April. Admiral Sharp observed, however, that if USAF aircraft in Thailand continued to be unavailable for use in South Vietnam, Navy aircraft would be needed.[11]

Concerning the operations against the North, the Defense Secretary said that a "doughnut" area around Hanoi-Haiphong complex and Phuc Yen airfield would continue to be exempt from attack. He favored at least one VNAF Rolling Thunder mission per week with USAF support but no combined VNAF-Navy missions. No decision was reached on JCS requests to attack SA-2 missile sites. Concerning Laos, McNamara asserted that USAF-Navy Barrel Roll and Steel Tiger operations there had been wasteful since Navy sorties had produced few results.[12]

The SA-2 sites were a growing concern, Source: U.S. Air Force

Following these meetings McNamara outlined for the President a three-step program to bolster the 33,000 U.S. and 2,000 Korean military personnel now in

Air War – Vietnam

South Vietnam. (The Korean force, engaged largely in engineering tasks, began arriving in February 1965.) The first and only step he recommended for immediate action was the dispatch of 48,000 more U.S. and 5,250, Australian and Korean troops, plus three marine air squadrons. This force, which he proposed to deploy from May through August, would establish more enclaves, provide 20,000 men for logistic support, and conduct operations with Vietnamese units. A second step, to be considered later, called for deploying 56,000 more men, including an Army air mobile division, additional Marines, and Korean troops. The third step was not spelled out.

According to Secretary McNamara, the tempo of air strikes against North Vietnam was "about right" and had psychological as well as physical effects, although, he said, air attacks "cannot do the job alone." He concurred with Ambassador Taylor and others that the Hanoi-Haiphong area should not be hit since "we should not kill the hostage." All conferees agreed that air strikes should continue during any negotiation talks.

McNamara asked the President to consult again with the Australian and Korean governments about their proposed troop commitments. He recommended that the chief executive inform U.S. congressional leaders of the decisions to establish an international security force, deploy more troops and change the mission of U.S. forces.[13]

Korean troops arrived in Vietnam during 1965. By 1973 more than 300,000 had served there. Source: Republic of Korea

As a result of the Honolulu and Washington decisions, the JCS updated their 17 April deployment plan and sent it to McNamara at the end of the month. The Air Staff felt that the Army concept for deploying an air mobile division to the central highlands in South Vietnam had been overtaken by events.[14]

Speeding Unit Deployments

As the president weighed these recommendations, the JCS sent McNamara another air, ground, and naval plan for holding Southeast Asia. In addition to the 36,000 U.S. military men in South Vietnam on 30 April, the Joint Chiefs proposed adding 117,000 U.S. and 19,750 Korean, Australian, and New Zealand troops in subsequent weeks and months. They identified 12 USAF fighter-bomber, reconnaissance, and airlift squadrons and 4,813 USAF personnel to be deployed to South Vietnam, Taiwan, the Philippines, Japan and Okinawa. The proposal was

Plans and Operations 1965

not a unanimous one as general McConnell questioned the basic strategy it reflected. According to the USAF chief of staff:

The deployments and logistic actions imply a judgment that [the] United States should prepare to engage the Chinese Communists in a land battle in Asia under gravely disadvantaged conditions. Any planned commitment of U.S. manpower on [the] Asia land mass should continue to be [the] subject of deliberate and measured analysis of the near and long term objectives, capabilities, risks, and costs.

He urged more study of the impact of such deployments on NATO commitments, possible contingencies elsewhere, long term policy for Southeast Asia, and the imbalance and over commitment of U.S. military forces. [15]

Meanwhile, there were more setbacks in South Vietnam. The Viet Cong began a "monsoon" offensive with new weapons and appeared capable of launching large-size attacks anywhere in the country. In Saigon, following another coup attempt and demonstrations by Catholics who feared that the Quat government might make a neutralist settlement, that regime fell on 17 June. The military again took control, appointing the VNAF Commander, Marshal Ky, as Premier.

During this period, Westmoreland and Sharp asked for the immediate deployment of more U.S. troops, which triggered another intense debate in the JCS.[16] The Air Staff was deeply troubled by the U.S. drift toward an Asian land war and McConnell requested a special intelligence assessment of the need for still more ground deployments. He asserted that Army plans to dispatch an air mobile division which would be supported logistically by the Air Force had not been adequately examined. He declared that the increasing assistance by North Vietnam to the Viet Cong had added "a new dimension" to the war which required heavier air attacks on the north.[17]

Marshal Nguyen Cao Ky. Source: National Archives

The intelligence community backed MAC/V's assessments and recommendations (USAF intelligence neither agreed nor disagreed). Ambassador Taylor dissented. He conceded the necessity for more U.S. troops but thought perhaps one third of the number requested would be enough for the duration of the monsoon season. He said that South Vietnam's problems were aggravated chiefly by factionalism, politics, and poor military leadership (especially in the I corps area) rather than by the annual Viet Cong offensive.[18]

The Westmoreland-Sharp views prevailed and the JCS on 11 June recommended deployment of 45,000 more U.S. military personnel (23 battalions, the army's air mobile division, and four USAF fighter squadrons) and nine Korean battalions. The Joint Chiefs recognized that more air base facilities were needed before the additional USAF squadrons could deploy. They urged heavier air strikes on

Air War – Vietnam

"important" targets and more armed reconnaissance in the North to demonstrate American determination. Increases in Vietnamese Army strength would be postponed until November while hard-hit units were reconstituted.[19] Ambassador Taylor, meanwhile, in answer to newsmen's queries, stated on 11 June that there was no immediate need for more U.S. troops and no prospect of a U.S. build-up to 300,000 men.[20]

On 15 June, McNamara approved the deployment of an air mobile division to South Vietnam (it was to arrive in South Vietnam by 1 September) and two days later, after important modifications, the rest of the JCS recommendations. Although he was against the decision on the air mobile division, McConnell met shortly with the Army chief to discuss USAF logistic support for it, estimated at about 800 short tons per day.[21]

U.S. Military Month End Strengths In South Vietnam 1965					
Month	Army	Navy*	Marines	Air Force	Total
Jan	14,752	1,103	891	7,112	23,858
Feb	15,201	1,131	1,447	7,158	24,937
Mar	15,592	1,271	4,721	7,527	29,111
Apr	16,192	1,561	8,944	9,324	36,021
May	22,588	2,912	16,265	9,963	51,178
Jun	27,350	3,756	18,112	10,703	59,921
Jul	39,650	4,646	25,533	11,593	81,422
Aug	48,077	5,324	34,227	18,719	100,347
Sep	76,179	6,039	36,442	13,637	132,297
Oct	92,755	8,529	36,788	15,207	153,279
Nov	104,508	8,869	37,897	18,297	169,571
Dec	116,755	8,749	38,190	20,620	184,314

* Includes Coast Guard - Source: HQ MAC/V

At this time, the administration planned to approve fewer additional U.S. units than recommended by the JCS. On 16 June McNamara announced that 21,000 more troops would shortly go to South Vietnam, raising the U.S. force there from about 54,000 to between 70,000 to 75,000 of which 21,000 would be combat troops. Publicly, the primary mission of U.S. troops was to secure and patrol important military installations. They could be used for combat with Vietnamese troops only in an emergency. More would be sent if needed. He said that the Viet Cong had 65,000 combat and combat support troops, 80,000 to 100,000 part-time guerrillas, and 30,000 political and propaganda workers. The South Vietnamese, with 574,000 regular and paramilitary forces, had less than a 4 to 1 advantage.

Plans and Operations 1965

Since a higher, ratio was needed to cope with the threat, more U.S. strength was required.[22]

Westmoreland and Sharp quickly insisted that the approved levels, less than half recommended by the JCS on 11 June, was insufficient to meet the critical situation in South Vietnam. During the debate which followed, both commanders, at White House request, sent additional assessments.[23] Westmoreland said that without nuclear weapons a quick victory was impossible. He warned of a long war of attrition and raised his demands. He asked for 44 U.S. battalions in 1965 and more in 1966 to relieve the war-weary Vietnamese forces. He also asked for more USAF aircraft and for 30 more Army and Marine helicopter units exclusive of the 27 authorized and those for the air mobile division. Sharp stated that more coastal enclaves were needed from Hue to Qui Nhon from which U.S. troops could expand. He expressed confidence that by working with Vietnamese units and by convincing rural Vietnamese of American support, the United States would succeed where France had failed.[24]

McConnell now supported the deployment of more ground units, but only in accordance with the enclave concept. He continued to stress the need for more air pressure on Hanoi, saying he was:

> more convinced then ever that these [air] operations cannot be divorced from and are the essential key to the eventual defeat of the Viet Cong. In November 1964, [the] JCS unanimously agreed that direct, decisive, action against the DRV was needed immediately. This course of action was not adopted and intelligence reports indicate that the current air strike program, while inconveniencing the DRV had done little to curtail or destroy their will and capability to support the insurgency, largely due to the restraints on the air strike program. In fact, the restraints have provided the DRV with the incentive and opportunity to strengthen both their offensive and defensive capabilities.
>
> So [the] C/S USAF considers an intensified application of air power against key industrial and military targets in North Vietnam essential to the result desired. During the period of time required to introduce more forces, any build-up of and support for the Viet Cong offensive should be denied. Failing this, more serious difficulties and casualties for U.S. and allied troops can be expected.

He again urged that the Air Force be allowed to strike targets in the 94-target list as well as others.[25]

The JCS, except for agreeing to some intensification of the air war against the North, did not adopt McConnell's views. On 2 July, the USAF Chief of Staff went along with a JCS recommendation to send more U.S. Army and Marine ground and support units to provide 34 "maneuver" battalions. The Joint Chiefs also asked for six to nine additional USAF squadrons (after the completion of more airfields). The new U.S. goal would be 175,000 military personnel for South Vietnam. Immediate and heavier air strikes on the North, they added, would constitute an "indispensable" part of the overall program but even as this recommendation reached the Defense Secretary, a further South Vietnamese military decline presaged still higher U.S. manpower needs.[26]

The Air War In South Vietnam (April-June)

With the approval by increments of larger American forces for South Vietnam, the United States increased its direct participation in the war. In the spring of 1965, however, air power still played the dominant role. U.S. Marine and Army ground units were committed primarily to the security of Da Nang AB and other installations in coastal areas. They engaged in small-scale actions until late June when the Army's 173d Airborne Brigade began its first large search and destroy operation in Zone "D" near Saigon.[27]

Navy A-4B Skyhawks joined Air Force bombers over South Vietnam.
Source: U.S. Navy

From April through June, the use of air power in Southeast Asia rose about 53 percent above the first three months of the year. In accordance with McNamara's orders, Westmoreland gave top priority to air strikes in South Vietnam. In April, after a fourth Navy carrier joined the Seventh Fleet, Navy and Marine aircraft began to supplement USAF-VNAF operations. In the largest single air effort of the war, nicknamed "Black Virgin" U.S. and VNAF aircraft flew 443 sorties on the 15th, dropping 900 tons of bombs during an attack against Viet Cong concentrations in a forest in Tay Ninh province. USAF planes flew 49 percent of these sorties.[28]

In May, Augmented by North Vietnamese units, the Viet Cong began their "monsoon" offensive and in subsequent weeks repeatedly engaged South Vietnamese forces. Some of the largest battles of the year were fought at Song Be (site of the first "monsoon" attack on 11 May) Ba Gia, Dong Xoai, and Cheo Reo. USAF and VNAF aircraft, despite bad weather, often staved off Vietnamese defeat by inflicting heavy casualties on the communists and causing enemy defections. MAC/V reports acknowledged the significant contribution of USAF strikes. Notwithstanding their losses, the communists often destroyed or seriously battered Vietnamese units whose strength was already undermined by desertions.[29]

The rising air activity taxed the resources of the 2d Air Division, with overcrowded bases and shortages of certain types of munitions being special problems. McNamara asked the Air Force if it had the resources to expand airfield facilities quickly. Although Secretary Zuckert advocated continued reliance on Army construction units, a study was initiated to determine whether USAF units could do this work in operational areas. Also in short supply were aircraft for

Plans and Operations 1965

forward air control operations. However, in late May and early June, the first of three additional O-1 USAF squadrons (approved by the President on 15 March) began to arrive. By the end of July all three USAF squadrons, flying aircraft obtained from the Army, and three Army O-1 companies were in place. But a new problem arose when it was found that the radio equipment in some of the O-1s was incompatible with that of USAF fighter aircraft.[30]

On 15 May, there was a serious USAF-VNAF setback when an accidental explosion at Bien Hoa AB destroyed 14 aircraft (10 USAF B-57s and 1 A-1E, 1 Navy F8U, and 2 VNAF A-1Hs) and damaged 31 (1 USAF H-43 and 30 VNAF A-lHs). Twenty-seven USAF officers and men were killed and 77 wounded. Ten vehicles, buildings, a fuel dump, and other facilities also were lost. An investigating team led by Lt. Gen. William K. Martin, the Inspector General, Headquarters USAF, concluded that the explosion was caused by a malfunctioning fuse on a bomb in a B-57.[31]

To make up for these losses, a Navy aircraft carrier on 16 May began "Dixie Station" duty for South Vietnamese operations. On 11 June, after a decision to maintain a fifth aircraft carrier with the Seventh Fleet, the Dixie Station duty became permanent.[32]

In a major administration decision, and despite misgivings by the Air Staff and the SAC commander, B-52 bombers originally scheduled for use only over North Vietnam were assigned saturation bombing missions in the South. This decision came after the "Black Virgin" forest attack of 11 April. The Air Force had considered the strike relatively successful but Westmoreland thought the results showed that the tactical aircraft could not conduct pattern bombing over a large area in a short period of time. In the first B-52 attack (Arc Light 1) on 18 June, 27 aircraft hit the "Zone D" area near Saigon, a Viet Cong stronghold. Although 30 bombers took off, two collided during refueling maneuvers and were lost as were eight of 12 crew-men. A third bomber aborted. On 4 and 7 July the B-52s hit the same area.[33]

When all else fails, trust the Gray Lady. B-52s proved an extremely effective means of delivering saturation attacks. Source: U.S. Air Force.

An analysis of the first three strikes suggested that they provided valuable training in conventional bombing but did not prove B-52s could destroy Viet Cong capabilities. Intelligence for spotting targets was poor and without follow-up ground attacks, the bombings appeared wasteful. Some members of the Congress and the press also questioned the effectiveness of the bombings. But as additional strikes demonstrated their value, their frequency was increased. General

Air War – Vietnam

McConnell later described the B-52 effort as "strategic persuasion" to encourage the communists to cease their aggression. He also noted that a few of the bombers could saturate very accurately a large enemy area in a few minutes and their use freed many tactical aircraft for other tasks. McNamara, Westmoreland, and Sharp also strongly backed the use of B-52s. A high level committee with representatives from the White House, the State and Defense Departments, and the JCS exercised careful control of the bombings.[34]

By mid-1965, the U.S. air effort in the South was reaching formidable proportions. In January, USAF combat advisory sorties totaled 2,392; in June combat sorties totaled 7,382. Westmoreland desired still more air power and asked to use USAF aircraft in Thailand for attacks in the South. Sharp doubted that such attacks would be effective. More USAF strikes from Thailand on North Vietnam and Laos would also be needed, and he did not wish to jeopardize this effort by asking the Thailand government to approve such USAF attacks on South Vietnam.[35]

The Air War In North Vietnam And Laos (April-June)

The number of USAF-VNAF-Navy strikes against North Vietnam also rose steadily. On 10 April the JCS authorized the use of 10 KC-135 Arc Light tankers each day for fighter-bomber and reconnaissance sorties. Combined U.S.-VNAF combat sorties totaled about 3,600 in April, 4,800 in June. USAF aircraft flew less than half the missions. But an analysis by JCS chairman Wheeler on 1 April and another by the CIA and the Defense Intelligence Agency (DIA) early in July showed that the strikes had not reduced appreciably North Vietnam's ability to defend its homeland, train its forces, and infiltrate men and supplies into South Vietnam and Laos.[36] In fact, there was evidence that Hanoi would try harder to defend itself since more MiGs and Il-28 bombers had arrived on its airfields, and SA-2 antiaircraft sites had been built to protect the small industrial resources in the Hanoi-Haiphong area. The Air Staff, while accepting the Analysis, noted it had failed to take into consideration the political restraints which had hampered, U.S. operations.[37]

On 3 April, in the first enemy air attack of the war, MiGs intercepted a Nay F-8E near Hainan Island, and Navy pilots claimed a "possible" first kill. In another surprise attack on the 4th, four MiG-15s and -17s shot down two USAF F-105s on a bombing mission over the North, the first U.S. losses to enemy aircraft. To improve air defense warning against the MiGs, the Air Force sent seven EC-121s to Tainan AB, Taiwan and then to South Vietnam bases for operations over the North as necessary. The Air Staff also pressed for the deployment of another F-104 squadron to the Western Pacific.

On 12 May air assaults on the North halted as the United States explored the possibility of negotiations with Hanoi. When there was no satisfactory response, the bombings resumed on the 18th.[38]

Meanwhile, on 14 April, the JCS urged McNamara to approve air attacks on SA-2 sites as they became operational.[39] On General McConnell's initiative, the JCS resubmitted a recommendation on 7 June to "eliminate" the Il-28 bomber "threat."

Plans and Operations 1965

On 3 July, it also recommended strikes against the SA-2 sites. On the 7th McConnell said that reconnaissance showed that three SA-2 sites would soon have a limited capability.[40]

The MiG problem seemed well in hand. Source: U.S. Air Force

Neither the Secretaries of Defense or State, the U.S. intelligence community, nor Westmoreland shared the JCS view on the gravity of the situation. They doubted that the IL-28s would hit the South and concluded that the SA-2 sites had not yet interfered with Rolling Thunder operations. Meanwhile, the MiG problem appeared well in hand. Navy Phantoms downed two MiGs on 17 June, and a Navy Skyraider another on the 20th. On 10 July USAF F-4Cs destroyed two MiGs with Sidewinder missiles. Five enemy aircraft were now destroyed and a sixth possibly so.[41]

The Joint Chiefs continued studies, begun earlier in the year, on aerial mining of key North Vietnam ports and a naval blockade. They withheld recommending such action since the administration did not wish to increase the danger of hostilities between the United States and third nation suppliers of Hanoi, especially the Soviet Union and China. The JCS did propose reprisal air strikes for the assassination or kidnapping of key U.S. officials and, at McNamara's request, Sharp sent a list of suitable targets to be attacked within 18 hours after Washington's approval.[42]

In April, air activity increased in Laos, and about 2,000 USAF-Navy combat sorties were flown (about half by the Air Force), but in subsequent months the number fell below this figure. MAC/V was the coordinating authority but McConnell thought that MAC/V did not have enough qualified air experts and Sharp should exercise control through his component commanders of the Pacific Air Forces and the Seventh fleet. McConnell, however, was unable to persuade the JCS to alter the command arrangements.[43]

In accordance with General Johnson's recommendations of 14 March, the JCS on 3 April ordered the inauguration of "Steel Tiger" armed reconnaissance over Laos to insure heavier strikes against enemy personnel and equipment on infiltration routes south of the 17th parallel.[44]

Meanwhile, Barrel Roll operations began concentrating solely on providing Combat support for Lao ground forces against communist Pathet Lao and North Vietnamese units. On the 29th some restrictions on Barrel Roll missions were relaxed, and on 9 May, USAF F-4Cs in Thailand were placed on daily "Bango" alert to hit targets of opportunity. Later, USAF F-105s were placed on "Whiplash" alert for the same purpose.[45]

The Lao government insisted on stringent rules to govern U.S. activities, and this created a tortuous and time-consuming target-approval procedure. The chain ran from CINCPAC to MAC/V to Ubon AB, Thailand, to Vientiane, Laos, and reverse. Washington authorities and the U.S. ambassadors in Laos and Thailand were all deeply involved. Consequently, days and sometimes weeks passed before pilots were permitted to hit certain targets. After an alleged U.S. air strike on friendly Lao personnel on 22 May, U.S. officials suspended Steel Tiger operations until 7 June, and the Lao Government imposed more rules for an area where communist infiltration into South Vietnam was believed to be heaviest, Brig. Gen. Phai Ma, the Lao Air Force commander did not accept the U.S. estimates of the infiltration problem. [46] By the end of June some rules had been relaxed and control from Washington reduced, but Barrel Roll and Steel Tiger operations remained less effective than U.S. military officials desired.[47]

New Command Arrangements

The period also witnessed important command realignments. On 25 June General Memore, the 2d air division commander, was given the additional responsibility of MAC/V's deputy commander for air operations and raised to the rank of Lieutenant General. Long discussed by the JCS and backed by the Army, the change was approved by McNamara although General McConnell and the Marine Corps chief believed that the "two hat" arrangement was inappropriate and organizationally unsound as it would divide Memore's efforts between two locations.[48]

Meanwhile, there were plans to separate U.S. military activities in Thailand from headquarters MAC/V. This had been advocated by the U.S. Ambassador in Bangkok, Graham A. Martin, to allay the concern of the Thai Government about becoming too closely identified with the war in South Vietnam. The JCS also split over this issue: the Army opposed but the Air Force, Navy and Marine Corps favored separation, since it would permit MAC/V to concentrate on defeating the Viet Cong. In addition, the three services apparently were concerned lest there be established eventually a larger Army-dominated Southeast Asia Command. The majority believed that a three-star USAF general should head the new command in Thailand.[49]

On 30 April, McNamara approved the separation but accepted the recommendation of the JCS chairman that a two-star Army general be Commander and a one-star Air Force general be deputy commander of the new command. Called U.S. Military Assistance Command, Thailand (USMAC/THAI) with headquarters in Bangkok, it was established on 10 July 1965 almost simultaneously with another organizational change. This was the reassignment two days earlier of the 2d Air Division from Headquarters 13th Air Force, Clark AB, the Philippines, to Headquarters PACAF in order to streamline and make more effective command and control procedures for the expanding tactical air operations. Air force units and six bases in Thailand remained assigned to the 13th Air Force but operational control was exercised by the 2d Air Division through the deputy commander in Thailand.[50]

Plans and Operations 1965

IV. PLANNING NEW DEPLOYMENTS

In July, the Vietnamese political situation under Premier Ky appeared more hopeful, as the mounting U.S. air strikes and the start of large-scale American ground sweeps helped restore momentarily the morale of friendly forces. Unfortunately, this favorable change was offset by the loss of additional Vietnamese territory in the II and III Corps areas, which produced an increasing number of refugees.[1]

A Larger Force For Southeast Asia

From 16 to 20 July McNamara, Wheeler, and other officials met in Saigon to assess the war effort and examine in detail Westmoreland's June proposals for sending more U.S. manpower to South Vietnam. On 8 July the White House announced that Henry Cabot Lodge would replace Taylor as ambassador. Taylor left Saigon on 30 July. Officially, Lodge became ambassador on 25 August. Ambassador Taylor described the most recent Vietnamese setbacks and the current military situation. He said the monsoons had made close air support unpredictable and reduced logistic support up to 30 percent. Air transport was the only reliable and, in several instances, the only means of reaching some provinces. He thought that Viet Cong willingness to come to terms would be dependent on the Rolling Thunder operations, the Saigon government's stability and capacity to administer a cleared area, U.S. determination, and the attitude of Hanoi and Peking.

Henry Cabot Lodge Jr
Source: U.S. Govt

Westmoreland and his aides outlined a proposed U.S. build-up in two phases. Phase I would require, by the end of 1965, 44 U.S. and allied battalions, 30 helicopter units, 20 USAF squadrons, and 6 Marine Corps squadrons for a total of 176,162 men. The ground forces would number 154,662, the Air Force 17,500, and the Navy 4,000. Twenty USAF squadrons would have the following aircraft:

Type Aircraft	No. Squadrons	Type Aircraft	No. Squadrons
B-57	1	Fighters (unspec)	6
F-100	4	A-1E	2
F-102	1	AC-47	1
F-104	1	C-130	4

Air War – Vietnam

During Phase I USAF combat capability would rise to about 16,750 sorties per month by the end of the year. Phase II deployments in calendar year 1966 would add 24 battalions, 18 helicopter units, 7 tactical fighter squadrons, and 2 transport squadrons. The 94,810 men would consist of 91,810 ground troops, 2,400 airmen, and 600 Navy personnel. At the end of Phase II, U.S. Forces would total about 270,972.

At that meeting, McNamara reaffirmed MAC/V's first claim to air resources, promising, if needed, additional aircraft carriers. He directed the Air Force to plan for a rate of 12 sorties per aircraft per day. Hz favored but made no firm decision on boosting the B-52 effort to 800 sorties per month as proposed by Westmoreland and Sharp in late June. He also supported modifying additional B-52s to obtain the 82 needed to achieve this sortie rate. He promised more engineering battalions to insure timely expansion of airfields and facilities, AM-2 airfield matting, he said, was being produced at a rate of one and one-half airfields per month or sufficient for 10 airfields by January 1965 (with about three million square feet per average airfield). [However] USAF information indicated sufficient production for only one airfield every two months.

The Defense Secretary was concerned about the Air Force-Army split in controlling aircraft in South Vietnam but did not dwell on the subject. General Memore, 2d Air Division commander, assured McNamara that all valid close air support requests for U.S. troops were being met. Memore emphasized the need for careful targeting of B-52 strikes to avoid wasting their expensive ordnance loads.[2]

The B-52 missions were becoming increasingly important but the desired number and force levels were disputed.
Source: U.S. Air Force

Some of the proposals and decisions were not fully in consonance with the views of the Air Staff. It believed that no ground forces should be sent in 1966 until air and naval power had hurt North Vietnam more severely, and that a maximum of 50 rather than 82 B-52s should be employed to provide 600 rather than 800 sorties per month. Although no decision was made on control of air resources, the Air Staff adhered to the belief that this problem could be resolved only by centralizing all air operations in South Vietnam under the 2d Air Division. But there was little prospect of JCS agreement on these issues.[3]

Meanwhile, on 19 July the JCS agreed on the construction or expansion of eight airfields in Southeast Asia and the western Pacific. Their location and suggested operational dates were:[4]

Plans and Operations 1965

Location	Number	Operating Date
Da Nang, South Vietnam	1	1 Sep 1965
Qui Nhon, South Vietnam	1	1 Nov 1965
Phan Rang, South Vietnam	1	1 Jan 1966
Sattahip, Thailand	1	1 Feb 1966
Unspecified, Thailand	1	1 Mar 1966
Unspecified, Western Pacific	3	Not given

The JCS also reviewed a "shopping list" of additional military requirements that Westmoreland gave to McNamara. The 1965 Phase I requirements were raised by about 20,000 men and an updated program was sent to the defense secretary on 30 July. Phase I now called for a U.S. force of 195,800 personnel with 34 maneuver battalions, 23 fighter squadrons, and 53 helicopter companies, and 22,250 allied personnel with 10 battalions. Official JCS approval of Phase I was delayed until August as estimated needs continued to increase.* McConnell supported the build-up but insisted that before confirming Phase II needs, the JCS should approve an overall strategy for the Western Pacific.[5]

Almost simultaneously President Johnson approved the dispatch of more Phase I-marked units. He announced on 28 July that U.S. strength in South Vietnam would rise almost immediately from 75,000 to 125,000 men, the maximum allowed until 1 September. It would provide 28 combat battalions and include the Army's air mobile division and appropriate air and logistic units. More troops would be sent later. The President pledged again America's determination to prevent the communist domination of Vietnam and Asia. Other officials said there would be no major change in the U.S. combat role. Vietnamese troops would bear the brunt of the fighting while U.S. units would guard U.S. bases and be available for emergency assistance.[6]

Impact On The Air Force

The spiraling U.S. military requirements for Southeast Asia, with costs expected to reach an estimated $10 to $12 billion per year, had a significant impact on force structures. On 23 July, Wheeler directed the JCS to review American world-wide military posture, and by early August the Air Staff was deeply involved in the evaluation, especially commitments to the North Atlantic Treaty Organization (NATO) and Cuban contingency plans. On the 5th, McNamara announced that because of Vietnam and other possible requirements, U.S. military strength would rise by 340,000 men to 2,992,000. The Air Force would increase from 809,000 to 849,000, largely to support stepped-up B-52, tactical, airlift, and logistic activities in Southeast Asia.[7]

To resolve urgent problems associated with USAF participation in the war, McConnell on 2 August designated the Air Staff Board as the principal coordinating agency in the Headquarters. In August, top USAF officials headed by Secretary Zuckert met in Honolulu to examine deployment, personnel, equipment, construction, and other matters incident to the approved and projected build-up of forces. With respect to the Phase I build-up they decided to convert all Air Force units already in place from temporary duty (TDY) to permanent change of station (PCS) and to ensure that additional units moving from the United States to Southeast Asia would be in PCS status. In the same month, USAF personnel were assigned to an OSD logistic task force created by McNamara on 31 July to expedite supplies to South Vietnam.[8]

The hike in the U.S. force goal in July prompted McConnell to press the JCS to appraise the military situation, state U.S. objectives, and prescribe a course of action for attaining them. Largely on his insistence, the Joint Chiefs on 27 August prepared a concept for Vietnam that singled out three basic military tasks, all of equal priority: (1) to force Hanoi to end its support of the Viet Cong; (2) to defeat the Viet Cong and extend control of the Saigon government over all of South Vietnam; and (3) to deter the Chinese communists and, if they intervened, to defeat them. The broad military strategy prescribed in the document which supported an intensified air and naval effort against North Vietnam contained many Air Force views. After studying the concept, McNamara sent the document to the State Department and the White House for use in further deliberations and informed the JCS that their recommendations on future operations in Southeast Asia would be considered on an individual basis.[9]

Meanwhile, the demand for more ground troops continued to increase. The last Phase I estimate had called for 195,800 U.S. military men for South Vietnam, but after July new assessments by the JCS and field commanders pushed the figure to 210,000, of which 34,500 were Air Force. The increase reflected requirements for more airlift, strike aircraft, air defense, airfield construction, artillery, support, and personnel for advisory, intelligence, communication, and security duties. On 23 August, the JCS recommended approval of the new Phase I figure, and in September the Defense Secretary sent the request with his endorsement to the President.[10]

McConnell was increasingly troubled by the impact of the projected deployments. He informed Gen. Hunter Harris, jr., PACAF Commander, that upon completion of Phase I, 67 percent of the Air Force's tactical fighters, 87 percent of its tactical reconnaissance, and 62 percent of its tactical airlift squadrons would be overseas. The Air Force could change unit missions or transfer units, but this would not provide either adequate rotational training in the United States or a sufficient number of units for deployment to meet NATO and other commitments. The Army too, he observed, was finding it more difficult to fulfill its needs.

McConnell reiterated his belief that only proper use of air power could simultaneously deter the Chinese Communists and minimize the growing imbalance in the U.S. military posture. "If air power is not used to greatest advantage" he advised Harris, "and our military and civilian leaders are not

Plans and Operations 1965

convinced of this advantage, I foresee a virtually endless requirement for more and more ground forces in Southeast Asia reacting to whatever strategy the Viet Cong, DRV, and CHICOMS wish to impose." [11]

The effect of the U.S. Commitment on Air Force resources was becoming increasingly manifest. On 4 October, in a major decision, the Air Staff converted 13 USAF fighter squadrons (3 F-100, 4 F-105 and 6 F-4C) in the United States from tactical missions to replacement training to meet anticipated combat aircrew requirements.[12]

New Agreements At Honolulu (27 September -7 October)

From 27 September to 7 October military planners again met in Honolulu, primarily to determine the military units and movement schedules for the 1966 Phase II forces. McConnell instructed Harris, the chief USAF Representative, to impress upon the conferees the impact of the increased Phase I forces on the U.S. military posture and the importance of evaluating this impact before recommending more deployments under Phase II and not set arbitrary dates for unit arrivals.[13]

At the conference Admiral Sharp asked for 19,954 more Phase I personnel above those requested in August. For Phase II, the conferees agreed on the need for three more USAF tactical fighter squadrons above the seven believed necessary previously. The additional Air Force, Army, and Navy units and personnel selected were approved for planning purposes only. While the conferees agreed that the logistic structure would not fully support either Phase I or II deployments, the serious military situation dictated deploying as many combat units as possible even if support were marginal and combat capability reduced. Sharp's report of the conference emphasized the need for the United States to maintain military "momentum" as there was now a "clear and unmistakable" surge of Vietnamese hope and confidence stemming from the presence and performance of U.S. forces.[14]

Reviewing Sharp's revised manpower request for Phase II, the Air Staff considered the figure too high because it included a demand for units not yet in existence or which could not be deployed for 18 months. On 14 October the JCS recommended to McNamara 12,000 more Phase I personnel, 934 of whom would be Air Force.[15]

Refining strategic and deployment plans after the Honolulu Conference, the JCS on 10 November updated the concept for integrating U.S., allied, and Vietnamese forces to destroy the Viet Cong and pacify South Vietnam. This included, again, an extended Rolling Thunder program against the North that would achieve a level of destruction that the Hanoi regime could not accept.

The JCS paper on deployments showed that the completion of Phase I would place 219,000 U.S. personnel in South Vietnam. Completion of Phase II would bring the total to 359,000 since current resources could not meet this and other U.S. military obligations, the joint chiefs asked the Secretary of Defense for immediate approval to establish a broader base for service manpower training and

Air War – Vietnam

rotation and authority to call up selected reserve personnel and units, activate new units, and extend tours of duty. To rebuild its military strength after Phase I was completed, the Air Force said it would require four more tactical fighter and three more tactical reconnaissance squadrons (150 aircraft). After Phase II, it would need four more tactical fighter squadrons (96 aircraft). The other services also described their larger force structure needs.[16]

The Air War In South Vietnam (July-November)

The need for more U.S. forces ix South Vietnam was apparent from reports from the field of battle. The fighting grew in intensity even as larger numbers of American military personnel were arriving after mid-1965. U.S. forces, largely Marine and Army ground troops, totaled 59,921 at the end of June and 153,279 at the end of October. Because airfield space in South Vietnam was limited the Air Force had to rely increasingly on facilities in Thailand, the Philippines, Okinawa and Japan. As a result, USAF personnel increases in South Vietnam were modest, rising from 10,703 to 15,207. Although USAF and VNAF units shared air activity with the U.S. Marines, Navy and Army, the Air Force performed a majority of air strikes.[17]

While restraints on the use of air power were fewer and the rate of enemy "killed" rose, concern about the fate of Vietnamese non-combatants increased. Or 7 July Westmoreland instructed all commanders to minimize civilian casualties. McConnell indorsed the letter but continued to feel that too many rules interfered with effective operations. He favored permitting unified commanders maximum latitude, in accordance with national policy, in planning and executing the air effort.[18]

In 1965, the B-57B force was hard at work providing ground support for U.S., allied and Vietnamese forces. Source: U.S. Air Force.

Plans and Operations 1965

September witnessed the beginning of larger-scale ground and air action. U.S. and South Vietnamese marines launched Operation Piranha while army units of the two countries in Operation Gibraltar attacked Viet Cong-North Vietnamese forces now estimated to exceed 200,000, including political cadres. From October through the end of the year multi-battalion forces were engaged in the central highlands in the heaviest fighting of the war. In a major air-ground battle at Plei Mei from 19 to 29 October, USAF B-57s, A-1Es and F-100s played a key role in breaking up the communist attacks. From 9 to 28 November a second major battle, Operation Silver Bayonet, was fought in nearby Idrang Valley. It was highlighted by the first use and USAF support of the Army's air mobile 1st Cavalry division which had arrived in September-October. On 16 November this operation saw the first B-52 close support strike of the war. [19]

In these and other U.S.-Vietnamese campaigns, communist forces were thrown back with heavy casualties. The operations demonstrated the effectiveness of close air support. U.S. Army and Marine commanders and the U.S. Embassy in Saigon on frequent occasions testified that airmen had given indispensable assistance to ground troops and praised highly the exploits of USAF strike and FAC pilots. [20]

Of major USAF interest was its support of the Army's air mobile 1st Cavalry Division. Although the division destroyed many Viet Cong soldiers, its operations created severe supply problems and strained the entire U.S. logistic system in South Vietnam. McConnell believed that initial reports justified his earlier warning against employing a division near Pleiku in the central highlands without first securing properly ground and air lines of communications. He thought that more heliborne units, if deployed, would demand greater tactical, B-52, and airlift support than had been envisaged by either CINCPAC or MAC/V. "I still believe" he informed Harris, "that a combination of regular Army division and tactical air can provide the most potent forces as demonstrated in the recent Goldfire exercises," (these exercises, held in 1964, tested Air Force tactical support of ground forces) but unless OSD could be convinced of this, he expected more Army air mobile divisions to become part of the U.S. military structure. He instructed Harris to document thoroughly the recent USAF experience with the 1st Cavalry division. General Harris's report, forwarded on 1 December, confirmed the need for very extensive Air Force close air and logistic support for the division. [21]

The increase in U.S. ground operations coincided with a build-up of USAF strength from October through the end of the year. Phase I Units poured into South Vietnamese and other Asian bases. Four F-100 squadrons went to Bien Hoa and Tan Son Nhut ABs, four F-4C squadrons deployed to Can Ranh

Combat evaluation of the F-5A started in late 1965. Source: U.S. Air Force

Air War – Vietnam

Bay AB which opened in November, and one RF-4C and RF-101 squadron each deployed to Tan Son Nhut. Special air units also reached South Vietnam. In October an F-5 "Skoshi Tiger" squadron with 12 aircraft arrived for combat evaluation, and in November an AC-47 "Puff, the Magic Dragon" squadron, the first of its kind, was deployed with 20 aircraft. The aircraft had previously undergone successful combat evaluation. Also deployed was a psychological warfare squadron with 4 C-47s and 15 U-10s for stepping up leaflet and loudspeaker missions approved in April,* and three spray-equipped C-123 "Ranch Hand" defoliation aircraft. Other Phase I squadrons with F-105, F-100, F-4C, RB-66, and C-130 aircraft deployed to bases in Thailand, the Philippines, Taiwan, Okinawa, and Japan.[22]

Most B-52s operated out of Guam in 1965. However, a search for additional bases was under way. Source: U.S. Air Force

The stringent Washington controls over B-52 operations moved the JCS, in August, to ask McNamara to authorize five "free bomb" zones. According to the Joint Chiefs, this would insure attacks on the communists in all types of weather, make more aircraft available for other tactical missions, and provide more stable air crew, maintenance, and logistic support for the bombers. When McNamara approved the recommendation on 29 September he stressed the importance of avoiding casualties among Vietnamese civilians.[23] In September, B-52 tactics were changed from "maximum effort" missions to a combination of more frequent strikes using fewer aircraft. More than 300 sorties were flown that month and that level was maintained through the end of the year.[24]

When doubts about the value of B-52 bombings continued to be expressed, General Westmoreland, in an August press conference, strongly defended their effectiveness. However, the lack of adequate "exploitation" by ground forces of areas bombed troubled the Air Force. By 3 October only 10 of the 37 missions flown had been followed up on the ground and in only two instances was there evidence of significant damage to the communists. Secretary of the Air Force Harold Brown, shortly after assuming this post succeeding Zuckert on 1 October 1965, asked the Air Staff for a study of the bombings. Its reply showed that the B-52s prevented concentration of enemy forces, often forced their withdrawal, instituted great fear, effectively destroyed major targets and boosted lagging South Vietnamese morale. The study also pointed to the need for better targeting. Brown considered the study sufficiently important to send copies to McNamara, who, in turn, sent them to the State Department and the White House.[25]

Plans and Operations 1965

Some thoughts were given to deploying the B-52s closer to the combat theater, say in the Ryukus, but this raised serious political questions. Thus, in July when B-52s launched a mission from Okinawa (where they had flown because of a storm in the Guam area), both the governments of Japan and the Ryukyu Islands protected vigorously, alleging such missions endangered Japan's neutrality. The U.S. ambassador to Japan, Edwin 0. Reischauer, also objected, warning that further flights from the islands could endanger U.S.-Japanese negotiations beginning in 1967 on the renewal of base rights in 1970. On 31 July 1965, Under Secretary of State (George W. Ball) asked McNamara for a ruling on the need for Okinawa for B-52 operations.[26]

The JCS quickly counseled more restraint in publicly confirming the operations, believing that this would decrease left-wing pressure in Japan against them. Backed by Gen. John D. Ryan, the SAC commander, and Admiral Sharp, the JCS stressed the importance of the island for U.S. contingency planning and asked for "unswerving" U.S. support for its use without hindrance. But Okinawa was not used again by the B-52s during the rest of the year. Meanwhile, new proposals were studied for basing the big aircraft in Thailand, the Philippines, or Taiwan.[27]

Airfield Expansion And Security

Airfield construction moved at a feverish pace in South Vietnam. Work began in May on a new airfield at Chu Lai and in June on another at Can Ranh Bay. In July construction was approved for a new airfield at Phan Rang and for additional work at Qui Nhon and Da Nang. In late August the JCS forecast a slippage in the schedule for the last three sites of from three to eight weeks. An Air Staff study identified the major problems as inadequate engineering units, poor construction methods, and lagging production of AM-2 airfield matting.[28]

To spur airfield expansion, Brown informed McNamara in October that while the Air Force would continue to rely largely on Army Engineers for air base work, it would use its own resources to activate two heavy repair units that would be mobile, flexible, and located so that they could respond rapidly when needed.[29]

At the end of 1965 construction was under way on three new airfields at Chu Lai, Cam Ranh Bay, and Phan Rang in South Vietnam, and eight others were being expanded. In Thailand, Sattahip AB also was undergoing major expansion.[30]

Air base security remained a problem since the Viet Cong made the bases prime targets and attacked them frequently, often with great success. On 1July a 14-man Viet Cong sabotage team unleashed a mortar and rifle attack on Da Nang, killing 1

Aftermath of a Viet Cong attack on Bien Hoa. Source: U.S. Air Force

airman, destroying 3 aircraft, virtually destroying 3 others, and damaging 4. The total monetary loss was estimated at $5 million. On 24 August a mortar and 105-mm howitzer strike at Bien Hoa wounded 9 Americans and at least 20 Vietnamese and damaged 22 USAF aircraft and 8 Army helicopters. On 27 October the Viet Cong attacked Marine Corps installations at Da Nang and Chu Lai, destroying 22 helicopters and 2 A-4s and damaging 18 helicopters and 5 A-4s. American personnel losses were 3 killed and 83 wounded. [31]

Air base vulnerability was attributed largely to lack of cooperation between Vietnamese Army and VNAF Commanders and their refusal to accept U.S. advice. A headquarters USAF inspection team, after visiting Da Nang, Bien Hoa, Tan Son Nhut, and Nha Trang in early September, believed that 1,381 more air police plus additional vehicles and radar equipment would strengthen internal air base defense. Its major recommendation was that U.S. Air Force assume from the Vietnamese responsibility for perimeter defense (except at Da Nang where U.S. Marines guarded the base), and that the JCS approve 33,600 more military spaces for this purpose. Unless this were done, the team predicted more Viet Cong attacks and USAF losses of personnel, aircraft, equipment and facilities.[32]

In a JCS review of the subject, McConnell observed that not all U.S. and allied troops had been used to secure U.S. bases as the JCS initially intended. The Army and Navy chiefs opposed any action however, that appeared to criticize CINCPAC and COMUSMAC/V in October, while visiting South Vietnam, McConnell reviewed the matter with Westmoreland, who indicated he did not plan to ask for a sizeable increment of troops solely for base security. As a consequence, McConnell made no further effort within the JCS to obtain combat forces for the poorly protected air bases. To help reduce their vulnerability to attack, he directed that all USAF aircraft in Southeast Asia be parked in revetments as soon as possible.[33]

The Air War In North Vietnam And Laos (July-November)

In the Rolling Thunder attacks on North Vietnamese targets in the latter half of 1965, enemy anti-aircraft fire took an increasingly heavy toll. The threat from SA-2 missile sites was of particular concern. After an SA-2 missile on 24 July downed a USAF F-4C, the first such U.S. loss, the administration allowed USAF-Navy aircraft to attack the missile site. The mission was carried out on 26-27 July, but was unsuccessful. On 9 August, 12 USAF aircraft hit another site, but it was later found to have been unoccupied. In the same month the JCS enlarged somewhat the boundary for permissible U.S. air operations against the sites. It also inaugurated two programs for locating and destroying the sites: "Iron Hand" and "Left Hook" with the latter employing electronic intelligence (ELINT) aircraft, reconnaissance drones, and other measures. [34]

In August the Air staff convened a study group to examine the SA-2 problem. One result of the groups work was McNamara's approval in October of the transfer of five USAF B-66Bs from Europe to Southeast Asia to Augment PACAFs electronic countermeasures capability. In the same month, he approved an "Iron

Plans and Operations 1965

hand" strike on another SA-2 site, and five Navy aircraft destroyed it on the 17th. There were also successful strikes on 31 October and 7 November.[35]

The JCS continued to chafe under the remaining restraints against hitting SA-2 and other more important targets, especially those in the Hanoi-Haiphong area in August. It proposed aerial mining and a blockade of major northern ports, a course of action long under study. In September the Joint Chiefs again recommended as a matter of "military urgency" air attacks on Phuc Yen airfield to destroy the Il-28 bombers there and other attacks on the SA-2 sites which were increasing in number. They also urged hitting other antiaircraft emplacements, four power plants, fuel storage facilities at Haiphong, and rail, highway, and waterway traffic between the Hanoi-Haiphong area and southern China.[36]

In reply, McNamara expressed doubts that the gains from more bombings would outweigh the risks. Intelligence estimates, he observed, indicated that heavier air strikes, especially in the Hanoi-Haiphong area, would not persuade Hanoi that the "price" for aiding the Viet Cong was too high. They might, in fact, induce North Vietnam to step up its assistance. Increased pressure could also trigger an enemy air strike on Da Nang or result in a confrontation between the United States and China. Like McNamara, the State Department's Assistant Secretary for Far Eastern Affairs, William P. Bundy, did not think that bombing the Hanoi-Haiphong area would force the North to accept a negotiated solution to the war.[37]

At Brown's request, the Air Staff in October made a special study of the effectiveness of USAF armed reconnaissance in North Vietnam. Its report, issued on the 29th, substantiated previous observations that traffic on main transportation routes and traffic support had been disrupted and that the transit time for supplies had increased. But the study concluded that the North's ability to resupply communist forces in the south had not yet been seriously impaired.[38]

On 10 November the JCS again recommended an enlarged air attack program on North Vietnam and Laos that would try to destroy 13 sites in the North which contained about 97 percent of that country's fuel storage capacity. The Air Staff especially considered it necessary to destroy these sites. The joint chiefs said that 446 aircraft, including 336 for the strikes and 80 for flak suppression, would be needed. They doubted that Hanoi would retaliate in any way. Although this request was not approved, the Service Chiefs were authorized for the first time to hit certain transportation targets connecting the major North Vietnamese industrial areas.[39]

In Laos, USAF-Navy combat sorties from July through November ranged from about 1,000 to 1,500 per month. Westmoreland continued to search for more effective means to apply air power. In July, in addition to a request for more sorties, he asked permission to launch small, air-supported, ground operations from South Vietnam into Laos to hit infiltration targets. Such operations had been supported by the JCS in 1964 but administration approval was withheld until September for these small "Shining Brass" attacks. The ground forces penetrated up to 20 kilometers into two Southern Laos provinces. Thai-based USAF aircraft supported the operations and several initial air strikes were successful.[40]

In the same month Westmoreland asked authority to use more South Vietnamese-based aircraft to supplement USAF-Navy efforts in Laos. USAF activities had been limited by the insufficient number of KC-135 tankers for in-flight refueling of the F-105s from Thailand and the F-100s from South Vietnam, and by Thai government reluctance to allow more USAF aircraft to engage in operations outside the country. Admiral Sharp approved this request in October subject to final concurrence by the U.S. Embassy in Vientiane.[41]

In mid-November, Gen. Phai Ma, Chief of the Lao Air Force, relaxed some of the severe restrictions previously imposed on Steel Tiger operations in southern Laos, scene of some of the heaviest communist infiltration. Also in November, the U.S. ambassador to Laos, William H. Sullivan, agreed to Westmoreland's request to use B-52s for strikes along the Lao-South Vietnamese border.[42]

To improve the coordination of USAF operations flown over North Vietnam and Laos from Thailand, PACAF on 23 November established the post of deputy commander, 2d Air Division and 13th Air Force. Brig. Gen. Charles S. Bond, jr., was named to fill the post beginning 7 January 1966, succeeding Brig. Gen. John R. Murphy who had served only as deputy commander for the 2d Air Division. General Bond was to transfer his headquarters from Udorn to Korat, Thailand, as soon as possible.[43]

Plans and Operations 1965

V. COMMUNIST GAINS AND U.S. RESPONSE

A Viet Cong unit displays its equipment to a friendly camera. The presence of AK-47s, mortars and an 82mm recoilless rifle indicate the growing power of insurgent units. Source: Vietnamese People's Liberation Army

By the end of October, more than 153,000 U.S. military personnel were in South Vietnam, 15,207 of them Air Force. Large-scale U.S.-Vietnamese air and ground operations since July had averted a Viet Cong take-over of the country. The communists had suffered 3,000 to 4,000 killed each month, and in November their losses were even heavier. For the first time since 1963 U.S. officials began to feel "optimistic" and Sharp publicly asserted that "we have stopped losing the war."[1]

But victory was not yet on the horizon. North Vietnam countered the U.S. build-up by further escalating its strength in the south. On 21 November Westmoreland alerted Sharp and the JCS to the fact that communist infiltration was at more than twice the rate previously estimated. Relative "force ratios" previously expected to rise to 3.3 to 1 by the end of 1965 in favor of the Vietnamese were down to 2.8 to 1 and threatened to fall to 2.2 to 1 by the end of 1966 even if all U.S. Phase II forces arrived as scheduled.[2]

The Saigon Conference in November.

During another high-level conference in Saigon in late November attended by McNamara and Wheeler, Westmoreland and his aides told top officials that the problems arising from the North's growing involvement were compounded by deepening Vietnamese weaknesses. The armed forces of the Republic of Vietnam were unable to cope with the communist threat and had lost the initiative. The people, in turn, had lost confidence in the Saigon Government's ability to prevent Viet Cong attacks and hold rural areas and lines of communication. Only more U.S. and allied forces could arrest this trend. If these were not forthcoming, the government would become weaker while the odds against success would become even greater.

MAC/V force estimates habitually referred to Viet Cong unit strengths in men. However, a significant proportion of the Viet Cong personnel, including this mortar crew, were women. Source: Vietnamese People's Liberation Army

MAC/V intelligence said the Communist had more than 220,000 men, including 113 combat battalions (85 Viet Cong, 27 North Vietnamese) and political cadres. They could assemble 155 battalions by the end of 1966, by drawing upon about 526,000 males in the south and 1,800,000 males in the North. Logistically, they needed 234 tons a day in the south, and they brought in about 190 through Laos, 25 through Cambodia, and 14 by sea. About 300 tons a day could enter in the seven month dry season and 50 tons a day in the five-month wet season.

Allowing for a possible increase of 30,00 Vietnamese regular forces and Augmentation of paramilitary and allied units, MAC/V concluded that additional U.S. personnel would be needed beyond the last estimate for Phase II. Total U.S. personnel in South Vietnam would rise to 389,544. USAF requirements would include possibly five more fighter squadrons, a C-130 squadron, and a new airfield. In view of communist manpower increases, Westmoreland urged quick approval and accelerated deployment of all the Phase II forces previously approved for planning purposes only, and certain "add-on" logistic units. The manpower goals were as follows:[3]

	Phase I	Phase II	Phase II add-on (also designated IIA)	Total
Army	133,916	82,106	52,000	268,022
Navy	9,905	1,961	200	12,066
Marines	40,770	24,417	0	65,187
Air Force	35,428	4,341	4,500	44,269
Total	220,019	112,825	56,700	389,544

Westmoreland praised highly the B-52 operations, saying they demoralized the enemy, boosted allied morale, and encouraged Vietnamese forces to enter the bombed areas. There were more targets than B-52s could bomb. He asked for

Plans and Operations 1965

simpler approval procedures to assure faster response in hitting targets. He also favored using B-52s to hit targets in Laos.[4]

In Laos, Westmoreland said that under the USAF-Navy Barrel Roll and Steel Tiger programs there had been about 2,700 sorties per month and he required 4,500. Barrel Roll attacks had succeeded in containing some of the infiltration of men and supplies, but the Steel Tiger effort in the southeastern part of the panhandle was less effective primarily because of rigid and tine-consuming restrictions imposed by the Lao Government. Bad weather and some diversion of the effort to the Rolling Thunder strikes against North Vietnam also had affected the program adversely.[5]

The MAC/V commander proposed an operational concept patterned after earlier U.S. experience in South Vietnam to assure more rapid approval for hitting fixed and other targets. U.S. FACs flying 0-1s and familiar with the area would be accompanied by Lao observers with authority to approve strikes. A better communication net would cut the time used by the Lao and U.S. governments for coordinating air activities. With the approval of the Lao government and the U.S. ambassador, William H. Sullivan in Vientiane, some acceleration of air attacks had already begun. To assure support for these strikes, nicknamed "Tiger Hound," Westmoreland asked for the immediate reallocation from use in South Vietnam to use in Laos of 20 Army 0-1s direction-finding, infrared, and other aircraft, and their replacement as soon as possible.[6]

After the conference ended, McNamara announced that while the allied forces had stopped "losing the war" and denied the Viet Cong a victory, Hanoi had made a "clear decision" to both "escalate the level of infiltration and, the level of conflict." Recent infiltration into the south, estimated at 1,500 men per month, would probably rise to 4,500 in the dry season. More U.S.-allied forces would be needed to oppose this build-up and he forecast "a long war."[7]

The defense secretary said he was "immensely impressed" with the effectiveness of the Army's 1st Cavalry Division around Plei Mei in South Vietnam. The concept of increased mobility and firepower had "proven out" and he planned to add another air cavalry division to the U.S. Army [thus foreclosing, it appeared, the USAF hope of limiting further Army heliborne expansion]. On the air war on the North, he reaffirmed U.S. policy of hitting infiltration routes rather than such strategic targets as Haiphong, since it was not the U.S. objective to destroy the Hanoi government.[8]

Air Cavalry in Action.
Source: U.S. Army

Air War – Vietnam

The Follow-Up

In reviewing Phase IIA USAF requirements with Brown in early December, McNamara said that the Air Force's supplemental appropriation request for fiscal year 1966 should provide for about 4,500 more men to support five additional fighter squadrons and one transport squadron, and for the building of two more airfields, one in South Vietnam and the other in Thailand. More O-1 and OV-1 reconnaissance aircraft probably were needed. The BarrelRolland Steel Tiger programs in Laos should be stepped up to 50 and 100 sorties per day, respectively.

The destruction caused by B-52 Arc Light missions could be spectacular.
Source: U.S. Air Force

The B-52 sortie rate should reach 800 sorties per month in about six months. To support this rate, McNamara approved enlargement of Andersen AFB, Guam, directed that Sattahip AB, in Thailand, be improved to accommodate the B-52s, and asked for further study of the need for basing the heavy bombers in Taiwan. He indicated that the SAC airborne alert might be reduced to help attain the higher sortie rate. On 29 November McNamara directed the discontinuance of the SAC airborne alert on 1 July 1966.[9]

The defense secretary said that Phase IIA would also require more U.S., Korean, and Australian ground units. The administration did not contemplate calling up U.S. reserves, and he said that the services should review their contingency capabilities without them. He asked for a "Red Ball" air express system, as Westmoreland had requested, to speed the flow of spare parts for helicopters, tanks, bulldozers, and other equipment. The remaining Deployment details would be worked out at another conference in Honolulu, scheduled for January 1966.[10]

Acting quickly, McConnell proposed on 5 December that the JCS agree to the new B-52 sortie rate. He said that about 70 bombers would be needed. He suggested that the JCS recommend basing some of the bombers on Kung Kuang AB, Taiwan. On the same day he directed the Military Air Transport Service to establish a "Red Ball" express and the first flights began on the 7th.[11]

After meeting with the President and other officials, McNamara on 11 December approved a speed-up in the deployment of the specific units requested by Westmoreland. Some additional Army units arrived by the end of the month. Except for some logistic units and four USAF tactical fighter Squadrons, virtually all Phase I elements scheduled for 1965 had reached their Southeast Asia and Western Pacific destinations.[12]

Plans and Operations 1965

Before December ended, however, there were prospects that still higher manpower goals might be set than had been contemplated by Westmoreland. On the 16th Sharp sent the JCS a plan for a further increase in Phase IIA goals for South Vietnam. His proposal called for the following manpower totals:

	Phase I	Phase II	Phase IIA	Total
Army	133,400	82,500	77,100	293,000
Navy	11,300	3,400	7,800	22,500
Marines	40,700	22,000	8,300	71,000
Air Force	35,400	4,500	15,500	55,400
Allied	21,100	--	23,500	46,600
Total	241,900	112,400	132,200	486,500

He also asked for 169,000 other U.S. military personnel to provide direct and indirect support of the war in other areas of Southeast Asia and the Western Pacific. This would raise the total number of combat and support personnel to about 655,500 by the end of 1965.

USAF Phase IIA requirements would rise from one troop carrier and 5 tactical fighter squadrons to 13 tactical fighter and 2 troop carrier squadrons. There would be 65 reconnaissance aircraft. Sharp also asked for the quick deployment of the remaining USAF Phase I squadrons and some Phase II and IIA squadrons in the first quarter of calendar year 1966. If directed, the Air Staff thought it could fulfill the higher USAF goals by drawing on its world-wide manpower resources and by transferring aircraft from Europe. However, it continued to believe that, if the wraps were taken off air and naval power, the deployment of such large ground forces would be unnecessary.[13]

The Air War in December

The number of U.S. combat sorties in South Vietnam in the last month of 1965 remained high. The ground war featured U.S. Marine Corps-Vietnamese operation "Harvest Moon" from 8 to 18 December in Quang Tin province. Combined air and ground action killed an estimated 400 communists and wounded 100. The Air Force believed, however, that planning for close air support in this operation was inadequate. Because a USAF liaison officer was not included in the initial planning, air-ground coordination was poor during the operation. USAF FACs, who were familiar with the area, were not asked to support Vietnamese Army or U.S. Marine units until an emergency arose. Nor were they given sufficient credit in U.S. Army and Marine after-action reports. The operation highlighted the difficulties of fighting with two distinct systems of air control, and the 2d Air Division again recommended adoption of a single, unified, tactical air control system.[14]

B-52s flew 307 sorties in December, including three close air support missions on the 12th, 13th, and 14th for the Harvest Moon operation. By year's end, 1,572 sorties had been flown. Although controversial when first employed on 18 June, the military value of super-bomber strikes was now highly praised by the Air Staff as well as by McNamara and Army and Marine Corps field commanders.[15]

Meanwhile, in connection with proposals to increase B-52 capability by moving the aircraft to either Thailand, the Philippines, or Taiwan, the State Department on 15 December emphasized the serious obstacles. In Thailand, bases were overcrowded and it was considered best not to raise the issue until the expansion of Sattahip AB was completed in 12 to 15 months. In the Philippines, the government was new and many Filipino congressmen were opposed to a plan to send troops to Vietnam. It appeared desirable not to broach the subject until the government had dispatched its "task force," possibly in March or April 1966. In Taiwan, the presence of B-52s would create the "serious risk of Chinese communist reaction against the island."[16]

F-105Ds were the backbone of the Rolling Thunder strikes. Source: U.S. Air Force

Rolling Thunder strikes against North Vietnam were maintained at a high rate until 24 December when a bombing truce began. On 1 December, about seven million leaflets were dropped over the North in the largest single leaflet operation of the war. On the 9th, 150 U.S. aircraft, 115 of them USAF, hit numerous targets in the largest single strike operation to that date. By the 24th, USAF aircraft had flown 26,154 sorties, of which 10,750 were strike and armed reconnaissance. Navy aircraft compiled a slightly higher total.

Air operations remained under many important restraints. Exempt from attack were the Hanoi-Haiphong area, airfields, and most SA-2 sites. Specifically, no strikes were allowed within 30 nautical miles of Hanoi, 10 of Haiphong, 25 of the Chinese border from the coast to 106 degrees East, and 30 of the Chinese border from 106 degrees east to the Laos border. Targeting was planned in advance for a two-week period and operations were under tight Washington control.[18]

Most of the 84 USAF aircraft lost in combat during 1965 were downed by antiaircraft fire. At least 56 SA-2 sites, 8 installations, and 1 support facility had been found in the North, the "threat" had diminished somewhat by year's end. Although 125 SA-2 missiles were observed in flight (presumably, many others were not observed), they had downed only 5 USAF (2 F-4Cs and 3 F-105s) and 5 Navy aircraft. Much of the effectiveness of these Soviet-built missiles had been nullified by their relatively poor guidance system, U.S. electronic countermeasures, and the evasive tactics of U.S. pilots. On the other hand, they forced pilots to fly in lower

Plans and Operations 1965

to hit targets, making them more vulnerable to ground fire and thus indirectly increasing U.S. losses.[19]

A year-end CIA-DIA analysis of the air attacks on the North since they began on 7 February 1965 indicated that they had inflicted about $28.5 million worth of damage. Despite strikes on a few key targets such as six electric power plants constituting 27 percent of total national capacity, the North's economy showed no sign of disintegrating. Economic life was disrupted [but] not crippled, and Hanoi's ability to supply communist forces in South Vietnam and Laos had not been reduced. In fact, the transportation system appeared to have carried as much tonnage in 1965 as in 1964, and there was less evidence of shortages than earlier in the year. The North Vietnamese had proved very resourceful. They had also received greater quantities of aid from the Soviet Union, Hungary, East Germany, Romania, and China. December witnessed a new high in imports. The bombing pause, beginning on 24 December and extending into January 1965, did not change the pattern of infiltration, training, and repair of communication lines. It enabled the North Vietnamese to move their supplies in daytime as well as at night.[20]

AC-47s proved their worth supporting friendly forces. Source: U.S. Air Force

In Laos, AC-47s were used for the first time on 1 December, and on the 10th, B-52s conducted their first Laotian strike. On the same day, Westmoreland, to help step up attacks in that country, delegated to General Memore, 2d Air Division commander, complete responsibility for planning. coordinating, and executing all USAF-Navy air operations.

In December, a new program, Tiger Hound, was added to the three (Yankee Team, Barrel Roll, and Steel Tiger) already being conducted in Laos. Tiger Hound missions, which began on a limited basis on the 5th, featured the use of U.S.-piloted 0-1s for visual reconnaissance and Forward Air Control and for airborne

Air War – Vietnam

command posts. The rules permitted unlimited armed reconnaissance along motor roads in a specified area in the panhandle, but allowed air strikes on targets of opportunity only within 200 yards of all other roads. Beyond this distance and outside the specified areas fliers could attack only targets approved previously or marked by Lao FACs as soon as they began these duties. Infiltration trails or way stations could not be attacked and napalm could not be used. When additional air resources became available, this program would receive the most emphasis.[21]

At The End Of The Year: The Air Force View

As 1965 neared its end, U.S.-Vietnamese officials were preparing to deal with the anticipated "ceasefire offensive" of the communists during Christmas and the Tet lunar holiday in January. An agreement was reached for a short truce on Christmas. A Tet policy was more difficult. The U.S. mission council in Saigon favored stopping only the ground war, not the air attacks in South and North Vietnam and Laos. The State Department in Washington, however, called for a suspension of all air and ground activity in South and North Vietnam. The Air Staff strongly supported the position adopted by the JCS on 27 December which opposed a "stand-down" of the war for Tet similar to one adopted for Christmas.[22]

Events overtook the recommendation. During Christmas there was a 30-hour truce in the fighting in South Vietnam (marked by many Viet Cong violations) and a suspension of bombing in the North. At the end of this period, fighting resumed in the South but the bombing pause in the north continued because President Johnson had undertaken a major peace offensive that was still continuing as the new year began. As part of the peace offensive, U.S. forces began applying more military pressure on the Communists in both South Vietnam and Laos.[23]

At the close of 1965, the United States had 184,314 military personnel in South Vietnam, 20,620 of then in the Air Force. An additional 14,177 military personnel were in Thailand, including 9,117 Air Force. There were 719 USAF aircraft in the two countries, including 15 tactical fighter squadrons (F-4Cs, F-100s, F-105s), 8 air commando squadrons (A-1Es, AC-47s, U-10s, C-47s C-123s) and 57 reconnaissance aircraft (RF-4Cs, RB-66s, RB-57s, RF-101s). Many backup units were in the Philippines, Okinawa, Taiwan, and Japan. There was also a formidable array of Army, Navy, and Marine Corps strength.[24]

The RF-101 fleet conducted long-range photographic reconnaissance missions and brought back imagery other aircraft could not equal.

Source: U.S. Air Force.

Plans and Operations 1965

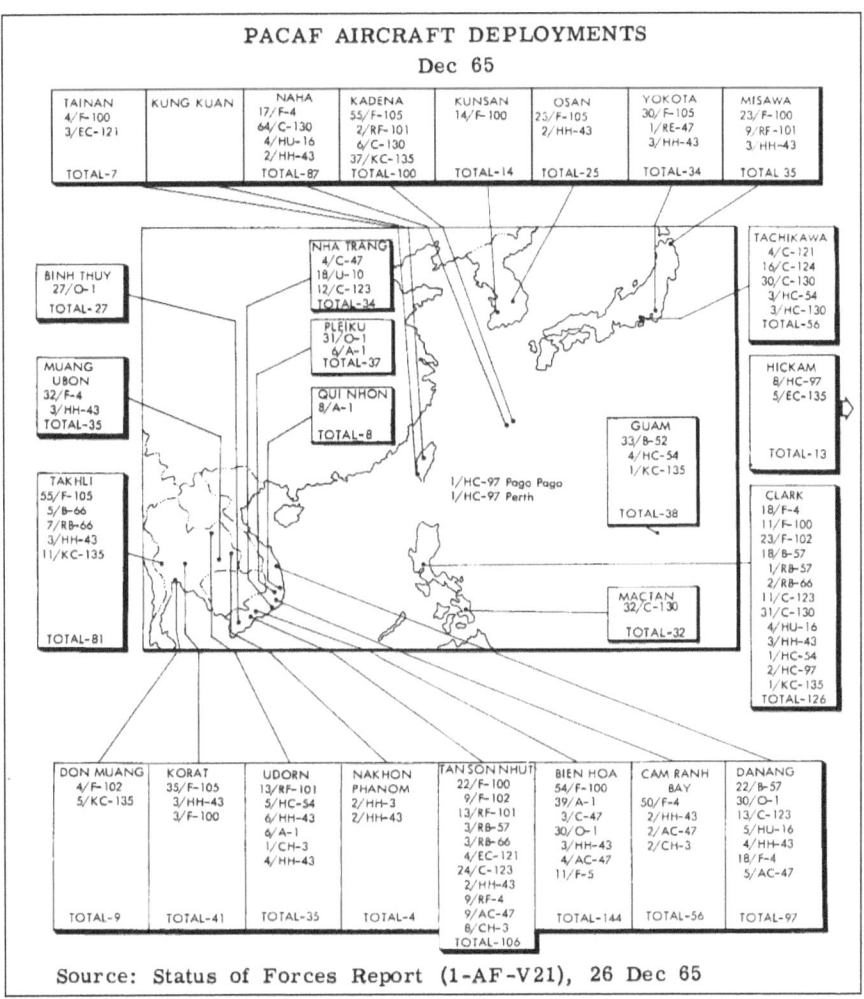

Source: Status of Forces Report (1-AF-V21), 26 Dec 65

The progressive build-up of U.S. power reflected the continuing military crisis in South Vietnam. With the friendly Vietnamese effort diminishing and the Viet Cong-North Vietnamese forces growing in size and aggressiveness, only a basic change in U.S. assistance during the year from a largely advisory and support mission to open combat operations had saved the Saigon Government from certain defeat. The rising tempo of the war was reflected in the combat statistics. South Vietnamese forces lost 11,333 killed, the Communists 36,925. U.S. operational and advisory losses for the year were 1,389 killed in action, of which 43 were Air Force. U.S. wounded in action totaled 5,984, of which 155 were Air Force.[25]

Although victory was not yet in sight, the services agreed that U.S. and allied forces had prevented a communist take-over of the country. However, they disagreed on the merits of the strategy followed in 1965 and planned for 1966. McConnell and the Air Staff, gravely concerned about the trend of the war, believed the failures of the past stemmed largely from a desire of the United States to achieve its objectives with small risks and minimum commitment. With the country now faced with the problem of spiraling military requirements, the Air Force foresaw the need for national mobilization to support a ground-oriented war of attrition in the south, while in the air campaign in the North, the United States would suffer the loss of expensive aircraft engaged in striking mostly insignificant targets.

The implications of the conflict were serious: the American People, faced with fighting a long war that would cost more than the Korean Conflict, might despair of victory, and the war itself could end in a stalemate. Meanwhile, the United States was reducing the amount of military power that it might need to apply in Europe and other areas where contingencies might arise. Insurgencies might occur elsewhere than Southeast Asia as the Communists became convinced they could wage "wars of liberation" without undue risk. There was also the possibility that Communist China might intervene directly in the Southeast Asia conflict.

The Air Force believed its position on the war had been consistent. Instead of a piecemeal build-up and a gradual application of military power that probably could neither gain national objectives in South Vietnam nor deter the Chinese, the United States should focus on North Vietnam, the source of the insurgency. It should employ quickly substantial air and Naval forces against primary targets such as fuel sites and facilities, power plants, and war industries, and conduct heavier interdiction strikes of roads, railroads, and canals. Although other service chiefs in varying degree and at different times had supported these views, McConnell was the only JCS member who believed that the United States should not deploy considerably more ground forces in South Vietnam until the North was isolated by air and naval power. This also placed him in disagreement with Admiral Sharp and General Westmoreland.

The Army consistently argued for more ground troops and near the end of 1965 Westmoreland proposed a total U.S. commitment of 389,544 men in South Vietnam. As 1966 began, the Marine Corps maintained that at least 500,000 troops would be needed in the South for at least five years, an estimate initially made 18 months earlier. The Navy believed that at least 600,000 men were needed and that delay in building up to this total would only increase U.S. casualties. But McConnell doubted whether even this number could drive all the communists out of South Vietnam and keep them out.[26]

The administration clearly was committed to using more ground troops, however, and to restricted bombing of North Vietnam. Testifying before a House committee early in 1965, McNamara asserted that it was his "strong personal opinion" that the United States could not end the war solely by bombing the north, even to the point of obliterating for all practical purposes the entire country. Most of the arms and ammunition used by the Viet Cong and the North Vietnamese, he observed, came from other communist nations. Wheeler supported him, saying that both the current concept for the war and Westmoreland's latest proposals for raging it were "correct." Thus the Air Force stood largely alone in its view of how best the war in Southeast Asia should be brought to a conclusion.[27]

Plans and Operations 1965

APPENDIX 1

U.S. Military Personnel In Vietnam and Thailand - 31 Dec 1965			
	Vietnam	Thailand	Total
Air Force	20,620	9,117	19,737
Army	116,755	4,765	121,520
Navy	8,446	185	8,631
Marine Corps	38,190	40	38,230
Coast Guard	303	0	303
Total	184,314	14,107	198,421
SOURCE: USAF Mgt Summary, SEA, 25 Mar 66, p 81 (S)			

APPENDIX 2

U.S. Casualties, 1 Jan 61 to 3 Jan 1966				
Deaths Due To Hostile Action				
	Vietnam	Laos	Thailand	Total
Air Force	121	8	0	129
Army	1,096	5	0	2,001
Navy	77	1	0	78
Marine Corps	349	0	0	349
Total	2,643	14	0	2,557
Deaths Due To Non-Hostile Action				
Air Force	81	1	27	109
Army	232	0	0	232
Navy	30	0	0	30
Marine Corps	124	0	0	124
Total	467	1	27	495
SOURCE: USAF Mgt Summary, SEA, 7 Jan 66, p 8 – 9 (S)				

APPENDIX 3

U.S. Aircraft Inventory - 3 Jan 1966				
	Vietnam	Thailand	On Carriers	Total
Air Force	514	205	0	719
Army*	1,614	0	0	1,614
Navy-Marine	332	0	223	555
Total	2,460	205	223	2,888
* Composed of 369 fixed-wing and 1,245 rotary				
SOURCE: USAF Mgt Summary, SEA, 7 Jan 66, p 19 (S)				

APPENDIX 4

USAF Combat and Combat Support Sorties In South Vietnam 1965			
Month	Sorties*	Month	Sorties*
Jan	8,523	**Jul**	14,259
Feb	8,714	**Aug**	15,634
Mar	10,840	**Sep**	17,943
Apr	10,911	**Pct**	18,041
May	12,970	**Nov**	21,127
June	12,690	**Dec**	22,004
		TOTAL	173, 656
* Includes 30 B-52 sorties first flown on 18 Jun 1965 and subsequent sorties			
SOURCE: USAF Mgt Summaries, 30 Jul 65 p 16 and 21 Jan 66, p 25 (S)			

Plans and Operations 1965

APPENDIX 5

Month	Bomber sorties scheduled	Bomber sorties over target	Bombers Lost
B-52 Sorties - 1965			
Jun*	30	27	2
Jul	149	147	
Aug	177	165	
Sep	327	322	
Oct	297	294	
Nov	312	310	
Dec	315	307	
Total	1,607	1,572	2

* The first sortie was flown on 18 June. Two bombers were lost in an air refueling mishap

SOURCE: Daily Staff Digest No.17 Hq USAF, 25 Jan 66 (C)

APPENDIX 6

USAF Combat and Combat Support Sorties In North Vietnam 1965			
Month	Sorties*	Month	Sorties*
Jan	0	Jul	2,342
Feb	52	Aug	3,136
Mar	390	Sep	4,141
Apr	1,468	Pct	3,486
May	2,819	Nov	3,330
June	2,358	Dec	2,632
		TOTAL	26, 154

* First Navy strike made on 7 Feb 65; first USAF strike on 8 Feb 65

SOURCE: Data Control Br, Sys Div, Dir of Ops, DCS/P&O (S)

APPENDIX 7

Total USAF Sorties In Laos - 1965			
Month	Sorties in Northern Laos	Sorties in Southern Laos	Total
Jan	56		56
Feb	310		310
Mar	426		426
Apr	1,042		1,042
May	739		739
Jun*	560		560
Jul	957		957
Aug	664		664
Sep	871		871
Oct	509	291*	800
Nov	527	774	1,301
Dec	694	1,392	2,086
Total	7,355	2,457	9,812
* First breakdown between sorties in northern and southern Laos			
SOURCE: Data Control Br, Sys Div, Dir of Ops, DCS/P&O (S)			

Plans and Operations 1965

APPENDIX 8

	USAF Combat Losses In South East Asia 1965				
Month	South Vietnam	North Vietnam	Laos	Thailand	Total
Jan	2	0	2	0	4
Feb	1	0	1	0	2
Mar	2	7	0	0	9
Apr	2	7	2	0	11
May	1	5	2	0	8
June	7	6	1	0	14
July	11*	9	1	0	**21**
Aug	4	9	0	0	**13**
Sep	6	15	1	0	**22**
Oct	9	10	0	0	**19**
Nov	11**	9	0	0	**20**
Dec	10	7	1	0	**18**
Total	66	84	11	0	**161**

* Three C-130s and three F-102s destroyed on the ground by Viet Cong mortar attack
** Five O-1Es destroyed on the ground by Viet Cong mortar attack

SOURCE: Data Control Br, Sys Div, Dir of Ops, DCS/P&O (S)

Air War – Vietnam

\multicolumn{6}{c}{USAF Operational Losses In South East Asia 1965}
Month	South Vietnam	North Vietnam	Laos	Thailand	Total
Jan	1	0	0	1	2
Feb	1	0	0	0	1
Mar	3	0	0	0	3
Apr	1	0	0	1	2
May	13*	0	0	2	15
June	7**	1	0	0	8
July	4	0	1	0	5
Aug	5	0	0	1	6
Sep	5	0	0	2	7
Oct	5	0	0	0	5
Nov	3	0	0	1	4
Dec	4	0	0	2	6
Total	52	1	1	10	64

* Ten B-57s destroyed on the ground by accidental bomb explosion
** Two B-52s destroyed in accidental mid-air collision on first mission

SOURCE: USAF Mgt Summaries, 30 Jul 65 p 16 and 21 Jan 66, p 25 (S)

Plans and Operations 1965

APPENDIX 9

USAF Aircraft Losses in Southeast Asia 1 Jan 62 – 3 Jan 66			
Type Aircraft	Hostile Losses	Other Operational Losses	Total
A-1E	21	13	34
B-26	9	1	10
B-52	0	2	2
B-57	14	13	27
C-47	1	2	3
C-123	4	12	16
C-130	4	4	8
CH-3C	1	0	1
F-4C	12	1	13
F-5	1	0	1
F-100	23	5	28
F-102	4	1	5
F-104	2	3	5
F-105	61	8	69
HU-16	0	1	1
H-43	3	0	3
KB-50	0	1	1
O-1E/F	23	13	36
RB-57	1	0	1
RB-66	1	0	1
RF-101	10	0	10
T-28	13	3	16
U-10	1	1	2
AC-47	2	0	2
TOTAL	211*	84**	295
* Includes 17 destroyed on the ground; ** Includes 10 destroyed on the ground			
SOURCE: USAF Mgt Summary SEA, 7 Jan 66, p 19 (S)			

Air War – Vietnam

APPENDIX 10

Total U.S. Aircraft Losses In South East Asia - to Jan 66			
Month	Hostile Action	Other Operational Losses	Total
Air Force	211*	84****	295
Army	73**		73
Navy-Marine	139***	79	218
TOTAL	423	163	586

* Includes 17 destroyed on the ground
** Consisted of 8 fixed-wing and 65 rotary. In addition, Army claimed 220 fixed wing and 769 damaged
*** Includes 22 destroyed on the ground
**** Includes 10 destroyed on the ground

Air Force data as of 3 Jan 66, Navy and marine data as of 5 Jan 66. Army data as of 31 Dec 65,

SOURCE: USAF Mgt Summary SEA, 7 Jan 66, p 19 (S) and 2 Mar 66, p 20 (S)

APPENDIX 11

Total Vietnamese Air Force Aircraft			
Type	Squadrons	Number Possessed	
A-1E	6	134	
C-47	1	33	
O-1A/U-17	4	107*	
O-6	(Flt)	9	
CH/UH-34	4	60	
TOTAL		343	

* Does not include 16 U-17s assigned to the 12th School Squadron

SOURCE: USAF Mgt Summary SEA, 7 Jan 66, p 87 (S)

APPENDIX 12

Vietnamese Air Force Combat Sorties 1965			
Month	Sorties	Month	Sorties
Jan	7,291	Jul	7,805
Feb	6,312	Aug	10,184
Mar	7,899	Sep	10,179
Apr	8,405	Pct	8,110
May	9,940	Nov	8,616
June	8,037	Dec	9,727
		TOTAL	102,505
SOURCE: USAF Mgt Summaries, 23 Jul 65, p 33 (S) and 7 Jan 66 p 64 (S)			

APPENDIX 13

Vietnamese Air Force Combat Losses 1962 - 65			
Type	Hostile Action	Accident	Total
Total aircraft	49	64	113
SOURCE: USAF Mgt Summary, 11 Feb 66, p 65 (S)			

APPENDIX 14

Type	South Vietnam	Viet Cong*
Strength	651,885**	229,757
Killed	11,333***	36,925***
Desertions	113,462***	11,000

* Includes North Vietnamese units
** Includes all regular, paramilitary and special forces and police
*** During 1965

SOURCE: Hq MAC/V Comd Hist, 1965, pp 268, 270, 272 and 283; N.Y. Times 24 Feb 66.

PART FOUR

AIR OPERATIONS 1966

Air War – Vietnam

Air Operations 1966

I. OBJECTIVES OF THE AIR WAR AGAINST NORTH VIETNAM

From its inception, the "out-of-country" air campaign in Southeast Asia, that is, against targets in North Vietnam and Laos, was limited in scope and objective. The first air strikes against North Vietnam were conducted on 5 August 1964 by Navy aircraft in retaliation for Communist attacks on U.S. ships in the Gulf of Tonkin. The next ones occurred on 7-8 and 11 February 1965 when USAF and Navy aircraft flew "Flaming Dart" I and II missions in retaliation for Viet Cong assaults on U.S. military bases in South Vietnam. These were followed by an air program against selected North Vietnamese targets in order to exert, slowly and progressively, more military pressure on the Hanoi regime. Designated "Rolling Thunder, " it began on 2 March 1965. As explained by Secretary of Defense Robert S. McNamara, the air attacks had three main purposes: raise South Vietnamese morale, reduce the infiltration of men and supplies to South Vietnam and increase its cost, and force the Communists at some point to the negotiating table.

Background to Rolling Thunder

The Rolling Thunder program was basically a USAF-Navy air effort but included occasional token sorties by the Vietnamese Air Force (VNAF). Adm. U.S. Grant Sharp, Commander-in-Chief, Pacific (CINCPAC), Honolulu, exercised operational control through the commanders of the Pacific Air Forces (PACAF), the Seventh Fleet, and the Military Assistance, Command, Vietnam (MACV). Co-ordination control was assigned to the PACAF commander with the tacit understanding that it would be further delegated to Maj. Gen. Joseph H. Moore, Jr., commander of the 2d Air Division (predecessor of the Seventh Air Force) in South Vietnam. Both the Air Staff and the PACAF commander considered this arrangement inefficient, believing that air assets in Southeast Asia, with few exceptions, should be under the control of a single Air Force commander.

With the air program carefully circumscribed, the North Vietnamese initially enjoyed extensive sanctuaries. These included the Hanoi-Haiphong area and the northeastern and northwestern portions of the country closest to China. Targets were selected by the Joint Chiefs of Staff (JCS) after considering the recommendations of Admiral Sharp and the MACV commander, Gen. William C. Westmoreland, the decisions being based on intelligence from the war theater and in Washington. The Secretary of Defense reviewed the recommendations and then submitted them to the President for final approval. Special targeting committees performed this vital task. [2]

Rolling Thunder at first was characterized by individually approved air strikes but, as the campaign progressed, the high authorities approved one- and two-week target "packages" in advance and also gradually expanded the bombing area. In August 1965 they narrowed North Vietnam's sanctuaries to a 30-nautical mile radius of Hanoi, a 10-nautical mile radius of Haiphong, a 25-nautical mile "buffer"

Air War – Vietnam

near the Chinese border extending from the coast to longitude 106° E. and a 30-nautical mile buffer from longitude 106° E. westward to the Laos border. By early September armed reconnaissance sorties had reached a rate of about 600 per week and did not rise above this figure during the remainder of the year. There was a reduction in the number of fixed targets that could be hit (however the list of 220 fixed targets as of 20 September was not reduced) and no extension of the bombing area. Poor weather contributed to the static sortie rate after September.[3]

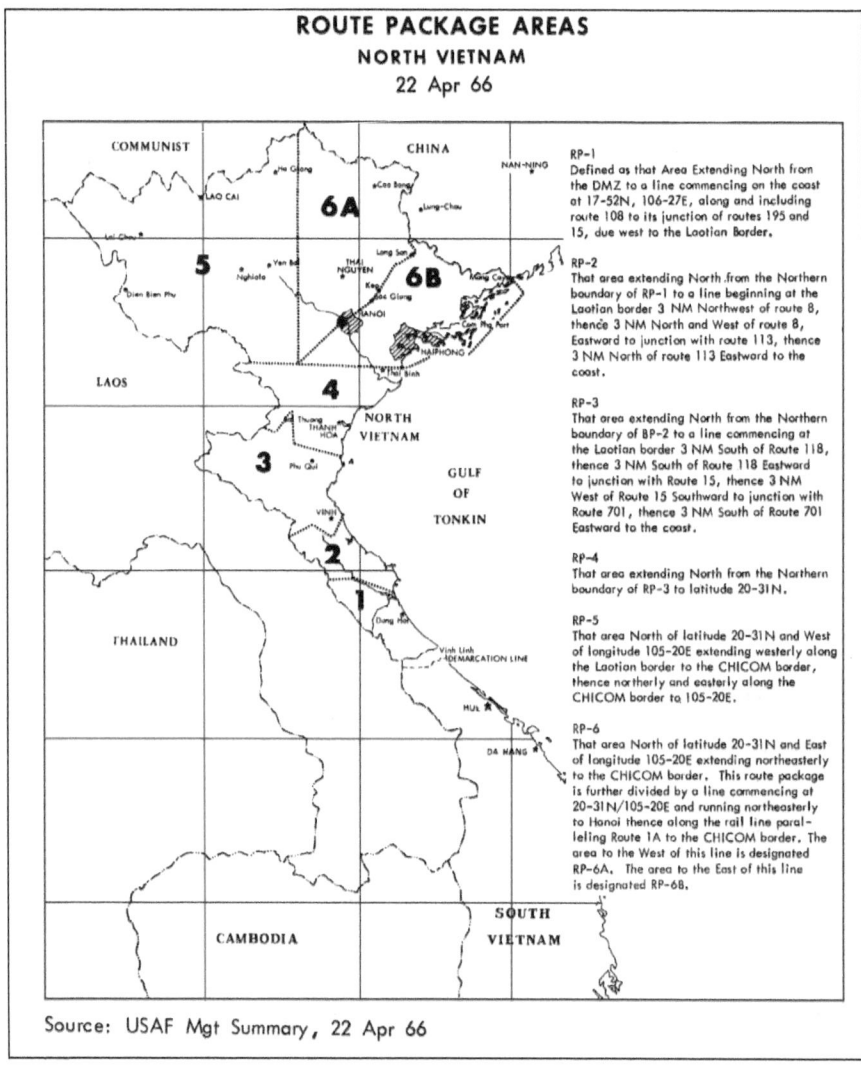

In November 1965, there was an important change in bombing procedure when Admiral Sharp, at the Navy's request, divided North Vietnam into six principal "route packages." Each included lines of communication (LOCs) and other

Air Operations 1966

targets suitable for armed reconnaissance strikes and were to be assigned to the Air Force or Navy for a two-week period, the duration of specific Rolling Thunder programs at that time. (Service air strikes against fixed JCS-numbered targets were excepted and took precedence over armed reconnaissance operations. Starting 10 December, the Air Force began armed reconnaissance flights in route packages II, IV and V, and the Navy in route packages I and III. With variations, the rotation policy continued until April 1966. General Moore, commander of the 2d Air Division, was dissatisfied with this split system of air responsibility. He felt it continues to forfeit the advantages of centralized air control under which the complementing capabilities of Air Force and Navy aircraft could be better coordinated.[4]

On 24 December 1965, the Americans began a two-day Christmas bombing pause in the air campaign against the North which eventually grew into a 37-day moratorium as the U.S. government made a major effort to find a basis for negotiating an end to the war. The limited bombing of targets in Laos and the air and ground war in South Vietnam continued, however.[5]

The Air Force and JCS Urge Early Renewed Bombing

Both the Air Staff and the USAF Chief of Staff, Gen. John P. McConnell, were deeply troubled by the bombing moratorium. Testifying before Senate committees early in January 1966, General McConnell observed that it enabled Hanoi to move men, supplies, and equipment around the clock and to restore its lines of communication. A delay in resuming attacks could prove costly in lives. Concerned about the relative ineffectiveness of the 1965 bombing effort, he favored removing political restraints on the use of air power to allow heavier strikes before a major U.S. and allied force buildup, then under consideration by the administration, was approved. He thought that the military effort against North Vietnam should have a priority equal to that given by the administration to the war in the South.[6]

Other service chiefs supported General McConnell's recommendations to resume and intensify the bombing of the North. On 8 January 1966 they informed Secretary McNamara that the bombing pause was greatly weakening the U.S. negotiating "leverage" and proving advantageous to Hanoi, permitting it to reconstitute its forces and continue infiltration through Laos into South Vietnam. They recommended renewed bombing 48 hours after a Soviet delegation, then in Hanoi, returned to Moscow. Concerned about a possible Communist misinterpretation of U.S. resolve, the Joint Chiefs wanted to insure that any peace negotiations were pursued from a position of strength.[7]

After a Central Intelligence Agency (CIA) and Defense Intelligence Agency (DIA) analysis confirmed that the 1965 bombings had failed to halt the resupply of Communist forces, the JCS prepared another recommendation for Secretary McNamara. On 18 January it urged, again in accordance with General McConnell's view, that the bombing moratorium end with a "sharp blow" followed by expanded air operations throughout the North. It suggested reducing the "sanctuary" areas to a 10-nautical-mile radius of Hanoi and Phuc Yen airfield, a 4-nautical mile radius of Haiphong, and a 20-nautical-mile "buffer" zone in the northeast and northwest areas near the Chinese border. The JCS also called for closing the major seaports (by mining) and removing other political restraints against striking important targets.[8]

On 25 January, in answer to a query from Secretary McNamara, the JCS proposed three alternate ways to resume the bombing. One would use all Thai-based USAF aircraft and planes from three Navy carriers, flying 450 sorties per day for 72 hours, hitting all land and water targets (vehicles, ferries, pontoon bridges, etc.) outside of the sanctuary areas. The second would use the same aircraft flying armed reconnaissance against all LOC and petroleum, oil, and lubricants (POL) targets for 24 to 72 hours with follow-on attacks in accordance with the first alternative. The third called for 600 armed reconnaissance sorties per week in southern North Vietnam with the tempo being increased until the target program recommended on 18 January was reached.[9]

In addition to their proposals to renew the bombing, the Joint Chiefs examined ways to improve air activity. They sent Admiral Sharp guidance on making more effective air strikes against watercraft on inland waterways in the North. Until the bombing halt, more watercraft had been observed as air attacks on the road and rail network had forced the North Vietnamese to rely increasingly on water transportation. The Joint Chiefs concluded that better air-delivered mines should be developed and asked the Chief of Naval Operations (CNO) to give special attention to this matter. [10]

The JCS also examined the problem of closing down the 124-mile rail link between Hanoi and Lao Cai. This and the Hanoi Dong Dang line were the two principal rail arteries to the Chinese border. Secretary McNamara had expressed surprise that the Hanoi-Lao Cai segment was still in service despite repeated air strikes by USAF aircraft before the bombing pause. On 22 January, the JCS chairman, Gen. Earle G. Wheeler responded that there were two reasons why it remained open: frequent aborts because of weather during December 1965 amounting to 37 percent of the planned sorties that month and the arrival of Chinese railway engineering personnel that substantially augmented the North Vietnamese repair capability. Estimates on the size of air defense and repair crews varied widely during 1966. To keep the line closed, said General Wheeler, would require the destruction of three bridges, at least 100 armed reconnaissance sorties per week, and the use of reliable, long-delay bomb fuses and seismic fuse antirailroad mines, both still under development.

Secretary McNamara's Views

The administration moved cautiously toward a decision on whether to renew the bombing of the North. On 19 January, Secretary McNamara informed the Joint Chiefs that their views on this matter were under constant study by the State Department. On the 26th, in a summation of the 1965 Rolling Thunder program, the Defense Secretary told a House subcommittee.[12]

> It was clearly recognized that this pressure, by itself, would not ever be sufficient to cause North Vietnam to move toward negotiation unless it

Air Operations 1966

were accompanied by military action in South Vietnam that proved to the North that they could not win there. These were our objectives then; they are our objectives now. A corollary of these objectives is the avoidance of unnecessary military risk. We, therefore, have directed the bombing against the military targets, primarily routes of infiltration.

We have not bombed Hanoi, we have not bombed Haiphong. We have not bombed certain petroleum supplies which are important. We have not mined the Haiphong port. We have gradually evolved from last February to mid-December, a target system that included all of North Vietnam except certain specified locations.

The targets were very carefully chosen and the rate at which the bombing program grew was very carefully controlled, all for the purpose of trying to achieve our limited objective without widening the conflict.

It was also Secretary McNamara's "strong personal opinion" that the war in South Vietnam could not be won solely by bombing the North and that the northern air campaign should be essentially a "supplement" to military action in the South. [13]

Although the air war was carefully limited, the Defense Secretary informed the President that it had already achieved the objective of raising the cost of infiltration. Air attacks had reduced the amount of enemy supplies reaching the South, carried mostly by trucks over greatly improved routes, from about 400 to 200 tons per day. Moreover, they had diverted 50,000 to 100,000 personnel to air defense and repair work, hampered the mobility of the populace, forced decentralization of government activities thus creating more inefficiency and political risk, and reduced North Vietnam's activities in Laos.

For 1966, Secretary McNamara thought that the bombing "at a minimum" should include 4,000 attack sorties per month consisting of day and night armed reconnaissance against rail and road targets and POL storage sites except in cities and the buffer zone near the Chinese border. He proposed more intense bombing of targets in Laos, along the Bassac and Mekong rivers running into South Vietnam from Cambodia, and better surveillance of the sea approaches. In the South there should be more harassment of enemy LOCs and destruction of his bases.

Recognizing that estimates of enemy needs and capabilities and the results of air action "could be wrong by a factor of two either way, " the Secretary advised the President that unless studies under way indicated otherwise, heavier bombing probably would not put a tight ceiling on the enemy's activities in South Vietnam. However, he thought it would reduce the flow of Communist supplies and limit the enemy's flexibility to undertake frequent offensive action or to defend himself adequately against U.S. allied, and South Vietnamese troops. Mr. McNamara suggested two possible by-products of the bombing effort: it should help to condition Hanoi toward negotiation and an acceptable end to the war and it would maintain the morale of the South Vietnamese armed forces. The defense chief also outlined for the President the 1966 military objectives for South Vietnam.14

These objectives were formalized during between a meeting between President Johnson, and South Vietnamese Prime Minister, Nguyen Cao Ky at Honolulu from 6 to 8 February. They agreed to try to: (i) raise the casualty rate of Viet Cong-North Vietnamese forces to a level equal to their capability to put new men in the field; (2) increase the areas denied to the Communists from 10 to 20 percent to 40 to 50 percent; (3) increase the population in secure areas from 50 to 60 percent; (4) pacify four high-priority areas containing the following population: Da Nang, 387,000; Qui Nhon, 650,000; Hoa Hao, 800,000, and Saigon, 3, 500,000; (5) increase from 30 to 50 percent the roads and rail lines open for use; and (6) insure the defense of all military bases, political and population centers, and food-producing areas under the control of the Saigon government

The Bombing Resumes and Further Air Planning

Having received no acceptable response from Hanoi to his peace overtures, President Johnson on 31 January ordered resumption of the bombing of North Vietnam. It began the same day. "Our air strikes from the beginning," the President announced, "have been aimed at military targets and controlled with great care. Those who direct and supply the aggression have no claim to immunity from military reply." Other officials told newsmen that the United States would continue to limit bombing of the North but intensify other aspects of the war, including more use of B-52 bombers and ground artillery in South Vietnam.[15]

As anticipated, the bombing moratorium had in fact benefited the North Vietnamese. USAF reconnaissance revealed that supplies had moved by truck and rail 24 hours per day and that repairs and new construction on the road and rail net likewise had proceeded on "round-the-clock" basis. General McConnell believed that the moratorium had permitted the North to strengthen its antiaircraft defenses, including expansion of its SA-2 system from about 50 to 60 sites. Admiral Sharp reported the enemy had deployed about 40 more air defense positions in the northwest rail line area and 26 more guns to protect routes south of Vinh.[16]

When the aerial attacks resumed as Rolling Thunder program 48, allied air strength in South Vietnam and Thailand consisted of about 689 U.S. and 125 Vietnamese Air Force tactical combat aircraft. The number of U.S. tactical combat aircraft by service were: Air Force, 355; Navy (three carriers), 209; and Marine Corps, 125. In addition the Air Force had 30 B-52's in Guam. At that time, North Vietnam possessed about 75 MiGs. More would arrive in subsequent months. The limitations placed on the renewed bombing effort disappointed the Joint Chiefs, especially since none of their recommendations had been accepted, In fact, the program was more restrictive than before the bombing pause. Armed reconnaissance during February was limited to 300 sorties per day and almost solely to the four route package areas south of Hanoi. Only one JCS target, Dien Bien Phu airfield, was hit several times, Poor weather forced the cancellation of many strikes and others were diverted to targets in Laos. A Pacific Command (PACOM) assessment indicated that the renewed air effort was producing few important results as compared to those attained during 1965 against trucks, railroad rolling stock, and watercraft.[17]

Air Operations 1966

Meanwhile, the bombing policy remained under intensive review. At the request of Secretary McNamara, General Wheeler on 1 February asked the service chiefs to establish a joint study group which would examine again the Rolling Thunder program and produce data that could serve as a basis for future JCS recommendations. They quickly organized the group under the leadership of Brig. Gen. Jammie M. Philpott, Director of Intelligence, Strategic Air Command (SAC). Its report was not issued until April. [18]

On 8 February, following a three-week conference of service officials in Honolulu to plan U.S. and allied air and ground deployments through fiscal year 1968, Admiral Sharp and his staff briefed Secretary McNamara on the results of their deliberations. They proposed a program of stepped up air attacks in the North and in Laos with the immediate goal of destroying Communist resources contributing to the aggression, and of harassing, disrupting, and impeding the movement of men and materiel. Admiral Sharp advocated 7,100 combat sorties per month for the North and 3,000 per month for the South. [19]

Secretary McNamara did not immediately respond to these sortie proposals. However, he approved, with certain modifications, CINCPAC's recommended schedule for additional air and ground forces. These deployments promised to strain severely the resources of the services, especially those of the Air Force and the Army. Concerned about, the impact on the Air Force's "roles and missions, force structure, overall posture and research and development needs, Lt. Gen. H. T. Wheless, Assistant Vice Chief of Staff on 18 February directed Headquarters USAF's Operations Analysis Office to undertake a "vigorous" analysis and asked all Air Staff offices to support the effort. Its major purpose was to develop a more comprehensive data base on the use of air power in Southeast Asia. [20]

Because of the decision to deploy more forces and the likelihood of stepped up air and ground operations, General McConnell decided a number of organizational changes were necessary. He directed the Air Staff to replace the 2nd Air Division with a numbered Air Force, upgrade the commander of the Thirteenth Air Force in the Philippines to three-star rank, and formalize USAF-Army airlift arrangements in the theater. [21]

With the air campaign continuing at a low tempo, the JCS, with Air Staff support, reaffirmed its prior recommendation to Secretary McNamara for accelerated air operations against the North and to strike all targets still under administration wraps. If this could not be approved, the JCS urged extending operations at least to the previously authorized areas. The Joint Chiefs warned that if more remunerative targets could not be hit to compensate for the handicaps imposed by operational restraints, more air sorties should be flown elsewhere. They also raised their estimated sortie requirement for the northern campaign from 7,100 to 7,400 per month, citing Admiral Sharp's newly acquired intelligence which confirmed additional enemy deployments of SA-2 missiles and possible Chinese antiaircraft artillery units in the northeast region. 22

Secretary McNamara informed the JCS that the political atmosphere was not favorable for implementing these recommendations, Some Air Staff members

Air War – Vietnam

attributed the administration's cautiousness to the Senate Foreign Relations Committee hearings on the war, which began 4 February under the chairmanship of Senator J. William Fulbright. In addition, the Defense Secretary was known to believe that there were limitations to what air power could do in the type of war being waged in Southeast Asia. Mr. McNamara thought that even the obliteration of North Vietnam would not completely end that country's support of enemy operations in the South since most of the arms and ammunition came from other Communist nations. He firmly believed that the war would have to be won on the ground in South Vietnam.23

Secretary of the Air Force Harold Brown echoed this administration position, asserting publicly on 25 February that the destruction of the North's remaining industrial capacity would neither prevent the resupply of equipment and troops in the South nor end hostilities. He also said:[24]

> "should it appear that we were trying to destroy North Vietnam, the prospect of escalation by the other side would increase, and with it would increase the possibility of heavier U.S. casualties and an even harder and longer war; our objective is not to destroy North Vietnam. It is to stop aggression against South Vietnam at the lowest feasible cost in lives and property. We should take the course that is most likely to bring a satisfactory outcome at a comparatively low risk and low cost to ourselves. Our course is to apply increasing pressure in South Vietnam, both by ground and supporting air attacks; to make it clear to the North Vietnamese and Viet Cong forces, that life is going to get more difficult for them and that war is expensive and dangerous."

Thus, for the time being, the JCS-recommended program for an accelerated air campaign against North Vietnam had no chance of receiving administration approval.

Air Operations 1966

II. INCREASING THE AIR PRESSURE ON NORTH VIETNAM

On 1 March the JCS generally endorsed Admiral Sharp's "Case I" air, ground,. and naval deployment program leading to stepped-up operations against the Communists in North and South Vietnam and Laos. Case I called for deployment of a total of 413,557 U.S. personnel in South Vietnam by the end of calendar year 1966. It also recommended again that the war be fought in accordance with the Concept for Vietnam paper which it had approved on 27 August 1965 and later amended. This paper called for air strikes against the North's war-supporting industries in the Hanoi-Haiphong area, aerial mining of the ports, additional interdiction of inland and coastal waterways, and special air and ground operations in Laos, all recommended many times in various ways. But administration authorities continued to favor a more modest air effort against the Hanoi regime.[1]

Air Operations and Analyses

The new Rolling Thunder program number 49 was ushered in on 1 March. It was still limited to armed reconnaissance of the North but the administration had broadened the authorized attack area to include coastal regions and had eased restrictions to permit the use of air power up to the level existing when bombing ceased on 24 December 1965. The Air Force and Navy were allocated a total of 5,100 armed reconnaissance sorties (and 3, 000 for Laos), with the number to be flown by each contingent on weather and other operational factors. Poor weather, however, limited their sorties to 4,491 during the month. The Air Force concentrated its efforts against targets in route packages I, III, and VIA, the Navy in route packages II and IV and against coastal targets in route package I through IV. The VNAF flew token sorties in route package I under the protection of U.S. Marine Corps electronic and escort aircraft. On 10 March the JCS again pressed for its proposed accelerated air program with early attacks on POL sites, the main rail system running from China, and the mining of deep water ports. Again the recommendation was not acted upon.[2]

Meanwhile, the North's air defense system began to pose a greater threat to USAF and Navy operations. On 3 March photo reconnaissance aircraft discovered about 25 MIG-21 fuselage crates at Phuc Yen airfield near Hanoi. USAF " Big Eye" EC-121D aircraft also detected airborne MiGs about 55 times during March, although there were no engagements. General Curtis E LeMay, former CSAF, first recommended striking the North's airfields on 10 August 1964 and the JCS sent its first recommendation to do so on 14 November 1964. By 1 March 1966 the JCS had made a total of 11 such recommendations but the administration had approved strikes on only three small airfields at Vinh, Dong Hoi, and Dien Bien Phu in May 1965, June 1965, and February 1966. respectively. Admiral Sharp directed the PACAF and Seventh Fleet. commanders to prepare for counter-air operations and

Air War – Vietnam

the SAC commander to submit a plan for a B-52 strike, if necessary, against Phuc Yen and Kep airfields. He asked for additional electronically equipped USAF EB-66 aircraft to reduce the effectiveness of the SA-2 missiles and the anti-aircraft guns. "'Jamming" was thought to have already reduced the usefulness of enemy air defenses.[3]

Aircraft losses to enemy ground fire continued to cause much concern. A Joint Staff study of the problem during March showed that 199 American aircraft had been lost over North Vietnam since the bombings began on 7 February 1965, sixteen of them by SA-2 missiles. The aircraft loss rate was six times higher in the northeast, the most heavily defended area, than in the rest of North Vietnam. Headquarters USAF estimated the North's antiaircraft strength at 2,525 guns.[4] (Estimates of North Vietnam's antiaircraft gun inventory varied considerably during 1966)

To improve its analysis of aircraft losses and other operational data, the Air Staff on 26 March established an *ad hoc* study group in the Directorate of Operations. In the same month the Chief of Operations Analysis, in response to General Wheless' directive of 17 February, completed an initial study on the effectiveness of air interdiction in Southeast Asia;. It summarized the enemy's supply requirements, his capability to transport supplies by land or sea, and the extent air strikes had hampered such activities. One conclusion was that air attacks had not yet decreased the movement of men and supplies from the North through Laos to South Vietnam. They had, however, inflicted about $15 to $16 million direct and $8 million indirect damage on the North's economy and forced Hanoi to recruit 30,000 more personnel, in addition to local forces, to perform repair work. An analysis of one route from Vinh to Muang Phine suggested that air attacks had caused the Communists to increase their truck inventory by one-third and their transport time by two-thirds. 5

Another Operations Analysis interdiction study listed enemy targets destroyed or damaged in North Vietnam and Laos through March 1966 as follows:

	North Vietnam			Laos		
	Destroyed	Damaged	Total	Destroyed	Damaged	Total
Transportation vehicles	1, 537	2,500	4,307	515	485	1,000
LOC network	546	4,381	4,927	398	4,886	5,284
Counter-air	134	189	323	145	67	145
All other	3,681	4,196	7,877	2,783	1,259	3,991
Total	5,898	11, 266	17,164	3,841	6,697	10,426

Air Operations 1966

Transportation vehicles included bridges, road cuts, rail cuts and ferry ships.

Counter-air included aircraft, runways, antiaircraft sites, SA-2 sites, and radar sites.

All other included buildings, POL tanks, power plants, locks and dams.

Concerning the Communist effort to fill craters and repair roads damaged by air attacks, there were indications that only one man-day of direct productive effort per attack sortie was needed to perform this task. "At this rate, " the Operations Analysis study observed, "a few hundred sorties per day would only make enough work for a few hundred men."

As for Communist supplies, the study estimated that in 1965 they averaged 51 tons per day across the North Vietnamese-Laos border and 16 tons per day across the Laos-South Vietnamese border. For 1966 (through March), the figures were 70 and 35 tons respectively. The Laos panhandle infiltration routes in themselves appeared to be capable, despite air attacks, of supporting the current low-level combat by Viet Cong and North Vietnamese forces. To support a higher combat level, for example, one day in seven, the Communists would have to use other supply channels or dip into South Vietnamese stockpiles, either of which would complicate their distribution problems.[6]

The Beginning Of Rolling Thunder Program 50

Concurrently, there was planning for the next Rolling Thunder program. In meetings with General Wheeler on 21 and 23 March, Secretary McNamara set forth certain guidelines for stepping up air strikes in the northeast and hitting additional JCS targets. The Joint Chiefs quickly responded by proposing Rolling Thunder program 50. It called for launching 900 attack sorties against major lines of communication and striking nine POL storage areas, six bridges, one iron and steel plant, one early warning and ground control intercept (EW/GCI) site, and one cement plant, the latter in Haiphong. Admiral Sharp planned to conduct this program within an allocation of 8,100 sorties (5,100 for North Vietnam, 3, 000 for Laos),[7]

Administration authorities approved this program, which began on 1 April. For the first time in 1966 armed reconnaissance was authorized over the far northeast and four new JCS targets (all rail and highway bridges) were cleared for interdiction. However, some time before program 50 ended on 9 July, permission to strike the other JCS-recommended targets was withdrawn. Dissatisfied with the restrictions, General McConnell and the Marine Corps chief jointly advised the JCS that "sound military judgment" dictated that all the targets be hit immediately. Higher administration officials withheld consent, however, principally because of the unstable South Vietnamese political situation which developed after the ruling junta's ouster on 10 March of Lt. Gen. Nguyen Chanh Thi, the I Corps commander.[8]

Poor weather in April again limited the number of attack sorties flown against the North and delayed until 5 May the completion of strikes against the four authorized JCS targets. Other air operations included armed reconnaissance against roads, rail lines, watercraft and similar LOC targets. April also saw several important developments: establishment of the Seventh Air Force, the first B-52 strike in North Vietnam, a marked step-up in Hanoi's air defense effort that resulted in a U.S. downing of the first MiG-21, a change in the command and control of route package I, and the beginning of a study on increasing air pressure to offset civil disturbances in South Vietnam.[9]

The establishment of the Seventh Air Force, effective 8 April, followed General McConnell's successful efforts to raise the stature of the major USAF operational command in the theater. General Moore continued to serve as its chief with no change in his relationship with other commanders. Also, in accordance with General McConnell's wishes, the commander of the Thirteenth Air Force in the Philippines was raised to three-star rank on 1 July.[10]

SAC made the first B-52 strike against the North on 12 April when 30 bombers dropped 7,000 tons of 750- and 1,000-pound bombs on a road segment of [the] Mugia Pass near the Laotian border. It was believed to be the single greatest air attack on a target since World War II. Initial reports indicated that "route 15" had been "definitely closed" by a landslide as had been hoped; however, 26.5 hours later reconnaissance photos showed all the craters filled in and the road appeared serviceable, attesting to the quick repair capability of the North Vietnamese. A second strike by 15 B-52's on 26 April on a road segment six kilometers north of Mugia blocked the road for only 18 hours. The apparent inability of the B-52's to close down the road expressed by the Secretary of State and other officials and a Seventh Air Force report of an SA-2 site near Mugia, prompted Admiral Sharp on 30 April to recommend to the JCS no further attacks on the pass. In fact, the bombers were not again used near North Vietnam until 30 July.[11]

Towards the end of April Hanoi stepped up its air defense activity, dispatching 29 to 31 MiGs against USAF and Navy aircraft. In nine separate engagements in five days, six MiGs were destroyed, all by USAF F-4Cs which suffered no losses. The first MiG-21 was downed on 26 April by two F-4Cs. Antiaircraft fire continued to account for most American aircraft combat losses with 31 downed (14 USAF, 17 Navy), while two F-102 and a Navy A-1H were struck by SA-2 missiles.[12]

Meanwhile, a change in command and control of air operations in route package I followed a meeting on 28 March between Admiral Sharp and the JCS. The PACOM commander recommended that General Westmoreland's request for partial operational control of this area be approved and that the sector be accorded the same priority as for South Vietnam and Laotian "Tiger Hound" air operations. General Westmoreland urgently desired more air power to hit enemy approaches to the battlefield area near the Demilitarized Zone (DMZ) for which he was responsible. Admiral Sharp thought that 3,500 sorties a month was warranted alone for route package I.[13]

USAF commanders and the Air Staff objected to the proposed change, feeling that MACV's command authority should be limited to South Vietnam. They believed that the PACAF commander should remain the sole coordinating authority for the

Air Operations 1966

Rolling Thunder program. Nevertheless, Secretary McNamara approved the change on 14 April and the JCS endorsed it on the 20th. To allay any doubts where he thought the war's emphasis should be, the defense chief said that air operations north of route package I could be carried out only if they did not penalize air operations in the "extended battlefield, " that is, in South Vietnam, the Tiger Hound area of Laos, and route package area I. Under this change Admiral Sharp still retained partial operational control of route package I. General Westmoreland's authority was limited to armed photo reconnaissance and intelligence analysis of Rolling Thunder and "Iron Hand" operations. Simultaneously, the Air Force-Navy rotational bombing procedure in other route packages, in effect since late 1966, also ended. [14]

The civil disturbances and reduced U.S. and allied military activity in both South and North Vietnam that followed General Thi's dismissal prompted the Joint Staff on 14 April to recommend a step-up in the attacks in accordance with the JCS proposals of 18 January. It thought this might help arrest the deteriorating situation. A special Joint Staff study of the problem also examined the possibility that a government coming to power in Saigon might wish to end the war and ask U.S. and allied forces to leave. [15]

The Air Staff generally supported the Joint Staff's recommendation for an intensified air offensive against the North and withdrawal of U.S. forces if a local fait accompli left the United States, and, its allies no choice. But the Army's Chief of Staff doubted that heavier air strikes could resolve the political situation in South Vietnam. Observing that Admiral Sharp already possessed authority to execute some of the recommended strikes, he opposed sending the Joint Staff's study to Secretary McNamara on the grounds that if U.S. strategy was to be reevaluated, it should be by separate action. General McConnell suggested, and the JCS agreed, to consider alternate ways of withdrawing part or all of the U.S. forces from South Vietnam should this be necessary. Reviews were begun but in subsequent weeks, after political stability was gradually restored, the need to consider withdrawal action lessened and no final decisions were taken.

The Rolling Thunder Study of 6 April

April also witnessed the completion of the special joint report on the Rolling Thunder program requested by Secretary McNamara in February. Prepared under the direction of General Philpott, it was based on all data available in Washington plus information collected by staff members who visited PACOM, MACV, the 2d Air Division, and the Seventh Fleet.

Completed on 6 April, the Philpott report reviewed the results of one year of Rolling Thunder operations (2 March 1965 - 2 March 1966). During this period U.S. and VNAF aircraft had flown about 45,000 combat and 20, 000 combat support sorties, damaging or destroying 6,100 "fixed" targets (bridges, ferry facilities, military barracks, supply depots, etc.), and 3,400 "mobile" targets (trucks, railroad rolling stock, and watercraft). American combat losses totaled about 185 aircraft.

Air War – Vietnam

The report touched briefly on Laos where the air effort consisted primarily of armed reconnaissance in two principal areas designated as "Barrel Roll and Steel Tiger. It noted that the effectiveness of USAF strikes in Laos was limited because of small fixed targets, high jungle growth, and mountainous terrain that hampered target location and identification. Also, important targets were normally transitory and had to be confirmed carefully before they could be attacked. The operations in North Vietnam and Laos, said the report:

> have achieved a degree of success within the parameters of imposed restrictions. However, the restricted scope of operations, the restraints and piecemealing effort, have degraded program effectiveness to a level well below the optimum. Because of this, the enemy has received war-supporting materiel from external sources, through routes of ingress, which for the most part have been immune from attack, and has dispersed and stored this materiel in politically assured sanctuaries. Although air operations caused significant disruption prior to the standdown, there has been an increase in the North Vietnamese logistic infiltration program, indicating a much greater requirement for supplies in South Vietnam.

Of a total of 236 "JCS numbered" targets in North Vietnam, 134 had been struck, including 42 bridges. Among the 102 untouched targets, 90 were in the northeast area and, of these, 70 were in the sanctuary zones of Hanoi, Haiphong, and the buffer territory near China. Elsewhere in the North 86 percent of the JCS targets had been hit. The report further asserted:

> The less than optimum air campaign, and the uninterrupted receipt of supplies from Russia, China, satellite countries, and certain elements of the free world have undoubtedly contributed to Hanoi's belief in ultimate victory. Therefore, the Study Group considers it essential that the air campaign be redirected against specific target systems, critical to the capability and important to the will of North Vietnam to continue aggression and support insurgency.

It consequently proposed a three-phase strategy. In Phase I, over a period of four to six weeks, the United States would expand the armed reconnaissance effort over the North except for the sanctuary areas and again attack previously struck JCS-numbered targets in the northeast. Air units also would strike 11 more JCS-numbered bridges, and the Thai Nguyen railroad yards and shops; perform armed reconnaissance over Kep airfield; strike 30 more JCS numbered targets, 14 headquarters/barracks, four ammunition and two supply depots, five POL storage areas, one airfield, two naval bases, and one radar site.

In Phase II, a period of somewhat less duration than Phase I, American aircraft would attack 12 military and war-supporting targets within the reduced sanctuary areas, consisting of two bridges, three POL storage areas, two railroad shops and yards, three supply and storage depots, one machine tool plant, and one airfield. During Phase III all remaining JCS-numbered targets (now totaling 43)would be attacked, including six bridges, seven ports and naval bases, six industrial pl-ants,

Air Operations 1966

seven locks, 10 thermal/hydroelectric plants, the headquarters of the North Vietnamese ministries of national and air defense, and specified railroad, supply, radio, and transformer stations.

Concurrent with this program, the study group proposed three attack options that could be executed at any time: Option A, strike the Haiphong POL center; Option B, mine the channel approaches to Haiphong, Hon Gai, and Cam Pha; and Option C, strike four jet airfields at Phuc Yen, Hanoi, and Haiphong.

Finally, it proposed that Admiral Sharp should determine when to hit the targets in each of the three phases, the weight of the air attacks, and the tactics to be employed.[17]

General Wheeler, who was briefed on the report on 9 April, called it a "fine professional approach," a "good job," and endorsed it. The manner in which it should be sent to Secretary McNamara created difficulties, however. General McConnell suggested that the Joint Staff prepare "positive" recommendations for the implementation of the report's air program, stating that if this were not done, it would not receive the attention it deserved. But strong service support was lacking for that approach. An agreement eventually was reached to send the report to secretary McNamara with the Joint chiefs "noting" it. They advised him it was fully responsive to his request, was in consonance with the JCS recommendations of 18 January 1966, and would be useful in considering future recommendations of the Rolling Thunder program.[18]

Air Operations in May: Beginning of "Gate Guard"

The Rolling Thunder study had no immediate impact on air operations. In fact, Secretary Brown on 22 May publicly affirmed the administration's decision not to expand significantly attacks on new targets. He said such action would not cut off infiltration but would raise the danger of a "wider war". Not stated by Secretary Brown was the fact that civil disturbances in South Vietnam triggered by the dismissal of General Thi on 10 March still prompted the administration not to risk escalation of the war at this time.[19]

Thus the authorized level of 5,100 sorties for North Vietnam remained unchanged in May and only a few important attacks on fixed targets were approved. The principal operation was against seven targets within the Yen Bai logistic center which were struck by 70 USAF sorties. Monsoon weather again plagued the air campaign, causing the cancellation of 2,972 USAF-Navy sorties or about 32 percent of those scheduled. USAF sortie cancellations amounted to 40 percent.

Heavier North Vietnamese infiltration toward the DMZ as indicated by more truck sightings led to a change in tactics. Beginning on 1 May, a special air effort called "Gate Guard" was initiated in the northern part of the Steel Tiger area in Laos and then shifted into route package I when the monsoons hit the Laotian region, utilizing many of the "integrated interdiction" tactics developed in Laos earlier in the year, Gate Guard involved stepped-up air strikes on a series of routes or "belts" running east to west. Many special USAF aircraft were used: C-130 airborne command and control centers, C-130 flare aircraft, EB-66s for ECM, and

Air War – Vietnam

RF-101s. Attack aircraft interdicted selected points in daytime and destroyed "fleeting targets" at night.

During the month there were few MiG sightings and only one was destroyed. Heavy antiaircraft fire accounted for most of the 20 U.S. aircraft (13 USAF, six Navy, one Marine) that were downed. USAF losses included seven F-105's in the northeast. The enemy's ground fire, General McConnell informed a Senate subcommittee during the month, was "the only thing we are not able to cope with." whereas the SA-2's which were deployed at about 103 sites had destroyed only five USAF and two Navy aircraft. The SA-2's were countered by decoys, jamming techniques, and evasive aircraft tactics. Air Force confidence in the value of anti-SA-2 operations was challenged in a Seventh Fleet study, dated 12 July 1966 and based on SA-2 USAF and Navy firing reports. It asserted that the value of ECM and ,other jamming techniques was uncertain as aircraft with deception devices normally sought to evade the missiles when fired upon. [22]

During May the Air Staff began a study effort to establish requirements for a suitable, night, all-weather aircraft interdiction system using the latest munitions, sensors, and guidance equipment to provide an "aerial blockade" against infiltrating men and supplies. This followed an expression of frustration by high State Department and White House officials in late April about the inability of air power to halt these movements into the South. As part of this study, the Air Staff solicited the views of PACAF, SAC, and other commands, advising them of the need for a solution within existing bombing restraints. Recommendations to "strike the source" of Communist supplies, they were informed, were politically unacceptable and likely to remain so. [23]

In a joint reply on 24 May, the commanders-in-chief of PACAF and SAC, Generals Hunter Hamis, Jr. and John D. Ryan, pointed to improved results from air operations in route package I and in parts of Laos. They said that interdiction could become even more effective by greater use of air-delivered mines (against ferries), "denial" munitions with delayed fuses insuring "longevity" up to 30 days, around-the-clock air strikes on selected routes south of Vinh, special strikes against Mugia Pass, and improved air-ground activity in Laos, They also proposed the use of low-volatile chemical-biological agents to contaminate terrain and surface bursts of nuclear weapons. The latter would "dramatically" create "barriers" in areas difficult to by-pass. To implement these measures, General Harris again stressed the need for centralized control of air resources, asserting it should be a "high priority" Air Force objective. But most of these suggestions could not or would not be implemented in the immediate future. [24]

Highlights of June Operations

June witnessed another step-up in air activity over North Vietnam, the major highlight being USAF-Navy strikes, beginning 21 June, against previously exempt POL storage sites and culminating in major POL strikes in Hanoi and Haiphong on the 29th. (See details in Chapter III.)

Other targets continued to be hit, such as the Hanoi-Lao Cai and Hanoi-Dong Dang rail lines, but most USAF sorties concentrated on route package I targets

Air Operations 1966

which absorbed about 93 percent of the total flown in the North that month. These strikes reflected the importance General Westmoreland placed on curbing the flow of enemy troops and supplies toward and into the DMZ, Gate Guard targets were hit hard and, after the introduction of USAF MSQ-77 "Skyspot" radars for greater bombing accuracy, the infiltration "gates" were "guarded" virtually around the clock.

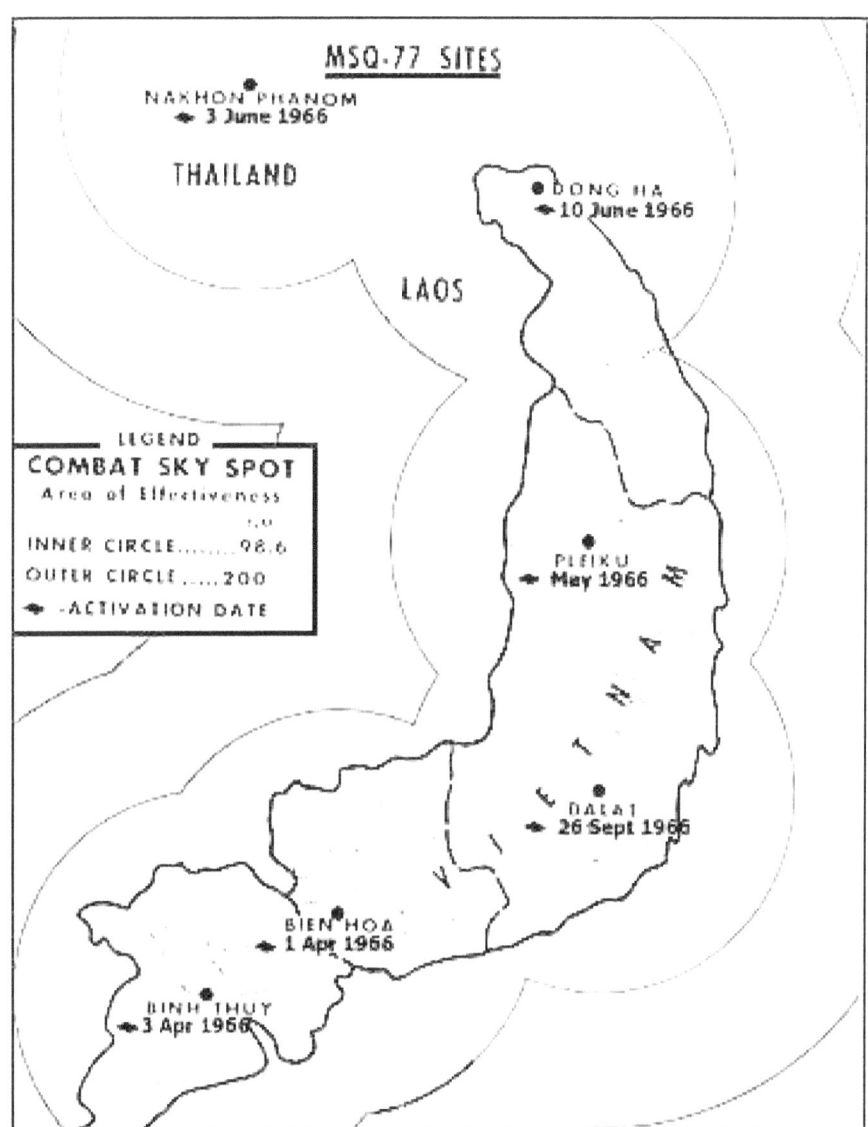

The initial MSQ-77 radar was placed at Bien Hoa, South Vietnam on 1 April 1966, and the second one at Pleiku in May. With the installation of the third and fourth radars at Nakhon Phanom, Thailand and Dong Ha, South Vietnam on 3 and

Air War – Vietnam

12 June, respectively, the system could be used for air strikes in route package I. A fifth radar was placed at Dalat, South Vietnam on 26 September. The MSQ-77 was an MSQ-35 bomb-scoring radar converted into a bomb-directing radar with a range of 200 nautical miles.

About 97 percent of the Navy effort was concentrated along the coast in route packages II, III, and IV. The VNAF flew 266 sorties in route package I, its highest total against the North in 12 months. [25]

The Gate Guard campaign seemed to confirm the value of night air attacks. By 7 July the nighttime missions had achieved better results than those in daytime, 164 trucks being destroyed and 265 damaged compared with the daytime toll of 154 destroyed and 126 damaged. [26]

Despite these successes, Gate Guard operations faced certain handicaps. During daylight hours USAF 0-1 forward air control (FAC) aircraft used to support U.S. strikes were highly vulnerable to the heavy ground fire and, when forced to fly higher, became less effective. Also, interdiction points, often on flat terrain, were easy to repair or by-pass. And the North Vietnamese could store and service their trucks in numerous small villages, secure in the knowledge that U.S. aircraft would not attack civilian areas. Events finally overtook the Gate Guard effort. Continued infiltration through the DMZ prompted Headquarters MACV to develop a "Tally-Ho" air program; a more ambitious effort to block, if possible, a large-scale invasion by North Vietnamese troops through the DMZ into South Vietnam's northernmost province.[27]

Air Operations 1966

III. THE POL STRIKES AND ROLLING THUNDER PROGRAM 51

As indicated, the highlight of the air war and of the Rolling Thunder program since its inception were the POL strikes in June 1966. General McConnell and the other service chiefs had long urged the destruction of North Vietnam's major POL sites but the administration did not seriously consider attacking them until March.

Background of the POL Air Strikes

Some months before, in December 1965, a CIA study had concluded that the destruction of the North's POL facilities would substantially increase Hanoi's logistic problems by requiring alternate import and distributing channels and the use of more rail cars, drums, and other storage items. CIA analysts recognized that the North Vietnamese probably anticipated such attacks and that the POL facilities near Haiphong, a major port city, politically were sensitive targets. Assessing the consequences of a POL air campaign, they further concluded it would (1) not change Hanoi's policy either toward negotiation or toward sharply entering the war; (2) probably result in more Soviet pressure on the regime to negotiate; (3) force Hanoi to ask for and receive more supply and transport aid from China and air defense aid from the Soviet Union; (4) aggravate Soviet-Chinese relations, and (5) cause further deterioration of U.S. - Soviet relations, especially if a Soviet ship were hit. Soviet counteraction was thought possible and might take the form of attacks on U.S. ferret aircraft or interference with U.S. access to West Berlin. Chinese Communist intervention in the war, while possible, was considered unlikely.[1]

In March another CIA study predicted that the destruction of POL sites (and a cement plant in Haiphong) would severely strain the North's transportation system. It was one of the most influential documents to bear on the subject. On 23 March Secretary McNamara informed General Wheeler that a new Rolling Thunder program directed against POL storage and distribution targets might be favorably received. On 25 April, Deputy Secretary of Defense Cyrus R. Vance assured the JCS that its 1965 POL studies were now receiving full consideration. On 6 May, a White House aide, Walt W. Rostow, recalling the impact of oil strikes on Germany in World War II, suggested to the Secretaries of State and Defense that systematic and sustained bombing of POL targets might have more prompt and decisive results on Hanoi's transportation system than conventional intelligence indicated.[2] Mr. Rostow observed that in 1965 U.S. estimates showed that 60 percent of the North's POL was for military purposes and 40 percent for civilian needs. The current ratio was now placed at 80 and 20 percent, respectively.

On 31 May, although a final decision to hit the major facilities had not been made, Admiral Sharp was authorized to attack certain POL-associated targets in the

Air War – Vietnam

northeast along with five small route targets. On 6 June General Westmoreland advised CINCPAC that an improving political situation in South Vietnam (since civil disturbances began on 10 March) was causing Hanoi much disappointment and dismay. Noting this circumstance and the heavy toll inflicted by the air campaign over North Vietnam and Laos, he recommended that these psychological and military gains be "parlayed into dividends" by hitting the POL storage sites. To do so later, he warned, would be less effective because of dispersal work already under way.[3]

Support continued to build up. Admiral Sharp quickly endorsed General Westmoreland's views and, on 8 June, the U.S. Ambassador to South Vietnam, Henry Cabot Lodge suggested that intensified bombing was the most effective way to get Hanoi to the negotiating table. General McConnell, who had long supported such action, told a Senate subcommittee that hitting POL targets would have a "substantial" effect on the amount of supplies the Communists could send to their forces in South Vietnam. An Air Staff intelligence report asserted that hitting the sites would have "a most profound" impact on Hanoi's infiltration activities and expressed confidence it could be done without causing severe civilian casualties.[4]

The Strikes of 29 June

The administration now moved toward its decision. In a preliminary action, the JCS on 16 June authorized Admiral Sharp to hit all of the POL dispersal sites listed in the current Rolling Thunder program except those within a 30-nautical-mile radius of Hanoi, a 10-nautical-mile radius of Haiphong, and 25 nautical miles from the Chinese border east of longitude 105° 20' E. and 30 nautical miles west of longitude 105° 20' E. On 21 June, USAF jets struck gasoline and oil depot sites ranging from 28 to 40 miles from Hanoi. Several other sites, previously exempt from attack, were hit in ensuing days outside the Hanoi-Haiphong area. 5

In addition, extraordinary steps were taken to prepare for the attacks on POL targets in the two main cities of North Vietnam. On 23 June, after Secretary McNamara and General Wheeler had informed President Johnson of their precautionary measures to avoid attacks on civilian areas and foreign merchant ships. Nine rules were laid down:

- use of pilots most experienced with operations in the target areas,
- weather conditions permitting visual target identification,
- avoiding to the extent possible populated areas,
- minimum pilot distraction to improve delivery accuracy,
- use of munitions assuring highest precision consistent with mission objectives,
- attacks on air defenses only in sparsely populated areas,
- special security precautions concerning the proposed operations, and

Air Operations 1966

- personal attention by commanders to the operations.

The JCS authorized Admiral Sharp to strike early on the 24th seven POL storage facilities and a radar site at Kep, northeast of Hanoi. Although special security precautions surrounded the planning, the news media soon reported the essential details of the operation. This forced the administration to postpone it and deny any decision had been made.

The strike was rescheduled and took place on 29 June. A USAF force of 24 F-105s, 8 F-105 "Iron Hands", 4 EB-66s plus 24 F-4Cs and 2 F-104s for MIG "cap" and escort hit a 32-tank farm about three-and-a-half miles from Hanoi. Approximately 95 percent of the target area, comprising about 20 percent of the North's oil storage facilities, was damaged or destroyed. Simultaneously, Navy A-4 and A-6 aircraft hit a large POL storage area two miles northwest of Haiphong. This facility, containing an estimated 40 percent of the North's fuel storage capacity and 95 percent of its unloading equipment, was about 80 percent destroyed. One USAF F-105 was lost to ground fire. Four MiG-l7s challenged the raiders and one was probably shot down by an Iron Hand F-105. No SA-2 missiles were observed. Maj Gen. Gilbert L. Myers, deputy commander of the Seventh Air Force

F-105s formed the backbone of the force used for the POL strikes. Source: U.S. Air Force

termed the raids "the most significant, the most important strike of the war'" Secretary McNamara subsequently called the USAF-Navy strike "a superb professional job," although he was highly incensed over the security leaks that preceded the attacks.[7]

In a press conference the next day, the defense chief said the strikes were made "to counter a mounting reliance by North Vietnam on the use of trucks and powered junks to facilitate the infiltration of men and equipment from North Vietnam to South Vietnam. " He explained that truck movements in the first five months of 1966 had doubled, and that daily supply tonnage and troop infiltration over the "Ho Chi Minh trail" were up 150 percent and 120 percent, respectively, over 1965. Further, the enemy had built new roads and its truck inventory by December 1966 was expected to be double that of January 1965. This would require a 50- to 70-percent increase in oil imports over 1965. The Secretary also justified the timing of the strikes, asserting that the "perishable" nature of POL targets made it more desirable to attack them now than earlier in the year.[8]

President Johnson said that the air strikes on military targets in North Vietnam "will continue to impose a growing burden and a high price on those who wage war against the freedom of others. " He directed that in the forthcoming weeks first priority be given to "strangling" the remainder of Hanoi's POL system except for that portion in areas still exempt from air attack. He also wanted more bombing of the two main rail lines running between Hanoi and China.[9]

The Mid-1966 Assessment

Shortly after the 29 June POL strikes, another major conference took place in Honolulu to review the war and plan additional U.S. and allied air, ground, and naval deployments. A mid-year assessment of the war, contained in a letter from Admiral Sharp to the JCS and the Office of the Secretary of Defense (OSD), was expanded in briefings for Mr. McNamara in Honolulu on 8 July. The PACOM commander said that he considered the air program for North Vietnam still inadequate, observing that previous recommendations to hit major ports of entry, logistic targets leading from China, and certain POL sites (in addition to those struck on 29 June) had not been approved. He thought it impossible to prevent the enemy from moving supplies from North to South and thus to "isolate the battlefield"; rather, the "highest task" was route interdiction and striking new targets as they were uncovered. Recent intelligence showed that the air campaign was hurting Hanoi. Its repair and reconstruction force now totaled about 500,000 and the morale of the government and troops was declining. To raise the cost of infiltration, he proposed striking as soon as possible 33 important exempted targets and more of the enemy's supplies, road and rail repair centers, and military training areas.[10]

Trucks proved to be hellishly difficult targets to find and kill.
Source: People's Liberation Army of Vietnam

Admiral Sharp pointed to Hanoi's greater effort to hide and disperse its logistic supplies because of the air attacks. As a result there was greater U.S. effort in the first six months of the year to uncover more of the following types of targets:

Type of Target	1 Jan 66	1 Jul 66	Total New Targets
Truck Parks	55	126	121
Military Storage Facilities	316	969	380
POL	38	180	142
Military Installations	680	939	259
Transshipment Points	7	65	65
Total	**1,096**	**2,006**	**967**

Air Operations 1966

The table showed an increase of 90 percent in significant targets since 1 January 1966 with the major portion consisting of truck parks, military storage facilities, and transshipment points.

During the first half of the year, Admiral Sharp continued, Rolling Thunder strikes had destroyed or damaged 1,076 trucks, 900 pieces of rolling stock, and 3,304 watercraft. A total of 2,771 trucks were destroyed or damaged in Laos. Discussing the North's air defense system, he said that Hanoi's antiaircraft gun inventory had increased from about 859 in February 1965 (when the bombings began) to more than 4,200, an average increase of about 205 guns per month. The North also possessed 20 to 25 active SA-2 battalions, good early warning, ground control interception equipment, and a respectable MiG force.[11]

SA-2 defenses were spreading all the time. Source: U.S. Air Force

In reply, Secretary McNamara reported that President Johnson had accorded first priority to "strangulation" of the North's POL system. Thus, it was essential to determine Hanoi's land and sea distribution system, categorize the targets, and then render them ineffective. The Secretary also pointed out the need for increased interdiction of railroad lines, particularly bridges in the northeast and northwest leading to China. Expressing concern over U.S. aircraft attrition, he said OSD was working with the services on ways to reduce it.[12]

The Beginning of Rolling Thunder Program 51

The strangulation campaign was incorporated into a new Rolling Thunder program, number 51. It was authorized by the JCS on 6 July and went into effect on the 9th. Armed reconnaissance could now encompass all of North Vietnam except for the established sanctuary areas (i. e., a 30-nautical-mile radius of Hanoi, a 10-nautical-mile radius of Haiphong, and 25 to 30-nautical-mile buffer area adjacent to China). Admiral Sharp assigned PACAF specific responsibility for halting all rail traffic in the northeast and northwest sectors. In addition, the JCS on 9 July authorized an increase in attack sorties for North Vietnam and Laos from 8,100 to 10,100 per month.[13]

Because of the high priority assigned to the strangulation effort, and in response also to Secretary McNamara's direction, the Air Staff on 16 July established an Operation Combat Strangler task force headed by Maj Gen Woodrow P. Swancutt, Director of Operations, Headquarters USAF. Its immediate objective was to evaluate POL strangulation and LOC interdiction plans prepared by the Seventh Air Force and PACAF. simultaneously, the Air Staff established an Operations Review Group within the Directorate of Operations under Col. LeRoy

J. Manor, an enlarged and reorganized successor to the ad hoc study group formed on 26 March 1965. It examined the effectiveness of combat and combat support operations in southeast Asia as well as the activities of USAF worldwide operational forces.14

Under Rolling Thunder program 51, USAF aircraft initially concentrated on route packages I, V, and VIA and the Navy on the others. Then on 20 July, at the direction of General Westmoreland, the Air Force inaugurated a "Tally-Ho" air campaign in route package I in a renewed effort, somewhat similar to Gate Guard, to curb Communist infiltration into and through the DMZ. Also, on 6 August at General Westmoreland's request and by the decision of Admiral Sharp, the "Dixie station" aircraft carrier used for air operations in South Vietnam was moved to "Yankee Station," thereby providing three rather than two carriers for the stepped up air activities against the North. Another important change was an agreement between the Seventh Air Force and Seventh Fleet commanders whereby the former would provide about 1,500 sorties per month in the normally Navy-dominated route packages II, III, and IV. By September USAF aircraft generally were covering 46,265 square miles or 77 percent of the land area of North Vietnam. The Navy, by comparison, was covering 13,891 square miles or about 23 percent of the land area. The Air Staff and General Harris considered the arrangement better than the relatively rigid delineation of service air responsibility for the North that had existed previously. Although the agreement took effect on 4 September, restrictions on air operations prevented its full realization until the restrictions were eased in December 1966.15

Bad weather was a serious problem. One solution was to use an EB-66 to spot targets on radar with the F-105s dropping on command. Source: U.S. Air Force.

The immediate priority, of course, was given to POL sites. The campaign increased in momentum until the week of 13-19 August when 140 attack sorties were flown against POL targets. Thereafter the sortie rate dropped. By the end of August an estimated 68 percent of known POL storage capacity in route packages I, V, and VI had been destroyed. On 19 September the remaining POL capacity in the North was placed at about 69,650 metric tons, of which 18,526 metric tons were not yet authorized for destruction.16

By the end of September it was apparent that the POL strikes were becoming less productive. There had been no let-up in Soviet deliveries of POL supplies and the North Vietnamese continued their dispersal efforts. Supported by Combat Strangler analyses, PACAF considered

Air Operations 1966

the benefits derived from attacking the scattered sites no longer worth the cost in aircraft lost. In a report to Secretary Brown on 14 October, PACAF stated that the POL campaign had reached the point of diminishing returns and that the Soviet Union and China could adequately supply the North with POL products. Also, U.S. air power could best force changes in POL handling and distribution by striking targets listed in Rolling Thunder program 52 proposed by the JCS on 22 August. This program called for 872 sorties over 19 new targets. This would constitute, PACAF felt, the best kind of "strategic persuasion" before Hanoi could devise countermeasures.[17]

The railroad strangulation effort, particularly against the Hanoi-Lao Cai and the Hanoi-Dong Dang lines running to China and located in route packages V and VI A, was not especially productive because of bad weather and the ability of the North Vietnamese to repair the lines quickly. In fact, PACAF believed it was virtually impossible to maintain an effective air program against them. Weather problems in the two route packages forced the cancellation or diversion of about 70 and 81 percent of the attack sorties scheduled for July and August, respectively. The weather improved in September but turned poor again in October.[18]

Enemy antiaircraft defense, including additional SA-2s also added to the difficulty in interdicting the two main rail lines. As American aircraft losses rose, Admiral Sharp on 20 September ordered a reduction of about one-third of the air strikes in route package VIA until measures could be devised to reduce the toll. For example, on 7 August anti-aircraft guns knocked down seven U.S. aircraft (six USAF, one Navy), the highest one-day total since 13 August 1965 when six were shot down. American combat losses in the North during the third quarter of the year were: 41 in July, 37 in August, and 26 in September. Eighty of these were USAF aircraft. In October combat losses declined to 23, only nine of them USAF.[19]

MiG pilots also became increasingly aggressive. Fifteen "incidents" in July resulted in two MiG-21s and one MiG-17 being shot down against the loss of one USAF F-105 and one Navy F-8. During an engagement on 7 July, two MiG-21s for the first time in the war fired air-to-air missiles against two F-105s but failed to score. Another milestone in the air war occurred on 21 September when the biggest air-to-

Vietnamese MiG-17 Pilots Scrambling. U.S. reluctance to attack ground control centers for fear of killing Russian and Chinese "advisors" gave the MiG pilots a priceless advantage. Nguyen Van Bay was the leading North Vietnamese MiG-17 ace with five kills, 2 F-8s, 1 F-4B, 1 A-4C and 1 F-105D. Source: People's Liberation AF of Vietnam

air battle to date was fought over the North. In seven separate encounters USAF pilots downed two MiG-17s, probably a third, and damaged a MiG-21 without suffering any losses.[20]

The Tally-Ho Campaign.

In terms of total sorties flown, the largest portion of the USAF effort, as in previous months, was concentrated in route package I which included the DMZ, the area of the greatest enemy threat. Intelligence believed that about 5,000 North Vietnamese had infiltrated through the zone in June. PACAF speculated that these enemy movements may have been due to the recent success of Tiger Hound air operations in Laos which, together with monsoon weather, had virtually blocked certain logistic routes in that country.[21]

As more enemy troops pressed toward the DMZ and intelligence reported that the North's 324B Division of 8,000 to 10,000 men, had crossed over into the I Corps area of South Vietnam, General Westmoreland asked Lt. Gen. William W. Momyer, who succeeded General Moore as Seventh Air Force commander on I July, to prepare an air program similar to Tiger Hound in Laos for the most southern part of route package I including the zone. Already under way just south of the DMZ was a combined U.S. Marine and South Vietnamese Army and Marine air and ground effort called Operation Hastings. General Momyer quickly outlined a "Tally-Ho" air campaign against enemy targets in an area about 30 miles inside North Vietnam from the Dai Giang river below Dong Hoi through the DMZ to its southern border. The first Tally-Ho air strike was made on 20 July by USAF and Marine aircraft, the latter beginning regular operations in the North for the first time. Previously Marine Corps activities in the North consisted of eight sorties in April and two sorties in June. Like Gate Guard, C-130 airborne control was employed and, for the first time, USAF O-1 FACs flew into North Vietnam to help find targets. To sustain Tally-Ho, Tiger Hound activity in Laos was scaled down.[22]

Although Tally-Ho included the DMZ, military operations within the zone were not conducted immediately. The political problems associated with such action had been under study for some time. On 20 July, the day Tally-Ho began, the JCS finally authorized Admiral Sharp to launch air or artillery strikes in the southern half of the zone. This followed protracted State and Defense Department negotiations which resulted in State's approval if the allies had concrete evidence that the North was using the zone for infiltrating men and materiel, if there existed an adequate record of the Saigon government's protest to the International Control Commission (The ICC composed or representatives from India, Canada, and Poland was established in July 1954 as a result of the Geneva conference that ended the French-Indochina war. Its primary function was to supervise the 1954 Geneva agreements.) concerning Hanoi's violation of the zone, and if an appropriate public affairs program was begun prior to military action in the zone.[23]

After these conditions were fulfilled, the JCS on 28 July specifically authorized B-52 strikes in the southern portion of the DMZ in support of U.S. South Vietnamese "self-defense" operations. In their first attack there, on 30 July, 15 B-

Air Operations 1966

52's dropped bombs on ammunition dumps, gun positions, and weapon staging targets. In August B-52's returned there several times.[24]

On 22 August General McConnell informed Secretaries Vance and McNamara of a rising trend in USAF out-of-country night operations, especially in North Vietnam, and of his expectation that the trend would continue in the Tally-Ho campaign. But shortly thereafter the hazards of antiaircraft fire and inadequate aircraft control forced a reduction in the use of USAF 0-1 FACs and, consequently, of other combat aircraft. In fact, the night attack effort, despite General McConnell's hopes, did not show a significant rise again until December.[25]

In September the advent of better weather and better results with the use of MSQ-77 radar permitted intensification of the Tally-Ho operations. Many secondary explosions often followed USAF-Marine corps air strikes. The first B-52 strike in the northern portion of the DMZ was made on 16 September and others soon followed until 26 September when they were halted in the zone east of route package I to permit ICC inspection of North Vietnamese troop infiltration. As the Communists continued to use this area, administration authorities on 13 October rescinded the prohibition against air and artillery strikes. On the 14th B-52 strikes were stopped in the zone, this time because of the danger from suspected SA-2 sites.[26]

Tally-Ho continued through October and into November. As in the Gate Guard operations, Tally-Ho FAC pilots often were forced up to 1,500 feet by ground fire, thus reducing the value of visual reconnaissance. They also experienced severe turbulence over mountainous terrain and poor weather added to their difficulties.[27]

The Tally-Ho program remained under constant review. Initial evidence appeared to show that its operations destroyed many enemy structures, supplies, antiaircraft positions, and vehicles, and that it hampered, but did not stop infiltration on foot through the DMZ. On I0 October, during a briefing for Secretary McNamara and other top officials who were visiting Saigon, Brig. Gen. Carlos M. Talbott of the Seventh Air Force indicated that Tally-Ho and other air activities possibly had caused the enemy to reach the limit of his supply capability. PACAF officials thought that Tally-Ho and U.S.-South Vietnamese "spoiling" attacks in and below the DMZ had thwarted a major offensive planned by the North Vietnamese into the I corps. On the 13th, the JCS, in answer to a White House request for an assessment of the enemy threat in the zone, likewise reported that spoiling attacks and tactical and B-52 air strikes in and near the demilitarized area had defeated the North Vietnamese and prevented them from seizing the initiative. But the service chiefs warned that the enemy still retained considerable offensive capability and that U.S. reinforcements should be sent to that region.[28]

However, these were general observations. The USAF Vice Chief of Staff, Gen. Bruce K. Holloway, when pressed by Secretary Brown on the effect of the air effort on North Vietnamese movement through the DM.Z, was less certain about the results of Tally-Ho operations. He replied: "I do not know what the effect is and nobody else seems to know," adding that there was much "speculation and

Air War – Vietnam

excuses why it's hard to determine." He said that there were several actions under way to improve data-gathering in the DMZ area. These included establishing a tactical air support analysis team (TASAT) composed of 20 Air Force and Army personnel to insure systematic data-reporting, forming a similar USAF-Army team to assess B-52 strikes, inviting the Army and Navy to join the Air Force Combat Strangler task force in assessing the results of the air campaign, and organizing an air weapon survey board.[29]

The need for more reliable information on Tally-Ho activities near the DMZ was also reflected in the observation of a USAF intelligence officer in South Vietnam who was associated with the air campaign. "We don't know how effective we were," he commented, "for we don't know what we stopped or the amount of flow." He thought the program could be made more productive by defoliating the terrain and by improving intelligence, targeting, and communication procedures. Subsequently, a list of targets believed to have been damaged or destroyed by the Tally-Ho program was compiled.[30]

Air Operations 1966

IV. ANALYSES OF THE AIR CAMPAIGN

The beginning of Rolling Thunder program 51 also witnessed the start of a greater Air Staff effort to analyze the effectiveness of USAF operations in Southeast Asia, particularly in North Vietnam. With the assignment of more personnel in July to the Operations Review Group under Colonel Manor and Operation Combat Strangler under General Swancutt, the Air Force improved its ability to collect and evaluate operational data and to respond to requests from higher authorities for information on different aspects of the air war.

Operational Studies

One of the early important products of the Swancutt task force was its analysis of the Seventh Air Force POL and LOC air campaign against North Vietnam. Completed on 30 August, it pointed to the inflexibility of air operations in the North. This situation was attributed to seven main factors: air restrictions that reduced aircraft maneuver, the prohibition against striking certain target areas, the "route package" system that divided into relatively independent regions the USAF and Navy target areas of responsibility, a targeting system that had the effect of concentrating air power and thus "telegraphing" U.S. intentions to the enemy, bad weather and anti-aircraft defenses that left little choice in tactics, the existence of few profitable targets, and fragmented command and control of air activities.

Based upon its analysis, the task force recommended two primary changes: a broadened target base to allow an increase in the tempo of air operations and a single centralized command and control system for air. It also began assembling a complete statistical record of aircraft losses, ordnance expended, results of air strikes, and tactics employed (because of the inordinately high aircraft losses in route packages V and VIA), and analyzing Seventh Air Force and PACAF plans weekly. The group also proposed that the Air Force seek permission for its aircraft to hit targets in the Navy-dominated route packages II, III, and IV when weather forced diversionary strikes, and it recommended more night air operations. Agreements subsequently were reached to allow USAF units to make diversionary air strikes in the Navy areas, the new policy becoming effective on 4 September.[1]

Also in August the Air Staff examined the value of air attacks on North Vietnamese watercraft. This was in response to a query from Secretary Brown who observed that Admiral Sharp, in his briefing of 8 July in Honolulu, had indicated that 2,358 watercraft had been attacked by air to that time.[2] General Holloway advised on 22 August that in Admiral Sharp's view, air strikes on largely coastal watercraft through mid-1966 had not always been worth the effort, although they did have a harassing effect on the North Vietnamese. Since July, because of the stepped up air operations on land transportation routes, a larger volume of barge traffic had appeared on inland waterways. In the Thanh Hoa and Vinh areas, watercraft construction was exceeding civilian needs. Some watercraft carried POL drums, tanks, and ammunition, and there were more attempts to

Air War – Vietnam

camouflage them. Thus, said General Holloway, Admiral Sharp now believed that they were worthwhile air targets.[3]

On 13 September, again at the request of Secretary Brown, the Air Staff undertook a detailed study of the types of target systems in North Vietnam. The approach included an examination of the cost and the length of time needed to destroy a part or all of each target, and the effect its loss would have on Hanoi's ability to continue hostilities. The primary target systems being studied were electric power, maritime ports, airfields, navigation locks and dams, industrial facilities, command and control sites, extractive industries, military installations, and LOCs. The project had not been completed by the end of the year.[4]

The Effectiveness of Air Power

The Air Staff also assembled data to reply to numerous questions raised by Secretary McNamara on the effectiveness of air power. On 2 September, during a meeting with Air Force, Navy, and other officials, the defense chief asked the Air Force to examine the combat use of F-4C and F-105 aircraft. He wished to determine whether F-4Cs should fly most of the sorties against North Vietnam, especially against "fleeting" night targets, and whether F-105s should be employed in South rather than North Vietnam. He also asked for a comparative study of the performance of propeller and jet aircraft in night operations over route packages I and II. From the Navy, Secretary McNamara wanted recommendations on how to increase the number of night sorties over North Vietnam.[5]

A Vietnam Veteran. The three victory stars on this F-4C were scored on May 12, 1966, by Maj. W.R. Dudley (pilot) and 1Lt. I. Kreingelis (WSO) flying for the 390th TFS, 35th TFW using an AIM-9 Sidewinder against a MiG-17; May 14, 1967, by Maj. J.A. Hargrove (pilot) and 1Lt. S.H. Demuth (WSO) flying for the 480th TFS, 366th TFW using a pod-mounted 20mm cannon against a MiG-17; and June 5, 1967, by Maj. D.K Preister (pilot) and Capt. J.E. Pankhurst (WSO) flying for the 480th TFS, 366th TFW using a pod-mounted 20mm cannon against a MiG-17. Source: U.S. Air Force.

Air Operations 1966

On the basis of data collected by the Air Staff, Secretary Brown advised the defense chief on 28 September that while the F-4C and F-105 aircraft were both suited for daytime attack missions, the F-4C was more effective at night, principally because it carried two pilots. This permitted better target-finding, better radar-controlled formations (by the rear pilot), and more protection for pilots against "spatial disorientation/vertigo." Although a switch in the use of the F-105 from North to South Vietnam would reduce its losses, other reasons militated against such a change. It would affect the logistical base of the two aircraft, probably not reduce aircraft attrition in route package areas V and VI (where enemy defenses were heaviest), and create an aircrew replacement problem. He supported the assigned missions of the two aircraft and the practice of "attriting" the F-105s first in order to conserve the F-4Cs.

SAFD Secretary Brown reported that comparisons between propeller and jet aircraft in night operations were inconclusive because of vast differences in their use. In North Vietnam the Air Force used its A-1s in less defended areas while the Navy did not employ its A-1s until an area was first tested by A-4s. In Laos, Air Force A-1 losses were higher because of lower attack speed or more ordnance-delaying passes against targets. [6]

The study requested by Secretary McNamara on stepping up night operations over North Vietnam was submitted by Navy Secretary Paul M. Nitze. He said more night sorties would cause a drop of about 15 percent in Navy attack efforts, reduce effectiveness by about 50 percent compared with daytime strikes, result in more civilian casualties, and double operational aircraft losses, although combat losses would remain about the same. In view of these findings, and because he believed it was necessary to maintain pressure on the North "around the clock", Secretary Nitze recommended no change in the current "mix" of day and night sorties. [7]

Secretary McNamara also expressed dissatisfaction with the level of air analysis performed by the services, pointing to the differences between the estimates made in several studies on the effects that the POL strikes would have on North Vietnamese infiltration and those that actually occurred. He asked the Navy Secretary especially to review past CIA, DIA, and other reports on this matter as well as analyze the general subject of aircraft losses. He enjoined the Air Force to make more "sophisticated" analyses of the conflict, asserting that this was one of the "most important" things that it could do.[8]

On 3 November Secretary Nitze sent Mr. McNamara an initial report on the Navy's most recent air studies. The findings, and admissions, were unusual. He said the report showed that (1) there was insufficient intelligence data to produce a viable assessment of past or projected air campaigns; (2) North Vietnam's logistic requirements for forces in the south, compared with its capabilities, were small, thus permitting Hanoi to adjust the level of conflict to its available supplies; and (3) North Vietnam's estimated economic loss of $125 million versus $350 million of Soviet and Chinese aid taken alone, was a "poor trade-off" when compared with the cost of achieving the end product. The first two factors, the Navy Secretary observed, emphasized the magnitude of the task of disrupting North Vietnamese infiltration.

Admittedly, he continued, air attacks had produced some results such as requiring North Vietnam to provide for an air defense system and to maintain a 300,000-man road and bridge repair force that reduced resources available for infiltration into South Vietnam. And prisoner of war and defector reports testified to some success of the air and ground campaign in the South. Nevertheless, because of the inadequacy of available data, analysts were unable to develop a logical case for or against the current air campaign at either a higher or lower level. "This is not a criticism of the analytical effort," said Mr. Nitze, "rather, it is a reflection of the degree to which decisions in this area must be dependent on judgments in the absence of hard intelligence."

The Nitze report included a review of studies including the March 1966 CIA study which preceded and led to the U.S. decision to attack North Vietnam's POL system. The overall purpose of the air strikes had been to strain Hanoi's transportation system. Interviews with CIA analysts disclosed that many of their assumptions were based on certain estimates of the logistic capacity of the Hanoi-Dong Dang rail line, the amount of seaborne imports, the impact of hitting a cement plant in Haiphong, and other data. In retrospect, other factors also bore, or could bear on the effectiveness of air operations against the enemy's logistic capability and resources, such as the existence of a road system parallel to the Hanoi-Dong Dang rail line, the construction by the Chinese of a new internal transport link to Lao Cai, the transport capacity of the Red River from Lao Cai to Hanoi, and the capability of the North Vietnamese to continue, although less efficiently, to produce cement in small, dispersed furnaces if the plant in Haiphong were destroyed. As the Haiphong plant was the only such facility in the North, the Air staff had seriously questioned the ability of the North Vietnamese to produce cement if it was destroyed. There were indications that the analysts' use of 1965 average import statistics to project future North Vietnamese requirements resulted in an overstatement of Hanoi's needs. These, and other examples, showed the inadequacy of the information base for evaluating the effectiveness of airstrike programs planned for North Vietnam.

To obtain better analyses for predicting the results of air strikes, the Nitze report indicated that the Chief of Naval Operations was establishing a special branch in the Navy's System Analysis Division to perform this vital task. [9]

Secretary Brown, in a reply to Mr. McNamara on 10 November, summarized current efforts to improve USAF analysis of the effectiveness of air interdiction. He cited the establishment in July of the Operation Combat Strangler task force and expansion of its functions to include development of a computer model to simulate air campaigns against North Vietnamese targets. The Air Force also was analyzing daily the air operations over North Vietnam, reviewing and evaluating major target systems including the anticipated effect of air attacks on the North's economy and on infiltration into the South, and studying the length of time required to destroy a given percentage of target systems and the cost of striking them in terms of sorties, munitions, and aircraft. This effort had been assigned top priority and the necessary resources. In addition to briefing the Air Staff, the task force made the various analyses available to the Joint Staff and OSD and posted pertinent data in a special situation room.

Air Operations 1966

The Secretary of the Air Force also advised that the USAF study of major target systems in North Vietnam was 50 percent complete and would be finished early in 1967, after which a second analysis would "interface" all target systems to determine the cumulative effect of the destruction of several complimentary target systems. In addition, a special analysis of night operations was under way.[10]

Studies on Aircraft Attrition

Another problem area that received increased attention after mid-1966 was aircraft attrition. Following a USAF briefing on this subject on 6 June, Secretary McNamara asked the Air Force for a detailed analysis of losses.[11]

On 19 July Secretary Brown submitted coordinated USAF-Navy reply. Over North Vietnam, he said, the majority of aircraft losses (74 percent) were due to automatic weapon and light antiaircraft guns, and most aircraft (77.1 percent) were hit below 4,000 feet. The losses were distributed fairly evenly over the route packages, with no meaningful differences in the loss rates by routes. He said an apparent USAF aircraft loss rate amounting to "three times" that of the Navy's was due principally to the lack of a clear definition of strike sorties, the limitations of the joint reporting system, and frequent diversion of sorties. Overall Air Force and Navy aircraft losses were quite similar, amounting to 3.96 and 4.32 aircraft per 1,000 sorties, respectively. He reported there was no data on the frequency of aircraft exposure to antiaircraft weapons at different altitudes, the proportion of losses sustained on each segment of an attack area, and the extent of increasing aircraft exposure to ground fire induced by avoiding SA-2 missiles.

An analysis of operational data for the period 1 October 1965 through 31 May 1966 by cause of loss, including "take-off" for combat missions, the Air Force Secretary continued, showed that by far most of the operational losses were due to aircraft system failures. The ratio of system failures to total operational losses in this period were by service: Air Force, 23 of 44; Navy, 10 of 29; and Marine Corps, three of nine. Of the 36 system failures, 22 involved aircraft engines, five were due to flight control problems, and the remainder were random system failures which occurred only once or twice. In addition, the Navy lost nine aircraft in carrier landings.

Compared with normal peacetime attrition, Secretary Brown added, actual operational losses in Southeast Asia for fiscal year 1966 were below predicted figures for USAF F-100s, F-104s, F-4Cs and F-5s. Only F-105 losses were higher than expected and several efforts were under way, including a study by the Air Force Systems Command, to modify the aircraft in order to reduce combat losses. In addition to air crews, hydraulic-pneumatic systems (such as fuel and flight control) and aircraft engines were most vulnerable to enemy fire.[12]

At the request of Deputy Secretary Vance, the Air Force also made a special study of aircraft losses during night missions over North Vietnam and Laos. Reports submitted by Secretary Brown and General McConnell on 24 and 25 August showed that for the period 1 January - 31 July 1966, the aircraft loss rate per 1,000 sorties for night armed reconnaissance sorties averaged 0.84 compared to 4.27 for day armed reconnaissance. Night sorties were considerably less hazardous, primarily because North Vietnam's air defense weapons were largely optically directed. [13]

Aircraft losses remained of particular concern to the Air Staff since they threatened the Air Force's planned buildup to 86 tactical fighter squadrons by June 1968. On 29 August, General Holloway, the vice chief of staff, sent a report to General Wheeler on the effect of the losses on the Air Force's capabilities. It showed that at current aircraft loss rates the Air Force would be short five tactical fighter squadrons at the mid-point of fiscal year 1968 and three squadrons short at the end of the fiscal year. The approved squadron goal might not be reached until after the third quarter of fiscal year 1969. The report also indicated that an OSD-prepared aircraft "attrition model" needed adjustment to reflect more clearly sorties programmed for North Vietnam. It was on the basis of this model that OSD on 19 November 1965 had approved additional production of 141 F-4s to offset attrition. General Holloway said that the Air staff would continue its analysis of this problem, [14]

Aircraft attrition was, of course, being followed closely by administration officials and congressional critics. In recognition of the problem Secretary McNamara on 22 September announced plans to procure in fiscal year 1968, 280 additional largely combat-type aircraft costing $700 million. Although the largest number were earmarked for the Navy, the Air Force would receive a substantial portion of the total. [15]

The Hise Report

Meanwhile, on 26 September, a Joint Staff study group completed a more detailed examination of aircraft attrition. Its findings were contained in the "Hise Report", named after the group's director, Marine Col. Henry W. Hise, whom General Wheeler had designated on 28 July to perform this task. Some of the ground work of the Hise Report had been done by a study group headed by USAF Brig Gen. R. G. Owen at the request of General Wheeler on 25 April. The Hise study group consisted of four representatives, one from each of the services, including USAF Col. C. L. Daniel and one representative from the DIA.

The Hise group studied all factors affecting aircraft losses using data from joint operational reports, the DIA, and interviews with Air Force, Navy, and Marine commanders and airmen at Headquarters PACOM and in Southeast Asia. It covered all aircraft losses, whatever the cause, from January 1962 through August 1966. Totaling 814, the aircraft were lost in the following areas: North Vietnam, 363; Laos, 74; and South Vietnam, 377. The report analyzed the main factors affecting aircraft losses: time, enemy defenses, tactics, targeting, weather, sortie requirements, ordnance, aircrews, and stereotyped air operations.

The report's major conclusion was that North Vietnam had been given an opportunity to build up a formidable air defense system and noted, in support, General Momyer's recent observation: "in the past three months the enemy has moved to a new plateau of [air defense] capability. He now has a fully integrated air defense system controlled from a central point in Hanoi. " Both the antiaircraft

Air Operations 1966

guns and S,A-2 missiles, according to the Hise Report, had had a "crippling effect on air operations. The vast majority of aircraft losses were attributed to ground fire, with 85 percent of all "hits" being scored when the aircraft were below 4,500 feet. If Hanoi were permitted to continue its buildup of air defense weapons, the United States eventually would face a choice of supporting an adequate air campaign to destroy them, accepting high aircraft losses, or terminating air operations over the North.

The report also pointed to a number of other problems. It said that between 1 July and 15 September 1.966 USAF's 354th TFS had experienced an inordinately high aircraft loss rate. Additionally, some pilots in the theater were overworked, several squadrons had fewer than authorized pilots, F-105 pilots had "low survivability" in route packages V and VIA, stereotyped operations contributed to air losses, and a larger stock of ordnance was needed to provide for a more intense anti-flak program.[16]

The F-105 Wild Weasels External stores include QRC-380 blisters, AGM-45 Shrike and AGM-78B Standard ARM (Anti-Radiation Missile). Source: U.S. Air Force.

General Harris on 20 October forwarded the PACAF-Seventh Air Force assessment of the Hise Report to General McConnell. He generally agreed with the report's conclusions about the buildup of the North's anti-aircraft defenses and the need to broaden the target base. But he thought the report added little to a fundamental discussion of aircraft losses since it cited largely a number of well known facts. General Harris modified or took exception to a number of points raised. Concerning the effect of SA-2 missiles (which forced pilots down to within range of antiaircraft guns), he said that Air Force "Wild Weasel" and "Iron Hand" forces were mitigating the effect of the SA-2's on tactics, although a major

Air War – Vietnam

development effort was still needed in this area.. Wild Weasel aircraft, largely F-100F's and F-l05F's, were specially equipped with electronic countermeasures (ECM) equipment for anti-SA-2 operations. Iron Hand was the operational code name for attacks on SA-2 sites.

In bad weather it was the lack of an all-weather bombing system that limited operations rather than SA-2's. The Soviet-made missiles merely complicated bombings, making it difficult for aircraft to fly higher lest they become vulnerable to a missile hit. [17]

With respect to high losses incurred by the 354th TFS, General Harris attributed this primarily to aggressive leadership, accidents, and misfortunes in only one squadron, something that often happened in peace as well as in war without identifiable causes. Nor did he consider overwork or fatigue of pilots a factor in aircraft losses. F-105 pilots at Takhli and Korat Air Bases in Thailand, for example, in July flew an average of 56.7 and 43.9 hours respectively. In August they flew 48.2 and 36.5 hours respectively. Although aircraft often flew twice in one day, pilots seldom did except during "peak loads" and this was an infrequent requirement.

General Harris also took issue with a statistical interpretation showing that F-105 pilots flying 100 missions over route packages V and VIA would suffer excessive losses. Although the figures (based on July and August data) were approximately correct, they represented the greatest attrition rate in a period of maximum losses in the highest risk area in Southeast Asia. Seventh Air Force records showed that only 25 percent of pilot missions were in high risk areas. Thus, in a 100-mission tour, an F-105 pilot would not lose his aircraft over enemy or friendly territory as often as alleged. He further observed that the F-4C loss rate was about one-fourth that of the F-105 rate. He conceded that some squadrons at Takhli and Korat Air Bases had been below authorized pilot strength during the June-September period.

2.75" rockets were in short supply. Source: U.S. Air Force.

Air Operations 1966

The PACAF commander also agreed that, to some extent, there was a tendency to use standard or "stereotyped" tactics because of the need for efficient air scheduling and to meet JCS objectives. But it was North Vietnam's effective early warning and ground control interception system rather than stereotyped tactics that aided the enemy and provided him with nearly total information on U.S. air operations. The advantages of existing air scheduling, he thought, far exceeded the disadvantages. [18]

The Air Staff and General McConnell considered the data in the Hise Report as accurate and generally accepted the findings. On 10 October the JCS informed Secretary McNamara that, to the extent possible, Admiral Sharp and the services had taken several steps to ameliorate the aircraft loss rate. But certain other measures would require administration approval, particularly increased production of specific types of munitions for more effective suppression of enemy air defenses. There included 2.75" rockets with M-151 heads, Shrikes, CBU-24s, and 2,000- and 3,000-pound bombs. The Joint Chiefs reaffirmed their recommendation of 22 August that Rolling Thunder program 52 be adopted to broaden the target base over North Vietnam and make possible increased destruction of enemy air defense sites.[19]

The Hise Report findings prompted Dr. Brown and Deputy Secretary of Defense Vance to seek clarification of certain aspects of aircraft attrition. Detailed replies subsequently were incorporated into a JCS paper in which the service chiefs also cited two major policy handicaps of the air war that contributed to aircraft losses. These were the administration's restrictive targeting policies and its observance of the sanctuary areas around Hanoi, Haiphong, and in the buffer zone adjacent to China. They endorsed the Hise Report finding that North Vietnam's air defense system eventually could make air attacks unprofitable and reaffirmed the need for more ECM equipment and suitable ordnance. They disagreed with the report's belief that pilot fatigue contributed to losses, but conceded some pilots had been over-worked because occasionally there were insufficient numbers of them. They pointed to Admiral Sharp's recent directive (of 2 October) stating that sorties allocated for North Vietnam and Laos were not mandatory figures to be achieved but were issued to indicate the weight of air effort that should go into certain areas. Air units were not to be pressed beyond a reasonable point. [20]

McNamara's Proposal to Reduce Aircraft Attrition

Meanwhile, based on a study by his Southeast Asia Program Division of 1965 aircraft loss rates, Secretary McNamara on 17 September sent the JCS a plan to reduce aircraft losses, particularly the Navy's. It took into consideration the Air Force's force structure which the division believed could absorb aircraft losses more easily. To reduce Navy losses, the Defense Secretary suggested shifting about 1,000 carrier sorties per month from North Vietnam and Laos to South Vietnam with the Air Force increasing its sortie activities in those two countries. He thought this might reduce Navy losses by about 59 aircraft during the next nine months. In absolute numbers, USAF losses had been less and Navy losses more than planned, in part because some "higher loss" targets initially planned for the

Air Force had been assigned to the Navy. Loss rates varied widely by target. Overall, Mr. McNamara saw no significant difference in the air performance of the two services, asserting that "I think they're both doing a magnificent job and I see no difference as measured by loss rates in their effectiveness in combat."[21]

Generals McConnell and Harris strongly opposed any change in sortie assignments. so did the JCS which on 6 October replied by noting that differences between projected and actual aircraft losses in December 1965 had stemmed primarily from the high level of air effort in route packages V and VIA and the significant increase in enemy air defenses. The Joint Chiefs also observed that OSD had underestimated both total combat sorties to be flown over North Vietnam and Navy's non-combat aircraft losses. A shift in sorties to reduce losses would pose considerable operational difficulties for the Air Force by requiring more flying time and air refueling missions in order to reach the northernmost targets. The Navy too would have to make important operational adjustments.[22]

Affirming that every effort was being made to reduce aircraft and aircrew losses, the JCS again recommended Rolling Thunder program 52 as the best solution. It also noted that, under current projections, even with the recently announced (22 September) procurement increase, new production would not equal aircraft losses.[23]

In view of this reply, Secretary McNamara abandoned plans to switch Air Force and Navy operational areas.

Air Operations 1966

V. THE AIR WAR AT YEAR'S END

While the Air Force concentrated on Tally-Ho strikes, the administration in late 1966 took another look at JCS proposals to increase the air pressure on North Vietnam. During a conference in October in Honolulu to review additional U.S. force deployments, Admiral Sharp proposed a revised strike program averaging 11,100 sorties per month against the North for 18 months beginning in January 196?. On 4 November the JCS endorsed both the deployment and sortie proposals and again advocated mining the sea approaches to North Vietnam's principal ports, as well as several other actions.[1]

On 8 November General Wheeler urged Secretary McNamara to approve the Rolling Thunder program 52 sent to him initially on 22 August. Except for some fixed targets, the program would prohibit armed reconnaissance within a 10-nautical-mile radius of Hanoi and Phuc Yen airfield and the Haiphong sanctuary would be limited to a radius of four nautical miles. The JCS chairman singled out a number of other major targets remaining in the North, commenting briefly on each. He proposed striking three SA-2 supply sites, observing that since 1 July 1965 at least 949 SA-2s had been launched against U.S. aircraft, destroying 32. He suggested attacks on certain POL storage facilities, estimating that 24,800 metric tons remained of an initial 132,000 metric tons of fixed POL storage capacity. Dispersed sites, he said, held about 42,500 metric tons. Other targets on his list included the Thai Nguyen steel plant, the Haiphong cement plant, two Haiphong power plants, four waterway locks (related to water transportation), and the port areas of Cam Pha and Haiphong.[2]

On 10 November Secretary Brown informed Secretary McNamara that he endorsed the proposed Rolling Thunder 52 program. It would include 472 strike sorties against selective targets (canal crater locks, POL storage areas, manufacturing and electric power plants and SA-2 support facilities) in route package areas V, VIA, and VIB. On the basis of 1 April - 30 September 1966 attrition rates, there would be a loss of eight aircraft. He thought the air strikes would reduce and discourage shipping operations, reduce POL storage, increase replenishment, repair, and construction problems, and make more difficult the resupply of Communist forces in the South.[3]

Approval of Rolling Thunder Program 52

The administration on 12 November approved a modified Rolling Thunder program 52. It contained 13 previously unauthorized JCS targets: a bridge, a railroad yard, a cement plant and two power plants in Haiphong, two POL facilities, two SA-2 supply sites, and selected elements of the Thai Nguyen steel plant. Ten vehicle depots also were earmarked for attack. To assure success of the overall program, the JCS raised the authorized attack sortie level to 13,200 per month for November. In separate but related planning action, Secretary McNamara limited the JCS-recommended air and ground deployment program

Air War – Vietnam

through June 1968 on the grounds that an excessively large buildup could jeopardize some recently achieved economic stability in South Vietnam.[4]

Despite the new attack sortie authorization, the northeast monsoons restricted program "52" operations for the remainder of 1966. Actual sorties flown in November totaled 7,252 (3,681 USAF) and in December, 6,732 (USAF 4,129). These figures compared with the year's high of 12,154 U.S. attack sorties flown against the North in September. A sudden administration decision in November to defer striking six of the approved JCS targets also affected the sortie rate.[5]

Among the authorized targets were the Hai Gai POL storage site, hit on 22 November by USAF F-4C's, and the Dap Cai railroad bridge, a holdover from program "51". Navy aircraft struck the Haiphong SA-2 supply complex and the Cam Thon POL storage area. On 2 December USAF aircraft hit the Hoa Gai site for a second time while Navy aircraft conducted a first strike against the Van Vien vehicle depot. The latter was subsequently hit six times through 14 December. USAF aircraft also hit Yen Vien railroad year for the first time twice on 4 December and conducted restrikes on 13 and 14 December. Both the vehicle depot and the railroad yard were heavily damaged.[6]

The Furor Over Air Strikes "On Hanoi"

The USAF and Navy strikes of 13 and 14 December against the Van Vien vehicle depot and the Yen Vien railroad yard had international repercussions. The depot was about five nautical miles south of Hanoi and the yard, a major junction of three rail lines with two of them connecting with China, about six nautical miles northeast of Hanoi. Both the North Vietnamese and Russians immediately charged that aircraft had struck residential areas of Hanoi, killing or wounding 100 civilians. Allegedly, several foreign embassies were also hit, including Communist Chinese. Headquarters MACV quickly asserted that only military targets were struck. The State Department conceded that the attacking aircraft might have accidentally hit residential areas but strongly suggested that Hanoi's antiaircraft fire and SA-2 missiles (of which more than 100 were fired during the two days, a record high) may have caused the civilian damage. 7

Debriefings of the crews of seven USAF flights participating in the 13 and 14 December strikes on the railroad yard indicated that two flights experienced problems. The crews of one had difficulty acquiring the target and were uncertain of the exact release coordinates because of clouds and a MiG attack. Although they thought the ordnance was released in the immediate target area, they conceded it might have fallen slightly southwest of a bridge located south of the railroad yard. Poor weather also prevented the crews of a second flight from seeing the railroad yard and bomb impact was not observed, although they thought the ordnance struck rolling stock.

The Communist allegations and the growing criticism by certain groups in the United States and abroad about the war's escalation prompted the administration on 16 December to suspend further attacks on the Yen Vien railroad yard. On the 23d Admiral Sharp advised all subordinate commands that until further notice no air attacks were authorized within 10 nautical miles of the center of Hanoi.

Air Operations 1966

Attacks on other fixed targets were also halted for the time being. On 26 December a New York Times correspondent, Harrison E. Salisbury, who arrived in Hanoi on the 23d reported on alleged eyewitness accounts of the 13 and 14 December airstrikes that resulted in civilian casualties and damage. The Defense Department on the same day acknowledged that some civilian areas may have been struck accidentally but reemphasized its policy to bomb only military targets in the North and to take all possible care to avoid civilian casualties. It was impossible, it said, to avoid some damage to civilian areas.[9]

Other Air Operations in November and December

Other air action in the last two months of 1966 included restrikes along the Hanoi-Lai Cai railroad line in route package V and continuation of the Tally-Ho air campaign in route package I. In fact, about 43 percent of the total U.S. air effort in the North, and 64 percent of the USAF effort was directed against targets in route package I. An Air Force compilation of the results of the Tally-Ho air campaign from 20 July through November showed the following:

Target	Destroyed	Damaged	Other
Trucks	72	61	
Structures	1,208	624	
Anti-aircraft and air warning positions	92	22	
Roads cut, cratered or seeded			339
Landslides			6
Secondary explosions			1,414

Nevertheless there was still considerable uncertainty as to the overall effect of this air program on North Vietnam's ability to resupply the South.

A limited number of USAF road cutting and other air strikes were also made in route packages II, III, and IV, There were no B-52 strikes in the North in November but in December 78 sorties were flown in the DMZ and 35 sorties slightly above the zone. From 12 April 1966 when the first strike was conducted against North Vietnam through the end of the year, B-52s flew 280 sorties including 104 sorties in "DMZ North." The major B-52 effort was directed against targets in South Vietnam. Year-end operations were also highlighted by 48-hour Christmas and New Year "truces". Although bombing ceased over the North during each truce period, USAF reconnaissance flights continued. USAF attack sorties for the year totaled 44,500, slightly more than 54 percent of the 81,948 attack sorties flown in the North by all U.S. and VNAF.[11]

Meanwhile, the JCS in November asked Admiral Sharp to comment on the "Combat Beaver" proposal that the Air Staff had developed in conjunction with the other services to support Secretary McNamara's proposed electronic and ground barrier between North and South Vietnam. Using Steel Tiger, Gate Guard, and Tally-Ho experience, Combat Beaver called for day and night air strikes on key logistic centers. This, it was hoped, would create new concentrations of backed-up enemy materiel and equipment suitable for air strikes. It would complement any ground barrier system and could begin immediately.[12]

Air War – Vietnam

Admiral Sharp's comments were critical. He said that with certain exceptions Combat Beaver was similar to the current air program. He thought that it overstressed the importance of air strikes in route packages II, III, and IV and would result in high aircraft losses. It would not, in his view, increase overall air effectiveness but, instead, disrupt the existing well-balanced air effort. Taking into account CINCPAC's comments and those of other agencies, the Air Staff reworked the proposal and, at the end of December, produced a new one, designating it the integrated strike and interdiction plan (ISIP).[13]

Assessment of Enemy Air Defenses

CHRONOLOGY OF THE GROWTH OF NORTH VIETNAM'S AIR DEFENSES 1964-1966	
Jul 1964	Air defense system based on obsolescent equipment. Anti-aircraft guns, 50; SA-2's,0; air defense radars, 24; fighter aircraft, 0.
Aug 1964	Introduction of MIG-15's.
Mar 1965	Introduction of improved air defense radars such as ground control intercept.
Apr 1965	First use of MIG fighter aircraft. Detection of first SA-2 site under construction.
Jun 1965	Increase in air defense radars to 41
Jul 1965	First SA-2 fired at U.S. aircraft. Introduction of 100mm antiaircraft guns.
Aug 1965	Significant increase in low-altitude air defense radar coverage. Increase in antiaircraft strength to about 3,000 guns.
Dec 1965	Introduction of MiG-21's. Beginning of emission control of air defense radar.
Mar 1966	Introduction of system for identification, friend or foe.
Jul 1966	First MiG use of air-to-air missiles.
Aug 1966	Completion of a sophisticated air defense system. Anti-aircraft guns, 4,400; SA-2's, 20 to 25 firing battalions; air defense radars, 271; fighter aircraft, 65.
Dec 1966	Air defense system includes: light and medium antiaircraft guns, 6,398; SA-2 sites, 151; SA-2 firing battalions, 25; MiG-15s and -17s, 32; MiG-21s,15; use of air-to-air missiles.
SOURCE: Briefing Report on Factors Affecting A/C Losses in SEA, 26 Sep 66, prepared by Col. H.W. Hise, JCS (TS); USAF Mgt Summary (S), 6 Jan 67; p 70; Ops Review Gp, Dir/Ops, Hq USAF; *N. Y. Times*, Jul 66.	

Air Operations 1966

By the end of 1966 the overwhelming number of U.S. combat aircraft losses in the North was still caused by conventional antiaircraft fire. The Seventh Air Force estimated the enemy's antiaircraft strength had grown from 5,000 to 7,400 guns during the year. Nevertheless, U.S. aircraft losses were decreasing with 17 downed in November and 20 in December. The Air Force lost 24, 12 in each of the two months.[14]

The MIG threat increased in December, apparently in response to the latest U.S. attacks on important targets. During 35 encounters and 16 engagements two F-105!s were lost as against one MiG. one of the losses, on 14 December, was the first one attributed to a MiG-21 air-to-air missile. Other air-to-air missiles were fired on at least five occasions during the month, but U.S. air superiority was easily maintained. Between 3 April 1965, when the MiGs first entered the war, and 31 December 1966 there were a total of 179 encounters and 93 engagements. The aerial battles cost the enemy 28 MiGs as against 9 U.S. aircraft, a ratio of 1 to 2.8. Of the nine losses, seven were USAF and two were Navy. In addition, there were two "probable" USAF losses to MiGs. In December, the enemy's combat aircraft inventory, recently augmented by Soviet deliveries, was believed to consist of 32 MiG-15s and -17s, 15 MIG-21's, and six Il-28s, all at Phuc Yen airfield.[15]

SA-2s continued to take a small but steady toll. They claimed one USAF aircraft in November and three in December. Because the missiles precluded the use of optimum air tactics, Admiral Sharp on 22 November proposed to the JCS a major effort to solve the SA-2 problem. He placed the current SA-2 strength at 28 to 32 firing battalions (the year-end estimate was 25 battalions) and warned that the number would increase unless air restrictions were eased. Already a shortage of special munitions and properly equipped aircraft prevented a large-scale attack on these mobile, well-camouflaged units. Only a "blitzkrieg" type of attack could prevent their movement.[16]

For the short term, Admiral Sharp recommended the use of all available aircraft to detect SA-2 sites, revision of the current targeting system to include SA-2 assembly and storage areas regardless of location, a priority intelligence effort to locate key SA-2 control facilities, and attacks on high priority targets in the North in random fashion to avoid establishing a predictable pattern of attack. He also urged steps to increase Shrike production, assure positive control and tracking of all U.S. aircraft through the USAF "Big Eye" EC-121 program, improve distribution of SA-2 data, exploit more fully color photography in penetrating camouflage, and equip all aircraft with ECM, chaff, homing radars, and warning receivers.

Further, the State and Defense Departments should release statements to discourage the Soviets from deploying additional SA-2 systems by pointing to the danger of escalation, and the "intelligence community" should constantly review and distribute all relevant SA-2 information.

For the long term, Admiral Sharp said there was a need to expedite procurement of an anti-radiation missile, develop better warheads using the implosion

Air War – Vietnam

principle, employ beacons to aid in finding SA-2 emitters, provide VHF/UHF homing capabilities for Wild Weasel aircraft, and improve data exchange between the Rome Air Development Center and Southeast Asia operational activities. [17]

The Air Staff generally agreed with Admiral Sharp's recommendations. The JCS also concurred and directed General McConnell to procure and deploy adequate numbers of anti-SA-2 devices and equipment. The Joint Chiefs were still undecided at the end of the year whether to recommend to Secretary McNamara an all-out campaign against the SA-2's in the immediate future. [18]

Assessments of the Air War Against North Vietnam

As 1966 ended, General McConnell and the Air Staff remained convinced that greater use of air power, especially in North Vietnam, was the only alternative to a long, costly war of attrition. They also thought it would make unnecessary the massive buildup of U.S. and allied ground forces still under way. Although the combined air and ground effort in South-east Asia had prevented a Communist takeover of South Vietnam, one Air Staff assessment found no significant trend toward the attainment of other U.S. objectives in that country.[19]

Within the JCS General McConnell continued to support recommendations to reduce operational restrictions and expand target coverage in the North. The level of air effort was less than he desired, but he believed air power had shown how it could be tailored to the geography of a country and, by the selection of weapons and mode of air attack, be responsive to political and psychological considerations. In some instances, it was clear, the Vietnam experience ran counter to conventional air power concepts. As he had observed in May, "tactical bombing" in South Vietnam was being conducted in part by "strategic" B-52 bombers and "strategic" bombing of the North was being conducted largely by "tactical bombers". [20]

Any evaluation of the effect of air power, especially in the North, had to consider political factors which limited military activity. To deal with this circumstance, General McConnell offered the following dictum: "Since air power, like our other military forces, serves a political objective, it is also subject to political restraints. Therefore, we must qualify any assessments of air power's effectiveness on the basis of limitations that govern its application." [21]

General Harris, the PACAF commander, singled out three principal factors hampering the air campaign against North Vietnam: political restraints and geographical sanctuaries that precluded striking more lucrative targets, poor weather for prolonged periods of time, and Hanoi's ability to repair and reconstruct damaged target areas. With respect to the last, PACAF officials acknowledged the North Vietnamese had "exceptional" recuperative capabilities to counter air attacks on trucks, rolling stock, and the lines of communications. They had built road and rail by-passes and bridges in minimum time, dispersed POL by using pack animals, human porters and watercraft, and developed an effective air defense system. Infiltration through the DMZ, Laos, and Cambodia was placed at 7,000 to 9,000 men per month, and the enemy logistic system was supporting an estimated 128,000 combat and combat support personnel with out-

Air Operations 1966

of-country resources. MACV and DIA eventually estimated that about 81,000 North Vietnamese entered South Vietnam in 1966. The infiltration rate was high in the first half and dropped sharply in the second half of the year. General Harris thought that an important "lesson learned" was that the gradual, drawn-out air campaign had created very little psychological impact on Hanoi's leaders and the populace. He also continued to believe (as did the Air Staff and other Air Force commanders in Southeast Asia) that control of air operations in the North, as well as in Laos and South Vietnam, was too fragmented and should be centralized under a single air commander. [22]

Admiral Sharp's view of the air campaign against the North in 1966 was that little had been accomplished in preventing external assistance to the enemy. Except for the June strikes on POL targets in Haiphong (which handled 85 percent of the North's imports during the year), the port was almost undisturbed. Of the nearly 82,000 attack sorties flown during the year, less than one percent were against JCS-proposed targets. In the critical northeast area (route packages VIA and VIB), of 104 targets only 19 were hit in 1965 and 20 in 1966; the remaining 99 percent of attack sorties were armed reconnaissance and flown to harass, disrupt, and impede the movement of men and supplies on thousands of miles of roads, trails, and inland and coastal waterways. He noted that despite severe losses of vehicles, rolling stock, watercraft, supplies and men from air attack, the North Vietnamese were ingenious in hiding and dispersing their supplies and showed "remarkable" recuperative ability. He concluded that the overall amount of supplies and men moving through the DMZ, Laos, and Cambodia into South Vietnam probably was greater in 1966 than in 1965. [23]

Secretary Brown took a somewhat different view of the air campaign believing it had inflicted "serious" logistic losses on the North. From 2 March 1965 (when the Rolling Thunder program began) through September 1966, air strikes had destroyed or damaged more than 7,000 trucks, 3,000 railway cars, 5,000 bridges, 15, 000 barges and boats, two-thirds of the POL storage capacity, and many ammunition sites and other facilities. He cited prisoner of war reports indicating that troops in the South received no more than 50 percent of daily supply requirements. In addition, the air war had diverted 200,000 to 300,000 personnel to road, rail, and bridge repair work, and combat troops for air defense. On 1 March 1967, Secretary McNamara estimated that Hanoi was using 125,000 men for its air defenses and "tens of thousands" of others for coastal defense. By December, military action in both North and South Vietnam had reduced battalion size attacks from seven to two per month and, in the past eight months, raised enemy casualties from 3,600 to 5,200 per month.

Although infiltration from the North continued, Secretary Brown said: "I do not believe that an air blockade of land and sea routes will ever be completely effective any more than a sea blockade can prevent all commerce from entering or leaving a country. " He thought the air attacks were becoming more effective due to improvements in intelligence, tactics, equipment, and techniques.

The Air Force secretary defended the administration's policy of exempting certain targets from air attack if they supported only the North's civilian economy, were close to urban areas and would cause civilian suffering if hit, and would not significantly affect in the short term the enemy's ability to continue fighting. He listed five criteria for judging whether to strike a target: its effect on infiltration from North to south, the extent of air defenses and possible U.S. aircraft losses, the degree of "penalty" inflicted on North Vietnam, the possibility of civilian casualties, and the danger of Soviet or Chinese intervention resulting in a larger war. He thought that a "Korean-type" victory, with the aggressor pushed back and shown that aggression did not pay, would meet U.S. objectives and make the war in Vietnam a "success." [24]

Secretary McNamara's views on the controlled use of air power against the North were well known. In a "deployment issue" paper sent to the JCS on 6 October in conjunction with deployment planning, he said that intelligence reports and aerial reconnaissance clearly showed how the air program against the North effectively harassed and delayed truck movements and materiel into the South but had no effect on troop infiltration moving along trails. He thought that the cost to the enemy to replace trucks and cargo as a result of stepped up air strikes would be negligible compared with the cost of greatly increased U.S. aircraft losses. In a summation of his views on the war before House Subcommittees in February 1967 he further stated:

> For those who thought that air attacks on North Vietnam would end the aggression in South Vietnam, the results from this phase of the operations have been disappointing. But for those who understood the political and economic structure of North Vietnam, the results have been satisfactory. Most of the war materiel sent from North Vietnam to South Vietnam is provided by other Communist countries and no amount of destruction of the industrial capacity can, by itself, eliminate this flow....

When the bombing campaign began he added, "we did not believe that air attacks on North Vietnam, by themselves, would bring its leaders to the conference table or break the morale of its people, and they have not done so."

The Defense Secretary also observed that although air strikes had destroyed two-thirds of their POL storage capacity, the North Vietnamese had continued to bring it in "over the beach" and disperse it. POL shortages did not appear to have greatly impeded the North's war effort. He reiterated the U.S. policy that "the bombing of the North is intended as a supplement to and not a substitute for the military operations in the South." [25]

Air Operations 1966

APPENDIX 1

U.S. and VNAF Attack Sorties in Southeast Asia: 1966					
	USAF	USN	USMC	VNAF	Total
North Vietnam	44,500	32,955	3,702	799	81,956
Laos	32,115	9,044	3,601	0	44,760
South Vietnam	70,367	21,729	37,602	32,033	161,731
TOTAL	146,982	63,728	44,905	32,832	288,447

SOURCE: Annual Supplement to Summary Air Ops, SEA, Cy 1966, prepared by Dir/Tac Eval, Hqs PACAF, 23 Jan 64; Ops Review Gp, Dir/Ops, Hq USAF.

Defense Lion notes that this table was badly corrupted in the original

APPENDIX 2

B-52 Sorties in SE Asia			
	1965	1966	Total
North Vietnam	0	176	176
South Vietnam	0	3,112	3,112
Laos	162	616	778
DMZ North	0	104	104
DMZ South	118	282	400
	280	4,290	4,570

SOURCE: Strat Ops Div, J-3, JCS; Ops Review Gp, Dir/Ops, Hq USAF

Defense Lion notes that this table was badly corrupted in the original

APPENDIX 3

U.S. and VNAF Attack Sorties in North Vietnam: 1966 By Month					
	USAF	USN	USMC	VNAF	Total
January*	57	80	0	0	137
February	1,547	1,265	0	0	2,812
March	2,559	1,919	0	0	4,478
April	2,477	2,818	8	144	5,447
May	1,794	2,568	0	103	4,465
June	4,442	3,078	2	266	7,788
July	6,170	3,416	370	243	10,199
August**	6,336	4,683	792	21	11,832
September	6,376	4,953	825	6	12,160
October	4,932	3,147	559	4	8,642
November	3,681	2,938	633	8	7,260
December	4,129	2,090	513	4	6,736
TOTAL	44,500	32,955	3,702	799	81,956

* Bombing of North Vietnam resumed on 31 January 1966.

** Reflects an increase from two to three aircraft carriers at "Yankee Station" beginning in August 1966.

SOURCE: Annual Supplement to Summary of Air Ops SEA, Cy 1966. Prepared by Dir/Tac Eval, Hqs PACAF, 28 Jan 67; Ops Review Gp, Dir/Ops, Hq USAF.

APPENDIX 4

U.S. Aircraft Losses in Southeast Asia
Hostile Causes

1965				
	North Vietnam	Laos	South Vietnam	Total
USAF	82	11	64	157
USN	85	8	6	99
USMC	3	3	0	6
TOTAL	170	22	70	262
1966				
USAF	172	48	76	296
USN	109	7	6	122
USMC	4	5	14	33
TOTAL	285	60	70	451

Operational Causes			
	1965	1966	Total
USAF	64	78	142
USN	27	40	67
USMC	10	12	22
Total	101	130	231

The above listing excludes helicopters but includes losses due to enemy mortar attacks.

USN and USMC figures subject to variations contingent on bookkeeping procedures.

SOURCE: Ops Review Gp, Dir/Ops, Hq USAF.

APPENDIX 5

USAF Combat Attrition in North Vietnam			
1965*			
Type of Sortie**	Sorties	Losses	Rate per 1,000 sorties
Attack	11,599	63	5.43
CAP/Escort	5, 675	7	1.23
Reconnaissance	3,294	9	2.73
Other	4,983	3	0.6
TOTAL	25,551	82	3.21
1966			
Type of Sortie**	Sorties	Losses	Rate per 1,000 sorties
Attack	44,482	138	3.10
CAP/Escort	9,041	6	0.66
Reconnaissance	7,910	19	2.40
Other	16,587	9	0.54
TOTAL	78,020	172	2.20

* Bombing of North Vietnam began on 7 February 1965.

** Excludes B-52 strikes.

SOURCE: Ops Review Gp, Dir/Ops, Hq USAF.

Air Operations 1966

APPENDIX 6

U.S. Aircraft Losses to SA-2s

Date	Missiles Fired	Confirmed Losses			Probable Losses			% Confirmed	% Effective
		USAF	USN	USMC	USAF	USN	USMC		
1965*	180	5	5	0	0	1	0	5.6	6.1
1966	1,057	13	7	0	5	6	0	1.9	2.9
Total	1,237	18	12	0	5	7	0	2.4	3.4

* The first SA-2 firings were sighted in July 1965.
SOURCE: Ops Review Gp, Dir/Ops, Hq USAF.

APPENDIX 7

SA-2 Sites in North Vietnam

	Jan	March	June*	Sep	Dec
1965	0	0	4	23	64
1966	64	100	115	144	151

* The first SA-2 site was detected in April 1965.
SOURCE: Ops Review Gp, Dir/Ops, Hq USAF.

APPENDIX 8

Light and Medium Antiaircraft Artillery Guns in North Vietnam

	Jan	Feb*	March	June*	Sep	Dec
1965	0	1,156	1,418	1,643	2,636	2,551
1966	2,884	3,092	3,159	4,123	5,009	6,398

* Bombing of North Vietnam began on 7 February 1965.

SOURCE: Ops Review Gp, Dir/Ops, Hq USAF.

APPENDIX 9

	U.S. Aircraft Losses in Aerial Combat			
	USAF	USN	USMC	Total
1965	2*	0	0	2
1966	5**	4***	0	9
Total	7	4	0	11

* Consisted of 2 F-105s.

** Consisted of 3 F-105's, 1 F-4C, 1 RC-47 and two "probables", 1 F-4C and 1 A-1.

*** Consisted of 3 F8s and 1 KA3. No "probables."

SOURCE: Ops Review Gp, Dir/Ops, Hq USAF.

APPENDIX 10

North Vietnamese Aircraft Losses in Aerial Combat				
1965				
Destroyed by	MiG-15s	MiG-17s	MiG-21s	Total*
USAF	0	2	0	2
USN	0	3	0	3
USMC	0	0	0	0
TOTAL	0	5	0	5
1966				
USAF	0	12	5	17
USN	0	4	2	6
USMC	0	0	0	0
TOTAL	0	16	7	23

* No "probables" listed,

SOURCE: Ops Review Gp, Dir/Ops, Hq USAF.

PART FIVE

THE SEARCH FOR MILITARY ALTERNATIVES 1967

Air War – Vietnam

Search For Alternatives 1967

I. THE SITUATION IN EARLY 1967

At the beginning of 1967, American officials were again fairly optimistic about the trend of the war in Southeast Asia. President Lyndon B. Johnson, in his State of the Union message on 10 January, declared that. while the end of the conflict was "not yet in sight." Gen. William C. Westmoreland, Commander of the U.S. Military Assistance Command, Vietnam (COMUSMACV) believed the enemy could "no longer succeed on the battlefield." The U.S. Ambassador to Saigon. Henry Cabot Lodge, on the 11th predicted that American war casualties would decrease in 1967 and that there would be "tremendous military progress." He emphasized the importance of continuing the bombing of North Vietnam. Gen. Maxwell D. Taylor, Special Consultant to the President, after a tour of the war zone. concluded at the end of January 1967: "I have a feeling that the Vietnamese situation may change drastically for the better by the end of 1967." [1]

Secretary of Defense Robert S. McNamara, in testimony before congressional committees. was also hopeful. He saw substantial military progress. with search and destroy operations an "unqualified success. " and air operations over South and North Vietnam producing good results. In addition. Mr. McNamara took comfort in such encouraging economic and political developments as South Vietnam's currency devaluation of 18 June 1966, which had arrested excessive inflation and promoted economic stability. and the election of 3 September 1966 in which 80 percent of the eligible voters cast ballots for a 117-man Constituent Assembly. This body would write a new constitution and help prepare for national elections leading to a new government. Another important event was the Manila Conference on 23-24 October attended by representatives of the seven principal allied nations participating in the war (Australia, New Zealand, Philippines, Thailand, Republic of Korea, South Vietnam, and the United States.) The conferees affirmed their determination "that the South Vietnamese people shall not be conquered by aggressive force and shall enjoy the inherent right to choose their own way of life and their own form of government." They also attested to the non-aggressive character of the seven-nation effort and promised withdrawal of allied military forces within six months or sooner if North Vietnam stopped its aggression and pulled back its troops. [2]

Secretary of Defense Robert Strange McNamara took the decisions that shaped the course and outcome of the war in Vietnam. Source: U.S. Library of Congress

The Joint Chiefs' and Air Force Views of the War

The Joint Chiefs of Staff (JCS) were not quite so sanguine. In a review of the war through 1966, they recognized several allied accomplishments: the prevention of a Communist takeover of South Vietnam. the infliction of heavy losses on the enemy, and the success of B-52 and ground forces "spoiling" operations. But the JCS found no "substantial trend" toward attaining the American objective of ending Communist efforts to conquer Southeast Asia. The Air Force view was especially sober. Its Chief of Staff. Gen. John P. McConnell, had generally supported JCS recommendations on building up U.S. forces in Southeast Asia, but he believed that more effective use should have been made of air power rather than to deploy so many ground troops. It was General McConnell's view that the military strategy being followed presaged a long, costly war of attrition and would require the use of even more troops. [3]

The Air Force argued strongly for a reduction of the restraints on the use of air power, especially against North Vietnam. General McConnell informed a House committee in March 1967 that three types of restraints had been imposed on the use of aircraft: the geographical areas in which they could operate, the ordnance they could carry, and the targets they could hit. It advocated striking at the remaining important war-supporting targets, particularly those in the "sanctuary" areas around Hanoi and Haiphong and in the "buffer" zone near China. It also wished to bomb or mine Haiphong harbor through which an estimated 85 percent of war-supporting imports entered from the Soviet Union. China, and other Communist (plus some non-Communist) states. With some exceptions, Adm. U.S. Grant Sharp. Commander in Chief, Pacific Command (CINCPAC), and the other members of the JCS shared these Air Force views. [4]

Other high officials. especially Secretary McNamara. disagreed. The Secretary was apprehensive, as were other administration leaders. that removal of bombing restraints might precipitate a wider war. He considered air attacks on the North as a supplement to and not a substitute for military operations in South Vietnam where, in the final analysis. the war had to be won. Secretary of the Air Force Harold Brown generally endorsed Mr. McNamara's position. [5]

U.S. and Allied Deployed Strength

The administration, partly for political reasons. had avoided a callup of reserves or extending service tours of duty and, except for the initial commitment of combat troops in 1965, had dispatched U.S. forces in accordance with specific deployment plans. The most recent, designated "Southeast Asia Deployment Program 4" and approved by Secretary McNamara on 18 November 1966, established U.S. manpower ceilings in South Vietnam of 439,500 by June 1967, 463,300 by December 1967, and 469,300 by June 1968.

The Air Force portion under this program was to remain stable, totaling 55,300, 55,400, 55,400, respectively. The June 1968 figure was about 52,900 less than the JCS had recommended. In addition to limiting the American commitment as much as possible, Mr. McNamara believed the manpower ceilings were needed to reduce excessive expenditures that might undermine South Vietnam's relative

Search For Alternatives 1967

price stability, achieved following the 1966 currency devaluation. He pointed out that inflation hit hardest the Vietnamese soldiers and civil servants on whom success in the war largely depended. [6]

At the end of 1966, American armed strength in South Vietnam stood at 390,568 (including 52,913 Air Force), and in Thailand, 34,489 (including 26,113 Air Force). In addition, the offshore U.S. Seventh Fleet possessed 36,300 personnel. South Vietnamese regular, regional, and popular forces numbered about 620,000 (including 15,070 in the Vietnamese Air Force). The strength of other allied forces, serving in South Vietnam at the end of 1966 was as follows: Australia, 4,533; New Zealand, 155; Philippines. 2,063; Republic of Korea, 45,605; and Thailand, 224 for a total of about 52,580. [7]

To prosecute the air war in South and North Vietnam and Laos, the United States had deployed more than 2, 000 combat and support aircraft and 1, 900 helicopters. Six hundred and thirty-nine combat aircraft belonged to the Air Force, 210 to the Navy, and 160 to the Marine Corps. Of the support aircraft, 534 were Air Force, 484 Army, 35 Navy, and 34 Marine Corps. The preponderant number of helicopters, 1,637, were operated by the Army. The Marine Corps possessed 229 and the Air Force 70. [8]

Augmenting these forces were 50 B-52 bombers on Guam and a Vietnamese Air Force (VNAF) with about 100 tactical and 152 support aircraft and 43 helicopters. American tactical aircraft and B-52's at the end of 1966 were flying about 25,000 attack sorties per month in Southeast Asia. In December, 13,246 of these sorties were flown in South Vietnam, 6,672 in North Vietnam, and 4,841 in Laos. Both the number of aircraft authorized and the attack sortie rate were below JCS recommendations. [9]

The bulk of the Air Force's combat aircraft were F-4s, F-105s, and F-100s with smaller numbers of A-1s and B-57s. Its principal reconnaissance aircraft were RF-4s and RF-101s. In addition, it employed a growing number of special air warfare (SAW) and specially equipped aircraft such as EC-121s, EB/RB-66s, and EC/RC-47s for electronic countermeasures operations and reconnaissance. Gun-firing AC-47s also were used in close support operations, and UC-123s for chemical operations to destroy jungle growth and crops in selected areas. [10]

The EC-121 Warning Stars provided radar coverage of the airspace over North Vietnam. Source: U.S. Air Force

Air War – Vietnam

Arrayed against American, South Vietnamese, and other allied forces were about 275,000 Viet Cong and North Vietnamese personnel in South Vietnam, the latter estimated at about 45,000. Enemy infiltration into the South was placed at 5, 300 to 9, 000 per month. Headquarters, Pacific Command (PACOM) early in 1967 estimated infiltration at 7,500 to 9,000 per month but Secretary McNamara thought this included confirmed and probable infiltrators. His "accepted" figure was an average of 5, 300 infiltrators monthly over the past nine months, and he pointed to a two to three-month lag in estimates.

North Vietnamese anti-aircraft gun crew. Source: Vietnamese People's Liberation Army

To defend its war supplies and lines of communications (LOC' s) from air attacks, the enemy had developed a highly sophisticated air defense system consisting of 37/57-mm and 85/100-mm antiaircraft weapons, surface-to-air missiles (SAMs), and a small air force. North Vietnam possessed about 54 MiG-15s and -17s and eight MiG-21's, plus a smaller number of MiG-15's and -17's and two IL-28 light bombers on the South China air bases of Yunnani and Peitan. Behind Hanoi's ground and air posture stood, of course, additional resources of North Vietnam and her principal suppliers, the Soviet Union, China, and other Communist states. [11]

Adjustments in Deployment Planning

On 16 - 19 January 1967, representatives from the Office of the Secretary of Defense (OSD) and the services met in Hawaii, reviewed Deployment Program 4 and agreed to raise the American military personnel ceiling in South Vietnam to 471,623 by 30 June. The revised service totals were: Air Force, 55,975; Navy (including Coast Guard), 28,431; Marine Corps, 71,000; and Army, 316,217. OSD confirmed the new ceiling on 31 March 1967. [12]

A-37 Dragonflys replaced the venerable A-1 Skyraiders in the VNAF. Source: U.S. Air Force

During the first six months of 1967, administration-imposed restrictions prevented deployment of additional Air Force units into the theater, but other actions led to an overall increase in USAF aircraft there. On 1 January, the Army transferred 83 C-7A Caribou transports in South Vietnam to the Air Force. Two other theater changes approved by OSD on 13 February were the transfer of six AC-47s from the Philippines to South Vietnam and the retention of an A-37

Search For Alternatives 1967

squadron (which had completed its "Combat Dragon" tests in December 1966) by USAF special air warfare forces until an evaluation of its operations were completed. The A-37 squadron. based at Bien Hoa AB, South Vietnam, began operational tests on 15 August 1967. Plans called for it to replace a USAF A-1 squadron in January 1968. Later that year, A-37s would begin entering the VNAF's inventory in lieu of A-1Es which would be returned to the Air Force. Early in 1967 the United States also welcomed the decision of the Australian government to send one Royal Australian Air Force (RAAF) Canberra squadron of eight aircraft to South Vietnam. The RAAF unit began combat operations on 23 April 1967.[13]

The beginning of 1967 also witnessed the redeployment of 25 USAF helicopters and 164 personnel from Thailand to South Vietnam. despite considerable Air Force and other opposition to the move. On 22 June 1966. Secretary McNamara had approved the temporary transfer of 10 USAF CH-3's and 11 UH-1Fs from South Vietnam to Nakhom Phanom AB. Thailand, to augment the Air Force's 606th Air Commando Squadron (ACS) which then possessed four UH-1Fs and other aircraft. On 12 January 1967. he ordered the return of all the helicopters to South Vietnam. His directive triggered a flurry of "reclamas" and compromise proposals by the Air Staff. JCS. PACOM. MACV. and the U.S. embassies in South Vietnam. Thailand. and Laos. Virtually all officials desired to keep the helicopters in Thailand in order to aid the Thai counterinsurgency program and to continue intelligence-gathering activities. Mr. McNamara. however. was opposed to more direct American involvement in the Thai program and. on 31 January, he reaffirmed his decision but allowed the four UH-1Fs. initially assigned to the 606th to remain in Thailand. However, he directed that they not be used in Thai counterinsurgency operations.[14]

The early months of 1967 also witnessed a decision to deploy Strategic Air Command (SAC) B-52 bombers to Thailand for the first time. The B-52 "Arc Light" program, approved by Secretary McNamara in late 1966, called for higher combat sortie rates than those recommended by the JCS. On 11 November 1966, the Secretary authorized 800 sorties per month beginning the following February. This decision

U-Tapao airfield started as a forward operating base for B-52s and quickly evolved into a strategic installation of major significance.
Source: Royal Thai Army

required the dispatch of 20 additional bombers to the theater to join the 50 already on Guam. Numerous studies had been made on the possible stationing of some B-52's in South Vietnam, the Philippines, Okinawa, Taiwan, or Thailand to improve "reaction time. " South Vietnamese bases, however, were deemed too insecure. Okinawa and Taiwan as B-52 base sites involved sensitive political questions. A site in the Philippines also posed political problems and would require costly and time-consuming base expansion. In the end, the planners were left with U-Tapao AB, Thailand, as the most promising location. The Air Staff, General McConnell and other JCS members strongly favored that base since the flying time and cost of bomber missions from there against targets in South Vietnam would be substantially less than from Guam or other bases under consideration.

After a preliminary White House decision in late 1966 to emplace the B-52's in Thailand, U.S. Ambassador Graham A. Martin began final negotiations with Thai officials. Extended talks, which included a quid pro quo in the form of additional military assistance to Thailand, culminated in agreement on 2 March to station 15 B-52's at U-Tapao. On 23 March Secretary of State Dean Rusk publicly announced the decision. Three bombers arrived at the Thai base on 10 April and flew their first mission the same day. Other bombers deployed in May and June and the last five (of 15) on 9 July. [15]

Search For Alternatives 1967

II. THE DEBATE OVER TROOP DEPLOYMENTS

While the services undertook to build up their forces to the approved level of 471,623 by 30 June 1968 and simultaneously began work on an anti-infiltration system in Vietnam, the President, under increasing fire from critics of U.S. involvement in the war, expressed a desire to accelerate military operations. On 17 February 1967, in response to the President's request, Gen. Earle G. Wheeler, Chairman of the JCS, directed the Joint Staff to prepare a range of military proposals that would assure "a definite and visible improvement" in the war by Christmas 1967.[1]

The Joint Staff quickly produced three proposals, each of which called for a progressive increase in air, ground, and naval pressure on the enemy. Concerning the air campaign against the North, the Air Staff prepared an integrated tactical target plan which outlined a series of attacks on certain key targets within the prohibited and restricted areas of Hanoi and Haiphong. In sending these proposals to higher authorities, the JCS recommended that operations begin at the onset of favorable weather in April to assure the desired results by the end of 1967.[2]

F-105s attack a bridge in southern North Vietnam, 1966. Source: U.S. Air Force

On 22 February, the administration approved a limited number of the suggested actions. In the North, it authorized a modest increase in Rolling Thunder strikes, the mining of designated inland waterways and estuaries, and naval operations against certain targets ashore or in coastal waterways. It also approved extending special air and ground operations into Laos, and the use of artillery fire from South Vietnam against enemy targets in the demilitarized zone (DMZ), North Vietnam, and Laos. This new authority for air strikes, while welcome, was still considered inadequate by the Air Staff which felt that the attenuated combat operations would not halt resurgent Viet Cong and North Vietnamese activities. The Air Staff's pessimism appeared to be justified when, in. March, General Westmoreland proposed a reexamination of the overall U.S. strategy for Southeast Asia and recommended the deployment of additional forces beyond those previously approved by the administration. [3]

General Westmoreland's Proposals

Beginning on 18 March, General Westmoreland forwarded to the JCS his latest assessments of the enemy's strength. He estimated that the Communists had ground forces in the field equivalent to 10 infantry divisions and four infantry brigades, one armored and eight independent regiments, one artillery division, one antiaircraft division, 80 independent antiaircraft regiments, 32 SAM battalions, and seven transport units. He credited North Vietnam's home-based air force with the capability to launch 87 jet fighter and six jet bomber sorties on a single mission while 26 to 32 more jet fighters and two light bombers belonging to the North Vietnamese were based in South China.

SA-2 missile batteries were spreading across North Vietnam.
Source: U.S. Air Force

The MACV commander also reported that the Communists had achieved a net gain of 50, 000 men, tantamount to an increase of nine to 12 divisions, despite battle losses of 127,000 between January 1966 and March 1967. Even with a projected loss of 140, 000 men in 1967. they could increase their numbers by 27,

Search For Alternatives 1967

000 through more infiltration and continued recruiting in the South, despite their diminishing success in obtaining new recruits. He believed that the Communists could sustain a sizable force until mid-1968 and conduct operations one day in 30 with maneuver battalions of 70 to 80 percent of regular strength.

According to General Westmoreland, the Communists had deployed one or two divisions north of the DMZ and other combat units available that could attack through Cambodia and Laos. In South China were seven (of 34) Chinese armies and 13 more could be dispatched southward. The Chinese also possessed 2,229 jet fighters and two medium and 235 light bombers. and both China and the Soviet Union showed every intention of supporting the conflict at current or even higher intensity.

Accordingly. the MACV commander asked for the additional deployment by 1 July 1968 of a "minimum essential force" comprising two and one-third divisions and four USAF tactical fighter squadrons (this was revised shortly to five USAF squadrons.) After 1 July 1968, to avoid a protracted war. he would further require an "optimum force" of four and two-thirds divisions and 10 USAF tactical fighter squadrons. Admiral Sharp generally concurred with the Westmoreland requests.[4]

Chinese aid to North Vietnam was more than just propaganda. In 2009, an official Chinese defense white paper revealed that 320,000 Chinese troops had served in North Vietnam, 5,000 of whom were killed. Source: Chinese Ministry of Defense White Paper 2009

Before forwarding these recommendations. General Wheeler asked the Joint Staff to analyze the proposed manpower increases under two "cases.' Case I would avoid a reserve callup or extension of service tours; Case II would require both. In either "case. 11 the services would be required to retain a capability to meet their North Atlantic Organization (NATO) commitments and make no change in the length of tours of duty in the war zone. Manpower deployments would be in addition to the latest Program 4 personnel authorization of 471,623 for South Vietnam by June 1968.[5]

Air Staff and Joint Staff studies showed that the minimum essential force would provide 21 maneuver battalions (from two and one-third divisions), five fighter squadrons, one C-130 squadron, four river assault squadrons, 59 river patrol boats, and associated engineer and construction battalions. It would require a rise in American personnel in South Vietnam of about 78, 000 and bring the 1967 total to 548,801. The optimum force after 1 July 1968 would provide 42 additional maneuver battalions (from four and two-thirds divisions), 10 USAF fighter

255

Air War – Vietnam

squadrons, and require another air base and a complete mobile riverine unit. It would result in the dispatch of about 122,000 additional American servicemen to South Vietnam for an overall total of 671,616. [6]

Early in April, at the request of Gen. John D. Ryan, Commander in Chief, Pacific Air Forces (CINCPACAF), who had succeeded Gen. Hunter Harris, Jr. on 1 February 1967, Admiral Sharp, and the Air Staff and the Joint Staff agreed to consider incorporating 12,009 more Air Force personnel in General Westmoreland's minimum essential force. A total of 7,989 would be deployed to South Vietnam to support additional UH-1 and CH-3 helicopters in operations outside the country and for general augmentation of other on-going activities. Another 4,020 would be sent to Thailand to support three more tactical fighter squadrons and to convert Nam Phong AB, which was in a "bare base" status, into a main operational base.

Air Staff-JCS Views of General Westmoreland's Requests

The initial Joint Staff paper advocating increasing the approved June 1968 manpower levels in South Vietnam by another 200,000 men deeply concerned the Air Staff and General McConnell. The Chief of Staff observed that U.S. strength during the past two years had risen "far in excess" of original requirements, yet the enemy was still a "potent threat. " As evidence, he cited the recent shift of I Corps units in "Operation Oregon" from one critical area to reinforce another and he noted that the JCS was weighing alternate measures to blunt the Communist thrusts, including a possible lodgment in North Vietnam. (Called Mule Shoe, the JCS study on a possible lodgment in North Vietnam was completed early in April.) Although current tactics might relieve pressure, General McConnell said they would not end Soviet or Chinese support of Hanoi. He believed that the fighting and staying power of the North Vietnamese and Viet Cong had been underestimated and he was not convinced that the addition of more troops, as contemplated by the Joint Staff, would bring about an early and decisive result.

However, the Air Force chief stated he would "reluctantly" support General Westmoreland's plea for more manpower and a possible reserve callup because of the situation in I Corps and because he was loathe to deny a field commander the forces he deemed essential. However, he said he would endorse the plan only if the JCS also recommend an immediate. expanded, air and naval campaign against the North to reduce or possibly obviate the need for more forces in the South. "The effective application of our superior air and sea power against North Vietnam's vulnerabilities, " he argued, "will cripple his capabilities to continue to support the war and will destroy his resolution to continue." [8]

The Joint Staff generally accepted General McConnell's suggestion and reworked its preliminary paper. Subsequently, on 20 April, the JCS recommended to Mr. McNamara that more American troops be dispatched to South Vietnam to maintain pressure against the enemy and that an expanded air campaign be authorized to further reduce the flow of men and supplies to the south. Specifically, the JCS proposed an increase of 127,111 "Case II" personnel in fiscal year 1968 above the number authorized in Deployment Program 4. The new total

Search For Alternatives 1967

would include 4,350 Air Force personnel to man five tactical fighter squadrons (F-100s and A-1s), one civil engineering squadron. and augmentation elements. The Army portion would total 71,200, the Marines 43,098 (consisting of one division/wing team plus augmentations), and the Navy 8,463.

Expanding air operations against North Vietnam included this USAF attack on the Thai Nguyen steel plant north of Hanoi, 1967. Source U.S. Air Force

In addition, the JCS recommended an increase of 10,288 "Case II" personnel for Thailand and other PACOM areas. These would be apportioned as follows: 4,025 Air Force for three tactical fighter squadrons, base augmentation, and other support; 3, 650 Navy to strengthen forces in the South China Sea and the Gulf of Tonkin; 1,690 Marines for air units on Okinawa and 923 Army for medical and other support in Japan. [9] To support these additional requirements, the JCS recommended a reserve callup for a minimum of 24 months. a 12-month extension of current service tours of duty. and asked for authority and funds to obtain the necessary equipment and other resources. [10]

OSD Request for Studies of Alternate Force Postures

There was no direct response to the JCS recommendations. Instead. on 26 April, Deputy Secretary of Defense Cyrus R. Vance asked the Joint Chiefs to examine as soon as they could, certain alternate force postures for Vietnam. One, which he listed as "course A," would add air, ground, and naval units totaling 250,000 men through fiscal year 1969 and possibly more later. This would permit greatly intensified military operations outside of South Vietnam to meet "ultimate" JCS requirements. The second. "course B." would add only 70,000 more troops during the next fiscal year. The Deputy Secretary requested an analysis of all aspects of course A: cost. reserve callups. service duty extensions. and military operations (the last to include possible Communist and free world reaction to an invasion of North Vietnam). He also asked for an analysis of bombing strategy for each course and desired special JCS consideration, under course B, of a bombing halt above the 20th parallel and of a complete end to the bombing of all of North Vietnam to "maximize" the possibility of ending the war. Finally. he solicited advice on strengthening the South Vietnamese Army. [11]

The rationale for limiting the bombing in North Vietnam to south of the 20th parallel originated in OSD. The Assistant Secretary of Defense for International Security Affairs asserted that at least as much destruction per sortie was possible by missions flown below the 20th parallel as above. He argued. for example. that it was probably 20 times more worthwhile to destroy a truck after it had traveled all the way to route package I near the DMZ than if it were destroyed further north in route package V. The Air Staff and JCS strongly disagreed.

Haiphong in North Vietnam was the primary point of entry for supplies. Closing a port by bombing it is hard. Mining, on the other hand, is very effective. Source: Vietnamese Govt

For the next several weeks. the Joint Staff worked to prepare recommendations. coordinating its effort with the services. Admiral Sharp. and General Westmoreland. Meanwhile. at the request of General McConnell. the JCS prepared a separate plan for submission to OSD which called for an accelerated air campaign to reduce "external" imports into North Vietnam. The Air Force Chief of Staff. disturbed over past JCS failures to convince OSD and high administration officials of the importance of such an effort. had brought to the chief's attention a special target study employing a new "econometric" technique and produced at his request. He said it showed "beyond doubt the necessity for a realistic air campaign. [12]

In forwarding the USAF proposal to Secretary McNamara on 20 May, the service chiefs cited the rise of war-sustaining imports into the North by sea, and the possibility they might soon include more advanced offensive and defensive weapons. The JCS indicated that about 85 percent of all war-sustaining materiel entered North Vietnam through its ports and about 15 percent by rail or road from China. The total volume had risen from about 800,000 metric tons in 1964 to more

Search For Alternatives 1967

than 1.3 million tons in 1966 and was still on the upswing in early 1967. The Joint Chiefs urged "neutralizing" enemy logistic bases in the Hanoi and Haiphong areas employing a "shouldering out" bombing method. This would consist of striking first at peripheral areas, then the port targets, then other logistic sites, followed by the mining of Haiphong harbor. Simultaneously, the USAF and Navy air arms would conduct an intensive campaign against roads and railways leading from China and the eight major North Vietnamese airfields (of which only three had been hit thus far). Calling the proposal "a matter of urgency," the service chiefs asked Secretary McNamara to transmit it to the President. [13]

The following day, 21 May, the JCS submitted to the Secretary its evaluation of the proposed courses "A" and "B" requested by Mr. Vance. For course A, the Joint Chiefs proposed a reserve callup, extension of terms of service, and adding 125,000 troops in fiscal years 1968 and 1969, respectively. In fiscal year 1968, they recommended adding to the Vietnam force three USAF tactical fighter squadrons, one and one-third Army division force equivalent, one Marine division wing-team (including two Marine tactical fighter squadrons), the remainder of the Navy's riverine mobile force, and other units. Outside of Vietnam, they proposed to increase military strength in Thailand by three USAF tactical fighter squadrons, and build up the Navy's Seventh Fleet by adding one cruiser, five destroyers, one assault patrol boat (APE), eight landing ship tanks (LST), and other support. In fiscal year 1969 the principal forces earmarked for Vietnam would consist of five USAF tactical fighter squadrons, two and one-third Army divisions and, off-shore, one battleship.

Riverine forces formed an increasingly important part of deployment packages.
Source: U.S. Navy

Under their proposed course B (providing 70,000 more men, the maximum possible without a reserve callup), augmentation in Vietnam would be limited to three USAF tactical fighter squadrons, one and one-third Army division force equivalent, the remainder of the Navy riverine mobile force, and other minor units. Deployments outside of Vietnam would consist of three USAF tactical fighter squadrons to Thailand and Seventh Fleet additions of one cruiser, five destroyers, one assault patrol boat, eight landing ship tanks, and other support. [14]

The JCS believed that course A would allow the allies to continue the initiative, provide a better posture for combat operations into Laos, Cambodia, or North Vietnam without reducing pacification and other programs, and hasten an end to the war. On the other hand, course B would permit only more "in-country" deployment of forces to the I Corps which might not suffice to sustain American and South Vietnamese operations beyond the immediate future. Under either "course," the Joint Chiefs urged expanded and intensified air action with emphasis on striking the Hanoi-Haiphong logistic base and import facilities and the aerial mining of specific inland waterways, ports, and coastal areas north of Haiphong.

This picture allegedly shows a downed U.S. Navy aircraft in North Vietnam. However, the aircraft is painted midnight blue which suggests the picture actually dates from the Korean War. Source: Vietnamese People's Liberation Army

Although heavier air and naval pressure against North Vietnam would lead to more Soviet and Chinese assistance to Hanoi, the JCS believed that neither Moscow nor Peking would intervene militarily. The Chinese could be expected to provide major reinforcements under three conditions: if requested by Hanoi's leaders, if the United States undertook a sizable ground invasion of North Vietnam, or if the Hanoi regime was in danger of collapse. General McConnell concurred with the above assessment but believed that the JCS strategy outlined on 20 April would provide more assurance for ending the war on terms favorable to the United States. [15]

In separate comments, General Wheeler urged "as a matter of high priority" the strengthening of South Vietnamese forces and renewed effort to obtain more free-world troops. although these steps would not lessen the need for additional American deployments. He also strongly opposed any partial or total bombing cessation of the North, arguing this would prove costly to the allies, prolong hostilities, and be interpreted by the Communists as an "aerial Dien Bien Phu." [16]

The Draft Memorandum to the President

In late May. the Air Staff and JCS were also asked to comment on a draft OSD memorandum to the President on future action in Vietnam. Prepared by a study group within the Office of the Assistant Secretary of Defense for International Security Affairs, this paper made an overall analysis of the war and the proposed courses "A" and "B" deployment plans (as modified).

The memorandum observed that the "big war" was going well. The enemy had suffered considerably and, beginning in March 1967 (according to General Westmoreland), the "cross-over" point was reached when his losses began to

Search For Alternatives 1967

exceed his replacements. Inflation was under control and the transition to responsible government in Saigon was proceeding as well as could be expected.

However, the "other war" was unsatisfactory. The Saigon government's real control was limited to enclaves. There was widespread corruption and little evidence of remedial action for social and economic ills or of momentum in the pacification program. In the Mekong delta, the tempo of operations was slow, the population apathetic, and many government officials seemed to have working arrangements with the Viet Cong. Imports into South Vietnam were still rising as rice deliveries from the delta decreased. The Communists held large parts of the countryside and believed the United States could not translate military success into political gain for the Saigon government.

With respect to increasing U.S. strength in Southeast Asia, the draft memorandum found "course A" unacceptable and unnecessary. It would introduce 200,000 or more troops into South Vietnam through fiscal year 1969, raising the total to about 670,000 in that country and to 770,000 within the theater. The additional cost in fiscal year 1968 alone would be $10 billion. General Westmoreland had said that without more U.S. forces above those authorized in Deployment Program 4, the war could go on for five years; with 100,000 more men, three years; and with 200,000 more men, two years. These estimates took into account a certain "degradation" of military effectiveness because of reserve callups, and morale and leadership problems.

Course A would also create "irresistible" U.S. pressure for ground action into Cambodia, Laos, and possibly North Vietnam, thereby risking Soviet, Chinese, and possibly North Vietnamese reaction to such moves, especially if accompanied by heavier American air attacks or mining of

McNamara's obsession with statistics led directly to a tightly-focused reliance on numbers such as the notorious "body count". These proved meaningless, an experience which impacts on U.S. Army policy to this day. Source: U.S. Army

harbors. The Soviets, for example, might send more and improved rockets, jet aircraft, and other equipment to the North. Also there was no indication that bombing thus far had reduced Hanoi's will to resist, its ability to resupply the South, or increased its willingness to negotiate. In addition, North Vietnamese morale was probably sustained by continued Soviet and Chinese aid and the expectation that American policy toward the war would change after the forthcoming Presidential election in November 1968.

The paper thus argued that course B deployments (as modified), providing a maximum of 30,000 to 50,000 more U.S. troops in South Vietnam by the end of

Air War – Vietnam

1968, would be more acceptable. Course B as described by Deputy Secretary Vance, however, would have provided a maximum of 70,000 more U.S. men for South Vietnam in fiscal year 1968. This restrained program would avoid extending the conflict, limit the bombing to south of the 20th parallel, improve prospects for negotiations, and contribute to advances in pacification that might follow adoption of a new Vietnamese constitution, national elections later in 1967, and an improved national reconciliation program.

The draft OSD memorandum emphasized the importance of narrowing and understanding the limited American objective of the war, which was to allow South Vietnam to determine its own future. This did not mean a U.S. effort out of proportion to the South's in the face of coups, corruption, and indications of lack of Vietnamese cooperation. Nor did it mean American insistence on the rule of the country by certain groups or a non-Communist government, although certain groups and types of government were preferred to others. [17]

Leading North Vietnamese Ace was Van Coc Nguyen of the 921st Fighter Regiment with nine kills, all scored using heat-seeking missiles. He retired in 2002. Source: Vietnamese People's Liberation Air Force.

The Air Staff's view was that the draft memorandum obviously was slanted toward a minimal buildup of U.S. forces and no significant step-up of military action. It called for little or no augmentation of air operations in South Vietnam and for more restraints on bombing in the North.[18] As the other services were equally critical of the document. the Joint Chiefs on 1 June informed Secretary McNamara that the draft memorandum did not address the implications of free-world failure in South Vietnam. Deployment of 200,000 more U.S. troops and a callup of reserves (course A). they said. would be supported by the American people who did not want "peace at any price. " nor would these two measures necessarily create an "irresistible drive" for military escalation. They also believed that an intensified air and naval effort against North Vietnam would not automatically result in a confrontation between the United States and the Soviet Union or China.

Course B deployments. the JCS continued. would prolong the war, reinforce Hanoi's belief in ultimate victory. and probably cost the United States much more in lives and money. The proposal to limit the bombing of the North to south of the 20th parallel would give the Hanoi government many advantages. induce it to redouble its efforts. and preclude a favorable end to the war. Observing that the OSD draft memorandum revealed an "alarming pattern" that augured a significant change in U.S. objectives for South Vietnam, the service chiefs reaffirmed their understanding of American policy as that embodied in national security action memorandum (NSAM) 288 of 17 March 1964. which called for a free.

Search For Alternatives 1967

independent. non-Communist Saigon regime. They recommended against sending the document to the President, giving it further serious consideration. and asked for the approval of the JCS proposals of 20 April. [19]

The U.S. Worldwide Military Posture

The JCS view on the draft memorandum to the President elicited no formal response from OSD. Meanwhile. the service chiefs on 20 May also had informed Secretary McNamara of their concern about the declining U.S. worldwide military posture. They foresaw a loss of the American initiative in Southeast Asia. decried the force limitations. and warned of a weakening capability to meet other contingencies and commitments.

To bomb or not to bomb, and if to bomb – where. That was the question in late 1967. Source: U.S. Air Force.

The incremental and restrained U.S. response in the war. they averred, made "highly possible" further involvement in Laos. Cambodia. Thailand. or Korea. The North Koreans. they observed. had recently committed a flagrant violation of the Korean Armistice Agreement of July 1953. Berlin. North Africa. and the Middle East were other trouble spots.

The Joint Chiefs affirmed the need to sustain CINCPAC's fiscal year 1968 forces and simultaneously NATO requirements. They proposed earmarking a contingency force of 10 tactical fighter squadrons and three division force equivalents (DFE) for Southeast Asia. and establishing separately a smaller contingency force of three tactical fighter squadrons and one DFE.

To regain the "strategic initiative. the JCS again recommended an expansion of American military strength beginning with a selected callup of reserves and extension by 12 months of service tours of duty. Concurrently. the allies in South Vietnam should apply more pressure against the Communists within that country and step up air and naval operations against North Vietnam. [20]

Secretary Brown's Views on Deployments and Bombing

Simultaneous with the above top-level deliberations, the Secretaries of the Air Force and Navy, the JCS, and Defense Intelligence Agency (DIA), at the request of Mr. McNamara, reviewed possible measures to reduce the flow of arms and men into South Vietnam. Observing the "considerable controversy" surrounding bombing policy, the Defense Secretary solicited comments on the two "most promising" alternatives. The first would concentrate the bombing on enemy lines of communication (LOC) in route packages I, II, and III (i.e., south of the 20th parallel) except for new or rebuilt targets. The second would emphasize bombing and armed reconnaissance on all LOCs in route packages VIA and VIE (i.e., primarily above the 20th parallel). Air strikes would be directed only against fixed targets associated with LOCs and MiG aircraft on airfields. A "sanctuary" running eight nautical miles outward from the center of both Hanoi and Haiphong would be exempt. In addition, two other alternatives might be considered: no bombing of the North's ports and port facilities, or using "every effort" to halt imports into the North (as recommended by the JCS on 20 May). Secretary McNamara also asked for estimates on aircraft losses and possible Soviet and Chinese reaction to these alternative courses of action. [21]

U.S. aircraft losses were mounting steadily. Source: Vietnamese People's Liberation Army

The Air Staff and JCS found both alternatives inadequate. The first would be most advantageous to the enemy and indicate a weakening of American resolve in the war. The service chiefs suggested a third alternative, permitting armed reconnaissance and strikes against all important fixed targets, including airfields, on LOCs in route packages VIA and VIE. and restricting the sanctuary area around Hanoi and Haiphong to eight and two nautical miles, respectively. There would be no bombing of the Haiphong wharf or mining of the harbor and commercial shipping waters north of the 20th parallel. But even these proposals, the JCS warned, would be insufficient to halt substantially Hanoi's imports and destroy other remaining resources in the North. Accordingly, the Joint Chiefs reaffirmed their recommendations of 20 May. [22]

On 3 June Secretary Brown sent his report to Mr. McNamara. It contained a partial reply to the draft memorandum to the President. Reviewing American and South Vietnamese objectives, he perceived three deployment options: adding 200,000 more troops in fiscal years 1968 and 1969; sending only 30,000 more troops in calendar year 1968; and withdrawing 100,000 troops per year to see "what would happen" (although this appeared out of the question unless another political coup occurred in Saigon during the coming U.S. presidential election).

The Air Force Secretary opposed deploying 200, 000 more troops, arguing that a force this size would neither accelerate the pacification effort nor make the North Vietnamese "fade away." In all probability, it would provoke Hanoi into a larger military response, raise American casualties, and convince the South Vietnamese, seeing 700,000 U.S. troops in their country, that this was not their war, a danger

Search For Alternatives 1967

that already existed. It would also create new pressure for expanded military operations in Laos, Cambodia, and North Vietnam and hazard an "unacceptable risk" of war with China. Secretary Brown favored his second option: providing only 30,000 more troops (which would add 10 maneuver battalions), possibly without a reserve callup or taking other mobilization actions. He thought this number could redress the military situation in I Corps and serve as a "buffer" while American troops conducted other operations in II and III Corps and the South Vietnamese combat and pacification activities in IV corps. This would avoid any great increase in U.S. casualties and the risk of further escalation in the war.

Concerning air strategy against the North, Dr. Brown favored current policy rather than adopting the alternatives of concentrating the bombing in the area south of the 20th parallel or on designated LOCs and ports. Any major diminution in air activity between route packages IV and VI (i.e., above the 20th parallel) would eventually require more U.S. troops in the South, raise allied casualties, and possibly inhibit South Vietnam's political evolution. On air effectiveness, the Secretary observed, it was more difficult to estimate the impact of air strikes on infiltration in North Vietnam and Laos than in South Vietnam. The only thing certain about the present level of out-country air strikes was that it caused a significant diversion of enemy manpower (only five percent of the population but many persons with skills to man air defenses and to make road, rail, and bridge repairs).

Bombing Lines of Communications was effective but arduous and costly.
Source: U.S. Air Force.

Secretary Brown's "optimum air strategy" called for maintaining the existing level of operations or reducing it somewhat where restrikes were unnecessary, using new air techniques. He was against striking the port of Haiphong, an act he felt would pose another "unacceptable risk." He also believed that the Joint Chiefs' recommended "shouldering out" air tactic and a proposed "power play" concept (sent by General McConnell to the Chief of Naval Operations and calling for sinking American ships in channels 18 miles from Haiphong), could have grave consequences and possibly evoke a Soviet or Chinese response. In sum, he visualized the air war against the North as similar to operations in the South, a "war of attrition." [23]

Secretary of the Navy Paul H. Nitze. after analyzing the bombing alternatives raised by Mr. McNamara, concluded that intensive bombing in southern North Vietnam (i.e., south of the 20th parallel) would reduce the enemy's capability to maintain a supply flow to the south, much more so than if bombing was concentrated in areas above the 20th parallel. [24]

As part of the high level consideration of future deployment and bombing policy. Secretary Brown continued to send to OSD Air Staff reports on the effectiveness of air power. These included evaluations by the Air Staff's "Combat Strangler" task force (established in July 1966) of the results of interdicting the North's petroleum. oil and lubricants (POL) system and lines of communications, and of Air Force weapons used. He also submitted summaries of air plans and target studies. notably, an integrated strike and interdiction plan against southern North Vietnam and Laos. an econometric study on the potential effectiveness of an intensified air campaign against more targets in the North, and a "Combat Alley" plan for destroying the North's MiG air bases. A number of these plans also were sent to the State Department. [25]

III. THE 525,000 U.S. TROOP CEILING FOR SOUTH VIETNAM

In July 1967, Mr. McNamara and other high Washington officials flew to Saigon to review the deployment issue and allied strategy in the war. It was the Defense Secretary's ninth trip in five years. The news media accurately revealed that the American troop increases under consideration ranged from a low of 35,000 to 70,000 to a high of 200,000. [1]

The Saigon Conference of 7 - 8 July

Upon arrival on 7 July. Secretary McNamara and his aides were briefed by the principal American civilian and military officials in South Vietnam. In contrast with General Westmoreland's dark forebodings in March about the war. the mid-year review contained a more hopeful tone. Ambassador Ellsworth Bunker, who had assumed his Saigon post on 4 April 1967, succeeding Ambassador Lodge. saw a greatly improved military situation following recent allied offensive operations and an encouraging South Vietnamese combat performance near the DMZ and elsewhere. Suffering from heavy casualties and a higher defection rate than in 1966, the Communist forces had failed to win a major victory. Politically, the Saigon government was moving toward a broader and more stable constitutional government. The pacification program was advancing faster and there was more economic activity and stability throughout the country.

Ellsworth Bunker.
Source: U.S. Govt

The American Ambassador admitted that serious difficulties remained. Enemy thrusts. while blunted. had not been stopped and infiltration, still the crux of the problem, was estimated at about 6, 500 personnel per month. Poor leadership still plagued Saigon's armed forces and South Vietnamese motivation and involvement was unsatisfactory due to apathy, inertia, widespread corruption, inadequate physical security, lack of social justice, and incompetence in civil administration. While there was no reliable opinion-taking organization such as in the Dominican Republic (Ambassador Bunker played a key role in American efforts to settle the revolution in the Dominican Republic in 1965-1966) to determine peoples attitudes outside of the cities, the Ambassador thought none of the problems were insuperable "if we stick with it long enough."

General Westmoreland likewise pointed to military progress and stressed the high cost being paid by the enemy. The growing success of the air and sea offensive, he thought, was being matched by the less dramatic gains of the ground campaign.

Air War – Vietnam

He urged stepped up military and pacification activities in South Vietnam, increased air pressure against the North, and new combat initiatives in Laos.

Gen. William D. Momyer, Commander of the Seventh Air Force, and Admiral Sharp gave briefings on the overall air effort. With respect to operations in the South, General Momyer observed that about 30,000 close air support sorties were flown in 18 ground campaigns and the combined air and ground fire killed 19,928 of the enemy. The largest action thus far, called "Junction City," occurred in the northwest corner of III Corps. B-52's were averaging 27 sorties per day from bases in Guam and Thailand with about 30 percent of the sorties flying in support of ground operations.

General Momyer reviewed the round-the-clock air operations in the North in the "Tally-Ho" area near the DMZ and in route package I, where about 18,500 attack sorties were flown from January through June 1967. With the arrival of better weather in May, a large portion of the air effort in Laos (i.e., the "Steel Tiger" and "Tiger Hound" programs) had been shifted to that region. Recent air strikes. he said, had depleted significantly the enemy's resources north of the DMZ. In the first six months of the year they had demolished or damaged 2,298 trucks, destroyed 6,297 tons of supplies. and caused 9,857 secondary fires or explosions which ruined another 1, 593 tons of supplies. In addition, the air campaign had kept the North's aircraft out of South Vietnam, and prevented the Communists from moving antiaircraft guns and SAMs to LOCs in southern North Vietnam.

Admiral Sharp reported that the enemy was "hurting" and thought the allies were "at an important point in the conflict." He recommended greater latitude for commanders in planning and executing air strikes in the remaining months of good weather, and opposed any further strictures such as limiting the bombing to south of the 20th parallel. This. he said, "would have adverse and disastrous effects." He reaffirmed the importance of bombing and mining the harbor at Haiphong and recommended attacks on six basic targets: electrical facilities, maritime ports, airfields, transportation, military complexes, and war-supporting industries. He called for integrated air strikes on all significant targets in North Vietnam and Laos, and singled out especially the need to reduce the size of prohibited areas around Hanoi and Haiphong.

The PACOM Commander pointed to a "significant" downward trend in Hanoi's ability to support the war because of more efficient U.S. air operations. He cited the enemy's high aircraft losses and his inability to use three airfields (because of bombings) which lessened the danger from MiGs. more SAM firings with faulty guidance, reluctance to fire SAMs in good weather for fear of allied detection of sites, decreasing antiaircraft fire along the northeast rail line and other sectors (also because of American bombings), and fewer U.S. aircraft losses to SAMs and antiaircraft fire in route packages VIA and VIB.

Discussing manpower needs, a MACV briefer noted that the latest approved deployment program authorized 483,222 spaces for South Vietnam (the approved manpower totals for South Vietnam were under constant revision) and that an additional request for 13,124 for fiscal year 1967 would raise the total to 496,346.

Search For Alternatives 1967

With respect to further troop increases as proposed in OSD's courses "A" and "B," he emphasized that the first, providing five tactical fighter squadrons and two and one-third divisions each for fiscal years 1968 and 1969, would provide greater assurance for maintaining pressure on the enemy and for shortening the war. Course B, allowing for only 70,000 more troops, would decrease American options and increase them for the Communists. [2]

MACV officials also presented to Mr. McNamara and his aides five force "package" programs, one of which contained new proposals to strengthen South Vietnamese forces, including the VNAF. [3]

Approval of the 525,000 U.S. Troop Ceiling.

On 11 July, after spending five days in South Vietnam, Mr. McNamara and his party departed for the United States. On the 12th, General Westmoreland flew home to attend the funeral of his mother in Columbia, S.C. He arrived in Washington and was invited to the White House, where he met again with the Defense Secretary and General Wheeler spent the night of 12 July at the White House and the morning of the 13th, during which time he and the President discussed Vietnam affairs.

In 1967, U.S. troop commitments to Vietnam passed the half million mark. Photo Source: U.S. Army

Without consulting the JCS, the three men agreed to establish the new U.S. troop ceiling in Vietnam at 525,000 men. This was about 45,000 above the currently authorized strength. President Johnson approved the figure the same day. There was no immediate public disclosure of it, although at a press conference on 13 July at the White House. attended by Generals Wheeler and Westmoreland and

Air War – Vietnam

Secretary McNamara, the President made it clear they had agreed on future plans for Vietnam. President Johnson did not disclose the new figure until 3 August when he announced that 45,000 to 50,000 more U.S. troops would go to South Vietnam by 30 June 1968. Subsequent planning called for the new U.S. manpower ceiling of 525,000 to be reached in March 1969.

The decision to limit the buildup of man-power, in contrast to the MACV commander's earlier desire for upwards of 200,000 men, apparently was based on a number of factors. They included the relative, if slow, success in the war (as described during the just completed Saigon conference), the administration's desire to avoid military or economic mobilization. concern about the inflationary impact of more troops on South Vietnam's fragile economy. and the possibility that an excessively large U.S. force would convince the South Vietnamese this was not their war and encourage military operations that might trigger Chinese intervention.

In public statements on 12 and 13 July. the President and Secretary McNamara jointly agreed that the additional U.S. manpower to be sent to Vietnam would not result in a reserve callup or an extension of tours of duty. The Defense Secretary also stressed the need for more effective use of the 1,300,000 American, South Vietnamese, and other allied troops already in South Vietnam. He said progress had been made in the military, political, and economic fields, but in a fourth area, pacification, it was still very slow. The President and his aids agreed that, despite considerable enemy infiltration, the war was not stalemated.

McNamara promised that new technology would help the hard-pressed F-105 force by reducing losses and enhancing capability.
Source: U.S. Air Force

In air operations. Secretary McNamara said some "very significant" changes in technology had greatly enhanced U.S. capability to make all-weather attacks on LOCs in South and North Vietnam. These changes. in conjunction with new weapons and ordnance. substantially improved the effectiveness of air strikes and reduced aircraft losses. But he reaffirmed his belief that air power alone against the main LOCs in the North could not stop the flow of men and supplies to the South, no matter how competently it was managed or directed. Rather, it could reduce the amount of supplies and make the war more costly to the enemy. The objective of penalizing the enemy was being met, he said, citing as evidence PACOM data that showed 400,000 to 500,000 North Vietnamese engaged in repairing the LOCs.[4]

Also on 13 July. Secretary McNamara orally informed the three service chiefs of the new manpower ceiling and asked them to submit a detailed troop list using the

Search For Alternatives 1967

five force "packages" prepared by General Creighton M. Abrams, Deputy Commander of MACV and his staff and given to him in Saigon. Based on the just completed briefings in Saigon and MACV's fiscal year 1968 force requirements. the Abrams packages suggested how the 1968 goals might be achieved without a callup of reserves. extending terms of service. and by employing only minimum additional troops. General Abrams presented alternate choices on how to limit a further U.S. military buildup, such as using more South Vietnamese or Korean manpower, or substituting civilian contractor or direct hire personnel. On

Vietnamese forces continued to expand. Photo Source: U.S. Marine Corps.

13 June, Secretary McNamara had asked the JCS to expand its study on combat support to include the possible use of more South Vietnamese civilians using as an example the Korean service corps. a quasi-military organization that worked for the Korean Army. In a separate action. General Westmoreland proposed increasing South Vietnam's regular, popular. and regional forces from 622,153 to 685,739 during fiscal year 1968. This was endorsed by Admiral Sharp on 29 July and approved by Mr. McNamara on 7 October. The packages incorporated General Westmoreland's additional fiscal year 1967 troop request into his fiscal year 1968 proposals. [5]

Refinements in the U.S. Troop List

In response to Mr. McNamara's request, the Joint Staff prepared and the JCS on 20 July sent to OSD a "troop list" proposing a "mix" of the following forces: 16 maneuver battalions (13 Army, three Marine), four USAF tactical fighter squadrons, the 9th Marine Amphibious Brigade (MAE) (with two fighter squadrons), more units for the Navy mobile riverine force (to expand "Game Warden" operations on inland waterways and "Market Time" coastal patrols), and certain logistic and advisory units and personnel. Using the currently authorized strengths as a base, the services would increase their total manpower in South Vietnam in 1968 from 484,472 to 537,545 (including 59,528 Air Force), but would stay within the 52,000-man ceiling by converting 12,545 military to civilian direct hire and contractor personnel. [6]

USAF strength would rise by 3,380 spaces and include two deployed A-1 fighter squadrons (963 personnel), two "ready-status" fighter squadrons (1,031 personnel) in the United States, 10 AC-47 and 22 0-2 aircraft plus crews, support personnel, and other augmentations (1, 386 personnel). "The other augmentations" were to include deployment of seven more UC-123 chemical defoliation aircraft, personnel to convert some [units equipped with] F-4Cs to F-4Ds and 0-1s to 0-2s,

Air War – Vietnam

a "Red Horse" civil engineering squadron, and more spaces for the public affairs office and communications center.

F-4Ds started to replace the F-4C and F-104 in Vietnam from 1967. Source: U.S. Air Force

Neither General McConnell, the Air Staff, nor the other service chiefs were satisfied with the troop list prepared, of course. under the guidelines laid down by the Defense Secretary. They believed all four USAF fighter squadrons were needed in South Vietnam and especially objected to including two squadrons scheduled to remain in the United States on a ready-status basis, within the 525,000 manpower ceiling. They were backed by General Momyer who, in reviewing his sortie needs, cited the Army's rising demand for preplanned close support sorties (i.e. to aid ground troops not in actual contact with the enemy). At present, he could fulfill only about 60 percent of the number requested. In addition, Secretary Brown previously had informed Mr. McNamara that, in the event the Marine Corps could not provide two more squadrons, the Air Force could make five available in fiscal year 1968: three A-l's, one F-4D and one F-100, the last from the European Command.

The O-2 was introduced to replace the old O-1 Forward Air Control aircraft. It was, in turn, replaced by the OV-10. Source: U.S. Air Force.

The JCS also objected to including elements of the 9th MAB, temporarily engaged in South Vietnam, in the 525,000-man ceiling since it was a PACOM reserve unit based in Japan, subject to deployment anywhere, and already accounted for in previous manpower totals. The service chiefs warned it would be difficult to substitute civilian contractor and local direct hire personnel in lieu

Search For Alternatives 1967

of U.S. military spaces. For the recruitment of suitable civilians would have to compete with Saigon government plans to draft more men for the South Vietnamese armed forces.

Although the units in the JCS troop list would contribute significantly to prosecuting the war, the service chiefs noted that they provided less manpower than was recommended on 20 April. They also reaffirmed the validity of their views of 20 May in which they addressed the nation's world-wide military posture.

After reviewing the troop list and JCS comments, Secretary McNamara on 21 July verbally directed the service chiefs to prepare, subject to OSD changes, the dispatch of additional forces in fiscal year 1968. The principal new Air Force units approved for deployment consisted of one F-4D ready-status squadron (to remain in the United States until needed), two A-1 squadrons. 10 AC-47s. and 22 0-2s.

More A-1E units were approved for deployment. The A-1 was proving to be an almost ideal aircraft for use over South Vietnam. Source: U.S. Air Force

Air Force personnel in these units. and those needed to convert some [units flying] F-4Cs to F-4Ds and 0-ls to 0-2s totaled 2,242. The important deletions consisted of one A-1 ready-status squadron (351 spaces) and a "Red Horse" engineering squadron (600 spaces).

Mr. McNamara deferred, pending more justification, deployment of additional UC-123 aircraft. He thought the inclusion of the three USAF fighter squadrons (two deployed, one in ready-status) and the two Marine fighter squadrons (with the 9th MAB in Japan) adequate for the present. The latter would be sent only if the Air Force failed to meet Marine Corps air needs. The Secretary reaffirmed his decision to include the one USAF ready-status squadron, the 9th MAB. and a Navy APB unit (transferred from the offshore Navy to South Vietnam) as part of the 525, 000 manpower program (which was designated Deployment Program 5 on 5 October). [7]

On 10 August McNamara, noting that some problems associated with the new Deployment Program 5 ceiling needed to be worked out, tentatively approved for planning purposes the following totals:

	Air Force	Navy	Marine Corps	Army	Total
Program 4	56,148	30,039	74,550	323,755	484,472
FY68 Added Forces	2,242	4,234	7,523	33,297	47,296
Total	58,390	34,273	82,073	357,052	531,768
Civilians	-542	-812	--	-5,414	-6,768
Final Total	57,848	33,461	82,073	351,538	525,000

UC-123s fitted with spray equipment executed Ranch Hand missions to remove cover around U.S. installations. It seemed like a good idea at the time. Source: U.S. Air Force.

With reference to JCS recommendations of 20 April, the Defense Secretary disapproved deploying additional "out-of-country" forces except five destroyers for gunfire support, to come from existing fleet resources, and said he was considering activating a battleship. He asked for another "refined" troop list by 15 September containing justification for more units that might be desired with "trade-offs" from military to civilian spaces. Other directives indicated that OSD was firmly resolved to restrict further U.S. military buildup and spending in order to control South Vietnam's piaster expenditures and inflation.[8]

To provide the refined troop list requested by Mr. McNamara, Admiral Sharp on 23 August convened a special five-day conference in Honolulu. attended by representatives of the Air Force and other services and OSD. Reviewing existing plans. the conferees determined that more than 5,400 military spaces could be saved in Deployment Program 4 by inactivations. reorganizations and strength adjustments. This saving, plus the conversion of 12,545 military to civilian spaces. would permit the deployment of an additional 50,000 American personnel to South Vietnam during fiscal year 1968, and allow the services to remain within the 525,000-man troop ceiling for that country.

The conferees also studied a new request for about 1,164 hospital and other medical personnel to assure more medical aid for South Vietnamese war

Search For Alternatives 1967

casualties. Plans to expand the treatment of South Vietnamese war casualties began following President Johnson's visit to Southeast Asia in March 1967. The program received impetus as a result of findings by a Senate subcommittee in August headed by Senator Edward M. Kennedy. (For a brief discussion of South Vietnamese casualties caused by both enemy and friendly forces, see MACV Command History (TS). 1967, Vol III Annex B.) Although this was principally an Army program, 32 USAF medical personnel were required. These and other adjustments were incorporated into a new troop list sent to Secretary McNamara on 15 September.

Service Troop List For Fiscal Year 1968 Deployments To South Vietnam 15 September 1967				
	Program 4 End Strengths	Program 4 In-country and ordered deployed	Program 4 Not ordered deployed	Total
Air Force	56,148	55,987	161	56,148
Navy	30,039	28,740	1,299	30,039
Marine Corps	81,270*	81,270	0	81,270
Army	323,735	309,417	9,693	319,110**
Total	491,192	475,414	11,153	486,567

	FY67 Additional and FY68 Adjusted Deployments	Totals	Civilianization	Grand Total
Air Force	3,161	59,309	-600	58,709
Navy	7,483	37,522	-2,050	35,472
Marine Corps	969	82,239	-300	81,939
Army	39,365	358,575	-9,595	348,880
Total	50,978	537,545	-12,545***	525,000

* Includes elements of the 9th Marine Amphibious Brigade with a strength of 6,720 men.

** Includes Army space-saving adjustments totaling 4,625 as a result of certain unit inactivations, net strength adjustments to program #4, and adjustments from in-country audits.

*** Tentative

Source: JCSM-505-67 (TS), 15 Sep 67

Air War – Vietnam

In forwarding the troop list to Secretary McNamara on 15 September, the JCS simultaneously expressed its reservations about some of its provisions. They said that a successful conversion of 12,545 military to civilian spaces was "highly conjectural" from the standpoint of civilian recruitment, reliability and financing. They opposed the Secretary's inclusion within the 525, 000 manpower ceiling, three non-deploying squadrons, a Marine unit temporarily assigned to South Vietnam, and new hospital spaces. Because additional manpower would have to come largely from the U.S. strategic reserve, the service chiefs indicated they had begun another study on how best to reconstitute it, observing that Mr. McNamara had not yet replied to their views of 20 May on the weakening U.S. worldwide military posture. [9]

Overriding all JCS objections, Secretary McNamara on 5 October approved the troop list with its provisions for civilianization of certain military spaces and additional deployments. He said Deployment Program 5 would be revised to reflect the manpower changes, and instructed the service chiefs to review continuously their forces, deleting those no longer required to reduce the impact of more U.S. troops on South Vietnam's economy. He said that requests to send more high priority units should contain appropriate "trade-offs" of civilians for military spaces to assure no breaching of the 525,000 military manpower ceiling. Costs and resources for additional deployments or adjustments should be included in revised service budget estimates for fiscal years 1968 and 1969 in accordance with established procedures. An initial report on service civilianization efforts was desired by the end of 1967. [10]

On 13 October, Secretary McNamara also imposed a ceiling of 45,724 U.S. military personnel in Thailand, asserting that number should suffice to meet foreseeable needs in that country. He stipulated that the "ground rules" for sending new units or augmentations into Thailand would be the same as those for South Vietnam. He also cited a recent study by OSD's Systems Analysis office showing that Air Force base support in Thailand could be reduced by 500 spaces below requests and was necessary to remain within the manpower ceiling. This conclusion was contrary to an Air Staff view. based on the findings of the office of Inspector General, that with few exceptions no reduction in air base support was possible. [11]

Plans to Increase South Vietnamese Forces

Concurrent with the above planning, the Defense Secretary on 7 October also approved a JCS recommendation, based on proposals submitted by General Westmoreland and Admiral Sharp, to boost the strength of South Vietnamese forces from 622,153 to 685,739 personnel by the end of fiscal year 1968. The Air Staff supported the increase, agreeing it was desirable to transfer to the Saigon government a large share of the military effort, improve the balance between combat and combat support elements, and provide more forces for the pacification and railway repair and security programs.

Service allocations for the 63,586-man raise in South Vietnamese military strength were as follows: VNAF, 761; regular army, 12,843; Marines, 131;

Search For Alternatives 1967

regional forces, 34,353; and popular forces, 15,610. The Navy would lose 112 spaces. The VNAF portion would be primarily for headquarters support, political warfare, counterintelligence, security, clandestine operations. helicopter maintenance, and for personnel and dependent's needs. Pending receipt of more information, Mr. McNamara deferred a decision on another proposal by Admiral Sharp and General Westmoreland to raise the strength of South Vietnamese regular, regional, and popular forces from 685,739 to 763, 953 in fiscal year 1969. [12]

Other Deployment Actions

There were, meanwhile, other significant developments affecting the Air Force. One followed a severe fire on the carrier *Forrestal* on 29 July which killed 133 personnel, destroyed 21 aircraft, damaged 30, and forced the carrier's return to the United States for repairs. To compensate for the loss of Navy air support, the Air Force was directed to deploy to South Vietnam six USAF F-l00s and 10 B-57s from the Philippines and the Marines were asked to send two squadrons from Japan. In addition, the carrier *Constellation* was temporarily assigned to "Yankee Station" off North Vietnam. Approving these changes on 13 September, Secretary McNamara directed, however, that the additional USAF and Marine aircraft could remain in South Vietnam only until 15 November. By that date, bad weather over the North would reduce combat requirements and the other air resources available to PACOM should enable Admiral Sharp to meet priority sortie needs. [13]

The summer months also witnessed accelerated planning for construction of a linear strong point obstacle system (SPOS) extending inland about 13 kilometers (later lengthened to 23 kilometers) from the South China Sea just below the DMZ, and an air-supported anti-infiltration system for Laos. Personnel and supporting aircraft for the Laos system would be based in Thailand. By 7 August, Mr. McNamara had approved the use of 11,567 U.S. military personnel already in the theater or newly deployed to build and support the two systems. Of these, 7, 822, largely Army, were earmarked for the SPOS in South Vietnam and 3,745, largely Air Force, were scheduled for the air-supported system in Laos. Still considered additive to Deployment Program 4, the manpower for the two projects was later included in the manning list for Deployment Program 5 (issued on 5 October).

Air Force Brigadier General William P. McBride was appointed manager of the air-supported anti-infiltration system (designated "Muscle Shoals" on 8 September) with headquarters at Nakhom Phanom, Royal Thai AFB. General McBride arrived at the base on 18 October as commander of the Seventh Air Force Task Force (unofficially called Task Force Alpha). He immediately began organizing a special unit to operate the Air Force-staffed anti-infiltration surveillance center (ISC). Several supporting air units, specially equipped for communications relay or for dropping sensing devices and special "gravel" munitions, arrived at Thai air bases between September and 20 December. They operated 21 USAF EC-121s at Korat, and eight Navy OP-2Es, 19 USAF A-1Es, and 12 USAF CH-3s at Nakhom Phanom. Twelve Army UH-1F armed helicopters

Air War – Vietnam

also arrived at Nakhom Phanom from South Vietnam to fly escort missions for the CH-3's. In addition, 18 F-4's were earmarked for stationing at Ubon on 1 March 1968. Approximately 400 USAF personnel arrived between October and December 1967 to staff the ISC and related operational, communication, and weather facilities. This figure was boosted to more than 500 in early 1968. [14]

The EB-66 was a vital link in providing electronic defense to the strike aircraft operating over North Vietnam. Source: U.S. Air Force

In another action under the aegis of the new deployment program 5, Secretary McNamara on 23 October approved the movement of 13 more USAF EB-66 aircraft and 902 personnel to Thailand. This increased the U.S. manpower ceiling of 45,724 set by Mr. McNamara 10 days earlier. Five hundred and ninety-two personnel were associated with the 13 aircraft and the rest were allocated to the expanding electronic warfare program in the theater. When deployed, 41 EB/RB-66 aircraft would be operating in Southeast Asia. The Air Force's electronic warfare and intelligence collecting capability was also expected to be enhanced by converting 11 more C-47 to EC-47 aircraft by June 1968, a decision to this effect being made in September. A total of 40 of these aircraft were in Southeast Asia at the end of December. [15]

IV. NEW STUDIES ON DEESCALATION & MILITARY ACTION

The administration's decision in mid-1967 to limit American strength in South Vietnam to 525,000 personnel coincided with another relatively "optimistic" period in the war. The hopeful briefings in July in Saigon for Secretary McNamara by air and ground commanders were followed, in August, by more publicized reports that "the pressures are beginning to tell on the enemy." In the same month, Gen. Harold K. Johnson, Army Chief of Staff, declared that the recently approved 45,000 U.S. personnel increase for South Vietnam would be adequate, and he foresaw some reduction in American forces in 18 months. However, by September, the intractable North Vietnamese launched new thrusts against allied troops in I Corps. [1]

The Threat in the Demilitarized Zone

On 12 September, after a White House conference of top officials examined a proposal from the U.S. Deputy Ambassador to Saigon, Eugene M. Locke, for improving the allied war effort, the President asked General Wheeler for a JCS list of actions which would increase military pressure against North Vietnam. Secretary McNamara specifically requested a plan for a 12-month air campaign against the North to begin on 1 November. [2]

Counter-Battery fire – Gray Lady Style. Source: U.S. Air Force

Meanwhile, the administration's immediate attention centered on developments in I Corps near the DMZ. There the enemy had built up his strength and launched artillery, rocket, and mortar bombardments of allied positions at Dong Ha, Con

Air War – Vietnam

Thien, and Gia Lien. The intensified attacks, General Westmoreland warned the JCS, threatened to halt construction work on the anti-infiltration strong point obstacle system. To silence the Communist batteries, the MACV commander on 11 September launched Operation Neutralize, using principally Seventh Air Force and Marine tactical air units and B-52's to knock out enemy gun positions. Meanwhile, a partially stepped up air campaign in the North, devised largely by the Air Staff and which began in August attempted to isolate Hanoi from Haiphong and both cities from the rest of the country. The Air Force also hit 10 new targets in the buffer zone near China. [3]

The enemy threat near the DMZ. however. did not abate and, with deteriorating weather tending to hinder air operations, the President on 20 September asked Gen. Wallace M. Greene. Commandant of the Marine Corps, whose troops were primarily responsible for the defense of I Corps, to suggest several courses of action to deal with the situation. General Greene proposed five possible actions. all using existing forces in Vietnam and requiring only a modest change in combat restraints: (1) continue operations with current strength (i.e. maintain the status quo); (2) attack north of the DMZ to destroy enemy positions; (3) reorient the allied strategy to a mobile defense; (4) reinforce I Corps by at least two regiments and concentrate on enemy battalions and firing positions; and (5) increase the effectiveness of air and naval gunfire north of the DMZ where the bulk of enemy infiltration. supplies, and firing positions were located. The Marine commandant recommended only the last two. but also asked for a Joint Staff study of the situation. [4]

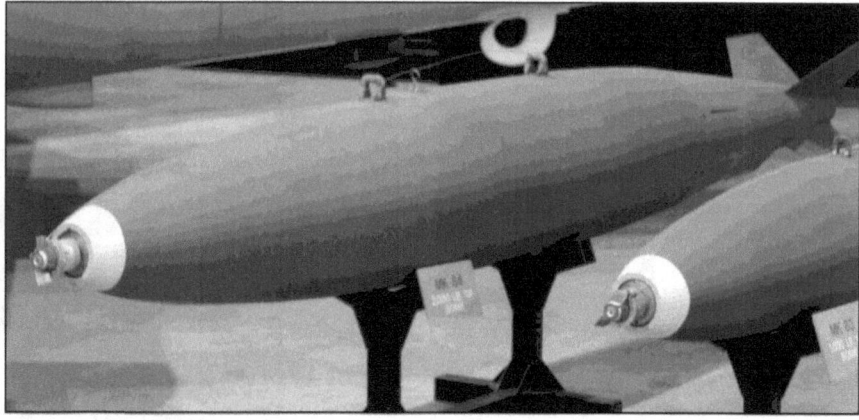

The Mk.84 2,000 pound bomb. Source: U.S. Air Force

In reply to still another White House request on 27 September, General Wheeler sent through Deputy Secretary Vance additional suggestions for dealing with the enemy in both North and South Vietnam. Some were long-term actions but all required White House approval. The JCS chairman proposed boosting B-52 sorties from 800 to 1,200 per month, authorizing B-52 overflights of Laos. employing 2,000-pound and heavier bombs such as the MK-84. permitting Air Force-Navy tests as soon as possible of MK-36 weapons. augmenting naval

Search For Alternatives 1967

gunfire and Army batteries in the DMZ area, accelerating the movement of units approved for Deployment Program 5, raising the level of South Vietnamese forces and equipping them more rapidly with the M-16 rifle, and intensifying research and development on finding concealed enemy artillery. He indicated preparatory steps had been taken to carry out these measures quickly, and awaited only official approval. [5]

The Air Staff generally supported these recommendations but expressed reservations about greater use of B-52s. In the absence of more precise targeting information, increasing the B-52 rate to 1,200 sorties per month or using larger bombs, which also could be carried by tactical aircraft, seemed an inefficient way to employ the SAC bombers. The Air Staff favored a modest increase in the monthly rate to 900 sorties and only a 48-hour "surge" capability of 1,200 sorties. [6]

OSD and the White House subsequently approved some of the recommendations made by Generals Greene and Wheeler. The MACV commander was authorized to reinforce Quang Tri Province in I Corps, and to augment his air, naval. and artillery firepower there including the use of more B-52 sorties. On 7 October, Secretary McNamara authorized a buildup of South Vietnamese forces from 622,153 to 685,739 by the end of fiscal year 1968. [7]

Deescalation Studies and Other Possible Actions

Concurrent with these military developments. Air Force officials noted indications of a possible shift in administration policy toward the war. On 21 September, the American Ambassador to the United Nations, Arthur J. Goldberg, in an address before the General Assembly. appeared to suggest that the United States might consider halting the bombing of North Vietnam if it could be assured of serious peace negotiations with Hanoi. This was followed, on 29 September, by a major address by President Johnson in San Antonio, Tex., in which he presented a "formula" for beginning negotiations with the Communists. [8] The President said in part: "As we have told Hanoi time and time again, the heart of the matter is this: The United States is willing immediately to stop all aerial and naval bombardment of North Vietnam when this will lead promptly to productive discussions. We would, of course, assume that while discussions proceed, North Vietnam will not take advantage of the bombing cessation or limitation." It was subsequently disclosed that the U.S. government sent the substance of the San Antonio formula secretly to Hanoi on 25 August.

These administration soundings triggered new "courses of action" studies by the Air Staff and Joint Staff. The principal ones concerned lowering the intensity of the fighting if the Communists reduced their activities or showed other evidence of weakening support for the war (i.e., "tacit deescalation"); lessening military activity because of congressional pressure, public debate. and other influences; and possible acceptance by Hanoi of President Johnson's San Antonio "formula" for ceasing air and naval bombardment of North Vietnam. Other studies centered on increasing military pressure on the Communists throughout Southeast Asia and launching a 12-month air campaign against the North or a four-month military campaign in Southeast Asia. All reflected the President's growing preoccupation with finding the right combination of political actions or military pressures to reduce the tempo of the war and to hasten its settlement. [9]

With respect to the first study, the Joint Staff prepared two "flimsies" or working papers on a possible American response to any tacit deescalation in fighting by the Communists, both providing, in effect, for a step-by-step decrease in hostilities. The Air Staff opposed this approach, believing it would be disadvantageous to the United States. It would permit the North Vietnamese to control the level and intensity of the war, possibly lessen allied air and ground activity. and negate the administration's objective of attaining peace in the shortest practicable time. The Air Staff also observed that tacit deescalation was but one of several alternatives open to Hanoi to reduce the tempo of the war. Since the Navy and Marine Corps endorsed the Air Force position, the Joint Staff decided to consider all of the alternatives that appeared open to North Vietnam in reducing military operations. No final action was taken on this subject by the end of 1967. [10]

In the second study, the Air Staff agreed with the Joint Staff that a lessening of allied activity could augur a major change in the conflict and possibly lead to a bombing halt of the North, signal other acts to decrease the fighting, and even result in a withdrawal of troops from South Vietnam. Accordingly, the service chiefs decided to review their major policy papers since November 1964 to determine if a lessening of warfare would permit the United States to achieve its goals or whether it would necessitate a change in them and, in turn, require the JCS to alter its strategy. [11]

The third study (on Hanoi's possible acceptance of the San Antonio "formula") was the most comprehensive examination to date of possible ways to negotiate an end to the war. Entitled "Sea Cabin," it was undertaken by an ad hoc group composed of Joint Staff. DIA. and OSD members and chaired by Lt. Gen. Andrew J. Goodpaster, Commandant of the National War College. A draft of the study was completed in December. Because an Air Staff analysis showed it included outdated intelligence, contained statements inconsistent with previous JCS judgments, and needed further review, General McConnell proposed. and the other service chiefs agreed, that the JCS merely note it and submit only preliminary comments to OSD. Accordingly. Secretary McNamara was advised that the study contained insufficient reliable intelligence on the overall impact of the air campaign on the North. The Joint Chiefs reaffirmed their judgment on how bombing could contribute to achieving American objectives. acknowledged the existence of diverse U.S. agency views on negotiating with the Communists while maintaining pressure on them. and suggested an inter-departmental examination of the problem with JCS participation. Deputy Defense Secretary Vance subsequently concurred with the last proposal and asked Secretary of State Rusk to establish an interdepartmental group. [12]

The fourth study on "increased pressure" combined earlier Joint Chiefs' views on possible "ultimate" U.S. military requirements as suggested by OSD with their response to the White House request of 12 September for a "pressure paper." General McConnell considered this study the proper "vehicle" for conveying the

Search For Alternatives 1967

position of the service chiefs to OSD and the President on further prosecution of the war. Observing that no one could predict how long it would take to defeat the Communists. he said it was now very evident that the strategy employed in the past three years had not produced the desired result. [13]

Sent to Secretary McNamara on 17 October and later to the White House, the document cited basic policy as outlined in NSAM-288, 17 March 1964 (calling for an independent. non-Communist South Vietnam). other policy guidelines. and the principal JCS recommendations for attaining American objectives. It also pointed to certain administration restraints on JCS action. such as requiring "graduated" pressure on the enemy, permitting "sanctuary" areas in North Vietnam (particularly around Hanoi and Haiphong and in the buffer zone near China). and limiting special operations in Laos and Cambodia.

Under current policy. the JCS said. North Vietnam was paying heavily for its aggression and had lost the initiative in the South. While the "trend" was with the free-world forces. South Vietnam was making slow military. political and economic advances. To accelerate the rate of progress called for more military pressure. The service chiefs advocated 10 major additional steps. none requiring an increase in U.S. deployments. Several pertained to removing restrictions on air operations in the North: reducing the size of "prohibited" areas around Hanoi and Haiphong to the cities proper. thus making more important targets available to air strikes; shrinking the "buffer" zone area near China to permit unrestricted air attacks on rail lines and roads up to five miles from the Chinese border; authorizing CINCPAC to strike or restrike all targets outside of newly defined restricted areas; and permitting the JCS to authorize air strikes within restricted areas such as Haiphong on a "case-by-case" basis.

The JCS further recommended the mining of deep water ports. inland waterways. and estuaries north of the 20th parallel. and extending naval (Sea Dragon) operations. They favored emplacing Talos surface-to-air missiles on U.S. ships. stepping up air strikes in Laos and along North Vietnam's borders. and establishing "saturation bombing" zones in certain areas of Laos. as in the region northwest of the DMZ. the Nape. and Mu Gia Pass. They urged eliminating restrictions on B-52 overflights of and air strikes in Laos and ending a "cover" requirement for air strikes in South Vietnam when the targets were in Laos.

The AC-119 proved to be a potent weapon against communist supply lines.
Source: U.S. Air Force

Air War – Vietnam

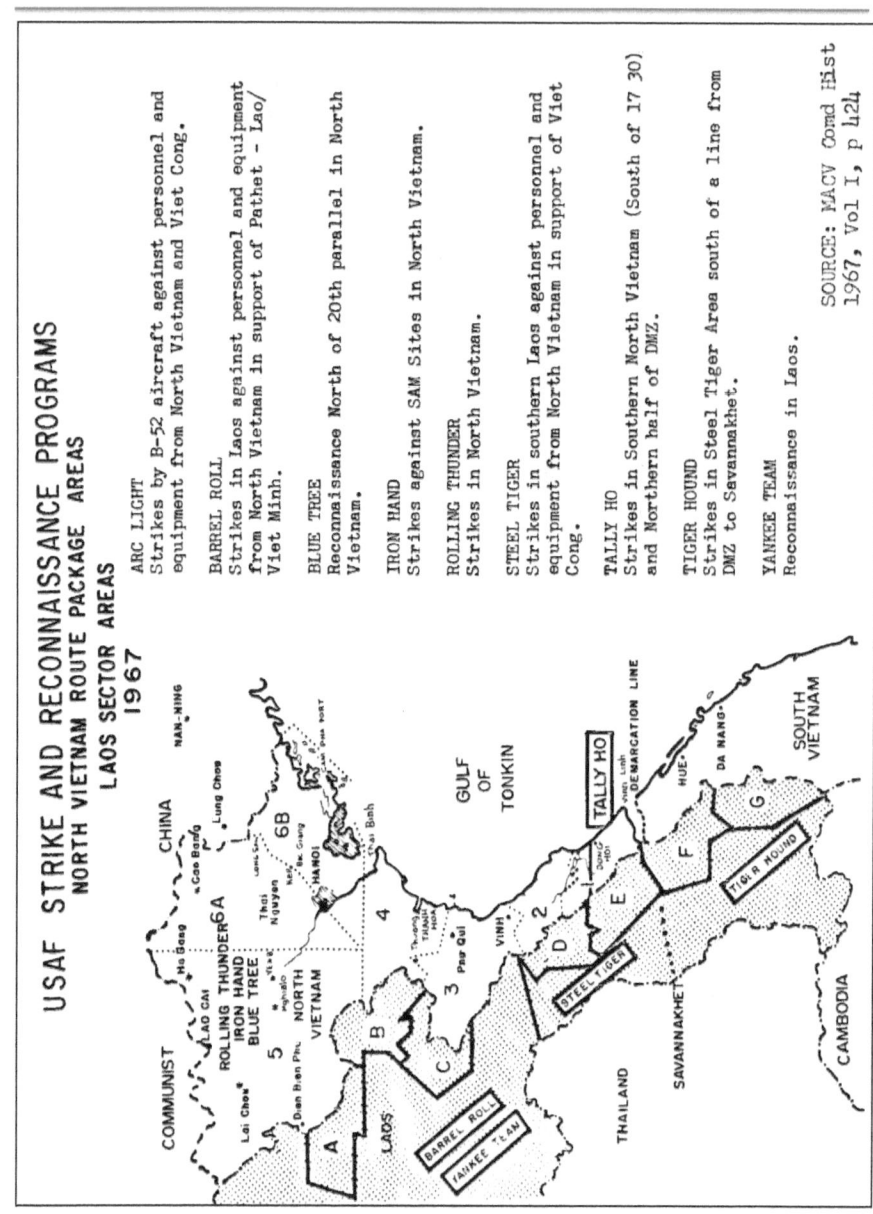

In addition. the JCS proposed expanding current "Daniel Boone" operations into Cambodia (to detect Communist activity) to the full length of the South Vietnamese-Cambodian border. allowing limited sabotage destruction activities, air strikes on border targets and unlimited helicopter missions near the border. and enlarging special programs in North Vietnam to improve the credibility of a resistance movement. The service chiefs believed that the major Soviet and Chinese reaction to all of these actions would be limited principally to providing more assistance to North Vietnam and to propaganda. [14]

Search For Alternatives 1967

High administration officials rejected these proposals, since, instead of offering "new thinking" for carrying the war against the enemy, principally within South Vietnam, they would greatly expand military operations in Laos, Cambodia, and North Vietnam which was contrary to Presidential policy. However, the administration was not yet through exploring alternate ways that could somehow accelerate allied progress in the war and simultaneously contribute to deescalating the fighting or negotiations to end it. Two more plans now came under JCS consideration. One concerned the 12-month air campaign against the North requested by Mr. McNamara and the other, requested by Secretary Rusk on 8 November, related to military operations in Southeast Asia in the immediate ensuing four-month period. [15]

Because the Defense Secretary wanted the 12-month campaign plan developed at the Washington level (only a concept had been prepared by PACOM in September), the JCS chairman on 14 October established a nine-man planning group headed by Air Force Lt. Gen. John C. Meyer, Director of Operations, Joint Staff. It included representatives from the Air Staff and other services, the Joint Staff, PACOM, and DIA. The group's terms of reference called for an air plan that would be an integral part of the over-all U.S. strategy in Southeast Asia. [16]

The draft of a plan acceptable to General McConnell and the Air Staff was completed on 29 November. It was a variation of many previous JCS recommendations. In essence, it emphasized air and naval operations in North Vietnam against port cities, materiel distribution lines (especially those running southward), and other targets. It would require only the use of currently deployed or programmed Air Force and Navy units but possibly more B-52 sorties. One of the plan's assumptions was that the stepped up operations would not trigger a Soviet or Chinese response. Transmission of the document to Mr. McNamara was delayed, however, until the service chiefs could complete the four-month plan for Southeast Asia requested by Secretary Rusk on 8 November. [17]

Mr. Rusk envisaged a State-Defense paper that would preclude the need for a weekly examination and approval of many small, short-range military operations and also accelerate in a very brief period allied progress in the war. The JCS input was sent through Secretary McNamara. Under current policy guidance, said the service chiefs, no new program could increase significantly the rate of allied progress in the near future. This was especially true with regards to efforts to expand the South Vietnamese armed forces and the pacification program and to improve the effectiveness of both. Taking a long view, they affirmed their belief that the present integrated military strategy for Southeast Asia (which they thought was generally being followed) was sound and would eventually achieve the objectives of NSAM-288, 17 March 1964, and those enunciated by the JCS on 1 June 1967. However, they thought there could be some improvement in the next four months if the United States avoided military truces (e. g., during the coming Christmas, New Year, and Tet holidays) and maintained pressure on the enemy. Again hoping to persuade administration authorities, the JCS also listed a series of measures for stepping up operations against the Communists, some of which were in the draft 12-month air campaign paper and in other JCS documents. These would require more action against North Vietnam in the form of air strikes on 24 important targets, mining the harbors of Haiphong, Hon Gai, and Cam Pha, ending bombing restrictions around Hanoi and Haiphong, allowing reconnaissance patrols in the northern half of the DMZ, launching Operation York II in the A-Shau Valley concurrently with limited South Vietnamese raids into Laos, and conducting other operations in both Laos and Cambodia. [18]

Although General McConnell had approved these recommendations, he and the Air Staff had misgivings about Operation York II and sending reconnaissance patrols into the Communist side of the DMZ as both might increase significantly manpower needs and require a major change in policy. [19]

Not unexpectedly, the administration disapproved the renewed JCS proposals for mining of harbors, striking targets in prohibited areas, or removing other major air restraints. In fact, the administration's response to the numerous "courses of action" papers indicated it desired no major change in military policy in Southeast Asia. Rather, it was moving toward making improvements in "in-country" programs, hoping this might contribute to de-escalation of the war and possibly negotiations. This trend became clearer in November when General Westmoreland and Ambassador Bunker and their staffs arrived in Washington to participate in another review of war policy. [20]

Search For Alternatives 1967

V. OTHER PROPOSALS TO SPEED UP PROGRESS IN THE WAR

By November 1967 the administration had additional reasons to adhere to its current military strategy in the war. From Saigon, it had received increasing optimistic reports which cited the high casualties suffered by the enemy, because of allied air and ground operations, and his failure to win any major battles. Political and economic conditions in South Vietnam also seemed much improved. In mid-November, General Westmoreland and Ambassador Bunker arrived in Washington to attest personally to the more favorable developments, to discuss new administration proposals to speed up allied progress and to seek approval of their own recommendations. [1]

The Administration's Eight Programs for South Vietnam

To prepare for the Westmoreland-Bunker visit, the JCS had informed MACV that the White House was considering giving top priority to eight programs in South Vietnam over the next six months. These would consist of: coordinated allied attacks on the Viet Cong infrastructure (including the construction of detention centers for 10,000 to 20,000 Communists); more integrated South Vietnamese-U.S. military operations; more South Vietnamese army search and destroy and security operations against Viet Cong battalions; more U.S. advisors for regional and popular forces; opening up and making the LOCs more secure; stepping up programs such as land reform, agricultural productivity, and universal education; encouraging more local government responsibility and attacking corruption; and employing locally trained personnel to support military research and development efforts. [2]

General Westmoreland subsequently added a ninth for "top priority" attention: improvement of South Vietnamese armed forces. He noted that each program would require additional authorizations from the JCS or other high officials with respect to personnel, equipment, funds, and adjustments in priorities. [3]

After studying the nine programs, the director of the Joint Staff foresaw some "maximum impact" arising from the quick dispatch of more American military advisors, greater effort in destroying the Viet Cong infra- structure, and building detention centers. But it would be more rewarding, he thought, to modernize South Vietnamese forces in order to accelerate the war's progress. However, this effort would take 12 months to gather "momentum" and would require, in allocating equipment, giving preference to South Vietnamese over American units. [4]

The Westmoreland-Bunker Briefings

In Washington, Ambassador Bunker and General Westmoreland participated in public as well as in White House, congressional and Pentagon briefings on the war. The Ambassador said that about 67 percent of the South Vietnamese people were now under Saigon's control (compared with 55 percent a year earlier), about

Air War – Vietnam

17 percent were under Viet Cong influence, and the remainder were in contested areas. He cited political gains, such as the inauguration of the Thieu government on 31 October, and reported that the South Vietnamese armed forces were improving, pacification was progressing, and the new government was taking steps against corruption. He believed that another bombing pause against the North should be contingent on some "reciprocity" by the Hanoi regime.

General Westmoreland said he had "never been more encouraged in my four years in Vietnam. " He saw the war entering a new phase and predicted, with continued military success, that the United States could begin shifting the burden of combat to the South Vietnamese in about two years. He opposed any lengthy bombing halt in either North or South Vietnam during the approaching holiday season, but said he could "live" with a short pause. 5

In a briefing for the JCS (similar to one given to Mr. McNamara), the MACV commander's "main theme" was on operations to improve the military situation in Southeast Asia during the next six to eight months. Stressing that real military pressure had been applied against the Communists for only one year, General Westmoreland outlined his current strategy. It consisted of "grinding down" guerrilla forces, driving main units into the jungles, mountains and border areas, and destroying enemy bases; opening roads for commerce and for Saigon's economic and social programs; blocking infiltration and bombing LOCs; forcing the North Vietnamese to divert more manpower to air defense and its transportation system; and preparing the South Vietnamese forces for a larger role in the war.

This strategy, General Westmoreland thought, had severely hurt the Communists, driven them to the border areas, and decreased recruitment which was down to 3,500 men per month compared with 7,000 per month a year earlier. Air and ground action had caused serious losses of personnel and supplies, and the Navy's sea blockade against infiltration had forced the enemy to use the treacherous land routes through Laos. Conversely, South Vietnamese forces were becoming more professional self-confident, and effective, and within the country there was political progress and some initial steps toward social reform. Roads were being opened.

For the future, the MACV commander advocated continuing the present policy at an accelerated rate, including the bombing of North Vietnam. He warned that "there was no better way to prolong the war than to stop the bombing of the North." In two years or less. he believed that South Vietnamese forces would be able to bear an increasing share of the war. thus permitting a phasing down of the American effort. He made three basic recommendations: modernize the South Vietnamese forces as rapidly as possible and as fast as they could receive equipment; send Deployment Program 5 forces as soon as possible; and increase B-52 sorties to 1,200 per month.

With respect to his first recommendation. General Westmoreland asked the JCS and OSD to approve his entire South Vietnamese program for fiscal years 1968 and 1969. This included accelerating shipments of M-16 rifles. M-60 machine

Search For Alternatives 1967

guns, M-29 18-mm mortars, M-79 grenade launchers,105-mm and 155-mm howitzers. AN/PRC-25 radios. trucks and other items, and assuring that the South Vietnamese possessed sufficient helicopters. With new and additional equipment the burden of the war in 1968, described as "Phase II" would shift more onto the shoulders of the South Vietnamese and they would assume a major share of front line defense of the DMZ area. although U.S. assistance in the delta region (IV Corps) would increase.   Under this program. General Westmoreland saw no need to raise the 525, 000 U.S. military ceiling for South Vietnam. The President concurred. [7]

As part of the Joint Staff's examination of the Saigon government's military needs. the Air Staff summarized approved VNAF programs and urged they be fully supported. These consisted mainly of aircraft conversion projects. Thus, one VNAF squadron of C-47s would convert to AC-47s and two C-47 squadrons to C-119's in fiscal year 1968; three A-1 squadrons to A-37s by the end of fiscal year 1969; and four H-34 helicopter squadrons to UH-1Hs by the end of fiscal year 1972. An important problem was finding enough H-34 aircraft to bring the VNAF's helicopter strength up to the authorized five squadrons plus "extras" for attrition. [8]

Secretary McNamara supported the stepped-up modernization of South Vietnamese forces "in principal" but asked for more data as soon as possible before giving final approval. [9]

In connection with General Westmoreland's second recommendation to speed up the movement of Deployment Program 5 forces, the JCS had anticipated it. On 9 October it had requested the services to again determine what units and personnel could be dispatched to Southeast Asia by 1 March 1968. In their replies, they reported on actions taken since 6 September to assure the movement of 18,000 additional troops (including 148 Air Force personnel and four UC-123s) to South Vietnam. However, another 27,900 troops remained to be deployed including 1,100 USAF officers and men. No estimate was available for 3,700 other Army and Navy personnel. The Air Staff, urged to reexamine its schedule, on 15 November determined that only 388 of the remaining USAF personnel could be sent by 1 March 1968. [10]

The expedited deployment of two brigades of the Army's 101st Airborne Division, approved by Secretary McNamara on 23 October, produced the largest single Air Force airlift of the war. This operation, designated Eagle Thrust, witnessed the air movement of 10,024 men and 5,357 tons of support equipment from Campbell Airfield, Ky., to Bien Hoa AB, South Vietnam, between 17 November and 29 December 1967. The entire operation required 369 C-141 and 22 C-133 missions. The two brigades arrived in the war theater about six weeks ahead of the original schedule. [11]

Concerning General Westmoreland's third recommendation, Secretary McNamara on 21 November, authorized an increase in the B-52 sortie rate from 800 to 1,200 per month. In April 1967, General Westmoreland had asked for a sustained 1,200 per month B-52 sortie rate as soon as possible. Subsequently. Secretary Brown informed Mr. McNamara that the SAC bombers could fly this rate. if necessary. starting 1 February 1968 from bases in Thailand and Guam. Upon completion of construction at U-Tapao AB in June 1968, the B-52's would be able to provide 750 sorties per month from that base alone. The ground work for this capability was laid on 6 November when the Defense Secretary. after obtaining Thai government approval sanctioned an increase from 15 to 25 B-52s at U-Tapao AB, Thailand (although the JCS recommended 30), plus about 1,000 additional military personnel. At that time he was responding to a JCS recommendation to consent to only a "surge" rate up to 1,200 sorties per month for 60 days. if necessary. with 72 hours advance notice. The service chiefs observed that the deployment of additional bombers and personnel to Thailand (by June 1968) would reduce somewhat the Air Force's capability to support the current strategic integrated plan (SIOP). The B-52 operations were eased by another administration decision on 5 December, which, with approval of the Lao government, authorized overflights of Laos. This change promised to save about $18 million per year, the difference in cost for 25 B-52s flying directly from U-Tapao to South and North Vietnamese targets versus detouring around Cambodia. [12]

U.S. Strategy and Strength at the End of 1967

Thus. at year's end the administration was engaged in stepping up its civilian and military programs within South Vietnam. Pacification and economic stability would continue to receive high priority. "In-country" military activity against the Communists would include more air-supported ground offensives, more B-52 sorties. and limited incursions into border areas. There would be a "crash" effort to complete the linear strong point obstacle system in South Vietnam and the air-supported anti-infiltration system in Laos. Portions of the 600-meter wide strong point obstacle system became operational in late 1967, and the anti-vehicular (Mud River) section of the air- supported anti-infiltration system attained an initial operational capability (IOC) on 1 December 1967. To strengthen South Vietnam's military posture. steps would be taken to accelerate the training and equipping of Saigon's regular, regional, and popular forces (to total 685,739 by June 1968), and most of the U.S. Deployment Program 5 forces would be sent by March 1968. American military strength in Thailand would be held to about 48,000.

To avoid precipitating a wider conflict. the air effort in North Vietnam or Laos would not be significantly intensified. Large-scale air and ground assaults against enemy troops in Laos. Cambodia or North Vietnam would be prohibited. By public statements and internal policy, the administration was exhibiting a greater desire to deescalate or negotiate a settlement of the war. During a brief Southeast Asian visit in December. President Johnson. while restating America's war objectives. also asserted that he now favored talks between the Saigon government and the Communist-led National Liberation Front. [13]

In seeking a lower military tempo and possibly negotiations. the administration was buoyed by reports of increasing losses of and strains on North Vietnam's military and civilian resources. Both Air Staff and MACV analyses of 1967 military operation in South and North Vietnam and Laos compared with those in 1966 showed considerably greater enemy casualties. MAC/V estimated overall

Search For Alternatives 1967

enemy losses for 1967 at 169,200. including 24,000 non-battle casualties. This contrasted with enemy manpower replacement by infiltration and recruitment of 113,700 for a net loss of 55,500. (In March 1967. General Westmoreland had anticipated a net increase in Communist strength by the end of the year. Air Staff figures showed an increase in enemy killed in action in South Vietnam from 55,524 in 1966 to 87,468 in 1967. [There was also an] apparent reductions toward year's end in the infiltration rate.

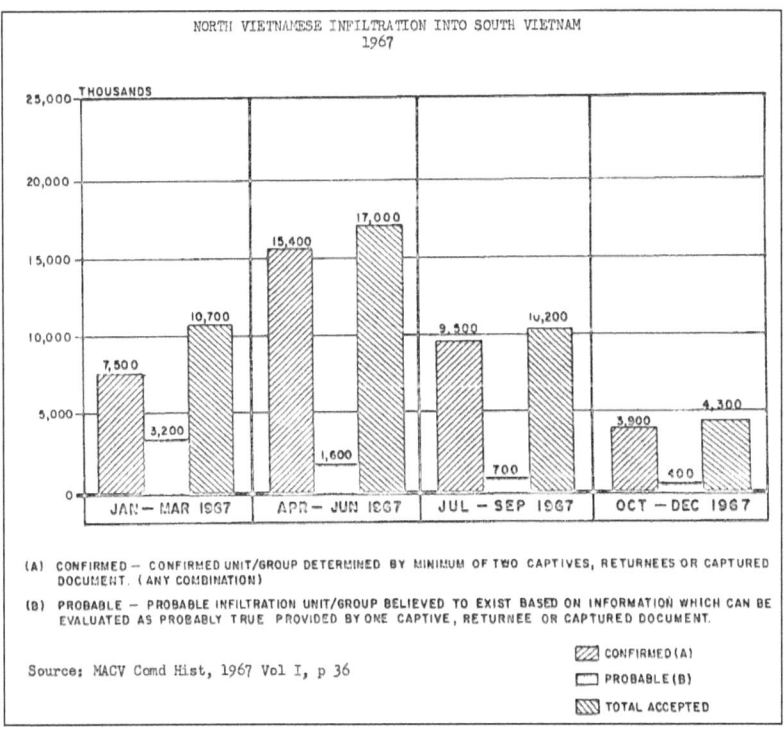

Although infiltration figures lagged by at least six months, MACV estimates showed a considerable drop between the second and fourth quarters of 1967. In January 1968 MACV estimated 1967 infiltration to have totaled about 54, 000 but expected the final total to be about equal to the 1966 total of 87,200. [There were also] higher truck, rolling stock, and watercraft losses, increased need for imports, and reduced war-supporting capacity. MACV estimated that air attacks in North Vietnam and Laos from 1 January through 20 December 1967 destroyed or damaged 5,261 motor vehicles, 2,475 pieces of rolling stock and 11,425 watercraft. The Air Staff concluded that 1967 witnessed for the first time. a net enemy loss of about 2,000 trucks above imports with about 9, 000 to 10,000 trucks still in North Vietnam's inventory. In addition, pacification reports were encouraging. [14]

However. there was also concern that past and current "progress" indicators were not sufficiently thorough or reliable. This was manifested in the Air Force by Secretary Brown's requests for better analyses of the results of the air effort. The

Air War – Vietnam

consequence was the issuance, beginning in September 1967, of a monthly publication entitled: "Trends, Indicators, and Analyses," by the Operations Review Group, Directorate of Operations. It sought to evaluate progress toward achieving the three basic objectives of the air war in North Vietnam. These were: reducing the flow of men and materiel moving from North to South Vietnam. increasing the cost of the war to the North, and convincing Hanoi it could not continue its aggression without incurring severe penalties. In the same month, also at Secretary Brown's request, the Air Staff formed a joint Operations Analysis-Rand Corporation study group to better pinpoint operational issues and analyze the effects of the air war in Southeast Asia. At a higher level. the White House on 25 October directed the creation of an interagency task force, chaired by the Central Intelligence Agency (CIA), to improve accuracy in estimating enemy casualties, weapon losses, extent of population control. the effect of the Chieu Hoi or "open arms" reconciliation program. and other "progress" indicators.

Both the strategy and deployment levels in South Vietnam and Thailand were, as has been noted, less than desired by General McConnell and the other service chiefs. In the JCS deliberations during the year, the Air Force Chief of Staff remained a consistent advocate of the use of more air power against North Vietnam, convinced this would minimize the need for more troops, decrease allied casualties, and shorten the war. In the absence of more authority for stronger air programs, he agreed with the other service chiefs on virtually all measures they mutually thought might shorten the conflict, such as mining or blocking Haiphong harbor, narrowing "sanctuary" areas to hit more war-supporting targets, calling up U.S. reserves, and modernizing Vietnamese forces.

Despite the Joint Staff's frequent advocacy of heavier air attacks against North Vietnam, the administration refused to alter its air policy. Testifying before a Senate committee early in 1968, Secretary McNamara asserted that few strategically important targets remained in the North and that the agrarian economy there could not be collapsed by bombing. Further, the enemy's low combat requirements precluded "pinching off" the flow of supplies to the South. Mr. McNamara emphasized .that, except for manpower, Hanoi's war effort was sustained principally by military and economic aid from Communist countries valued, in 1967, at about one billion dollars. [16]

As 1967 ended, 486, 600 American troops were in South Vietnam (including 55,900 Air Force) and 44,500 in Thailand (including 33,500 Air Force). This represented an increase during the year of 96,032 and 10,011personnel in the two countries, respectively. While Air Force deployments to South Vietnam were relatively small, amounting to only 2,987 personnel, they were substantial in Thailand where they increased by 7, 297. South Vietnamese regular, regional, and popular forces totaled 641,000 (including 16,253 VNAF), an augmentation of 21,000. There were also 42,000 in the civilian irregular defense group and 73,400 in the national police. The last was boosted by 13,400 men during the year to assure more internal security. Other allied forces totaled about 60,000, an increase of 7,678 including a Royal Australian Air Force Unit. Principal allied strength at the end of 1967 was as follows: Australia 6,600 (including one squadron of eight

Search For Alternatives 1967

B-57s); New Zealand, 500; Philippines, 2,000; Thailand, 2,400; and Republic of Korea, 48, 800.[17]

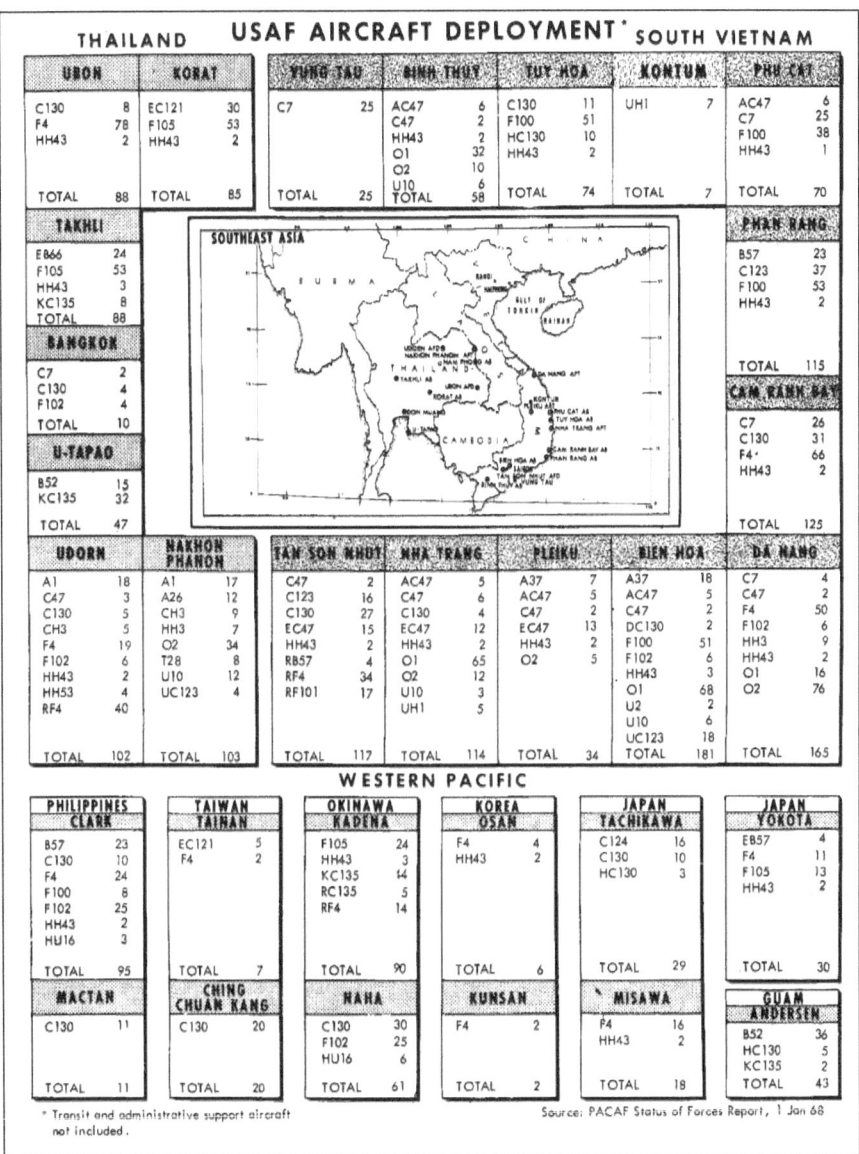

Although the number of American, South Vietnamese, and other allied personnel in South Vietnam and Thailand thus rose by 134,000 during the year, there was a slight decrease in U.S. combat aircraft in the two countries and with the Seventh Fleet. The total dropped from 1,009 (including 639 Air Force) at the end of 1966 to 922 (including 650 Air Force) at the end of 1967. Non-combat aircraft and

Air War – Vietnam

helicopter arrivals, on the other hand, rose substantially with the Army sending nearly 1,000 more helicopters to the theater during the year. USAF aircraft strength in 1967 was also changed by more modernization. A number of F-104s, F-105s and F-4C's were replaced by F-4D's, and FAC 0-ls by 0-2s. (U.S. manpower and aircraft strengths during 1967 and proposed strengths through 1969 are included in the appendix).

APPENDIX ONE
U.S. MILITARY & AIRCRAFT STRENGTH
20 December 1967

U.S. Military Strength in South Vietnam						
	Jun 67*	Dec 67*	Jun 68**	Dec 68**	Jun 69**	Dec 69**
USAF	55,700	55,900	57,900	58,000	58,700	-
USN/CG	29,000	32,200	33,100	34,800	35,500	-
USMC	78,400	78,000	81,800	81,800	81,900	-
USA	285,700	320,500	344,700	346,700	348,900	-
Total	448,800	486,600	517,500	521,300	525,000***	-

U.S. Military Strength in Thailand						
	Jun 67*	Dec 67*	Jun 68**	Dec 68**	Jun 69**	Dec 69**
USAF	28,300	33,400	35,300	35,300	35,300	-
USN-MC & CG	400	800	500	500	500	-
USA	10,300	10,300	12,200	12,200	12,200	-
Total	39,000	44,500	48,000	48,000	48,000	-

U.S. Military Strength Offshore						
	Jun 67*	Dec 67*	Jun 68**	Dec 68**	Jun 69**	Dec 69**
USN-MC & CG	41,300	36,500	42,000	42,000	42,000	-

Air War – Vietnam

U.S. Fighter and Attack Aircraft in Southeast Asia						
	Jun 67*	Dec 67*	Jun 68**	Dec 68**	Jun 69**	Dec 69**
USAF	621	650	704	704	698	698
USN/CG	182	188	203	198	212	213
USMC	153	154	170	170	170	170
USA	-	-	-	-	-	-
Total	956	992	1,077	1,072	1,080	1,081

B-52's						
	Jun 67*	Dec 67*	Jun 68**	Dec 68**	Jun 69**	Dec 69**
Thailand	10	15	25	25	25	25
Guam	54	36	****	****	****	****
Total	54	51	****	****	****	****

**** Undisclosed

Search For Alternatives 1967

Allied Fighter And Attack Aircraft in Southeast Asia						
	Jun 67*	Dec 67*	Jun 68**	Dec 68**	Jun 69**	Dec 69**
VNAF	108	90	72	99	108	108
RAAF	8	8	8	8	8	8
Total	116	98	80	107	116	116

U.S. Fixed-Wing Non-Attack Aircraft in Southeast Asia						
	Jun 67*	Dec 67*	Jun 68**	Dec 68**	Jun 69**	Dec 69**
USAF	662	793	851	854	854	854
USN	37	52	43	40	45	42
USMC	37	46	47	49	61	61
USA	384	550	594	609	609	609
Total	1,120	1,440	1,535	1,552	1,569	1,566

U.S. Helicopters in Southeast Asia						
	Jun 67*	Dec 67*	Jun 68**	Dec 68**	Jun 69**	Dec 69**
USAF	62	69	97	97	97	97
USN	0	0	0	0	0	0
USMC	301	296	304	292	280	280
USA	2,036	2,600	2,698	3,103	3,242	3,235
Total	2,399	2,965	3,369	3,492	3,619	3,612

Air War – Vietnam

USAF Fighter Attack Aircraft in Southeast Asia						
	Jun 67*	Dec 67*	Jun 68**	Dec 68**	Jun 69**	Dec 69**
A-1	31	41	61	61	61	61
A-26	8	12	12	12	12	12
B-57	21	23	24	24	0	0
F-100	199	197	198	198	198	180
F-102	24	24	30	30	30	30
F-104	16	0	0	0	0	0
F-105	132	107	108	90	72	54
F-4	179	213	234	252	288	324
F-5	0	0	0	0	0	0
T-28	11	8	12	12	12	12
A-37	0	25	25	25	25	25
Total	621	650	704	704	698	698

Search For Alternatives 1967

	USAF Fixed-Wing Non-Attack Aircraft in Southeast Asia					
	Jun 67*	Dec 67*	Jun 68**	Dec 68**	Jun 69**	Dec 69**
RB-57	3	3	4	4	4	4
EB-66	19	28	41	41	41	41
RF-4	60	74	76	76	76	76
RF-101	2.5	17	16	16	16	16
AC/C-47	3.5	40	46	46	46	46
EC-47	42	40	47	47	47	47
EC-121	6	30	27	27	27	27
C-123	84	7.5	88	91	91	91
WC/HC/C-130	16	17	17	17	17	17
KC-135	39	40	40	40	40	40
C-7A	84	8.5	96	96	96	96
0-1	202	182	103	.50	34	34
0-2	18	137	191	191	191	191
OV-10	0	0	27	80	96	96
U-10	2.5	2.5	32	32	32	32
HU-16	4	0	0	0	0	0
Total	662	793	851	854	854	854

	Helicopters in Southeast Asia					
	Jun 67*	Dec 67*	Jun 68**	Dec 68**	Jun 69**	Dec 69**
UH-1	13	12	1.5	1.5	1.5	1.5
HH-43	2.5	30	32	32	32	32
HH-53	0	4	8	10	10	10
HH-3	14	16	20	18	18	18
CH-3	10	7	22	22	22	22
Total	62	69	97	97	97	97

Air War – Vietnam

U.S. and VNAF Fighter and Attack Aircraft Sorties					
	Jun 67*	Dec 67*	Jun 68**	Dec 68**	Jun 69**
South Vietnam	16,544	15,200	15,256	14,893	15,347
Laos	1,441	6,700	1,874	6,340	1,807
North Vietnam	11,471	5,700	11,514	7,623	11,570
Total	29,456	27,600	28,644	28,856	28,724

B-52 Sorties				
Jun 67*	Dec 67*	Jun 68**	Dec 68**	Jun 69**
832	800	1,200	1,200	1,200

* Actual (Military strength figures include TDY personnel).

** Current Plan

*** The 525,000-man U.S. ceiling would be reached in March 1969.

Source: Memos (S), Asst SECDEF (SA) to Secys of Mil Depts et al, 29 Dec 67 and 15 Feb 68.

PART SIX

TOWARDS A BOMBING HALT

1968

Air War – Vietnam

USAF AIRCRAFT DEPLOYMENT IN ASIA
29 December 1968

SOUTH VIETNAM

BINH THUY		CAM RANH BAY		DA NANG		NHA TRANG		PHAN RANG		PHU CAT		PLEIKU	
AC47	3	C7	29	A1	2	AC119	4	AC47	3	AC47	3	A1	18
C47	1	C130	37	AC47	5	AC47	8	A37	14	C7	21	AC47	4
HH43	2	F4	49	C7	4	C47	3	C123	37	F100	65	C47	2
O1	38	HH43	2	C47	1	C130	3	F100	67	HH43	1	EC47	17
U10	8			F4	54	EC47	16	HH43	2			HH43	2
				F102	6	HH43	2	O1	6			O2	5
				HH3	10	O1	36	O2	12				
				HH43	2	O2	24						
				O1	15	U10	6						
				O2	59	UH1	8						
TOTAL	52	TOTAL	117	TOTAL	158	TOTAL	110	TOTAL	141	TOTAL	90	TOTAL	48

BIEN HOA						TAN SON NHUT	
A37	20					C123	16
AC47	5					C130	26
C47	3					EC47	15
DC130	2					HH43	2
F100	55					RB57	2
HH43	3					RF4	29
O1	62					RF101	15
O2	15					TOTAL	105
OV10	21					TUY HOA	
U10	9					C130	10
UC123	22					F100	75
WU2	3					HC130	9
						HH43	3
TOTAL	220					TOTAL	97

BAN ME THUOT						VUNG TAO	
UH-1	9					C7	30
TOTAL	9					TOTAL	30

THAILAND

BANGKOK		KORAT		NAKHON PHANOM		TAKHLI		UBON		UDORN		U-TAPAO	
C7	2	EC121	31	A1	40	EB66	29	AC130	4	C47	3	B52	34
C130	6	F4	20	A26	17	F105	56	C130	7	C130	6	KC135	41
F102	4	F105	35	C123	5	HH43	2	F4	72	CH3	8		
		HH43	2	CH3	10			HH43	3	F4	41		
				HH3	7					F102	6		
				O2	37					HH43	2		
				OV10	6					HH53	6		
				U10	13					RF4	34		
				UC123	2					UH1	4		
TOTAL	12	TOTAL	88	TOTAL	137	TOTAL	87	TOTAL	86	TOTAL	110	TOTAL	75

WESTERN PACIFIC

GUAM		JAPAN				KOREA				OKINAWA				PHILIPPINES		TAIWAN	
ANDERSEN		ITAZUKE		MISAWA		KUNSAN		KWANGIU		KADENA		NAHA		CLARK		CHING CHUAN KANG	
B52	55	EB66	4	F4	18	F4	8	F105	10	B52	19	C130	28	B57	4		
HC130	3	RF101	19	HH43	2	F100	50	HH43	2	C130	3	F102	15	C130	25	C130	26
KC135	2					HH43	2			F105	15	F106	1	F4	19		
								TOTAL	12	HH43	3	HH3	1	F102	30	KC135	20
		TOTAL	23	TOTAL	20	TOTAL	60	OSAN		KC-135	37			HC130	4	TOTAL	46
		TACHIKAWA		YOKOTA		F4	19	F102	12	RC135	8			HH3	2	TAINAN	
		C124	14	EB57	4	F106	17	HH43	2	RF4	13			HH43	2	EC121	5
		C130	20	F4	20	HH43	2	TOTAL	14	SR71	3					F4	4
		HC130	4	HH43	2			TAEGU									
								F4	11								
								HH43	2								
TOTAL	58	TOTAL	38	TOTAL	26	TOTAL	38	TOTAL	13	TOTAL	101	TOTAL	45	TOTAL	86	TOTAL	9

Transit and administrative support aircraft not included.

SOURCE: USAF Mgt Summary, Southeast Asia, 3 Jan 1969; p 22.

Towards A Bombing Halt 1968

I. MILITARY & POLITICAL SITUATION, EARLY 1968

When 1968 began, Washington officials were optimistic about the war in Southeast Asia since it seemed that -the Allies were closer to achieving their objectives. The armed forces of the free world had grown stronger during 1967, while those of the .North Vietnamese and Viet Cong had become weaker. "We feel," President Lyndon B. Johnson stated on New Year's Day, "that the enemy knows that he can no longer win a military victory in South Vietnam."[1] Studies for a de-escalation of the war were under way and there was new confidence that a cease-fire might be negotiated with the enemy.

President Lyndon Johnson.
Source: U.S. Govt

Evaluations of the War

During a visit to Washington in November 1967, Gen. William C. Westmoreland, Commander. U.S. Military. Assistance Vietnam (COMUSMACV) and the American Ambassador to Saigon, J. Ellsworth Bunker, had expressed considerable optimism. In meetings with the President, members of Congress. State and Defense Department officials, and in public statements, they declared that the United States and its allies were now winning the war. However,

Gen. William C. Westmoreland.
Source: U.S. Army

General Westmoreland, although he supported the President's military policy, desired to increase the pressure on the enemy. and warned against any letup in the bombing. of North Vietnam. He also wished to increase the number of B-52 sorties from 800 to 1,200 per month, modernize South Vietnamese forces as fast as they could absorb additional equipment and send the remainder of Deployment. Program 5 Vietnam as soon as possible. By the end of 1967, Secretary Robert S. McNamara had authorized an increase in B-52 sorties and a speed-up in the movement of Deployment Program 5 forces. He was studying a proposal to accelerate the modernization of South Vietnamese forces.[2]

Air War – Vietnam

Hopeful but somewhat less sanguine. as 1968 began was a report on the war's progress prepared by the Joint Staff of the Joint Chiefs of Staff (JCS). Pointing to the enemy's heavy losses in manpower, decline in population control, failures to launch major attacks and the need to operate from border areas, the report said that the military objectives were closer to attainment than at the beginning of 1967. It saw significant political gains in the South Vietnamese election of 3 September 1967 and the inauguration of Nguyen V Thieu on 31 October of the same year as the head of the new Saigon government. The report concluded that the air campaign against the North had reduced to "less than optimum" the number of troops and the quantity of supplies reaching the South. In fact, air operations had transformed North Vietnam's economy into little more than a distribution system. On the other hand. the Viet Cong and North Vietnamese were not yet "down and out." Dedicated and vigorous, they could transport the necessary military resources into South Vietnam, sustain current levels of operation, and commit major forces where there was a high probability of success. The North Vietnamese were able to adjust to selective bombings. Also, bad weather frequently halted or reduced air operations and the use of more antiaircraft weapons further degraded the effectiveness of Allied bombings.[3]

President Nguyen V Thieu. Source: Vietnamese Govt

More disquieting was an assessment by Maj. Gen. William E. Depuy, Special Assistant for Counterinsurgency and Special Activities, JCS, who warned Gen. John P. McConnell, Air Force Chief of Staff and other service chiefs that the new Thieu government still had to demonstrate its ability to govern. Despite plans for civil and military reform announced by President Thieu on 31 October, government ministries had not yet made civil changes nor had the Republic of Vietnam Armed Forces (RVNAF) reorganized in accordance with the recommendations of the U.S. Military Assistance Command. Vietnam (USMACV). Also. the powers of "war lord " corps commanders, who acted without regard to Saigon, had not been reduced. junior officers were increasingly restive about corruption, and morale was low among revolutionary development (pacification) cadres and province officials who lacked power to institute new programs. In short, there absence of "forward motion" which could put the government "on the road." General Depuy felt that "leverage" to force

Gen. John P. McConnell, Air Force. Source: U.S. Air Force

Towards A Bombing Halt 1968

changes in the lower echelons of the Saigon government was not possible unless it was applied first by Ambassador Bunker and General Westmoreland. Although Saigon'ss concern about "negotiations" made it difficult to apply pressure, General Depuy nevertheless believed that the "nettle had to be grasped." [4]

The status of the Saigon regime and U.S. strategy in the increasingly debated in Congress and the public media. In January 1968 Sen. Edward M. Kennedy (Dem, Mass.) proposed placing greater emphasis on effecting social and political reforms, reducing corruption in South Vietnam and adopting a "clear and hold" rather than a "search and destroy" strategy in the war. President Johnson his asked his service chiefs to comment on the Kennedy proposals.

In his reply to the President. General McConnell said that military success should take priority over efforts to achieve internal reforms in the Saigon government. He recognized the need for national stability in developing viable social and political institution, but opposed U.S. threats to withdraw forces until the Saigon regime reduced corruption, a condition not unique in South Vietnam although of special significance there. As for "clear and hold" military operations. he thought such a change in U.S. strategy would give the Communists more freedom to attack and inflict losses on Americans and would create demands for additional troops. The Air Force Chief of Staff felt that the administration should continue to explain to the complex problems of Vietnam did not lend themselves "simplistic solutions" but required military as well as nonmilitary actions. [5]

Robert McNamara
Source: U.S. DoD

Concerning military actions, the Air Force had long contended that the war could be shortened and won with fewer U.S. casualties and with acceptable risks. if the administration reduced its restrictions on bombing North Vietnam. But top officials including Secretary McNamara disagreed. They were convinced that bombing would not reduce significantly the enemy's minimal combat requirements and that such a policy might trigger a conflict with the Russians or Chinese. Further, they argued that the war had to be won in the South. (For a resume of the views of General McConnell and Secretary McNamara on bombing operations in Southeast Asia (SEA), see Hearings (22, 23, 25 Aug 1967) before the Senate Preparedness Investigating Subcommittee of the Armed Services Committee 90th Cong. 1st Sess, Air War Against North Vietnam. parts 3 and 4.)

Studies on a Bombing Halt and Negotiations To End The War

In early 1968 the Air Staff also continued to examine proposals for de-escalating or ending the war in Southeast Asia. One dealt with a 10 July 1967 plan developed by Representative E. Bradford Morse (Rep, Mass.) calling for a five-step de-escalation of the conflict. This would be achieved by reducing gradually the

Air War – Vietnam

bombing of North Vietnam southward toward the demilitarized zone (DMZ) while the Communist forces similarly deescalated. General McConnell sent his critique of the Morse plan to Air Force Secretary Harold Brown on 4 January 1968. He said the plan contained "serious pitfalls" and he particularly questioned its assumptions for attaining mutual de-escalation. Nevertheless, he believed the plan merited further analysis. [6]

Examining another proposal to achieve "tactical de-escalation of the war .. the Air Staff and Joint Staff agreed that in order to "tacitly" lower the tempo of the fighting, the administration's objective should not be less than a negotiated end to the war. And before the United States halted the bombing of the North, Hanoi should meet "minimum conditions" previously outlined by the JCS.   Although the services disagreed on some details of the tacit de-escalation proposal, they continued to examine its possibilities. [7]

The Air Staff also participated in a JCS study, "Sea Cabin" that explored President Johnson's 29 September 1967 San Antonio "formula" for ending the air. and naval bombardment of the North and negotiating an end to the war. (see Air War: Vietnam, Plans and Operations 1961 – 67, Part Five: The Search for Military Alternatives 1967 (Defense Lion Publications)). The JCS. Views generally in consonance with those held by the Air Force, were sent to Secretary McNamara at the end of January. The service chiefs recommended that the United States exact a stiff quid pro quo from Hanoi for a bombing halt. They felt strongly that bombing should be resumed if there were no serious discussions within seven days, if the enemy resumed major attacks. or if they concluded the bombing had given the enemy a substantial military advantage. [8]

Nguyen Duy Trinh
Source: Vietnamese Govt

Early in 1968 there was also considerable speculation about a statement on 29 December 1969 by North Vietnam's Foreign Minister Nguyen Duy Trinh, who suggested that negotiations might soon be possible. He indicated that his government's position had changed from "would talk" to "will talk" if the United States halted its attacks on the North. But President Johnson, Gen. Earl G. Wheeler (the JCS Chairman) and other Officials did not regard the statement as a "breakthrough: toward negotiations. The Hanoi regime, they believed, had not yet met their requirements for a bombing pause and would probably take advantage of a cessation of attack to strengthen its military posture. The new Secretary of Defense designate. Clark M. Clifford, informed a Senate committee on 25 January that he too opposed a bombing suspension, feeling it was premature. [9]

Notwithstanding these high-level views, the impact of new military crises in the Asian theater would soon alter fundamentally

Towards A Bombing Halt 1968

the administration's position concerning a bombing 'halt' and negotiations with Hanoi. Meanwhile. the United States and her allies continued to pit their combat strength and strategy against the Communists in the field.

U.S. and Allied Strength in Southeast Asia

In terms of manpower; the allies at the beginning of 1968 still enjoyed considerable superiority over the Communists, having fielded forces totaling more than 1,300,000 military personnel. Of this number, in South Vietnam 496,000 were American (including Air Force), and 641,000 were South Vietnamese (including regular, regional and popular force plus 16,253 in their air force). Saigon also could call upon a special South Vietnamese civilian irregular defense group of 42.000. In addition, other allied troops, mostly South Korean. were deployed in South Vietnam. Offshore 36,500 Americans manned the U.S. Seventh Fleet and, in Thailand, 45,500 U.S. personnel (including 33,400 Air Force) supported the air war in North Vietnam and Laos. American manpower was controlled tightly by the Office of Secretary of Defense through its Southeast Asia Deployment Program 5, issued on 5 October 1967. This document imposed a ceiling of 525,000 U.S. personnel South Vietnam and 45,724 in Thailand.

Reconnaissance aircraft such as this RF-4C played an important part in monitoring North Vietnamese activities but paid a heavy price for doing so.

Source: U.S. Air Force

Allied combat aircraft included 992 American, 90 Vietnamese Force (VNAF) and eight Royal Australian Air Force tactical fighters. Of the U.S. total 650 were Air Force. In addition, 51 Strategic Air Command (SAC) B-52 bombers were in Thailand and Guam to carry out Arc Light saturation bombing, mostly in South Vietnam. Also operating in the area were non-combat aircraft used for transport.

Air War – Vietnam

forward air control, reconnaissance., electronic,. and other support missions. Of these, the Air Force possessed more than half. Helicopters numbered 2,965 and belonged chiefly to the Army and Marine Corps, although the Air Force operated 69 on air rescue missions.

Arrayed against the free world forces were an estimated 200.000 Viet Cong and North Vietnamese soldiers, strongly backed by the resources of the Soviet Union, China, and other Communist states. As a result of allied air and ground operations in 1967, the Communists were believed to have suffered a net loss of 55,500 men. Air strikes had also discouraged Hanoi from maintaining many jet aircraft on its airfields. At the end of 1967 there were only 10 MiG-15s and eight MiG-21s stationed in North Vietnam, whereas 60 jets were on nearby South China bases. These consisted of 49 MiG-21s, three MiG-16 (Defense Lion notes that the original text says MiG-16 but the aircraft were actually MiG-15UTI) trainers, and eight IL-28 bombers.[10]

This order of battle, so heavily weighted in favor of the allies undoubtedly contributed to the aura of optimism about the beginning of the year.

Towards A Bombing Halt 1968

II. MILITARY CRISES LATE JANUARY - MARCH

The feelings of optimism expressed by military and civilian officials in Saigon began to fade in late January as a result of a series of unexpected events in Vietnam and Korea. On 21 January a specially trained team of 31 North Korean agents infiltrated into South Korea on a mission to assassinate President Park Chung Hee. Two days later the North Koreans seized the U.S. intelligence ship *Pueblo* about 13 miles off their coast. In South Vietnam, about this time the Communists completed an encirclement of the Marine base at Khe Sanh not far from the Laotian border and the DMZ. For 77 days they lay siege to about 6,000 Marines and a South Vietnamese Ranger battalion defending the post, while fears arose in the United that the enemy was trying to achieve another Dien Bien Phu. Although the enemy suffered huge casualties. he continued to ring the base, shelling it frequently while a major U.S. airlift replenished the stocks of the besieged Leathernecks. [1]

The intelligence ship USS Pueblo. Seized in January 1968, she remains in North Korean hands as of 2012. Photo Source: U.S. Navy

The most important enemy action, however, began in the early hours of 30 January. Under the cover of a military truce for South Vietnam, the Viet Cong and the North Vietnamese launched a month-long "Tet" offensive, attacking Saigon and key cities in I and II Corps in the north and numerous South Vietnamese and American military headquarters and airfields. The repercussions of these almost simultaneous blows throughout South Vietnam were far-reaching.

Crisis in Korea

The seizure of the Pueblo and other North Korean provocations along the Korean DMZ induced President Johnson on 25 January to order a limited callup of Reserve forces, a step long advocated by the JCS. Fourteen Air National Guard (ANG), eight Air Force Reserve (AFRES) and six Navy Reserve units totaling 14,878 personnel were called to active duty. Some Air Force units flew immediately to South Korea and a squadron of Air Force F-4s redeployed from Cam Ranh Bay in South Vietnam to South Korea. having been replaced Tactical Air Command (TAC) squadrons from the United States. (In May and June 1968 after further training, four of eight ANG F-100C squadrons recalled on 25 January were sent from the United States to South Vietnam.).

Air War – Vietnam

Crew of USS Pueblo in North Korean hands. Source: U.S. Navy

In a further show of force the next day, Secretary McNamara approved the movement of 26 addition B-52s and supporting tankers from the United States to the Pacific. In early February. under the code name Port Bow. Eleven bombers joined those already on Guam. Fifteen others along with nine KC-135 tankers went to Okinawa to be on hand if needed. [2]

Even as he was reacting to the *Pueblo* crisis, the President suspected, as he remarked in a brief report to the nation on 26 January, that the North Koreans might be trying to divert America's attention and energies from the Vietnam struggle. [3]

The Air Force at Khe Sanh

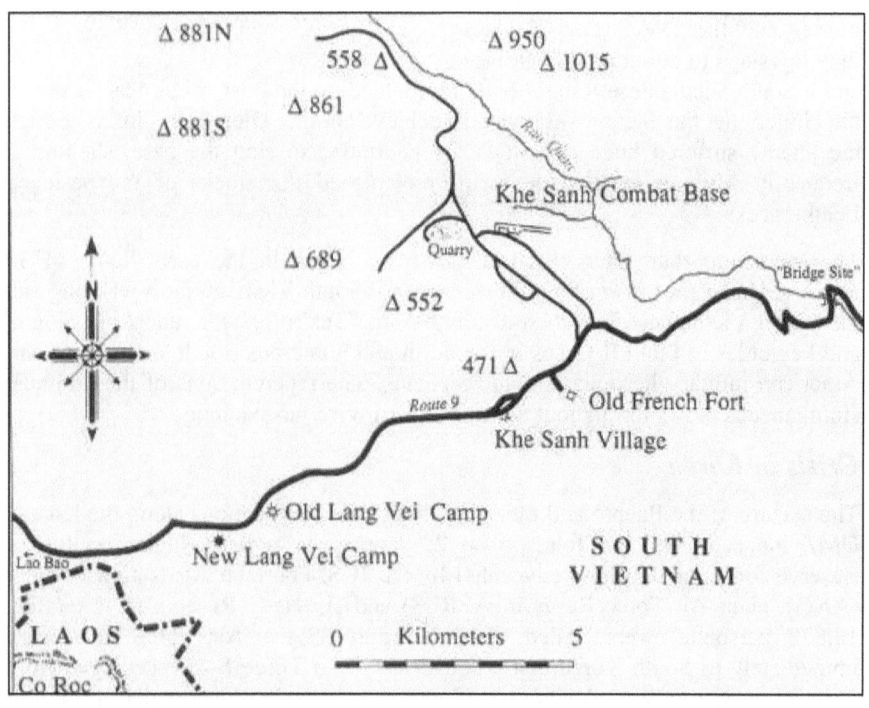

The Base At Khe Sanh. The defense of Khe Sanh was a remarkable demonstration of what tactical air power could achieve when the gloves came off. Source: U.S. Air Force.

Towards A Bombing Halt 1968

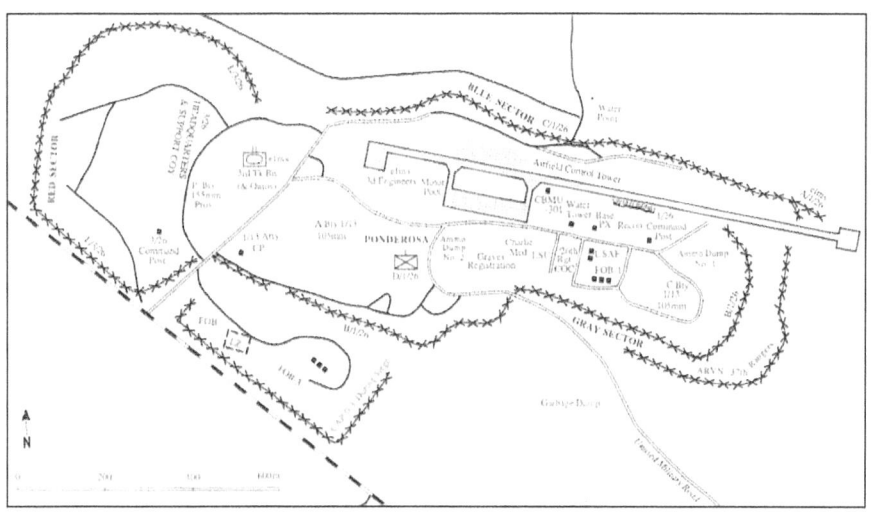

Detail Map Of The Khe Sanh Combat Base. Source: U.S. Marine Corps

Whether the President's supposition was correct or not, the fact was that the intractable Communists were preparing a major assault on Khe Sanh. Although there were strong arguments for abandoning the base. General Westmoreland, the JCS. and Secretary McNamara agreed it should be held for political and strategic reasons. General Wheeler termed Khe Sanh "the anchor of our whole defense of the northern portion of South Vietnam" and felt that defending it would tie down many North Vietnamese who otherwise would be free to attack elsewhere. Since the United States had the firepower and resupply skills to hold the base, military commanders saw opportunities for delivering a "severe" or "knockout" blow to the Communists. [4]

In an attempt to determine the size of the enemy forces in the area and to develop "target boxes" for B-52 strikes which were to follow, General Westmoreland in mid-January had launched an extensive air and ground intelligence search operation, known as Niagara I. On 22 January.. after the Communists had encircled Khe Sanh, the MACV commander began Niagara II operations. the greatest air offensive of the war. All available Air Force. Navy and Marine aircraft, SAC B-52s and Marine artillery were employed to prevent the garrisons capture. [5]

The Base At Khe Sanh Seen from Approaching Aircraft. Source: U.S. Army

Air War – Vietnam

Despite the confidence of military commanders that the Marine base could be held. there was considerable anxiety in Washington. In late January, General Westmoreland had warned that if the situation near the DMZ and at Khe Sanh worsened drastically, nuclear or chemical weapons might have to be used. This prompted General McConnell to press, although unsuccessfully, for JCS authority to request Pacific Command (PACOM) to prepare a plan for using low-yield nuclear weapons to prevent a catastrophic loss of the Marine base. [6]

Miss Buffy was a key player in the defense of Khe Sanh. Source: U.S. Air Force

At the White House the President made clear he did not wish to risk a defeat such as suffered by the French at Dien Bien Phu in 1954. Replying to a query from the President. General Wheeler assured him on 3 February that the military situation at Khe Sanh differed from that of the French in three ways: the United States had more and better equipped reconnaissance, all-weather fighter, and other tactical aircraft, plus B-52 bombers; there was reinforcing artillery from Marine positions east of the mountains; and there were vastly improved aerial techniques for resupply, medical evacuation, and for other needs. [7]

Official and public concern heightened when Communist forces overran the Special Forces camp at Long Vei near Khe Sanh on 6 - 7 February. This loss, accompanied by other temporary setbacks inflicted by the enemy's Tet offensive. raised probing Congressional inquiries about the war, evoked more criticism from the press. and led commanders in Saigon to request urgent reinforcements from the United States. [8]

Meanwhile, additional air power was brought to bear on Communist forces at Khe Sanh and related targets in South Vietnam and Laos. On 11 February the Joint Chiefs authorized the SAC and PACOM commanders to use the newly arrived Port Bow B-52s at Guam and Okinawa for these strikes. They also permitted an increase in the overall B-52 Arc Light sortie rate from 1,200 per month (originally scheduled to be attained by 1 February) to 1,800 per month. [9]

Mission count rose quickly. Source: U.S. Air Force

In mid-February another change in policy, long sought by the Air Force and other services, aided the defenders of Khe Sanh and other allied positions under attack by the Communists. It involved terminating the restriction, imposed by the Thai government in March 1965, on using Thai-based tactical aircraft for combat in South Vietnam. Heretofore, except for B-52s, Thai-based aircraft could be used only for operations in North Vietnam. The change permitted a more efficient use of existing theater resources. To accommodate more Air Force personnel the U S. military ceiling in Thailand was raised slightly to 47,461. [10] Since issuance of Deployment Program 5. 5 Oct 67 which authorized 45,724 U.S. military personnel in Thailand there had been only slight incremental increases in the ceiling.

The intensity of the fighting at Khe Sanh and elsewhere evoked new decisions on the use of equipment. A test of acoustic and seismic sensors to detect enemy vehicular traffic and troop movements. begun by the Air Force in its Muscle Shoals program (renamed Igloo White on 1 June 1968) in Laos in December 1967, was extended to the Khe Sanh area. Beginning 21 January the sensors were dropped by specially equipped Navy OP-2E aircraft and hand-launched from CH-3 helicopters based at Nakhon Phanom AB, Thailand. Enemy vehicular or personnel movements detected by sensors were relayed to orbiting USAF UC-121 aircraft which, in turn, further relayed the data to the USAF-operated Task Force Alpha infiltration surveillance center at Nakhon Phanom AB.USAF A-ls dropped special "gravel" munitions to impede movements. The tests soon demonstrated the usefulness of in gathering intelligence. [11] At Khe Sanh, an Air Force AC-130A Gunship II also went into action adding to the massive firepower poured down on the North Vietnamese troops surrounding the base. [12]

Dropping sensors over the jungle was a mission the designers of the P-2 Neptune ASW aircraft had never imagined.
Source: U.S. Navy

EC-121s served as relay points for sensor data. Source: U.S. Air Force

The heavy concentration of American and South Vietnamese ground and air units around Khe Sanh and in other parts of I Corps prompted General Westmoreland, on 8 March. to designate Gen William W Momyer, the Seventh Air Force commander and MACV's Deputy Commander for Air, as a "single manager" for air in that area. Within the JCS, this decision was contested by the other services, especially the Marines, who alleged the single manager concept for air threatened the "integrity" of Marine Corps air operations, and that it

Air War – Vietnam

established a precedent for centralized air control during periods of heavy combat. However, Generals McConnell and Wheeler supported Westmoreland's decision as did Admiral U.S. G. Sharp, the PACOM commander. The issue was eventually resolved by Deputy Defense Secretary Paul H. Nitze who upheld General Westmoreland's right to make the appointment. [13]

U.S. Attack Sorties In Defense Of Khe Sanh (Operation Niagara) 22 January – 31 March 1968				
	January	February	March	Total
USAF	2,143	3,911	3,630	9,684
USN	963	1,814	2,290	5,067
USMC	1,640	1,926	3,132	6,698
Total Tactical	4,746	7,651	9,052	21,449
SAC	334	1,057	1,176	2,567
Grand Total	5,080	8,708	10,228	24,016

Source: Project CHECO SEA Rprt (S) Subj: Khe Sanh (Operation Niagara) 22 Jan – 31 Mar 68, 13 Sep 68, pp 112-14

Meanwhile, the air effort to save Khe Sanh continued. Tactical air sorties for Niagara provided what the Marines called a "mammoth air umbrella" of fighter-bombers which covered Khe Sanh around the clock. From 2 January through 31 March Air Force, Navy, and Marine Corps fighter-bombers averaged more than 270 sorties each day responding to the needs of the surrounded Marines. The results were impressive: on an average, there were 87 secondary explosions and fires reported each day in March alone. During the entire period tactical aircraft touched off more than 4,700 secondary explosions and 1,910 secondary fires. They were credited with destroying much of the enemy's equipment. [14]

	Destroyed	Damaged
Trucks	250	50
Gun Positions	300	40
Bunkers	900	100

When fighter-bombers let up in their strikes against pinpointed targets, the B-52s of SAC unloaded their bombs on strongholds, troop concentrations and supply areas. Flying above 30,000 feet altitude and carrying a 27-ton payload of 108 500- and 750-pound bombs, they devastated their targets. Occasionally scoring hits within 1,000 meters of the Marine base. Photo reconnaissance of Khe Sanh subsequently revealed that the B-52s had destroyed more than 300 defensive positions, weapon sites and line of communications (LOC) targets; and triggered more than 100 secondary fires and 1,300 secondary explosions. Of 95,430 tons of air ordnance used during Operation Niagara in defense of the Marine Base (22 January-31 March). B-52's dropped 59,542 tons, tactical aircraft 14,724 tons. and Navy and Marine aircraft the remainder. As a result of combined B-52 and tactical air strikes, the enemy lost an estimated 10,000 troops, and his failure to over-run the base. according to General McConnell. "was directly related effectiveness of airpower." [15]

The effects of the B-52 raids also demoralized the enemy. Following one of the Arc Light strikes Marines reported that North Army (NVA) soldiers were found wandering in a daze and internally hemorrhaging. Explosions, reported a North Vietnamese diary, were "so strong that our lungs hurt." Fear of the B-52s also caused enemy desertions as in one instance when 300 Vietnamese troops en route to Khe Sanh fled from the ranks. [16]

C-130 doing a LAPES Drop at Khe Sanh. Source: U.S. Air Force.

While the B-52 raids and the tactical air strikes. Sustained the defense of Khe Sanh, USAF airlift assured the garrison's survival. Surpassed perhaps only by the Berlin airlift was the dramatic demonstration of aerial resupply for the surrounded garrison the Air Forcers 834th Air Division, and for two nearby outposts, hills "861" and "'881" by Marine helicopters. From 22 January until 8 April when "land route 9" was reopened to the base, USAF C-123s and C-130s made 447 landings and 576 airdrops. Of the latter, 15 were accomplished by the ground proximity extraction system (GPES). 58 by low-altitude parachute extraction system (LAPES), and 503 by the container delivery system. The GPES and the LAPES methods delivered supplies by approaching the delivery area slightly above ground level. With GPES the cargo was extracted by a hook attached to the cargo and extended from a boom at the rear of the aircraft. As the C-130 swooped low over the runway, the pilot tried to snag an arresting cable which upon engagement would jerk the pallets from the aircraft. With LAPES the cargo was snatched from the plane by the jolt of a blossoming parachute

In the final analysis it was the bravery of the cargo crews who flew into intense ground fire to supply the Marines holding the base that decided the issue. The North Vietnamese gunners had the runway and parking apron zeroed in and literally couldn't miss. The C-130 and C-123 crews went in anyway.
Source: U.S. Air Force.

C-7A Caribou aircraft were used on eight occasions. All casualty evacuations and personnel replacements also were made by air. Of the 12,430 tons of supplies delivered. 8,120 tons were airdropped and 4,320 tons were airlanded. General Westmoreland termed the resupply of Khe Sanh "the premier air logistical feat of the war." [17]

Elsewhere in South Vietnam during the 77-day siege, the Seventh Air Force and other Navy, Marine, and VNAF tactical aircraft flew thousands of close support, interdiction, reconnaissance, and electronic sorties in order to blunt another Communist gamble of the war, the Tet offensive. This massive onslaught over the length and breadth of South Vietnam would have a greater impact on American and allied policy than the enemy's effort to capture Khe Sanh.

The 1968 Tet Offensive

Despite their optimism about the war, as the new year began, allied commanders had anticipated another large-scale assault. As early as 9 January General Momyer was certain there would be a new offensive and doubted that it would be confined

Towards A Bombing Halt 1968

to the Khe Sanh area. When the siege of the base began, General Westmoreland requested authority to cancel a 36-hour military truce recently proclaimed by the allies (versus a one-week truce announced by the Communists) in recognition of the annual Vietnamese Lunar New Year. or Tet. beginning 30 January. His request was approved in Washington on the 29th. but it was to be applied only to South Vietnam's five northernmost provinces in I Corps. [18]

While the Americans awaited the enemy, the Saigon government generally disregarded the threat. Plans to celebrate the Tet holiday had not been interrupted, liberal military leaves and passes were granted, and on the eve of the enemy blitz. the South Vietnamese Army units outside of I Corps were only at 40 to 50 percent of their regular strength. Some units were in a state of alert. others were not. [19]

The Tet Offensive saw heavy fighting in Saigon. Source: U.S. Air Force

Consequently, when in the early hours of 30 January the Viet Cong and North Vietnamese attacked the capital and many other towns, as well as numerous South Vietnamese and American military bases and airfields. South Vietnam's Forces were unable to stem the enemy's surge. President Thieu quickly canceled the truce and placed his nation under martial law.

By the 31st the Communists were on the rampage throughout the country and within a few days had struck 36 of 45 provincial capitals, five of six autonomous cities, 64 of 242 district capitals and 50 hamlets. Viet Cong and North Vietnamese troops penetrated 11 cities in strength. In Saigon, where there was heavy fighting, 19 insurgents blasted a hole in the wall around the American Embassy, entered the grounds, and were finally killed trying to enter the building. Completely overrun and largely destroyed by air strikes and artillery fire, Vietnam's ancient capital Hue was not liberated by U.S. Marines and South Vietnamese troops until 25 February. The intensity of the far-flung assaults temporarily placed the allies on the defensive, forcing the troops to abandon much of the countryside in order to protect the cities. [20]

Blown up in its revetment, this poor RF-4C never stood a chance. Source: U.S. Air Force

Among the enemy's targets were 25 allied airfields. Communist troops at Hue overran the airstrip. destroying eight USAF O-1 and O-2 FAC aircraft. They also launched major assaults on the principal USAF-occupied airfields at Tan Son Nhut, Bien Hoa, and Binh Thuy but none were overrun or forced to discontinue operations. Some premature attacks had warned the allies, prompting General Momyer to declare a maximum alert for Air Force installations before the full impact of the enemy's offensive began. But the poor performance of some of the South Vietnamese assigned to air base security weakened the defense system. Between 30 January and 29 February, 25 Air Force aircraft were totally destroyed (by type they included seven O-1s, six O-2s, one F-4, four RF-4, one A-37, two F-84s, two F-100s, one RF-101 and one C-130) and 157 damaged on the ground by enemy rocket and mortar attacks. For the three services, the ground attacks in this period destroyed 53 aircraft of all types (30 fixed wing, 23 helicopters), and damaged 344. [21]

Meanwhile. all available aircraft were thrown into the battle. A substantial portion of tactical air was used for close air support of American and South Vietnamese troops. The Communist offensive warranted heavier bombing of supply routes in North Vietnam but air operations were handicapped throughout February by the worst weather since bombing began in 1965. [22] In fact, effective air operations over the North were greatly curtailed in the first three months of 1968 because of unprecedented weather. In the northern "route packages" (IV through VI), air strikes were possible only an average of three days per month

The VNAF's performance initially was poor. The policies adopted for the Tet holiday had left its units unprepared for sustained combat, and many flight line and cockpit jobs had to be filled temporarily by USAF advisors. By the end of the month, however, the situation had improved and the advisors rated the VNAF's performance as "highly satisfactory." [23]

General Westmoreland predicted on 2 February that the offensive "was about to run out of steam." He also anticipated a major attack on Khe Sanh, where the enemy had massed 20,000 troops and stated that his defeat there "may measurably shorten the war." [24]

While the destruction caused by the Tet offensive was extensive, the casualties suffered by the Viet Cong units that took part were so heavy that the Viet Cong never recovered. From Tet onwards, most of the fighting would be done by North Vietnamese regulars. The main target though was the U.S. public perception of the war's winnability and there, the North Vietnamese scored a decisive victory. Photo Source: U.S. Air Force.

Washington on the same day, administration officials viewed the Tet offensive with concern but not alarm. At a news conference President Johnson said that the enemy's objective was to demonstrate military and psychological success, to overthrow the Saigon government, and to prepare the way for a Communist

Air War – Vietnam

coalition government. Pointing to heavy Communist casualties he said the JCS, General Westmoreland and Secretary McNamara agreed that the Tet offensive was a military failure and there was no need to change basic strategy nor to increase the 525, 000 American military. personnel ceiling for South Vietnam. [25] Whether the Communists expected a general uprising, the overthrow of the Saigon government. and a "decisive" victory remains debatable. Many officials, including General Westmoreland, believed this was their true objective. A subsequent Defense Intelligence Agency analysis of the Tet offensive. however. states that "more persuasive evidence suggests the Communists were fully aware of the improbability of a full-scale military victory over the allies."

The JCS, however, were apprehensive. With the onset of the Tet attacks, General Wheeler directed the services to prepare a paper for Secretary McNamara demonstrating the need to fight North Vietnam on "a sound military basis." General McConnell, replying for the Air Force, favored an unrestricted air and naval campaign to destroy all military targets regardless of location. This meant full employment of B-52's and no strictures on the number of attack sorties to be flown or on the use of munitions. [26]

Approval for a "wraps off" air and naval effort was unlikely since the administration on 18 January. had imposed additional restraints on the bombing at the Hanoi-Haiphong area in response to a statement on 29 December 1967 by North Vietnam's Minister expressing an interest in negotiations. The administration had imposed a "prohibited" bombing area of ten and four nautical miles, respectively, from the center of each city. Selected air strikes were possible in the prohibited area with Washington approval.

Consequently, on 3 February. the JCS asked for authority to strike enemy targets up to three nautical miles of Hanoi and one and one-half of Haiphong. This would make more transportation staging areas, transshipment, road, railways, and waterways in the two cities vulnerable attack. Admiral Sharp would take measures to avoid striking populated areas and foreign shipping. The Joint Chiefs believed that Soviet and Chinese reaction probably would be limited to propaganda, diplomatic pressure and that air effectiveness could be increased without additional risk. In response OSD partially relaxed its on air strikes in the Hanoi-Haiphong area. [27]

Meanwhile, in ground fighting General Westmoreland pursued a strategy which assigned top priority to clearing the enemy from the cities of South Vietnam and second priority to denying him any territory of value. But to accomplish these twin objectives and to capitalize on the military opportunities open to him, the MACV commander needed more air and ground forces. [28].

III. WESTMORELAND SEEKS MORE TROOPS, AIRCRAFT & EQUIPMENT

In his request to Washington for more assistance in early February, General Westmoreland initially asked for one more USAF C-130 squadron (complete with ground handling equipment and maintenance crews), possibly a second squadron which he would keep on alert, more 0-1 FAC aircraft. helicopters, air drop equipment, and one naval mobile construction battalion for I Corps. He also desired faster distribution of M-16 rifles, M-60 machine-guns and M-29 mortars for the South Vietnamese Army. The Air Staff and other services immediately reviewed the impact that the emergency would have on deliveries previously scheduled under Deployment Program 5. [1]

Airlift augmentation was arranged promptly. The JCS approved temporary retention in PACOM of 16 C-130s scheduled for return, the deployment of a second squadron to the Pacific and the alerting of a third which followed shortly. Two of the squadrons would operate from Japan. In other actions, USAF UC-123 defoliation were pressed into airlift service for the remainder of February and a large part of the Air Force's transport fleet in the theater began flying on an emergency basis. [2]

Responding to additional appeals from Saigon. Secretary McNamara directed the immediate movement by air of 16 CH-34 and 30 CH-47 helicopters, 143 M-113 personnel carriers (to be taken from other military aid programs) and various arms including 20,000 M-16 rifles from Air Force units outside of Southeast Asia. Eighty five tanks would be shipped from Okinawa and the United States. [3]

USAF tactical transports also provided critical supplies to besieged forces using high-altitude drops.
Source: U.S. Air Force

Because of high losses and the need for more air sorties General Westmoreland asked for 119 additional FAC aircraft. The only available "extras" were to be found in Army National Guard and reserve units and these would have to be modified for use in Southeast Asia. On the recommendation of the JCS production of FAC O-2s was stepped up, although deliveries would take some time. [4]

Air War – Vietnam

A MACV request for the Air Force's low-altitude parachute extraction system to air drop supplies at Khe Sanh also posed a problem since the system, in early February, was still undergoing tests. The Air Staff finally concluded that it would be feasible to order sets for immediate production and simultaneously complete the tests and prepare operating instructions. Soon the LAPES and other air drop systems made signal contributions to resupplying the Marines at their besieged base. [5]

Meanwhile, General Westmoreland redeployed troops from other areas to I Corps and established there a MACV forward headquarters under his Deputy Commander, Gen. Creighton W. Abrams. These redeployments generated a need for more manpower as did the weakness of Saigon's forces and unexpected strength of the enemy. By 11th February the South Vietnamese had lost about 2,000 killed, 7,000 wounded and there were unknown numbers of absences from units. [6]

In many attacks, the Viet Cong failed to penetrate airfield defenses.
Source: U.S. Air Force

The cost to the enemy was considerable, totaling an estimated 32,000 killed and 5,000 captured and he also had lost more than 7,000 individual and "crew-served" weapons (those requiring more than one man to handle). However. he remained strong. The 84,000 troops believed committed to the Tet offensive represented only 20 to 25 percent of his strength. and most of his uncommitted manpower was still in I Corps. [7]

On 3 February, the ROK Defense Minister said that more U.S. aid was necessary to combat North Korean incursions, and without it his government might recall some of its troops in South Vietnam. In view of the crises and new assessments of enemy strength, General Westmoreland, through General Wheeler, asked Washington authorities to prevent any withdrawal of the present 49,000 man Republic of Korea (ROK) force in South Vietnam. If possible, he proposed to augment. allied forces with 11,000 more ROK troops and to expedite the deployment of a Thai infantry division (promised on 12 August 1967 by the Thai government for deployment in 1968). [8] Any increases in troop strength from these sources, he soon learned, would be too little and too late.

Plans to Speed Up Deployment of American Troops.

As there was little prospect of obtaining quickly more Thai troops, it became apparent that additional reinforcements would have to be American. On 9

Towards A Bombing Halt 1968

February Secretary McNamara asked the JCS for three alternate plans to reinforce MACV. These should include provisions for dispatching 150 more aircraft. However, he cautioned against recommendations requiring Congressional approval as this could trigger a further divisive debate on the war. [9]

Of the three plans hurriedly completed, the service proposed adopting the one that called for sending to South Vietnam the Army's 82d Airborne Division and six-ninths of a Marine Division Wing Team. Despite the apprehensions of the Defense Secretary, he should ask Congress for additional legislation to extend and recall more reservists. The Army and Marine force would also need support units. Although the Air Staff supported the JCS recommendations, it believed that it should have included a request for more support aircraft and heavier air attacks on North Vietnam.

The JCS cautioned, however, that the additional troops should not be sent until the military situation became clearer and manpower problems were resolved. "Deployable" forces (including the 82d and the Marine team), they noted, contained many personnel who had completed their Vietnam tours or were nearing the end of their military obligations. There also was a shortage of specialists for aircraft, helicopters, munitions, communications, and other jobs and it was important to continue to maintain an adequate training and rotation base. The JCS suggested that while readying the Army and Marine Force, certain reserve units should be recalled promptly and actions taken to alleviate shortages of aircraft, helicopters, and other important items. [10]

Before OSD acted on the JCS plans, General Westmoreland in concert with American Embassy officials in Saigon, asked the President to rush one brigade of the Army's 82d Airborne Division and one Marine Corps regiment to I Corps to preclude using troops needed elsewhere in South Vietnam. Although the Communists had been repulsed in other Corps area, more manpower was needed in I Corps to regain the initiative. The MACV commander said he could support logistically, would need them only for six months and planned to include them within the 525,000 U.S. manpower ceiling. [11]

President Johnson approved the request the same day (12 February), and on the 13th the Pentagon announced the decision to send 10,500 more men to South Vietnam, characterizing it as a "speed-up" in deployments authorized under the troop ceiling. Most of the 27th Regimental Landing Team (RLT) was already in the Pacific, deploying by ship from Okinawa. while Military Airlift Command's C-133's and C-141's flew the 82d's brigade and other personnel to I Corps by 26 February. The Saigon government, for its part, announced that it would add 65,000 more men to the Republic of Vietnam Armed Forces by June 1968. [12]

Air War – Vietnam

The President's decision led the JCS to request again a reserve callup. The service chiefs desired to mobilize immediately 46,000 Army, Navy and Marine reservists to support the 10,500-man deployment force and to rebuild the strategic reserve. Volunteer Air reservists could provide additional airlift needs. Anticipating further requests from General Westmoreland, the JCS warned that an additional Marine regiment and the remainder of the 82d would add to demands for support. To meet all pressing requirements, they desired to plan for the recall of an additional 137,000 reserve and national guard personnel (28,300 Air Force, 11,700 Navy, 39,000 Marine Corps and 58,000 Army). [13]

When Secretary McNamara solicited alternatives to these proposals, the service chiefs reaffirmed vigorously, the need to recall 46,500 reservists immediately and plan the recall of 137,000 more. Although their entreaties were not approved, the debate over mobilizing the reserve and national guard continued. [14]

General Westmoreland's Request for 206,000 More Troops

As retaking ground lost during the Tet offensive continued, troop shortages became apparent. Source: U.S. Marine Corps

As the Tet offensive continued into the second and third week of February. General Westmoreland advised Washington that he would need considerably more American manpower. To obtain a first-hand report of his requirements and the military situation, the President sent General Wheeler to Saigon to confer with the MACV commander. The JCS chairman returned to Washington in late February with a sobering report.

General Wheeler said that the Tet attacks had nearly succeeded in a dozen places and that defeat had been avoided only by the timely reaction of American forces. The revolutionary development program had suffered a severe setback and was aggravated by 474,000 more displaced personnel (by 1 March the estimate was 800,000). The urban people also reeled under the psychological blow of this harrowing month. With its effectiveness severely limited, the Saigon government had barely survived. Surprisingly, its army had withstood the initial assaults, but Vietnamese troops were now in a defensive posture around towns and cities. and there was concern about their steadfastness. MACV thought it would take the South Vietnamese army two to three months to recover from equipment losses and three to six months to recover its strength, although its problems were considered to be more psychological than physical. To be sure, the enemy suffered enormous casualties; nevertheless, he was operating in relative freedom in the countryside, recruiting heavily, while more North Vietnamese were infiltrating southward.

Towards A Bombing Halt 1968

In reporting his assessment. the JCS chairman said that despite considerable aircraft attrition, American air operations had lost none of their effectiveness. From 29 January to 21 February, Seventh Air Force increased its tactical fighter sorties by 8.5 percent and FAC sorties by 11 percent. Airlift resources, however, were strained from resupplying Khe Sanh and redeploying troops to 1 Corps, and because of enemy attacks against land and sea lines of communication in the Hue-Phu Bai area.

General Wheeler cited three major military problems that faced the command in Vietnam. First were the problems of logistics caused by bad weather, enemy action, and massive U.S. troop movements into the Da Nang-Hue area. Secondly, the poor defensive posture of the South Vietnamese Army had allowed the Viet Cong to enter pacified areas. Finally, insufficient forces outside of I Corps weakened the whole military structure. Moreover. there was the danger of synchronized enemy attacks on Khe Sanh, Quang Tri, the highlands, and around Saigon. which strained General Westmoreland's forces severely to meet all possible threats. He needed more troops.

The MACV commander's "stated requirement" was for 206,758 more U.S. military personnel which would boost the authorized total in South Vietnam to 731,756 by the end of 1968. This would increase Air Force personnel by about 22,000. The Army's increase would be 171,000 and the Navy would gain 13,000. The additional would provide 15 more tactical fighter squadrons and the equivalent of three U.S. ground divisions. General Westmoreland also desired one more ROK light division (about, 11,000 men). According to the proposed deployment schedule (including

Recovering ground lost in the cities was slow and immensely destructive. Hue City was almost totally destroyed by the fighting.
Source: U.S. Marine Corps

units previously approved in Deployment Program 5 but not yet deployed), eight of the Air Force's tactical fighter squadrons would deploy by 1 May 1968, four more by 1 September. and the final three by 31 December. [15]

Reviewing the requirements. Under Secretary of the Air Force Townsend Hoopes. assured OSD that the Air Force could meet General Westmoreland's proposed deployment schedule and, if necessary, deploy two squadrons within 48 hours. He warned, however, that the forces requested by the MACV commander would generate a need for more munitions and air bases (especially compel finishing

Nam Phong AB. Thailand). They would also require regular, national guard. and reserve units to provide additional reconnaissance, airlift and aeromedical support. [16] Mr. Hoopes also estimated that the costs for deploying and maintaining in Southeast Asia the new Air Force units would range from a minimum of $635. 5 million for the remainder of fiscal year 1968 to a maximum of $1.229 billion in fiscal year 1969.[17]

Meanwhile, the President had queried General Wheeler on the "maximum amount" of air power General Westmoreland could use "profitably" to carry out his mission. The JCS chairman's minimum estimate was 15 tactical fighter squadrons with deployments contingent on completing the present air base expansion program in Vietnam and Thailand. The MACV commander also needed, he said, two more C-130 airlift squadrons, 138 more FAC 0-ls 0-2s and OV-10s, more AC-47, AC-119 and AC-130 gunships. plus an increase in B-52 sorties from 1,800 to 2,250 per month.

Proposed naval air power augmentation included one carrier (which would result in a major change in U.S carrier deployments), more aircraft and helicopters for the Navy's water surveillance operations, and one Marine air group containing three helicopter squadrons. For the Army, General Westmoreland wanted substantially more helicopter assault and support units. These forces would have to be in place by the end of 1968. [18]

A New SecDef. Clark M Clifford. Source: U.S. Govt.

General Wheeler's report to the President was submitted by Deputy Defense Secretary Nitze, who observed that increased, air effectiveness might better be achieved by good target intelligence and accuracy in delivering munitions than by the number of sorties flown. He also said that the services desired more aircraft to support additional ground forces and that this problem was being studied. [19]

Before decisions were reached on these new military requirements there was a change in the leadership of the Department of Defense. On 1 March, Mr. Clark M. Clifford succeeded Mr. McNamara as Secretary, an event that promised to alter profoundly the strategy in the war.

IV. DEBATE OVER MORE DEPLOYMENTS & STRATEGY

General Westmoreland's request for a 40 percent increase in U.S. Forces in South Vietnam created much consternation in the administration, the Congress, and the public. The war was already the subject of violent debate in the nation. Nevertheless, the President asked his new Secretary of Defense to chair an *ad hoc*, cabinet level task force which would determine how General Westmoreland's needs could be met. The Air Force, the other services, and the Joint Chiefs were called upon to review or suggest alternate plans.

Three Air Force Strategies

As part of the review by Secretary Clifford's task force, Secretary Brown, Under Secretary Hoopes, and the Air Staff were jointly engaged in formulating three air strategies for prosecuting the war. On 4 March, Dr. Brown and Mr. Hoopes prepared a summary report to which were appended three pages describing the proposed air campaigns. Two had been prepared by the Air Staff and one by an *ad hoc* Operations Analysis-Rand study group. While task force was reviewing the papers, the Air Staff continued to refine details for the strategy in these studies. [1]

USAF attack on the Thai Nguyen steel plant. Source: U.S. Air Force

Under the first strategy, called Campaign I, the existing restrictions on bombing North Vietnam would be lifted to allow for more air strikes against a broader target base without regard to civilian damage or casualties. The principal targets

Air War – Vietnam

would include military headquarters, government control points population centers harboring vehicles and materiel, the ports of Haiphong, Cam Pha and Hon Gai (all three harbors would be mined), over-the-beach material centers, the northeast and northwest rail lines. and roads to the North Vietnamese-Chinese border.

Campaign I would focus on the North above the 20th parallel and consist of two types of operations: air harassment of enemy, to raise his defense costs, inflate manpower needs, reduce productivity, and cause problems in distribution, management, and other internal affairs; and heavier air attacks on significant to increase casualties, destroy more military potential, ruin rice crops and close ports and harbors. The North's road, rail, port capability, down from 15,000 to 8,000 short tons per day could be lowered to 4, 000 to 2,000 tons per day. Air harassment could reduce imports by about 25 percent, an amount probably insufficient to end the war decisively, whereas strikes on ports and mining of harbors would reduce imports by 75 to 90 percent. To accomplish these tasks, the Air Staff proposed a total of 170, 000 combat sorties annually: 120,000 by USAF tactical and B-52 aircraft, 35,000 by the Navy, and 15,000 by the Marines. The expected rise in aircraft losses and munitions expenditures would require an additional $2.5 billion. although the dollar outlays might be cut by using more guided bombs and substituting B-52's for tactical air strikes a one for 10 basis.

Secretary Brown feared that Soviet pilots might be sent to North Vietnam and Chinese may commit up to 50,000 troops. In reality, 320,000 Chinese troops were already serving in North Vietnam and Soviet pilots were already flying combat missions there. The leading Soviet pilot over North Vietnam was Vadim Petrovich Shchbakov, attached to the 921 Fighter Regiment, who was credited with six American aircraft shot down. Source: Russian Air Force.

Secretary Brown believed that Soviet reaction to Campaign I probably would consist of a hardened attitude toward the United States, some diversionary action against West Berlin and Korean DMZ and a step-up in the delivery of supplies, equipment and MiGs, including possibly Soviet pilots, to the North. The Chinese would also likely increase logistic and maintenance forces already in the North (estimated at 50,000) and occupy ports of North Vietnam if they felt that the bombing threatened the Hanoi government. However, Dr. Brown pointed out that more was needed on possible reaction of the Soviet Union, China, and other countries.

The Air Staff, on the other hand, believed that Moscow's response would be less severe than anticipated by Secretary Brown. It thought the Soviets might apply some pressure outside of Southeast Asia but probably would not use military forces to create a diversion. Thus Campaign I could force Hanoi to slow the tempo of fighting and eventually seek a compromise or to abandon the war. If it began in March the campaign's maximum effect would be felt by October when bad weather normally restricted the bombing and allowed the North Vietnamese to improve their transportation system.

The analysis for the second strategy Campaign II, was prepared by an Operations Analysis-Rand study group. It suggested various measures for exerting more pressure on the North Vietnamese-Laotian panhandles: diverting only USAF or all U.S. sorties from route package's IV through VI to route packages 1 through III in the Laos panhandle, interdicting selected LOC "belts" in southern North Vietnam, adding antipersonnel air strikes, using new land mines, and launching more B-52 attacks against LOC's in the Mu Gia and Ban Korai passes in Laos. The strategy further called for tripling the current sortie rate to produce a 10-fold increase, compared with 1967, in the destruction of trucks. Also proposed was stepped-up harassment of enemy repair crews and supply handlers to cause more delay in his transport of supplies.

Interdicting supplies moving through Laos was harder than it seemed.
Source: U.S. Air Force

The Air Force Secretary acknowledged the difficulty in limiting significantly the movement of Communist supplies by bombing. A study showed that the North Vietnamese had transported more goods than they required for operations despite "our most optimistic estimates of current damage, given the current rate of imports." To reduce the supply flow to a minimum meant improving air effectiveness "by a factor of four." He thought there was "an even chance" of achieving this by using new or improved sensors, aircraft and munitions, and by flying more sorties. If 120 trucks could be destroyed each day, Dr. Brown surmised. the Communists would find it most difficult to move many of them from China to the North Vietnamese and Laotian panhandles, refuel them en route. and provide the necessary support for 30,000 people manning the routes. [2]

The third strategy, Campaign III, called for a basic change in ground strategy and for more reliance on air power. It assumed that search and destroy operations had not given the South Vietnamese meaningful security and held no realistic promise of doing so. South Vietnamese and American ground forces would redeploy to give maximum protection to the heavily populated cities and adjacent rural areas. Once the population was secure, the Viet Cong infrastructure could be routed out. This was the "oil spot" concept tried earlier but never on a realistic basis. The free-fire zones outside of the secure areas would be subject to day and night air attacks by AC-47, AC-119, and AC-130 gunships, other tactical aircraft, and B-52s. Ground forces would hit main enemy units at a reasonable distance from the population centers.

The AC-130A would have been a key element in Campaign III. This is Azreal, Angel of Death, taken many years after the end of the Vietnam War. Source: U.S. Air Force.

The principal demands of. Campaign III would require an additional 126 FAC aircraft. 125 gunships, and 172 other tactical fighter aircraft to assure 24-hour surveillance and immediate air strikes. This strategy would accept somewhat higher aircraft attrition rates and relinquish territory to the enemy; as areas became secure, allied troops would move outward. Because of large casualties inflicted upon the enemy, his tempo of operations would slow down and eventually lead to tacit stabilization of the conflict at a lower level of intensity.

Campaign Ill required no increase in American ground troops. By safeguarding the population from terrorism. the Saigon government could concentrate on developing leadership and other programs that could generate enthusiastic support. In time, this scheme would enhance prospects for a compromise in a political settlement for all South Vietnam. [3]

As part of the Campaign III withdrawal of allied forces to secure limited areas, the Air Force proposed, as a beginning, the of Khe Sanh. Within six months the limited areas would be protected, then extended, and in 18 months the Saigon government would have sufficient control over most of the population and other

Towards A Bombing Halt 1968

resources in South Vietnam(in three-fourths of the country) to permit the initial departure of U.S. ground forces. This objective would call for tight population control, a necessity demonstrated in previous by the British in Malaya and the French in Indochina and Algeria.

In conjunction with expanding air action, the Air Staff recommended creating a center to consolidate the processing and evaluating of sensor data for "real time" evaluation of intelligence.

The Air Staff conceded there were risks in Campaign III. The communists probably would try to establish a government in the areas initially relinquished by the allies, although air attacks on and installations might prevent this. Also Hanoi might call for a cease-fire and propose a military status quo for both sides. If this were accepted, the United States would, of course, have to forego its objective of bringing all of South Vietnam under the control of the Saigon government.

Subsequent study led the Air Staff to conclude that the effective implementation of any of the three air campaigns would require a minimum of 1,101 USAF, Navy, Marine and VNAF aircraft, 105 B-52s and 104 gunships. This force could provide a total of 44,123 combat sorties per month as follows: tactical aircraft 39,720; B-52s 2,200; and gunships, 2,203.

Although developed separately, General McConnell felt the three campaigns should be combined into a single military concept "with a reasonable probability of providing the decisive impact required to achieve early settlement of the conflict." He solicited JCS support in requesting Secretary Clifford to recognize that, contrary to the administration's view, the war in South and North Vietnam was inseparable. The alternatives to "new and decisive emphasis on air operations against the North." he pointed out, were higher American costs for each cycle of enemy destructiveness leading eventually to a military standoff or a politically disadvantageous withdrawal of U.S. forces. General McConnell's proposal was made several days after the President had ordered a partial bombing halt and the JCS did not act upon it. [4]

At Secretary Clifford's request. another high-level appraisal of the three air strategies took place on 9 April when Air Staff representatives reviewed them with five members of the Presidents Science Advisory Committee (PSAC). The conferees agreed that Campaign I required the removal of air restrictions and better munitions. They also agreed that 20-mm cannons and incendiary had proved thus far to be the most effective weapons against enemy trucks. [5]

There was further study of the strategies, but on 10 May, U.S. and North Vietnamese representatives made an initial contact preparatory to peace talks. By then the adoption of the Air Force's three strategies. especially Campaign I, appeared remote. Mr. Hoopes, in fact, subsequently advised Secretary Clifford to resist pressures to resume the bombing in North Vietnam. He believed Hanoi's intransigence or its willingness to cooperate at the peace should dictate a ground strategy emphasizing shorter defense lines, better protection for the South Vietnamese people, and lower American casualties. He thought OSD should be ready. with a plan based on such a strategy. [6]

Air War – Vietnam

Response to the Westmoreland Troop Request

Meanwhile. the task force headed by Secretary Clifford had completed its initial review of General Westmoreland's request for 206,000 additional troops. To meet the MACV commander's most urgent needs, the task force proposed, in a memorandum to the President sending immediately 20,000 troops. It also approved calling up more reserves, larger draft calls and longer duty tours in Vietnam to provide the remaining 186,000 men desired. Simultaneously, it proposed stepped-up bombing of the North but not to the extent urged by Presidential consultant Gen. Maxwell D. Taylor, General Wheeler, and Walt W. Rostow. the President's Special Assistant for National Security Affairs. Except for reiterating the San Antonio formula of 29 September 1967, there would be no new initiative toward negotiations.

General Wheeler
Source: U.S. Army

On the issue of sending the full 206,000-man force, the task force was cautious. Such a step, it explained should be contingent on evidence of better performance by the Saigon government, the completion of new political and strategic studies to guide General Westmoreland, and week-by-week examination of the situation in Vietnam. New studies might show, for example, that MACV should not expect to destroy or rout all enemy forces from the South, that no number of allied forces could do this, and that the dispatch of more troops without substantial improvement in Saigon's armed forces might prove counterproductive. [7]

On 7 March, Secretary Clifford discussed the implications of the task force's memorandum at the White House. He informed the Chief Executive that he neither agreed nor disagreed with its recommendations. However. he expressed doubts about the efficacy of the present ground strategy, the bombing campaign. and the deployment of more American troops to Vietnam. The meeting assured further study of General Westmoreland's manpower requirement and the war's overall strategy. [8] In a Colombia Broadcasting System (CBS) television interview former President Johnson gave a considerably different account of high-level administration actions during this period than has been described by some of his principal officials at the time including Secretary Clifford. The issues which remain in controversy are over whether the President had asked for "recommendations" and "alternatives" to fulfill General Wheeler's "preliminary" manpower request who should be credited for proposing the partial bombing halt.

New Proposals for More U.S. and Allied Deployments

In the ensuing days, the MACV commander, still desiring reinforcements. asked for and the President tentatively agreed to send him at least 30,000 more troops as soon as possible of which 4,025 would come from the Air Force. The Air Force portion would include airlift, FAC, tactical air control, support personnel, as well as four tactical fighter squadrons, two of which had been included in Deployment Program 5. [9]

The Air Staff quickly assented to the proposal, believing that four more fighter squadrons would enable the Seventh Air Force to support the additional ground forces. But by the time a final decision was reached on 11 April, the figure had been reduced to 24,500. In the intervening period the administration, the JCS and the services had made an exhaustive review of the war, debating the cost and political impact of calling up more U.S. reserves, providing support forces, obtaining additional South Vietnamese or Korean units, finishing construction of Nam Phong AB and adopting one of the Air Force's three air strategies. [10]

Additional O-2 FAC aircraft were urgently needed. Source: U.S. Air Force

The debate on reserve callups was touched off in following the President's decision to send 10,500 troops to South Vietnam because of the Tet offensive. With the reservoir of trained military manpower rapidly depleting, new sources had to be tapped to support additional deployments and maintain an adequate reserve in the United States. Accordingly, the JCS on 15 March proposed three alternate national guard and reserve callup programs. The first required 39,677 personnel including 6,590 Air Force, the second 13,437 Army personnel plus an Army brigade to replace the Marine RLT (which had been sent to Vietnam in February only as an emergency reinforcement), and the third would alert 51,079 personnel, including 10,079 Air Force, for callup. OSD took no immediate action on these proposals. [11]

Another source for obtaining more troops was sought from Americas other allies in the war. On 15 March, the MACV commander proposed raising the strength of the South Vietnamese armed forces from 685,739 (approved on 7 October 1967) to 779,154 in fiscal year 1969 and to 801,215 in fiscal year 1970. The latter would include 5,124 more spaces for the VNAF, increasing its manpower to 21,572. OSD made no decision on these proposals until May. Meanwhile. to help speed the interim growth of Saigon's forces. General Wheeler proposed, and OSD agreed on 4 April, to add 31,475 "pipeline" spaces, including 750 for the VNAF, raising the authorized South Vietnamese strength to 717,214. In view of these actions the Air Staff asked the Seventh Air Force to consider a speedup in the training of the South Vietnamese air arm. [12]

Plans to obtain another ROK infantry division also deeply involved the Air Staff since the Seoul government desired as a quid pro quo for sending more troops, American support for a Air Force (ROKAF), squadron in South Vietnam. The JCS took the position, based on Air Staff and PACOM views, that unless the Koreans insisted on deploying a squadron, no action should be taken. If, on the other hand, a squadron was sent, it should be fully equipped with F-5 aircraft, pilots, support and maintenance personnel. An alternate plan called for using F-5's from U.S. sources with ROK manpower already trained to fly, support, and maintain the aircraft. In subsequent weeks, however, it became clear that no more Korean forces would be available. [13]

The proposed ROKAF Squadron for Vietnam would have flown F-5A aircraft. Source: U.S. Air Force.

Anticipating large troop augmentations in the war theater Secretary Brown and General McConnell renewed their efforts to OSD concurrence to complete Nam Phong AB, Thailand. On 16 September 1966, Secretary McNamara had approved only its "bare base" construction. But in the first 16 days of the Tet offensive (in February 1968), Dr. Brown observed, the financial loss arising from the destruction and damage of many aircraft manifestly justified the $14 million needed to make Nam Phong a main operating base. After completion of the base, estimated to take about 120 days, dispersal and safety of aircraft in Southeast Asia would be enhanced. [14]

These importunings again were to no avail. The Army did not wish to spend $7.5 million and assign 124 of its support personnel to help maintain Nam Phong and the Air Force would have to add 1,505 personnel after the base became operational. This would create a demand to raise a tight American manpower ceiling in Thailand. At the same time the U.S. Ambassador in Bangkok, Leonard Unger, was advising the State Department that more U.S. deployments to Thailand might exacerbate the political and military difficulties with that country. In the light of these problems. Deputy Defense Secretary Nitze informed Secretary Brown on 23 March that the administration should not proceed "at this time" with further construction of the base. [15]

Towards A Bombing Halt 1968

Air Staff Views of Other Proposals

While the administration was studying the Air Force's strategic views and other recommendations the Air Staff was reviewing other policy papers written for Secretary Clifford's task force. One of the papers prepared by Army planners, advocated a change in the objective of NSAM (National Security Action Memorandum) 288, 17 March 1964. which envisaged an independent, non-Communist South Vietnam. If this objective could not be achieved without an "all-out" military effort requiring large troop reinforcements, the revised NSAM 288 would call for an "honorable peace" and allow the South Vietnamese to devise their own political and economic system. The United States would negotiate with Hanoi unilaterally and depart from South Vietnam in a phased withdrawal over an 18-month period or longer. without achieving a decisive victory. Although such a course would damage American prestige the Army felt that there would be no serious long-term effect.

The Air Staff criticized the Army paper for its failure to consider that a basic change in strategy, (i.e. the use of more air power against North Vietnam) could attain the NSAM objective in the South. Moreover, in the eyes of the Air Staff, whatever strategy was adopted should permit the United States to extricate itself. Without jeopardy to its world position.16

Another paper reviewed by the. Air Staff was prepared by OSD's Office of International Security Affairs (ISA). It also recommended a revision in NSAM 288. The ISA office believed that the South Vietnamese Army had been greatly weakened and could not contribute substantially to allied progress in the ensuing months. Further, the Viet Cong and the North Vietnamese could maintain a state of "protracted conflict", offset any increase in American forces in the South, and were threatening allied forces in I Corps. There was danger of a collapse of Saigon's authority in the Mekong delta. The current strategy of destroying the enemy and driving him out of South Vietnam would require doubling the strength of American troops. But this would completely Americanize the war, totally frustrate the development of political and military strength in the South. and make impossible the attainment of U.S. objectives.

The ISA paper, sent as a draft memorandum to the President, also sought to achieve an honorable peace by permitting the South Vietnamese people to fashion their own political and economic institutions. It proposed sending U.S. military personnel to only the most populated areas of South Vietnam, stepping up of Saigon's armed forces, and warning the Thieu government to clean up corruption and improve its military forces. The paper assumed that the President would not authorize new military moves such as ground operations into Laos, Cambodia, or North Vietnam (including the northern half of the DMZ).. nor change the policy with respect to bombing the North and mining Haiphong harbor. [17]

The Air Staff disagreed with this paper, believing that the current NSAM 288 should not be revised or replaced and that withdrawals to populated areas in the South would not be in consonance with Presidential policy. It also noted that

Air War – Vietnam

General McConnell had, repeatedly pointed to the need to remove restrictions on air and naval operations against North Vietnam and. as the war continued there would be more compelling reasons to do so. [18]

A third paper, prepared by Secretary Clifford's task force, proposed a new in-depth study of American policy and strategy in the war. This might show, the paper conjectured that Westmoreland's request for massive reinforcements was no solution to his problem and that enemy forces could not be kept out of South Vietnam regardless of allied strength. Also. Better performance by the Saigon government's military units should deployment of more American troops. Parts of this paper were included in the initial report of Clifford's task force to the President on 7 March. The paper recommended a new NSAM limiting U.S. objectives to providing security for the South Vietnamese in populated areas rather than destroying. enemy forces, and leaving all of the populace in the South free to develop their own political system.[19]

The Air Staff objected to this paper as it too would rescind NSAM 288, and pointed to observations by Lt Gen. Glen W. Martin, Deputy Chief for Plans and Operations. on past American strategy on the war. General Martin noted that every time the military situation deteriorated, the authorities immediately looked for more ground troops while proposals to expand air operations received decreasing consideration. He emphasized the need, and the Air Staff agreed – for U.S. policy-makers to recognize the interrelationship of military operations between the two parts of Vietnam. This concept called for a single strategy and demanded decisive air action especially against the North in order to achieve allied objectives in the South. [20]

V. R&D FOR SOUTHEAST ASIA, 1968

Debate Over Air Force R&D

During an appearance before a Congressional committee on 28 February 1968, Gen. John P. McConnell, USAF Chief of Staff, stated that the "first mission" of the Air Force in South Vietnam was to support Allied ground forces "through close air support and interdiction," its secondary mission being the interdiction and destruction of Communist traffic "coming from North Vietnam into South Vietnam. " The Chief of Staff added that the Air Force had improved its interdiction operations but could not hope "to stop entirely the flow of supplies into South Vietnam. " The primary reason was the area's terrain and jungle, which made the Southeast Asia (SEA) war "a very difficult type of war to fight." In such an environment. accuracy had to be "very high in the delivery of weapons. We do not have enough good all-weather and night capability. We are gradually improving that. We should have had it before now. " [1]

General McConnell's comments on USAF operational deficiencies at the beginning of the year could have been echoed at its end. By December 1968 the Air Force still lacked an adequate night/all-weather attack capability and the accuracy of its weapon delivery systems remained poor. Although some research and development (R&D) projects had produced equipment of only marginal value (such as the Tropic Moon I and II systems and laser scan cameras), certain other equipment and munitions, newly modified or produced, had demonstrated outstanding capabilities in close support of friendly ground forces. Among these were the gunship, the B-52 used as a conventional bomber, and several types of ordnance.

As noted in an earlier historical study, the Air Force in early 1965 had been ill-prepared to conduct a tactical air campaign against the infiltration of enemy troops and supplies from North to South Vietnam. Despite the change in emphasis from nuclear to conventional "options" initiated by the Kennedy administration in 1961, the Air Force during the 1961-1965 period. had done little to improve its tactical capabilities. Even after the start of the air campaign

The A-26 (which was progressively redesignated from the original A-26 to B-26 to RB-26 then back to A-26) was the primary night attack interdictor as it had been in Korea a decade earlier. Source: U.S. Air Force.

against North Vietnam in February 1965, the Air Force took almost two years to deploy substantial quantities of new and modified equipment which significantly improved these operations. But major weaknesses remained. Thus, on 22 November 1967, the Air Staff's Tactical Panel noted that "The Air Force night attack capability was not good in World War II; little progress was made in Korea. Again, in SEA. there is the problem of stopping the enemy at night. Decisions have not been made on what aircraft and systems should be used. The Air Force must be careful not to lose the mission and opportunity to establish a permanent capability in the force structure." [2]

Dr. Harold Brown.
Source: U.S. Air Force.

Not only was the Air Force severely limited in its capability to locate and strike small, fleeting targets at night. but it also could not determine the success or failure of its interdiction efforts. Consequently, in late 1967, Secretary of the Air Force Harold Brown requested representatives from the Office of the Chief of Operations and Analysis and the RAND Corporation to analyze the effectiveness of the interdiction campaign. The group of about 25 was formally known as the AFGOA/RAND Southeast Asia Study group. The group's interim report, submitted in February 1968, stated that while the flow of materiel and personnel from North to South Vietnam had been impeded, the enemy continued to infiltrate sufficient supplies to sustain his war effort. This remained true despite the fact that the Air Force had improved its ability to interdict truck traffic on the Ho Chi Minh trail through Laos. Moreover, there was little evidence to indicate that "dramatic improvements" could be expected "unless the capabilities of our weapon systems are materially improved." [3]

According to the study group, improvements to electronic countermeasures (ECM) equipment and ordnance had led to better accuracy, but this still could not be considered an "accurate night and bad weather capability." Only a small part of the force was equipped to operate at night and the F-4C did not possess a computing sight for visual weapon delivery. The group thought that certain attitudes and administrative procedures had prevented speedy development of USAF weapons and suggested that measures could be taken to reduce the time from development to operational deployment. [4]

In its final report on the U.S. interdiction effort, on 1 July 1968, the study group concluded that air operations over North Vietnam and Laos "can only be assessed as inadequate. Ordnance delivery accuracy during day, night and 'weather' is inadequate; target acquisition at night is limited (whatever the state of the weather); and there are deficiencies in available ordnance. The group again noted that a major reason for these short- comings could be found in the USAF "decision-development-procurement" process, which had not adequately exploited technology nor satisfactorily responded to theater interdiction requirements.[5]

On 14 July 1968, in a separate report dealing with engineering development, the group reaffirmed that interdiction strikes were "no more than marginally effective; the inflicted damage was low, both absolutely and in terms of relative effectiveness." In terms of development effort. the group found "that for one reason or another some of the more effective steps that might have been taken have not and that the central cause is ambivalence in the decision process adversely affecting the introduction of innovations." No single office within the Air Force was responsible for equipping tactical aircraft to meet unique delivery conditions and attack difficult targets; neither was there "a sense of urgency regarding improvement in weapon delivery accuracies " Concluded the group: "We believe that the development procurement process for bombing effectiveness in SEA requires urgent review and some necessary modification. [6]

The group suggested that the technology for ameliorating near-term problems in SEA interdiction already existed, but was not always tapped. The many and complex difficulties involved organization, funding, and decision-making, with the last termed by the study members especially critical since it involved both the decision-making channels for requirements as well as the acquisition process. The decisions necessary for prompt and effective responses seemed almost impossible to obtain and to enforce.

The current structure was geared to long-term system development and could not deal effectively with short-term problems. To remove this major fault, the members suggested establishment of a small office at the highest level that would bypass the larger and more traditional R&D processes, and concentrate specifically on bombing and all other aspects of interdiction. This would also insure that priorities were assigned to promising short-term projects. [7]

This AC-130A, 54-1629, was one of the original six AC-130A Gunships deployed to SE Asia in the Fall of 1968. It carried 4x7.62 mm mini-guns and 4x20 mm Vulcan Gatling guns. This aircraft was lost to AAA on 24 May 1969.
Source: U.S. Air Force.

The group also recommended "crash" measures outside of regular channels to complete development and procurement of an advanced laser system for the F-4D to improve its visual bombing accuracy; develop and test a new bombing system using a forward-looking infrared radar (FLIR) sensor to improve night operations; increase and accelerate production of AC-130 and AC-119 gunships; and expedite tests of the Ka-band radar for the F-111 bombing system. It urged the Air Force to accelerate development of route-denial and "Paveway" ordnance, speed work on a 2,000- to 3,000-pound general purpose demolition warhead for the Walleye guided missile, and support the Navy's "Condor" program (since the Air Force lacked an accurate standoff weapon to use against heavily defended targets). [8]

Combat Lancer F-111As. Source: U.S. Air Force.

There was little immediate response to these recommendations. The first official comment came from Secretary Brown on 29 August when he suggested to Dr. Alexander H. Flax. Assistant Secretary for R&D. and General McConnell that the 14 July report appeared to be "very useful" and that the Air Force should find a way to speed the completion of critical projects. [9] Although the Chief of Staff felt that some portions of the reports could be helpful, he disagreed sharply with their views and conclusions. believing they were far too critical of the Air Force R&D effort. On 16 September, he stated bluntly to his Staff Directors that neither the summary report nor the supporting studies "should be construed to have the concurrence or endorsement of the Air Staff. the Chief of Staff, or the Secretary of the Air Force. [10]

The Air Staff nevertheless was disturbed by the thrust of the reports. although it was not convinced that the answer to the interdictions problem was to establish a single office of responsibility. Thus on 9 October 1968. Lt. Gen. Seth J. McKee. Assistant Vice Chief of Staff. asked Lt. Gen. Joseph R. Holzapple. Deputy Chief of Staff/R&D, to create an ad hoc committee to examine the proposals made in the 14 July report. The group's steering committee consisted of General Holzapple. DCS/R&D. chairman; Lt. Gen. Glen W. Martin. DCS/Plans and Operations; and Lt. Gen. Robert G. Ruegg. DCS/Systems and Logistics.

Even as he took this step. General McKee noted that the Air Force had previously investigated the validity of several projects listed in the report. It had found that a new laser bombing system would not improve the F-4Ds delivery accuracy. that development of route-denial and "Paveway" munitions had been stepped up, and that development of an angle-rate bombing system with a FLIR sensor had begun. Efforts were under way to develop a 3,000-pound laser guided bomb and an advanced air-to-surface missile. In addition. the Air Force had accelerated production of gunships and continued to explore the feasibility of a Ka-band radar for the F-111 bombing system. [12]

The addition of laser-guided Paveway bombs to the F-4D made a revolutionary difference to their ability to destroy small and/or very hard targets. Source: U.S. Air Force

The ad-hoc committee worked from October through December 1968 and, in January 1969. published two reports. The first, on 15 January. noted that "time. money. and testing compromises during the development program may also be major contributors to delay and unreliability." This report stated that projects which required urgent treatment definitely justified "higher risk management procedures." But special procedures to speed decisions could not be used for the majority of projects considered by the Air Staff. According to the committee, the Air Force lacked the experimental, design, and testing capacity to respond to many potentially productive ideas. It recommended that the Air Force improve its design and testing facilities and that the Air Staff Board panels, during their regular deliberations, identify any cases in which projects might merit special consideration. [13]

On 29 January 1969 the committee published its second report, a "White Paper" titled "Air Force Development/Procurement Actions in Response to SEA Problems." Considering development and procurement since 1965, the members insisted that the response had been effective. Although there had been some "temporary lapses in responsiveness," in general the innovation of Southeast Asia Operational Requirement (SEAOR) procedures had adequately served in pushing through the required short-term developments. [14] The committee listed many of the significant accomplishments in reconnaissance and electronic warfare. It cited the continued development of the RF-4C since 1962. This aircraft had been procured as a follow-on to the RF-101 and had also replaced the photo reconnaissance version of the RB-66. It noted the acquisition of forward and side-looking radar and pointed to work done on radar homing and warning (RHAW) equipment. which found expression in "Wild Weasel." The USAF schedule for

Air War – Vietnam

equipping the F-105F Wild Weasel III with an improved air-to-surface anti-radiation missile, the AGM-78B. was to be completed in March 1969. Electronic countermeasure pods had been developed and deployed to Southeast Asia to counter the enemy's surface-to-air missile (SAM) and anti-aircraft artillery (AAA) radars. In the area of night operations. the White Paper mentioned such improvements as the Tropic Moon and Black Spot aircraft and the Gunship II AC-130A's.[15]

This F-105D is a Thunderstick II conversion, which is easily distinguished from the standard -D model by the backbone electronics bay fairing from the cockpit to the vertical stabilizer. Thunderstick II allowed the aircraft to deliver ordnance under most flight conditions (speed, altitude and attitude) by automating weapons delivery and piloting. Source: U.S. Air Force

To improve visual bombing accuracy for the F-4C. a laser range finder had been developed and tested in the summer of 1968. Combat evaluation of the unit was scheduled for mid-1969. Laser technology also figured in the evolution of the laser guided bomb (early development had been particularly encouraging). laser target designator. and target seeker. all of which were being evaluated in the theater during 1968. As far as advancing adverse weather capabilities was concerned. the committee noted the contribution of the MSQ-77 "Combat Sky Spot"; the development (with F-105 retrofit in 1969) of the T-Stick TI advanced conventional weapons delivery system with long range navigation (LORAN); modifications to the F-111A delivery system prior to deployment to the theater in March 1968; development of the advanced Mark II conventional bombing system for the F-111D (delivery scheduled for early 1970); and the initiation of an over-the-horizon ground radar system ("Steer", a refinement of the radar bomb-directing technique which used two relay aircraft to control strike planes in low-altitude deliveries to ranges of 400 nautical miles from the ground terminal) and a F-111A radar correlation bombing system. to be placed under development in early 1969 with an initial operational capability (IOC) scheduled in mid-1971.[16]

According to the White Paper. all operational problem areas cited by the Operations Analysis I RAND reports had received attention. Delays in supplying equipment were traceable to lengthy development time required by certain items. but interim fixes had been provided in such cases. Concerning the suggestion that the Air Force set up a top level management group. the committee argued there were few projects that required special attention and that such a unit would put a strain on the USAF reserve of technically qualified officers. In summary. the White Paper concluded there was no evidence showing the lack of support for any

SEA problem. It said "There are, of course. differences of opinion concerning the management emphasis appropriate to any program. Opportunities for speeding the development of particular equipment always exist. The required commitment of resources must. however, be balanced against the total demand for development/procurement support." [17]

In January 1969, just after the Nixon administration took office. the outgoing Assistant Secretary of the Air Force (R&D) unburdened himself on the problem to Secretary Brown. who also was leaving the government. In retrospect. Dr. Flax said, between the mid-1950's and about 1963 the Air Force had concentrated its resources on nuclear weapons and the aircraft and equipment to deliver them. During the beginning of this period, he noted, "Air Force planning corresponded too literally and too narrowly to stated national policy." Even when the national policy changed in 1961, with its stress on R&D for limited war, for several years little change evolved in training, tactics, and equipment. [18]

According to Dr. Flax, the USAF Director of Requirements and many in the Air Staff were unalterably opposed to improving the F-4 aircraft. It had required pressure from his office. and from the President's Scientific Advisory Committee (PSAC). the Scientific Advisory Board (SAB). the Director of Defense Research and Engineering (DDR&E) and the Bureau of the Budget (BOB), before the Air Staff accepted a "weak compromise" with respect to F-4 conventional delivery modifications (the F-4D system). Because of this, development of a continuous-solution bombing computer and advanced bombing system along with improved sensors fell short of what could have been realized for the F-4D. The Air Force still did not accept the view, stated the Assistant Secretary of the Air Force. "that a substantial part of the tactical aircraft force should be equipped with accurate weapon delivery systems for conventional weapons. There are many in the Air Force who are still unreconstructed." [19]

Dr. Alexander H. Flax.
Source National
Reconnaissance Office

In the Air Staff. there were those who felt that present delivery accuracies were good enough for close support. However. Others held that. even if accuracy could be somewhat improved. the results would not justify the cost because there were no sensors available to detect small. fleeting targets in all kinds of weather. Too, there existed the view that available accuracies were not good enough to make all-weather delivery a cost-effective tactic to be widely employed and that improved guided bombs and missiles along with ground (MSQ-77) and air bombing control systems could provide sufficient improvement in delivery accuracy.

However. Dr. Flax said, there was an Air Force consensus that the F-111 should possess an accurate all-weather bombing system but disagreement over the

Air War – Vietnam

various ways of developing this capability. He also supported developing the "Paveway" series of electro-optical and infrared radar (IR) guided bombs.

Exploitation of technology. admitted Dr. Flax. was "not as good as it should be." but he was opposed to establishing new organizations to expedite such activity. The "quick reaction" programs would have to be treated as normal tasks by Air Staff and Air Force Systems Command (AFSC) management rather than by newly created special offices. As far as development of interdiction systems was concerned. there was no central office in either the Air Staff or AFSC "for doing a good job in this area. " [20]

There was another problem area Dr. Flax commented on. He said that, with regards to research and development for Southeast Asia, "we can lead the horse to water but we can't make him drink." No matter what new things the Air Force might develop or even produce. unless it could "promote approval of development" through the operational chain from the Joint Chiefs of Staff (JCS) down through the Military Assistance Command Vietnam (MACV) and Seventh Air Force, it could do no good for the forces in the field. "Examples abound," he said. "in which long delays in development or even failure to deploy potentially useful technological innovations must be attributed to resistance somewhere along this line of operational command."[21]

Examining The SEAOR System

Established in mid-1965, the Southeast Asia Operational Requirement system was designed to speed the identification of USAF equipment needs and to procure and introduce items more rapidly into the theater. By mid-1967, however, it became apparent that this system was not working as originally planned. Undoubtedly. part of the difficulty could be traced to U.S. defense planning, which had anticipated an early end to the war, perhaps within one or two years at most. Accordingly, the Air Force had continued to follow its peacetime R&D procedures with only slight modifications. [22]

USAF responsiveness appeared to be hampered by the requirement system itself, which a member of DDR&E's staff described as "awkward." [23] The theater commanders flooded the requirements channel and the number of approved SEAOR's, identified as either required operational capabilities (ROC's) or Class V modifications, clearly surpassed the Air Force ability to determine priorities and satisfactory funding sources.

Besides the lack of rigorous selectivity, some requirements which should have been completed in about 12 - 18 months, evolved into long-term development efforts. To compound the problem, the tremendous increase in requirements caused excessive specialization within the Air Staff which, in turn, led to duplication of effort. Also, the SEAOR system continued to be plagued by obsolete funding practices. [24]

In an effort to ameliorate the unsatisfactory funding process. improve the priority system. and reduce the time of equipment acquisition, a General Officers' SEAOR Review was held on 15-16 November 1967 at the Aeronautical Systems Division

(ASD), Wright-Patterson AFB, Ohio. Attending were senior representatives from Headquarters USAF. AFSC, PACAF, Tactical Air Command (TAC), Air Force Logistics Command (AFLC), and Seventh Air Force. Following this review, the Seventh Air Force and Headquarters. Pacific Air Forces (PACAF) were directed to devise, "to the extent practicable", a priority list of unfunded SEAOR's. Although no precise machinery for solving these difficulties was established, several procedures were agreed upon to improve coordination and increase the flow of pertinent information. [25]

With funding remaining one of the most serious problems, on 4 December Gen. James Ferguson. Commander, AFSC. Informed General McConnell that the lack of SEAOR resources had reached the critical stage and would become even worse as more requirements were received. The amount of money needed to complete already identified SEAOR's had already passed a half billion dollars in R&D and production funds as follows: [26]

Year	R&D	Production
FY 1968	$44,600,000	$98,039,000
FY 1969	$18,330,000	$391,181,000
Total	$62,930,000	$489,220,000

According to the AFSC Commander, since budgetary pressures could only become worse, the end result would be wasted effort in searching for technical solutions to SEAOR's for which there were no funds. Given the lack of funds, it would be much more productive to concentrate on those SEAOR's which could be seen through to completion. The Air Force, he said, must establish a priority system to satisfy the most critical requirements and also find more money for present and future SEAOR's. [27]

General Ferguson therefore suggested to the Chief of Staff that Headquarters USAF, in conjunction with the Seventh Air Force and PACAF, establish a ROC priority system similar to the Class V modification list. Under this procedure, unfunded and new SEAORs would be deferred until money became available. at which time Headquarters USAF would direct Systems Command to prepare (according to the priority list) an updated best preliminary estimate (BPE) for the next requirement. Also a review group should be established in the Air Staff to find money for new SEAOR's and for those critical requirements presently in need of funding. [28]

Acting on a directive from General McConnell. on 12 December 1967 General Holzapple established a SEAOR Review Board to analyze. approve, and fund Southeast Asia operational requirements. After receiving a best preliminary estimate from AFSC or AFLC, one of his aides (The Director of Operational Requirements and Development Plans) would recommend the requirement to the review board. A proposal would be presented only after an analysis of technical feasibility, determination whether the SEAOR could be completed within a reasonable time, and identification of a funding source. The review board would then decide whether the requirement should be pursued or canceled. [29]

Should it be canceled, the board would forward its rationale to General Holzapple. If he or his staff determined that the requirement could be satisfied (either in the near or long term), the board would then propose a funding source and an appropriate office to manage the SEAOR. Further, the board would decide which requirements could be consolidated and when a SEAOR would be considered completed. The board would also establish a priority list of all active SEAOR's, periodically review funding, and present recommendations to the Deputy Chief of Staff I R&D. [30]

The SEAOR Review Board – whose members consisted of the Director of Operational Requirements and Development Plans; Director of Development; Assistant for Reconnaissance; Assistant for R&D Programming; Director of Operations; and the Director of Maintenance Engineering, convened in early January 1968 to consider the entire range of critical problems that plagued the USAF requirements system for Southeast Asia. When this comprehensive review was completed, each SEAOR had been examined and almost two months had elapsed. Short and long-term requirements were identified; criteria for the required operational capabilities were developed; some SEAOR's were canceled or combined; and funding priorities were established. The SEAOR, newly-defined, was described as: "A Seventh Air Force requirement that can normally be satisfied by providing an initial operational capability (IOC) within 24 months after receipt of the BPE and Headquarters USAF approval. [31]

This marked an improvement over previous definitions but since more than 24 months was usually required to achieve an IOC. it still could not be considered either precise or binding.

The Air Staff Review Board agreed to consider approval of SEAOR's set priorities and look at the overall program. However, as it turned out, a semiannual General Officers' SEAOR Review Conference similar to the General Officers' Review of November 1967 took over many of the tasks of the board. It had been found that a periodic (every two months) and exhaustive review was impractical and, it was hoped, unnecessary. Held on 6-8 August at Headquarters USAF, the conference included representatives from the Air Staff,* Seventh Air Force, PACAF, TAC, AFSC, and AFLC. Requirements that needed funds or that had been plagued by technical problems were examined and views on current problems were exchanged. [32]

Despite the establishment of the semiannual general officers' conference, PACAF continued to develop its own quarterly list of funding priorities (for unfunded SEAOR's only) so that when money became available participating organizations could weigh the relative importance of SEA requirements. The Seventh Air Force also promulgated a list (not always in agreement with PACAF's), but the PACAF summary was the one sent to Headquarters USAF. SEAOR's were still approved by the Air Staff after it had received a best preliminary estimate, appropriate

Towards A Bombing Halt 1968

comments from the commands, and PACAF's validation. Nevertheless, it continued to review the requirements received and to study the SEAOR system. [33]

Subsequently. a task force that had studied Defense Department R&D procedures during the latter months of 1968 reported to the DDR&E that [34] "the present course of development of effective materials and techniques is particularly lengthy and its transfer to the field tortuous beyond necessity; this raises the question whether our Service R&D procedures are yet appropriate to the kind of real time. responsiveness of which the community is capable." Members of the task force included Dr. Gordon J. F. MacDonald. University of California (Santa Barbara). chairman; Dr. Chester Cooper. Institute of Defense Analysis; Dr. Richard L. Garwin. IBM; Dr. Murray Gell-Mann. California Institute of Technology; Dr. Marvin L. Goldberger. Princeton University; Dr. Harold Lewis. University of California (Santa Barbara); Dr. John L. McLucas. MITRE Corp; Dr. William A. Nierenberg. Scripps Institution of Oceanography; Dr. Guy J. Pauker. RAND Corp; Dr. Milton G. Wiener. RAND Corp; and Dr. Frederick Zachariasen. California Institute of Technology.

Thus. it appeared at year's end that the SEAORs system apparently was not working the way USAF planners had hoped when they established it in 1965 to meet critical combat needs. By 1968. this "crisis approach" had affected the overall ability of the Air Force to establish more orderly and cohesive R&D procedures. [35]

Countering The Enemy Defensive Threat

For several years the Air Force had studied ways to counter the growing North Vietnamese defensive threat. which comprised ground-based guns (including small arms. automatic weapons. and AAA), surface-to-air missiles (SAM's). and fighter aircraft. Although much attention focused on the SAM threat, about 75 percent of the USAF losses were caused by other types of ground fire. [36]

Between 1965-1967 USAF pilots maintained clear superiority over the Communist MiGs, but in late 1967 enemy tactics improved substantially. During September-December 1967, the United States lost 12 aircraft in air-to-air combat while downing 15 enemy planes. In contrast. over the first eight months of the year, 77 Communist planes were shot down with a loss of 24 U.S. craft. U.S. aircraft encountered SAM and AAA fire as soon as they flew over the coast line from the east or crossed the Red River from the west, evidence of coordination between the enemy's radar surveillance and his command

ZSU-23 23mm twin mount.
Source: National Archives

347

element. His MiG aircraft. which had been used very selectively, sought to interdict USAF planes in cloudy as well as clear weather employing tactics that indicated a radar-initiated intercept. The enemy's increased competence could be traced to a more effective use of ground control intercept (GCI) radars. As of August 1967, more than 200 early warning (EW) ground-controlled radars were deployed in North Vietnam along with AAA fire control and Fansong B missile control radars. [37]

A Bar Lock (P-35) GCI Radar.
Source: National Archives

The North Vietnamese deployed Bar Lock radars near Haiphong (with early warning coverage over the Gulf of Tonkin) and in the western part of Route Package 5 providing EW I GCI coverage for about 90 miles into Laos. Because they were mobile and camouflaged. it was difficult for USAF aircraft to locate and destroy these radars. Also. Air Force EB-66 electronic countermeasures (ECM) aircraft lacked adequate jamming power and maneuverability and consequently proved vulnerable in the so-called "high-threat" areas over North Vietnam. Even when the Bar Lock GCI radar was jammed. the enemy could track incoming planes successfully by employing other radars not affected by penetrator jamming. [38]

Disturbed by the overall improvement in the North's defense system. Gen. William W. Momyer, Seventh Air Force Commander. in January 1968 reported to Gen. John D. Ryan. Commander of the Pacific Air Forces. on the growing dangers to his strike aircraft. "We have made repeated attempts." said General Momyer. "to eliminate their GCI capability. with virtually no success." Not only were the radars mobile and well hidden. but in some cases they were located near population centers. thereby precluding attack. The MiG threat, he observed, was increasing more rapidly than the Air Force's ability to counter it. He recommended a crash program to deal with the situation. [39]

On 31 January General McConnell directed AFSC to determine how best to resolve the problem. "All aspects we're to be considered," said the Chief of Staff. including how to attack and destroy radars situated adjacent to population centers. [40] Acting on this directive. General Ferguson established a special task force (designated Have Dart) on 5 February 1968 to undertake an investigation and propose solutions. The AFSC commander recognized that not only had the enemy's GCI and overall defensive effectiveness caused an "unfavorable loss ratio of our aircraft," but they also affected the accuracy of their strikes. Many aircraft unloaded their ordnance prematurely in order to avoid Communist defensive fire. One of the continuing problems faced by the Air Force was unsatisfactory circular error probables (CEP's). Frequently, heavy ground fire compelled USAF pilots to release their weapons from inordinately high altitudes. For example. on dive-bombing missions weapons were released at about 8,000 feet in order to keep from going below 4,500 feet on pullout where heavy fire would be encountered.[41]

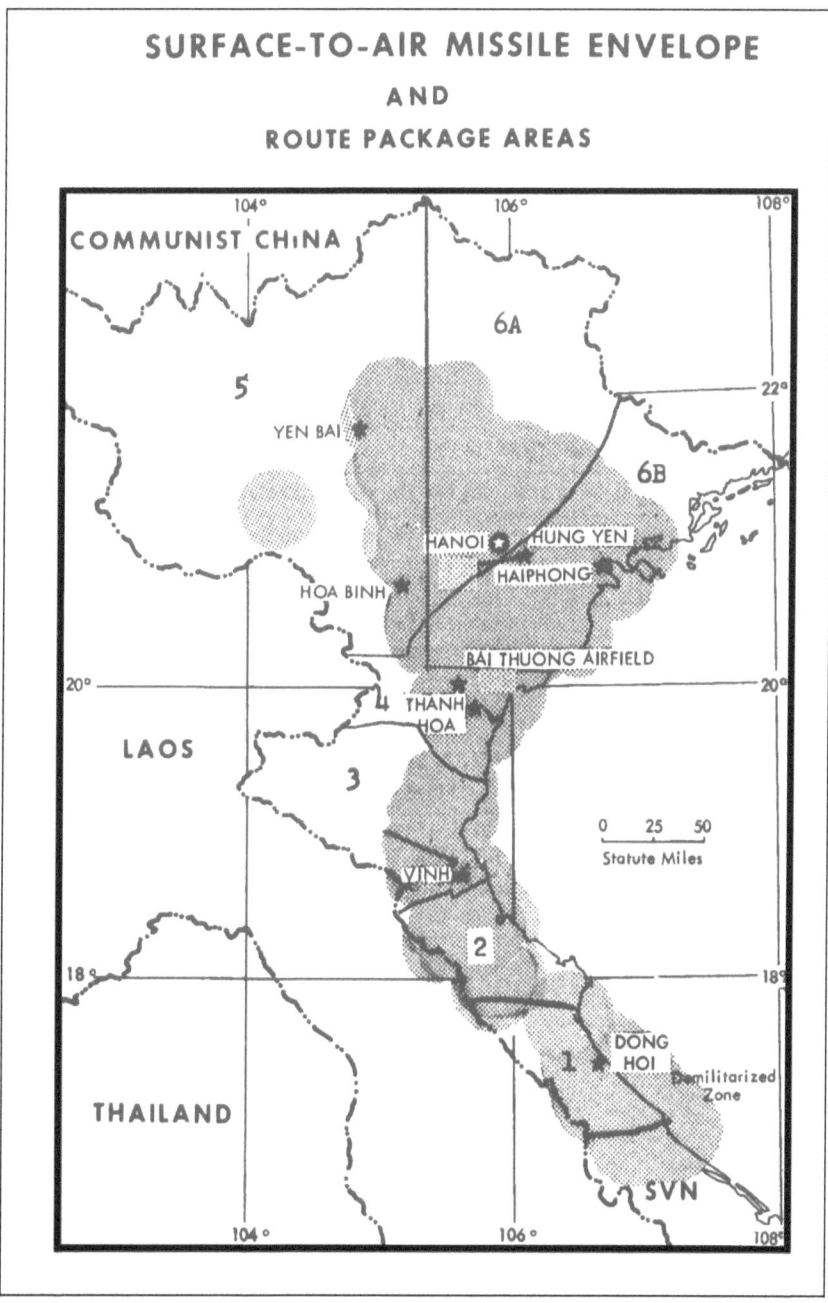

Air War – Vietnam

In a summary report completed on 13 March, the task force concluded that complete destruction of the GCI radars "appears unlikely" and that the enemy's Bar Lock radar appeared to be the most vulnerable to attack. However, to achieve a 500-foot CEP the F-4/F-105 had to drop 170 M-177 (750-pound) bombs to provide a 90 percent destruction probability against each GCI site. Since most of these sites were 'located within or near restricted areas, it was clear that the Air Force needed to acquire an accurate guided bomb. [42]

USAF officials further realized that Air Force electronic equipment was not adequate enough to counter enemy radars. To improve command and control in a hostile environment, the task force recommended improvements to EC-121 (College Eye), EB-66 and F-4D aircraft, deployment of a TPS-43 radar, and development of radar for helicopters. The information gathered by College Eye aircraft together with the Navy's positive identification and radar advisory zone (PIRAZ) ships could be used to produce controlled intercepts and positive identification for air-to-air missile launches without visual identification by the pilot. [43]

A Talos Missile fired from CG-10 Albany. On one occasion, a U.S. aircraft saw a MiG-21 taking off from Phuc Yen and alerted the Talos cruiser on duty in the Gulf. The cruiser fired a single Talos that hit the MiG before the pilot could retract his undercarriage. Source: U.S. Navy.

Also, the task force proposed that the Air Force employ the Talos missile against MiGs, develop day-visual radar acquisition and strike equipment along with a day/night/adverse weather integrated system of radar acquisition, give more attention to ECM jamming against EW/GCI radars, procure 1,000 Redeye missiles for use in air-to-air combat, and assign a high priority to guided bomb development. [44]

On 1 March, just prior to publication of the Have Dart report, the so-called Credible Comet study group, which included representatives from the Air Staff, TAC, ADC, AFLC and AFSC, reported that deficiencies in tactical electronic warfare (TEW) were adversely affecting air operations. This subject, which formed part of the Have Dart analysis, had long been of concern to operational commands and the Air Staff, since TEW constituted an integral part of air operations in any hostile environment. Without an effective electronic warfare capability, any plan for countering the North Vietnamese defensive threat would prove ineffective.

The Credible Comet group recommended that advanced TEW equipment be developed and installed on "all tactical aircraft exposed to a hostile electronic warfare environment." It suggested this include electronic attack devices to destroy the enemy's systems. ECM equipment. and electronic operational support (EOS) systems. To improve electronic warfare management. the group proposed a number of functional realignments within the Air Staff and a reorganization of operational and support commands. For research. development. and acquisition. a more cohesive and responsive cycle could be attained by clarifying R&D policies and procedures followed by the various commands during development of electronic systems. [45]

Reflecting concern with the entire research and development process. the report observed that the deficiencies uncovered during the Vietnamese war (which led to tactical shortcomings) necessitated a "broad reassessment" of organization and command responsibilities. For example, in Tactical Air Command some electronic warfare groups were assigned to reconnaissance. some to fighter wings. and others, such as Wild Weasel, to strike forces. Within the Air Force, noted the Credible Comet group, focal points "to accomplish the EW mission are dispersed, vague or nonexistent. The management picture also shows a lack of a total integrated systems approach with few clear-cut nodes of single authority or decision-making recognizable in the management network." [46]

An F-105 Wild Weasel. Source: U.S. Air Force

The group emphasized the importance of changing or even eliminating obsolete and time-consuming funding and procurement practices. Existing procedures for initiating AFLC's Class V electronic warfare modifications constituted, it said, a "less than efficient use of funds, manpower and facilities." Overall. a much more responsive RDT&E and acquisition cycle appeared necessary so that badly needed equipment could be produced more rapidly and in greater quantity. [47]

On 25 April 1968. shortly after the Have Dart and Credible Comet reports had been issued. the Joint Chiefs of Staff completed a review of the Night Song study. This study, initiated in January 1967 by Deputy Secretary of Defense Cyrus Vance in response to a marked improvement in Communist air defenses. was originally published in March 1967 with the proviso that it would be updated a year later. It recommended equipping USAF strike aircraft with an advanced radar homing and warning system and self-protection devices.

In its reappraisal of Night Song. the Joint Chiefs noted that the Communists' air defense system still depended on materiel shipped from the Soviet Union, China, and the eastern European bloc countries. As long as the North Vietnamese received equipment. they were capable of making their defense even more effective. Although the Air Force had improved its tactical strike craft since early 1967, the experience of the intervening period indicated that elimination of the

Air War – Vietnam

MiG threat was not feasible. As long as the enemy continued to use Chinese air bases near the North Vietnamese border, it would be impossible to remove the threat since it was U.S. policy not to strike within Chinese territory. [48]

Also. the JCS observed that the Air Force still did not have sufficient numbers of heavy bombs and needed more effective proximity and long-delay fuses. The Joint Chiefs hoped that eventual procurement of advanced fuses, electronic and infrared sensors. and laser equipment would enable pilots precisely to locate the enemy's radar. guns. and vehicles. [49]

The Night Song report reiterated that a "broad air campaign" was necessary to reduce the flow of materiel into North Vietnam, and it recommended strikes against additional military targets and war-supporting industry in the north. [50]

In late 1968, acting on these reports. the Air Force took steps to modify equipment and to develop new devices to deal with the defensive threat. USAF officials were optimistic that they could make inroads against the enemy's defensive system. although an early. complete solution was out of the question. However. with the cessation of the U.S. bombing campaign over the North. a final test of the additional offensive capabilities of the Air Force became a moot point.

Bombing, Interdiction, And Surveillance Operations

From the time it began full-scale operations in Southeast Asia in early 1965, the Air Force had sought to improve its bombing accuracy. Three years later. on 5 March 1968, General Holzapple admitted to a Congressional subcommittee that "we still have room to grow in terms of accurate delivery of ordnance." It was a very difficult problem. he said. "a problem inherent in any strike airplane." He advised the subcommittee that the Air Force planned to deploy the F-111 to Southeast Asia and predicted this new aircraft would lead to "a big step forward in the accurate delivery of ordnance." [51]

There were high hopes for the F-111A but its service debut was troubled.
Source: U.S. Air Force.

In October 1967 the USAF Combat Target Task Force, established by General McConnell to examine the problem of all-weather bombing, had recommended that six F-111As be deployed to Southeast Asia. For the long term. the task force suggested that a combat CEP of 200 feet or less be set as a criterion for such conventional all-weather bombing systems. Subsequently, in March 1968, the Air Force sent a small F-111A unit, designated Combat Lancer, to the war zone. Six aircraft, along with support personnel. arrived at Takhli AB, Thailand. on 18 March 1968. Nine F-111As were modified for Southeast Asia and six were originally deployed, of which three were lost in the first four weeks of operations. The cause of these crashes has been attributed to weld failure of the Bendix horizontal stabilizer link. Two additional aircraft deployed as replacements and one remained in the United States for testing.

Beginning on 25 March, they flew a total of 55 combat missions – averaging 2.46 hours, in Route Package #1 (North Vietnam from the demilitarized zone north to the 18th parallel). Low-level missions consisted of single-aircraft night flights. On several occasions between March and November 1968, Combat Lancer operations were suspended due to crashes, hydraulic system failure and metal fatigue of the wing carry-through structural box discovered at General Dynamics, San Diego. In late June, the F-111A's were restricted to flying without using the terrain following radar (TFR). After eight months in the theater, the unit returned to the United States on 23 November 1968. [52]

The F-111A Combat Lancer deployment showed that the F-111 was good but not good enough. This would change as the bugs were ironed out of the complex aircraft and it became a superlative strike aircraft. Source: U.S. Air Force.

A RAND Corporation analysis of Combat Lancer radar bombing completed prior to the termination of the F-111A operations found that the "verdict certainly must be 'not well enough' in terms of destruction of targets attacked." The basic Combat Lancer tactic included a low-altitude approach (200 to 1,000 feet) at night employing terrain masking, random headings, random release times and passive electronic countermeasures. The F-111A attained an overall CEP (in which bomb-miss distance was known) of 1,050 feet. By comparison the report observed that an F-105/F-4D radar bombing program (Commando Nail) showed an overall CEP of 2,000 to 3,000 feet with a 400 to 500 feet circular error for daylight dive-bombing over North Vietnam. However, because of a relatively high loss rate for

Air War – Vietnam

the F-105s and F-4s, these aircraft did "not appear to provide any overwhelming advantage over the F-111A." [53]

Therefore, the RAND report concluded that all major USAF aircraft left something to be desired as far as CEP was concerned, a conclusion previously reached by the Combat Target Task Force. The RAND analysis also indicated, however, that the F-111A showed promise for improved radar bombing. Substantially better results could be attained, it suggested, since Combat Lancer operations had been limited (crews could have been expected to improve with experience) and since Route Package #1 was a poor area for radar bombing. Considering the short operational period and the unfavorable conditions. the result according to RAND. could be construed as "fairly respectable." [This] conclusion [was] reported to Secretary of Defense Clark M. Clifford on 9 January 1969 by Secretary Brown. who emphasized that this finding was based on limited combat data (Memo (S). SAF to SECDEF. subj: COMBAT LANCER Preliminary Rprt. 9 Jan 69). [54]

A Combat Lancer final report subsequently issued by the USAF Tactical Fighter Weapons Center, noting that the F-111A's had dropped their bombs "with varying degrees of accuracy," estimated the planes had achieved a 400-foot overall CEP for radar bomb releases at 1,500 feet or less. Because of the brief duration of operations. the Center, like RAND, could not come to hard and fast conclusions. The concept of low-level F-111A penetration and attack during night and adverse weather," appeared valid." As for radar bombing results. the report stated that "the most critical factor affecting bombing accuracy was radar acquisition of the aimpoint." Results of combat showed that aimpoints with good radar return characteristics had a CEP of 233 feet while the CEP for ill-defined aimpoints was 2,304 feet. [55]

The dorsal hump that distinguished the T-Stick II conversions is clearly evident in this post-war photograph of an F-105D. Source: U.S. Air Force

As mentioned. the Air Force gave high. priority to the evolution of systems which would enable its combat aircraft to stay out of range of small arms and automatic weapons fire during daytime. and still operate under low overcast (2,000 to 3,000 feet) during the northeast monsoon season.   To acquire such a capability. the Air

Towards A Bombing Halt 1968

Force proposed several modifications to the F-105D/F. one of which was the T-Stick II/LORAN (long range navigation) weapon delivery system for which funds had been deferred by OSD. [56]

In August 1968. Adm. John S. McCain Jr, Commander in Chief. Pacific Command (CINCPAC) observed that U.S. aircraft still found it difficult to conduct air strikes at low altitudes and he requested a review of military R&D programs. He noted that a high percentage of planes had been lost to the enemy's automatic weapons fire at altitudes of 3,000 feet and under. What was needed. he said. were systems to keep planes out of the range of Communist guns and still enable pilots to accomplish their mission using improved navigation equipment and guided bombs. [57]

On 28 September 1968. the funds previously held by OSD were released by Deputy Secretary of Defense Paul H. Nitze who directed they be applied to the T-Stick II modifications. However. he required the Air Force to limit the work to one 18-unit equipment (UE) squadron (30 aircraft including training. support. and attrition aircraft) instead of the originally planned 65 airplanes. "It was necessary," noted Nitze, "that we obtain good data on the accuracies obtained with the LORAN bombing system to assess the desirability of providing other aircraft with this capability."

The avionics hump on the F-105D actually improved the aircraft's lateral stability. Source: U.S. Air Force.

As far as their oversea deployment was concerned, such a decision would depend on an evaluation of the modified aircraft. Extensive testing of early T-Stick II production models in the United States would be necessary with an original initial operational capability of mid-summer 1969. [58]

Ever since the administration committed substantial forces to the war. it had given high priority to interdiction of the enemy's supply and communications lines. And, on 4 March 1968. Admiral McCain reiterated that development and deployment of advanced interdiction systems and munitions were mandatory if the United States was to increase the pressure on the Communists. The Shed Light and Muscle Shoals (renamed Igloo White on 1 June 1968) programs, said Admiral McCain, should receive "full support." [59]

At the same time. Secretary Brown, concerned that the coverage of the enemy truck traffic in Laos was not intensive enough, suggested to Mr. Nitze that a combination of more sorties and more effective night operations would increase substantially the number of trucks destroyed. [60]

After the President on 31 March 1968 suspended U.S. bombing north of the 20th parallel (revised three days later to the 19th parallel), the major Air Force objective became the interdiction of the truck traffic in Laos, almost three-fourths of which operated in the area between Mu Gia pass and the demilitarized zone (DMZ). The infiltration through Laos remained substantial and as the dry season approached was expected to increase. On 2 July, the President's Scientific Advisory Committee (PSAC), concerned with the incessant movement of supplies into South Vietnam, recommended another special effort against the enemy's logistics. this time an intensified interdiction operation m Laos during the 1968-1969 northeast monsoon season (the dry season in Laos). Dr. Donald F. Hornig. the President's science adviser and chairman of the PSAC. met on 12 July with Defense Secretary Clifford. Deputy Secretary Nitze. and Dr. John S. Foster. DDR&E. to discuss the Advisory Committee's proposal. The group estimated that 60 percent of the materiel infiltrating into South Vietnam passed through Laos. most of it during the northeast monsoon. [61]

Modified EC-121s supported the Igloo White effort. Source: U.S. Air Force

Following this meeting, Secretary Clifford directed the preparation of an interdiction plan to attack enemy supply lines and evaluate Igloo White equipment. General McConnell assigned to Gen. Joseph J. Nazzaro. Commander in Chief. Pacific Air Forces (CINCPACAF), the task of establishing a group at Seventh Air Force headquarters to plan the interdiction campaign. Completed in late August and designated Commando Hunt, the plan envisioned the destruction of a greater number of trucks and supplies on the major infiltration routes in the Laotian panhandle. The proposed operations would hopefully tie down substantial enemy forces supporting the movement along the Ho Chi Minh trail while checking out the Igloo White sensors. The administration was especially anxious to strike key roads that the Communists had rebuilt over the past year. Intelligence indicated that the enemy's 559th Transportation Group, with about 50,000 personnel and well over 1,000 trucks, was located in the eastern part of the Laotian panhandle. [62]

Gen. Creighton Abrams. Commander. U.S. Military Assistance Command, Vietnam (COMUSMACV) approved the plan on 26 September 1968 and assigned it a high priority. Commando Hunt operations began on 15 November, with USAF Brig. Gen. William P. McBride, Commander of Task Force Alpha at the Infiltration Surveillance Center (ISC), Nakhon Phanom AB, Thailand, being responsible for integrated planning and control of Air Force, Navy. and Marine aircraft. General McBride's task force also directed the Igloo White air surveillance system (see discussion below) and was in a position to allocate Igloo White resources to the Commando Hunt project. The area of operations in the eastern segment of Steel Tiger extended from the Mu Gia pass to approximately

Towards A Bombing Halt 1968

six miles south of Tchepone, Laos, and covered about 1,700 square miles including 450 miles of primary roads. Information derived from Igloo White sensors was used as the primary intelligence base for locating truck concentrations. Also B-52 Arc Light aircraft were used against truck parks and supply storage areas as the Air Force increased the number of sorties allocated to strike in Laos. [63]

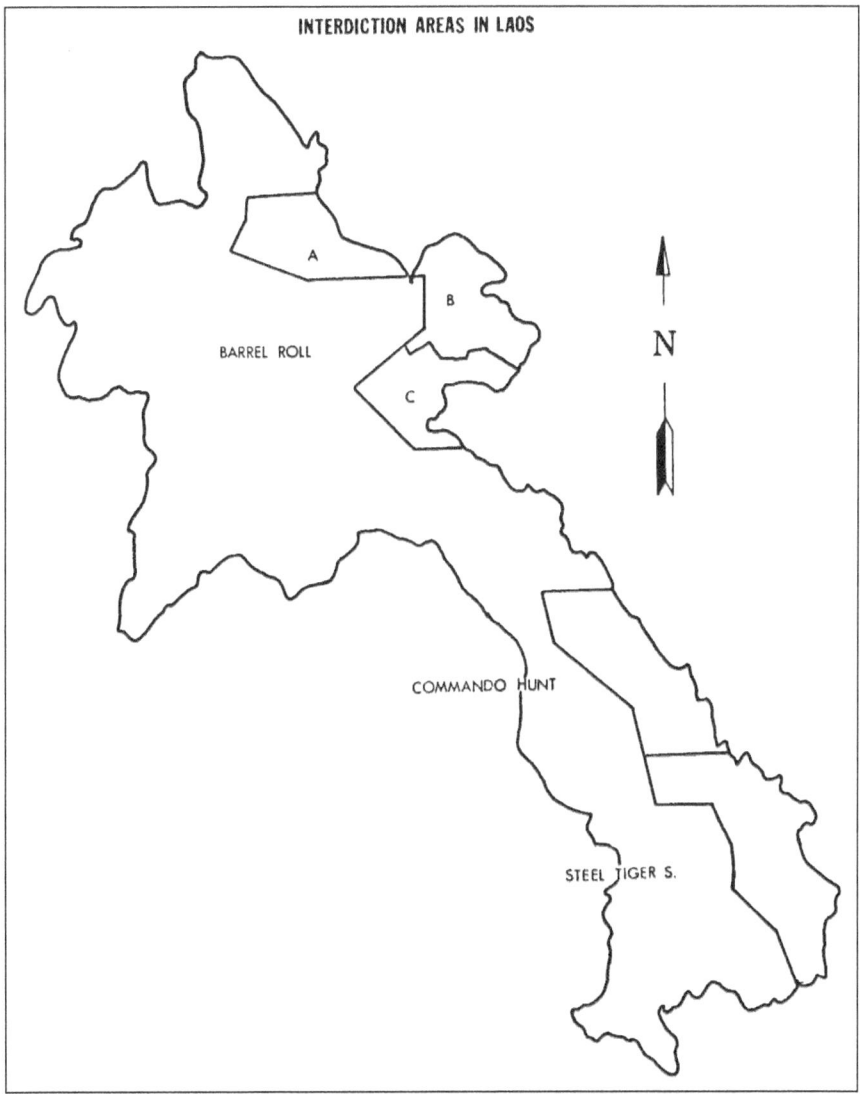

Between November 1968 and January 1969, the Air Force committed 40 percent of its SEA strike aircraft (including fighter-bombers, B-52s, and AC-130 gunships) against about 1,350 enemy trucks in Laos. Of the total, the North Vietnamese operated approximately 400 trucks per day although upon occasion the total was substantially higher. Commando Hunt emphasized attacks against roads and points which the Communists found difficult to bypass. When they completed their repairs, these same areas were hit again. During the Commando Hunt operation, the Air Force estimated that only 18 percent of the materiel entering Laos from North Vietnam actually arrived in South Vietnam, with 47 percent of it probably destroyed, 29 percent consumed, and six percent stored. [64]

The Air Force attributed the apparent success of Commando Hunt to several factors. First, the strikes were not arbitrarily limited in time and were expected to continue into June 1969, when the weather would make movement very difficult for the Communists. Second, Igloo White sensors had helped to locate interdiction points and areas. (The Air Force made a distinction between interdiction points and areas, using different tactics and munitions for each.) Also, the use of area denial munitions plus the effectiveness of the integrated command and control network under Task Force Alpha seemed to have made a difference. In the Air Force's view. Commando Hunt was perhaps the most effective American interdiction effort of the war excluding Khe Sanh. in which elements of interdiction. neutralization and even saturation bombing (especially by B-52's) combined to decimate the enemy and frustrate his objectives. [65]

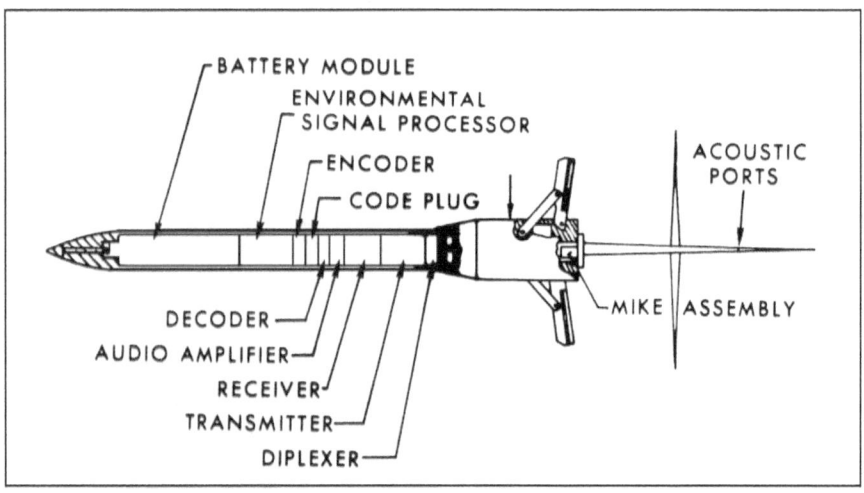

Cutaway Drawing of an ACOUSID III. This sensor could transmit sound from a built-in microphone. Over 20,000 sensors were dropped in Laos, and 80 percent of the sensors were operational after dropping. Source: U.S. Air Force

In 1967. when Secretary McNamara directed that a barrier system be constructed just below the DMZ and west to the mountain trails of Laos, the Air Force simultaneously began to deploy a complementary air surveillance system called Muscle Shoals. Redesignated Igloo White, the system became operational in December 1967. Its purpose was to gather intelligence on the enemy's personnel and vehicular movements through the use of a variety of sensors dropped over infiltration routes to provide 24-hour all-weather coverage. [66] Despite some technical problems. when initial operations began. it became clear that the Igloo White equipment was helpful in detecting enemy movements. The system proved

Towards A Bombing Halt 1968

sufficiently successful for General Abrams' staff to promulgate a plan (called Duck Blind and later Duffel Bag) designed to use sensors solely in South Vietnam to locate Communist base areas, truck parks. and possible ambushes as well as landing zone surveillance. [67]

On the basis of an Air Staff study. General McConnell recommended to Secretary Brown on 6 February 1968 that management of the Igloo White network be transferred from the Defense Communications Planning Group (DCPG), which had been responsible for setting it up, to the Air Force. He delineated two plans, the first calling for a phased transfer of responsibility from the DCPG to the Air Force, a process which would consume about five months after which time the DCPG would be disestablished. Subsequently. the Secretary of the Air Force would declare Igloo White a designated system and establish a system program office (SPO) to take over development responsibility. His second proposal envisioned Igloo White's immediate transfer as a so-called "designated system" with the SPO director being a member of the designated systems management group (DSMG) for the surveillance network. Secretary Brown agreed to support the latter option at the right time. [68]

In the meantime. an evaluation committee headed by Adm. James S. Russell (Ret.) concluded that. although Igloo White had not stopped infiltration (the Air Force emphasized that Igloo White was not an anti-infiltration system. but rather a surveillance system,) it showed "great promise for new and exciting military capabilities." Perhaps the major impetus for going ahead with Duffel Bag in-country development was the outstanding success achieved at Khe Sanh with battlefield sensor surveillance. Overall. the Russell committee felt that former Secretary of Defense McNamara had made a mistake when he placed "an untried infiltration-interdiction system in first national priority." It recommended the formation of a high-level committee reporting directly to the defense chief to study possible weaknesses in the military structure that led to the establishment of the DCPG. The highest national priority. declared the group. should be placed on development, production, and procurement of air munitions for interdiction. In general, the Russell committee concluded that the development of air munitions and delivery systems had been "inadequate. " Also, development and production of an all-weather. day-night aircraft should be accelerated. Finally. The committee proposed that the military coordinate plans to develop and deploy sensors so that the services might eventually take over DCPG responsibilities. [69]

On 7 November 1968 the Joint Chiefs endorsed the recommendation that the services make plans for a coordinated development of sensors and they agreed that greater stress should be placed on producing interdiction munitions and developing an effective tactical all-weather aircraft. The JCS also backed the Air Force's Commando Hunt interdiction plan. [70]

Acting on the Russell report and the recommendations of the JCS. Dr. Foster directed the Defense Communications Planning Group to transfer all procurement. systems engineering. and "operational interfaces" of the Igloo White system to the Air Force no later than July 1970. Although the Igloo White technology was "still in its infancy." said Dr. Foster. "I believe it is of national importance to continue

Air War – Vietnam

these developments with the same sense of urgency and dedication exercised by the DCPG over the last two years." [71]

Project Shed Light

Another project designed to provide the Air Force with the capability to find and" destroy the enemy and his supplies at night was Project Shed Light. established in March 1966. Although several Shed Light development projects were designed specifically to facilitate night operations and new or improved aircraft were required for the nighttime role. the fact remained that by 1968 the Air Force still had not developed a wholly satisfactory system. This was especially true for aircraft which could operate against the enemy over his own territory at night, and survive. [72]

Photographs of the Tropic Moon aircraft are very rare. This poor-quality illustration of a B-57G Tropic Moon III is one of the few in the public domain. Source: U.S. Air Force

Four major USAF systems, designated Gunship II, Black Spot, Tropic Moon I, and II, were deployed to Southeast Asia following their development under Project Shed Light. The first system, an AC-130A, was a self-contained. all-weather, night attack aircraft equipped with special sensors, four 7.62-mm "mini-guns" and four 20-mm gatling guns. In September 1967 a prototype model was deployed to Southeast Asia for evaluation and spent more than 10 months in combat before returning to the United States in November 1968. Between February 1968 and its last combat mission on 12 November. This Gunship II AC-130A [had flown] 151 missions [out of] 246 sorties with an average flying time per month of 111 hours. Bomb damage assessment was trucks sighted (1,000), trucks destroyed (228), trucks damaged (133), boats sighted (32), boats destroyed (9) and boats damaged (8). [73]

Based on the prototype's success, especially in support operations over areas where the enemy possessed only light anti-aircraft weapons, Secretary Brown approved the procurement of an additional eight AC-130s, 26 AC-119Gs, and 26 AC-119Ks. Four of the eight follow-on AC-130's were deployed to Ubon AB, Thailand, in December 1968 to fly interdiction missions over Laos. [74]

In August 1968 the Air Force deployed two night attack Black Spot C-123Ks initially to Osan AB, Korea, where they began surveillance operations in support of the Republic of Korea Navy against North Korean efforts to infiltrate South Korea by sea. The C-123s were equipped with forward-looking infrared radar moving target indicator (MTI), low light level television (LLLTV), and a laser ranger. In 28 sorties, the crews discovered they could detect water traffic with Black Spot equipment but were unable to identify which of the hundreds of vessels spotted were North Korean. In mid-November the two aircraft were sent to Phan Rang AB, South Vietnam, and on 1 February 1969 were redeployed to Ubon AB, Thailand, from where they operated against enemy lines of communications (LOC), logistic strong points, and trucks in the IV Corps and southern Laos. [75]

Black Spot operations in Vietnam and Thailand between 15 November 4 1968 and 13 March 1969			
	Attacked	Damaged	Destroyed
Trucks	727	156	255
Boats	103	24	55
Miscellaneous (Docks, buildings, camps).	138	78	27

The Air Force planned to return the C-123's to the United States in May 1969 for refurbishing and then redeploy them again to Southeast Asia as part of the permanent force. [76]

The Tropic Moon I program featured the development, testing. and deployment, in December 1967. of pod-mounted LLLTV night attack equipment on four A-lE aircraft. Based at Nakhon Phanom AB, Thailand, they began operations in the Steel Tiger area of Laos on 8 February 1968. In May, with the start of the rainy season in Laos, the Tropic Moon I planes moved to Bien Hoa AB, South Vietnam, for operations in the III and IV Corps. On 1 December 1968, the program was terminated, the LLLTV systems were removed and returned to the United States, and the A-1Es reverted to normal configuration, remaining in South Vietnam. [77]

Three Tropic Moon II B-57's deployed to Southeast Asia in December 1967. Based at Phan Rang AB. they started operations in the Steel Tiger area on 6 February 1968. During a 90-day combat evaluation that ended in May, these aircraft flew 116 sorties, detected 536 trucks, destroyed 31. and probably destroyed 43. They redeployed to the United States in July. Both Tropic Moon I and II programs proved disappointing. their effectiveness considered "marginal. " The major reason given for the failure of Tropic Moon II was that "the speed of the B-57 allowed insufficient time to identify targets." Also, the navigation equipment in the B-57 proved inadequate for the Tropic Moon II mission. [78]

In September 1967, two months before the operational deployment of the Tropic Moon I and II aircraft, a Shed Light General Officers conference had concluded that the B-57 was "the logical choice" for the Tropic Moon III mission of operating against small targets with a multi- sensor aircraft. On 28 November 1967. the Air Staff Board authorized the modification of 16 B-57's for the self-contained night attack role and OSD approved the program on 24 February 1968. [79]

The Tropic Moon III B-57's were to be equipped with low light level TV. forward-looking infrared radar with moving target indicator, and an advanced system for target detection. tracking and weapons delivery. In addition to radar homing and warning equipment and ECM. these aircraft were to have special ceramic armor to protect the crew and explosion-proof internal self-sealing fuel cells. [80]

Air Force officials visualized the modified B-57's as being able to perform the night attack role creditably after the war in Southeast Asia was over. Initial planning called for the development of two prototypes. Tropic Moon III contracts were let in late 1968 and training for crews and technicians began. The Air Force estimated that the 16 B-57's would be operational late in 1969. [81]

During their appearance before a House committee in February 1968, Secretary Brown and General McConnell emphasized the positive aspects of the Air Force's R&D programs for Southeast Asia. The Chief of Staff pointed out, for example, that over an 18-month period the Air Force had introduced into the operational inventory about 15 new air- deliverable weapons or major improvements in existing weapons. In this connection. Dr. Brown submitted to the committee a lengthy list of contributions of Air Force research to the Vietnam war. The items ranged from equipping USAF reconnaissance units with completely self-contained mobile photographic processing and interpretation facilities to ceramic armor kits for C-130 aircraft.

When a somewhat skeptical Congressman asked whether he wasn't being "overly-optimistic in what we expect of the developments and devices which become available each year." Secretary Brown admitted that such items "never perform in the field as they do on the test range." But, he argued. they always "perform better than last year's system." Further, he noted that the enemy also was developing systems, both in conventional war and in strategic war. "so we have to keep working on these things in order to stay ahead of the game." [82]

VI. THE PARTIAL BOMBING HALT & REASSESSMENT OF RESOURCES

As studies of alternate Air Force strategies. troop deployments, and administration policies were discussed and reviewed, Secretary Clifford concluded during his first month in office, that current American strategy in Southeast Asia was no longer justified and that some of the proposed alternatives were unlikely to attain U.S. objectives. Heavier bombing in North Vietnam and Laos could inflict heavier losses, but it would not stop the war. Dispatching the 206,000 more U.S. troops to South Vietnam desired by General Westmoreland also appeared untenable as it would require 280,000 more reservists. higher draft calls, and longer duty tours for most men in the services. An augmentation of that size, moreover, would cost $2 billion more in fiscal year 1968 and $10 to $12 billion in fiscal year 1969, would invite. domestic financial controls, would aggravate the balance of payments deficit by $500 million Further. there was no assurance that 206,000 more men would suffice. The enemy, who showed no diminution in his will to fight probably would respond to the American buildup and it was uncertain when South Vietnamese forces could "take over" the war.

As casualties mounted, the future became harder to predict.
Source: National Archives.

The problems facing the new Defense Secretary were apparent when he asked for "a military plan" for victory in the "historic American sense" and was told there was none. The lack of such a plan was attributed to three major political restrictions on waging the war: there could be no invasion of North Vietnam since it might trigger Hanoi's mutual assistance pact with Peking; there could be no mining of Haiphong harbor lest a Soviet ship be sunk; and there could be no pursuit of the enemy into Laos or Cambodia. These, and other constraints, he was told, precluded an all-out military effort. Since the Secretary and other high civilian officials had no intention of recommending their cancellation to the president, Mr. Clifford became "convinced that the military course we were pursuing was not only endless but hopeless," and that the primary U.S. goal "should be to level off our involvement and to work toward gradual disengagement." [1]

The President Decides to Halt the Bombing

After the President heard the views of Mr. Clifford. in the "closing hours" of March, he decided on a new course of action in an effort to end the war. As an initial step to entice Hanoi to the bargaining table, on the 31st he announced to the nation a partial halt to the bombing of North Vietnam. He said in part:

> Tonight I have ordered our aircraft and our naval vessels to. make no attacks on North Vietnam except in the area north of the demilitarized zone where the continuing enemy buildup directly threatens allied forward positions and where the movements of their troop and suppliers are clearly related to that threat.
>
> The area in which we are stopping our attacks includes almost 90 percent of North Vietnam's population. and most of its territory. Thus there will be no attacks around the principal populated areas. or in the food-producing areas of North Vietnam. [2]

The President limited the bombing of North to the area below the 20th parallel. On April 3, he changed the boundary to the 19th parallel. Urging Hanoi to join him in "a series of mutual moves toward peace," he asked for a prompt initiation of talks between the two sides and cautioned the Communists not to "Take advantage of our restraint." He renewed a pledge made in Manila with five allied nations on 24 October 1966 to withdraw all allied units (within six months) if North Vietnam disengaged from the war and the violence subsided.

Mining Haiphong Harbor was rejected on grounds a Soviet ship might be sunk. When the harbor was closed by mining in 1972, it proved an effective way of bringing Hanoi to the negotiating table. Source: U.S. Air Force.

Towards A Bombing Halt 1968

In addition to his new peace overture, the President made other major decisions. He limited U.S. military strength in South Vietnam to 549,500 troops and announced plans to accelerate the training and equipping of South Vietnamese forces so they could assume more combat responsibility and maintain military pressure on the enemy.[3]

ROUTE PACKAGES AND BOMBING LINES IN NORTH VIETNAM 1968

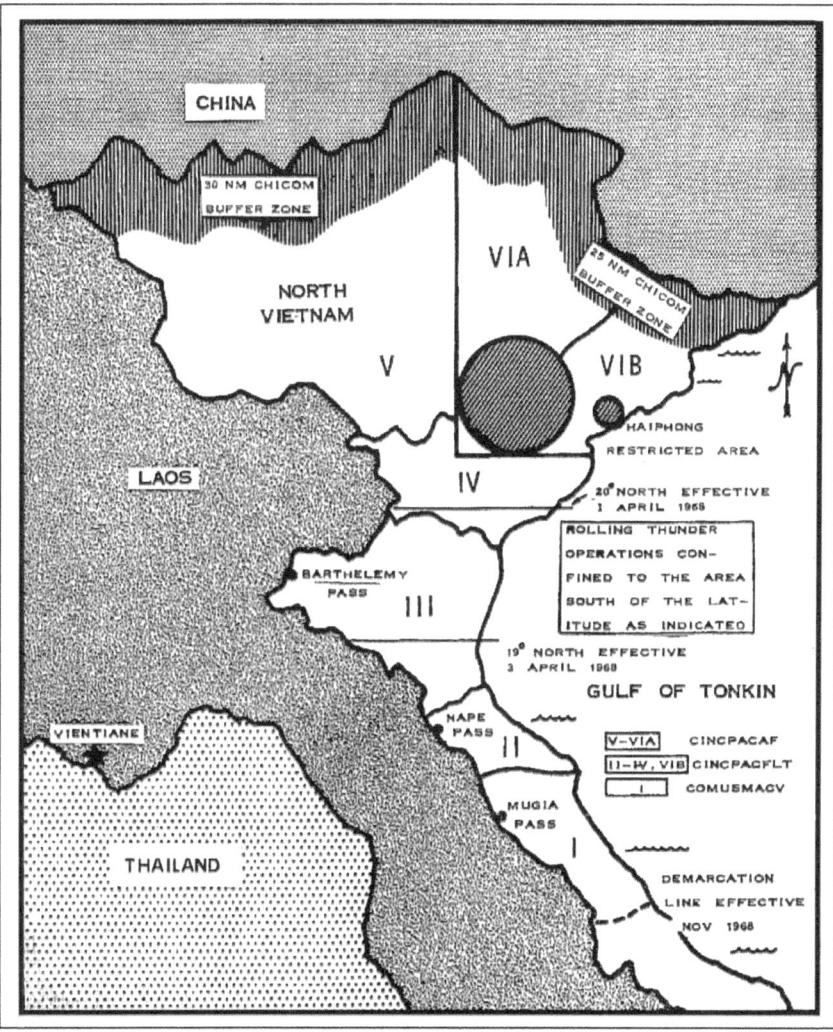

Service Views on the Bombing Halt

The service chiefs would, of course, support the President's decision, although they had opposed a bombing halt. General Wheeler had argued strongly against such a step on the grounds that it would prove costly to the allies, would prolong hostilities. and could be interpreted by the Communists as an "aerial Dien Bien

Air War – Vietnam

Phu." The Air Staff and the JCS had agreed that a bombing halt promised Hanoi's leaders many advantages, as they would interpret it as a weakening of American resolve. and be encouraged to redouble their war efforts. It would thus preclude a favorable outcome for the allies. General McConnell, in August 1967, had told a Senate subcommittee that limiting the bombing to below the 20th parallel would, after a short period. "certainly be disastrous." Throughout 1967 the JCS had expressed opinions that air and naval bombardment should be stepped up. not halted; that areas should be expanded, not narrowed. Consequently, in the eyes of the services, it appeared that the President sacrificed a tremendous military advantage as an enticement for peace. [4]

Phuc Yen Airfield. The speed with which the North Vietnamese exploited the partial bombing halt was entirely predictable.
Source: U.S Air Force

Two weeks after the 31 March decision, PACOM completed a study on the effect of the bombing halt. The study confirmed that Hanoi had gained a military advantage and that in light of apparent American and free world weakness. Communist intransigence in negotiations could be expected. Further restrictions on bombing below the 18th parallel or the DMZ would reduce the number of available targets in the North and expose allied positions in Vietnam to a greater danger from MiGs, artillery fire, and rockets. If the bomb line. extended to the provisional military demarcation line (PMDL) i.e. to all of North Vietnam, the enemy's position would be "militarily unacceptable."[5]

Nevertheless. a reorientation in bombing strategy was producing more salutary results. The administration had authorized unescorted photo and visual reconnaissance sorties to fly above the 19th parallel. Below it, however, Air Force, Marine, and Navy sorties nearly doubled over those flown during previous months, increasing truck "kills" nearly fourfold by May. Aircraft losses over the North decreased. Following a visit to South Vietnam, Secretary Brown reported that the "substantial increase in (bombing) effectiveness" in the North's panhandle below the 19th parallel was contrary to what he had anticipated. [6]

Further Debate on Reserve Callups

The Presidential decision to restrict the bombing did not end OSD-JCS debates on additional national guard and reserve callups. If anything, the arguments grew more contentious. On 2 April General Wheeler sought Secretary Clifford's approval to recall more reserves to support recently deployed forces and rebuild partially the strategic reserve. His request was for 56,877 assigned and 63,385 authorized reserve personnel including 6,435 and 7,685 Air Force personnel, respectively. Six ANG fighter groups (three F-84s, two F-86s, and one F-100) would enter active service. a part of which would be used to step up training of more FAC and air liaison officers (ALOs). As in the 15 March proposal, all of the

ANG and reserve units would be recalled in three increments by 30 June 1968. Additional costs were estimated at $180.1million in fiscal year 1968 and $530.4 million in fiscal year 1969. [7]

Concerned about the government's financial situation, Secretary Clifford advised the JCS Chairman that he was considering a less costly alternative. On 4 April he proposed a total callup of only 22,767 personnel (2,048 Air Force, 1,027 Navy, and 19,692 Army). Omitting the recall of six tactical fighter groups, he asked the Air Force to find a less expensive way of training FACs and ALOs, perhaps by substituting A-37, AT-33, F-5, or F-100 aircraft for the more sophisticated types. Mr. Clifford also indicated that he wished to limit the U.S. strategic reserve to six and two-thirds divisions and cancel the rotation part of the Pentagon's "Reforger" plan committing certain units exclusively to the North Atlantic Treaty Organization (NATO). This would make more units available, if necessary, outside of NATO. [8]

The A-37 offered a less expensive way of training FACs and ALOs
Source: U.S. Air Force

Plans to return the F-84 to the active inventory were discarded. Arguably, these aircraft may have served the VNAF at least as well as more expensive types.
Source: U.S. Air Force

With strong Air staff support. the JCS vigorously objected to calling up only 22,767 ANG and reserve personnel and to any change in the use of Reforger units. They agreed, however, to drop their request for three F-84 groups and the F-100 group. [9] Accordingly, on 6 April the JCS again asked for authority to call up the first increment of the total force proposed on 2 April and said they would review the requirement every 30 days. Four days later, in a new statement on America's worldwide military posture. They warned that reserve forces were inadequate against Communist threats facing NATO forces; the Asian countries of Laos, Cambodia and South Korea; and Latin America. [10]

Meanwhile, during the debate on the reserve callup, Secretary Brown sent the Defense Secretary a revised Air Force plan to ensure sufficient FAC-ALO training and to meet other pressing air unit needs. The two objectives, he reported, could be met by modifying or substituting other aircraft for those initially desired. and by recalling 3,489 ANG and AFRES personnel to man additional units. [11]

On 11 April, Secretary Clifford made his decision. He overruled JCS recommendations and announced a national guard and reserve callup of only 24,550 men. but he accepted Secretary Brown's revised proposals for the Air

Air War – Vietnam

Force. The service manpower allocations were as follows: Air Force, 3,488 (2,201 ANG and 1,287 AFRES); Navy, 1,028; and Army, 20,034. About 10,000 men would go to South Vietnam and the remaining 14,500 would be used to strengthen the strategic reserves. The Secretary confirmed that the President's decision raised the American troop ceiling for South Vietnam from 525,000 to 549,500 and reiterated the administration's policy to transfer gradually to the South Vietnamese the major responsibility for the war effort. In achieving this goal. President Thieu planned to add about 135,000 more men to his armed forces.

Southeast Asia Deployment Program 6

To reflect the change in the U.S. manpower ceiling for South Vietnam and other force structure adjustments in Southeast Asia, OSD on 4 April replaced Deployment Program 5 (issued 5 October 1967) with Southeast Asia Deployment Program 6. It called for deploying to South Vietnam four ANG F-100 squadrons and one Marine Squadron in May and June; deferring the deployment of one USAF F-4 squadron to Thailand (from February to June); deploying *(Note from Defense Lion Publications; the original text reads 'deploying' but this does not make sense in context. The intended word probably was 'delaying')* redeployment from Thailand to the United States of one USAF A-1 squadron and a Navy SP-2E unit; and extending the B-52 sortie rate of 1,800 per month from 15 February through June 1968 then dropping it to 1,400 sorties per month.

Above all, Vietnam was an infantry war. Source: National Archives.

Deployment Program 6 also called for an increase in South Vietnamese Army maneuver, artillery, and engineer battalions. New Army brigades would replace the 82d Airborne Division brigade and the Marine RLT 27. hurriedly sent to Vietnam in February. It also contained a new schedule, effective in September 1968, for converting 12,545 military to civilian spaces in South Vietnam, 600 of them Air Force, to preclude any overrun of the new 549,500 U.S. manpower ceiling. [13]

Deputy Defense Secretary Nitze questioned the necessity for five more squadrons in South Vietnam in view of the President's .decision to decrease the bombing of the North. He also asked the JCS for a plan to reduce the number of temporary duty units in South Korea, and wondered if the present force of 151 USAF tactical aircraft in that country would be needed through 1968. [14]

The JCS, strongly backed by the Pacific Air Forces PACOM commanders and the Air Staff, opposed any delay in the dispatch of the five squadrons to Vietnam. Foreseeing continued North Vietnamese infiltration and insurgency activity in South Vietnam, Laos, and Thailand, they emphasized the need for adequate air support for additional ground units. They also wanted no change in U.S. air strength in South Korea until the *Pueblo* crisis ended. [15]

Mr. Nitze subsequently withdrew his objections to the of the five fighter squadrons to South Vietnam, although he continued to question the requirement in view of plans to use more Air Force and Army gunships, and the availability in July of the Thai-based A-1 squadron. He indicated he might reopen the issue at a later date. [16]

Further Review of the B-52 Sortie Rate

The new restrictions on bombing North Vietnam and other considerations also had prompted Mr. Nitze on 14 April to ask for a review of the B-52 sortie rate (providing 1,800 sorties per month through June, and 1,400 per month thereafter). The cost of B-52 operations was high and the evidence of their effectiveness fragmentary. Moreover, the Army and the State Department desired to withdraw the bombers from Okinawa because of the forthcoming Ryukyuan elections on the island. The Japanese, who retained "residual sovereignty" over the island and the Ryukyuans, were fearful lest the flights of B-52s from Okinawa involve them in the war in Southeast Asia. The continued presence of the bombers could result in a decline in the Ryikyuan political party, with which the American High Commissioner (an Army General) could deal most easily. In view of these developments he asked for a report by 31 May. [17]

The aftermath of a B-52 strike. One of the problems in assessing the results of B-52 operations was that many of the benefits were intangible. Source: U.S. Air Force

In an initial reply to Mr. Nitze on 23 April, they urged the continuance of the 1,800 per month rate after June and pointed to the results achieved at Khe Sanh, where the enemy suffered a major defeat losing about one-half of his committed forces. Evidence of personnel, ammunition, equipment, and fortification losses found in prisoner of war reports attested to the importance of B-52 bombing. The JCS stated the 1,800 sortie rate was needed to support friendly ground operations against the enemy in the A Shau Valley and around Hue, to hit truck parks. troop concentrations and supply centers built up. since the partial bombing halt, and to meet another possible enemy offensive in June or July if peace negotiations were unsuccessful.[18]

Separately, General McConnell and Secretary Brown asked Mr. Nitze to delay the scheduled phaseout from the

SAC inventory in early fiscal year 1969 of four B-52 squadrons. The current high B-52 sortie rate, they said, wore out the bombers faster and speeded up their modification schedule. In addition, more B-52s were needed to handle other non-nuclear contingencies in Southeast Asia or elsewhere. Mr. Nitze disagreed. He said that the lack of funds necessitated the inactivation of the four B-52 squadrons and he foresaw no emergency requiring more B-52 non-nuclear bombers. If an emergency should arise present "surge" strength would suffice until stored B-52s could become operational.[19]

A second JCS report on the B-52 sortie rate requested by submitted late in May. It concluded that the bombers were effective and accomplished their task with the highest degree of accuracy and reliability. although bomb damage assessment was limited by bad weather, jungle canopy, terrain inaccessibility, and by insufficient follow-up by ground troops in the bombed areas. The 1,800 per month sortie rate should be maintained to assure striking all lucrative targets, and the bombers should not be removed from Okinawa for political reasons.

While bombing effectiveness could not be measured statistically, the report continued, there was evidence that the B-52 strikes forced the enemy to disperse, inhibited his speed of maneuver, compounded his command and control problems and shattered his morale. Furthermore. the bombers constituted a "dynamic reserve force" that could be used without incurring troop casualties.[20]

In this report to OSD. General Wheeler conveyed his and General Abram's concern about the administration's tendency to economize on the sortie rate, especially in light of the enemy's determination to achieve a major victory. Should funding cutbacks also hamper efforts to make the South Vietnamese forces more self-sufficient responsible in combat. American involvement in the war would be prolonged.[21]

The JCS arguments were successful, at least temporarily. On 22 June, Mr. Nitze approved "for planning purposes" continuation of the 1,800 sorties per month rate through December 1968, but said he would review his decision in 60 days and periodically thereafter. But in view of the State Department's belief that the Korean situation no longer warranted .bombers on Okinawa, he asked the JCS to determine if they could be added to the forces already in Thailand and Guam. Subject to the consent of the Thai government, he agreed to increase the number of B-52s at U-Tapao from 25 to 35 plus support personnel.[22]

Efforts to maintain the B-52 sortie rate were successful. Source: U.S. Air Force

The service chiefs replied that an 1,800 sortie rate could be sustained if 35 regular and four rotational bombers were stationed ay U-Tapao, 70 on Guam, and if six

Towards A Bombing Halt 1968

KC-135 tankers were moved from U-Tapao to Taiwan (for a total. of 21 tankers there). However,. they strongly urged retention the B-52s on Okinawa citing additional costs of the alternate plans and emphasizing "overriding military considerations." With the strong backing of PACOM, MACV and SAC commanders, the JCS persuaded OSD not to remove the B-52s from Okinawa, at least the remainder of 1968, although discussions of lowering the sortie continued. [23]

Faster Buildup of South Vietnamese Forces

The President's decision to speed up the training and South Vietnamese armed forces imposed additional work on the Air Staff and Joint Staff. Because peace talks in Paris could result in limiting the size of belligerent forces in Southeast Asia, General Wheeler on 10 April requested the Joint Staff to prepare a paper for the Defense Secretary soliciting his support for the largest RVN forces possible. Also, Deputy Defense Secretary Nitze asked for a plan which would assure South Vietnam's swift self-sufficiency in tactical air, logistics, and artillery. [24]

To fulfill General Wheeler's request, the JCS on 15 April proposed accelerating the buildup of Saigon1s forces to 801,215 men by the end of fiscal year 1969. The Air Force-supported VNAF would receive 5,124 more spaces for a total of 21,572. and there would be a step up in the distribution of M-16 and M-2 weapons to the regular and paramilitary units. respectively. [25]

In response to Mr. Nitze's request, the JCS on 23 May submitted a plan recommending faster delivery of arms and equipment to the Republic of Vietnam, partly by diverting military items from American units. This program would continue through fiscal 1973, although the total strength of RVN forces would remain at 801,215. VNAF units would expand to 11 tactical fighter, four gunship and 17 helicopter squadrons. Its personnel strength would increase to 36,855 with offsetting reductions in regular personnel. The cost of a larger VNAF for fiscal year 1968 through 1973 would total slightly more than $1 billion.

The service chiefs warned that the program to build up forces would encounter major obstacles. The South Vietnamese lacked trained manpower, particularly for the VNAF and the Vietnamese Navy (VNN); and there would be difficulties in diverting equipment including UH-1 Army helicopters to the VNAF and in obtaining supplemental appropriations from Congress. [26]

Modern aircraft are one thing; the trained personnel to support them quite another. Source: U.S. Air Force.

On 24 May Mr. Nitze approved the JCS plan of 15 April, but temporarily deferred its funding until an "action plan" and additional data on personnel "pipeline" increases were submitted. Then, on 25 June he approved the JCS plan of 23 May to accelerate the RVN military buildup, but only that portion of it pertaining to fiscal year 1969. This would give the VNAF two improved UH-1 helicopter in exchange for two older H-34 squadrons, more 105-mm 155-mm artillery battalions for the Army and Marines, and somewhat larger Regional and Popular Forces. [27]

	JCS PLAN FOR VNAF (23 May 1968)		
	Authorized FY68 Forces	Interim Force Structure End FY69	Expanded and improved Force Structure end FY73
Tactical Fighter Sqs	6	6	11
Helicopter Sqs	5	5	17
Liaison Sqs	4	4	7
Transport Sqs	3	2	4
Reconnaissance Sqs	1	1	1
Gunship Sqs	0	1	4
Training Sqs	1	1	1
HAWK Bns	0	0	2
AAW Batry	0	0	1
Air Bases	6	6	7
Source: JCSM-324-68 (S), 23 May 68			

Simultaneously, Mr. Nitze asked for a two-phase plan for expansion of South Vietnam's forces. Under "Phase I," Saigon's ground combat capability would be maximized and American participation in the war continued at the present level. "Phase II" would assume Saigon was self-sufficient to deal with any insurgency after U.S. and North Vietnamese troops withdrew from South Vietnam. The Deputy Defense Secretary asked for a preliminary report on Phase I by 15 August and a final report by 15 September. For Phase II he desired only a final report by 1 November. These dates subsequently were changed to 30 August, 30 September and 15 November respectively." [28]

Not all efforts to build up the South Vietnamese forces were physical; some were psychological. They desperately needed a boost in morale. To achieve this and encourage self-improvement for their Vietnamese ally, the Air Force and other services participated in Operation Limelight, a public affairs program designed to

Towards A Bombing Halt 1968

lift the RVNAFs esprit de corps of the troops and give more recognition to their performance and progress. The State and Defense Departments, PACOM. and MACV also contributed to this program. [29]

Air Staff / JCS Views on Negotiations

Following the President's 31 March address, the Air Staff shared in the preparation of a number of Joint Staff papers which incorporated the services' views on the impending negotiations. These had been solicited by General Wheeler and OSD. One paper called for by the Special Interdepartmental Group (SIG) of the 1954 and 1962 agreements on Vietnam and Laos to determine what provisions might be detrimental to American interests. A second contained data for negotiations (e.g. defining the meaning of. "preliminary talks," "de-escalation," and "cease-fire"), which Gen. Andrew J. Goodpaster would use in his role as Senior Military Representative to the U.S. negotiating team. A third. for Secretary Clifford, expounded a concept of negotiations. A fourth paper proposed a two-phase operational for redeploying certain forces in the event all bombing of North Vietnam ended, or for preparing to resume attacks quickly if necessary. [30]

While Washington and Hanoi sparred over a suitable place to begin peace talks (finally agreeing to hold them in Paris beginning 10 May) the Air Staff became concerned over the prevailing attitude in Washington which assumed that the negotiations would begin shortly and would be productive. Its apprehensions centered on the military drawbacks facing MACV. If negotiations proceeded swiftly, most of the reinforcements desired by General Westmoreland would not be sent. And, with the bombing of the North cut back to the 19th parallel. Hanoi clearly was "taking advantage" of the situation by increasing its infiltration to the South and by strengthening its air defenses. [31]

On 8 May the JCS sent two more papers to the Defense Secretary, both reflecting Air Staff views. The first addressed the negotiations for a complete bombing halt which the enemy insisted upon. The Air Staff believed that U.S. spokesmen in Paris should appreciate fully the impact of halting all attacks on North Vietnam. Though it would lessen domestic criticism of U.S. government policy, it would allow Hanoi to infiltrate more men and supplies, increase allied casualties and vitiate the effects of three years of bombing. The service chiefs concluded that:

> No combination of concessions which the North Vietnamese and National Liberation Front are likely to make unilaterally would afford the allied forces advantages commensurate with those afforded North Vietnam by cessation of bombardment. Maximum pressure should be applied at the negotiating table, therefore, in seeking to redress this initial disadvantage. Only if negotiations led to a cessation of hostilities in South Vietnam under conditions consistent with allied objectives will risks inherent in cessation of bombardment have been justified. [32]

The second JCS paper emphasized the importance of attaining U.S. objectives set forth in NSAM 288, 17 March 1964 calling for an independent, non-Communist South Vietnam. These required the withdrawal of all North Vietnamese troops and subversive elements from South Vietnam, Laos and Cambodia; restoration of effective inspection and verification of such withdrawals of the war in accordance. with the terms of the 1954 Geneva agreements. Prompt repatriation of prisoners of should be an important negotiating objective.

American concessions likely to prevent the United States from attaining its objectives. the JCS continued, would include the establishment of a coalition government with the National Front (NLF), agreement to an "in-place" cease-fire restricting the Saigon regime's freedom of action and representing a de-facto partition of the country, premature withdrawal of U.S. and forces from Southeast Asia, and cessation of air reconnaissance and coastal surveillance of North Vietnam and the DMZ.

The Joint Chiefs pointed to the absence of any Communist de-escalatory steps thus far which would correspond to the partial bombing halt, cited the stepped-up infiltration of men and supplies and warned of the possibility of another offensive against major urban centers. Although the United States was still negotiating from a position of strength, the JCS said they opposed any further reduction of military pressure against the North without substantial achievement of basic U.S. objectives in the war. [33]

On 10 May Deputy Defense Secretary Nitze sent both JCS papers to Secretary of State Dean Rusk for Ambassador W. Averell Harriman in Paris. He thought the JCS views were not inconsistent with those of the Ambassador and with other negotiating instructions. Meanwhile. in reply to a query from Mr. Harriman. General Wheeler sent him another paper, again stressing the importance of maintaining military pressure on the North during the negotiations. [34]

Late in May, the President and Secretary Clifford sought JCS advice on possible U.S. action if the Paris talks ended an unsatisfactory agreement or were abandoned. Addressing the first contingency, the JCS counseled against withdrawing any American forces and recommended continuing the war until the enemy became aware of the "inevitable destruction" of his capability. Military response should include air and naval attacks on the North with fewer restraints than existed on 31 March (when the partial bombing halt began). If the talks, were abandoned, air and naval attacks should resume (as indicated) and additional pressure put on the enemy through a series of air-supported small-scale overt and covert

Saigon Street Scene. By the time this picture was taken, the possibility of an American retreat from Vietnam was already looming. Source: National Archives

operations in Laos, Cambodia and the DMZ to aid military operations in South Vietnam. Other measures and their costs were also discussed. [35]

Not included in the JCS reply was an Air Staff judgment that partial bombing halt was not the "essential element" that brought Hanoi to the conference table. More plausible, it seemed, was Communist reasoning that, after inflicting many casualties on the Americans during the Tet offensive and with good weather making infiltration easier, it was time to talk and improve military positions. The Air Staff also believed that renewed bombing of the North would necessarily provoke Hanoi sufficiently to terminate the Paris talks. [36]

In a supplementary paper. the service chiefs reaffirmed their agreement with basic U.S. guidelines for the war (i.e avoid a wider conflict with the Soviets or China. do not invade North Vietnam or overthrow its government, and restore the principles of the 1954 and 1962 agreements). But they warned that the policy of gradual application of military power, restraints on attacking the North, and allowing protracted negotiations could result only in progressive decline of the allied capability to block attainment of Hanoi's goals in South Vietnam. [37]

In another action.. the Air Staff, with some exceptions, endorsed a JCS paper, prepared on 2 July for Ambassador Harriman, outlining requirements before the United States should consider a total bombing halt of North Vietnam. The service chiefs warned that Hanoi already was .using the partial bombing halt to strengthen its military position and that a renewal of attacks north of the parallel might be necessary. They recognized, however, that "overriding political considerations" might take precedence over JCS objectives. [38]

Meanwhile, on 1 June the JCS sent Mr. Nitze a two-phase plan for redeploying certain forces from Southeast Asia should all attacks on North Vietnam end. and then for resuming them if necessary. Phase I called for retaining, after a complete bombing halt, Air Force Marine, and Navy air units at their present locations, concentrating air operations in South Vietnam and Laos; preparing more aircraft to engage in combat operations (including against ground defenses and MiGs), and placing more aircraft on alert. They also recommended actions to assure the readiness of logistic, base construction, transportation, medical, and communications-electronic units.

Phase II provided four redeployment alternatives, each postulating the withdrawal of certain Air Force or Marine units in South Vietnam or Thailand to Japan, Okinawa, or the Philippines and withholding from combat a portion of or all Navy carrier aircraft. If necessary, these units could redeploy quickly to the war theater to resume operations. The JCS also restated its views concerning the advantages the Communists gained as a result of the bombing halt. [39]

The Air Force did not hide its skepticism of the enemy's intent in the months following the partial bombing halt. However, in view of the administration's determined effort to reduce the tempo of. The war and to achieve a political settlement, the Air Force, together with the other services, had no alternative but to reassess its role.

Air War – Vietnam

JCS REQUIREMENTS FOR A COMPLETE BOMBING HALT IN NORTH VIETNAM.

2 July 1968

1. Negotiating objectives

a. End to all infiltration.

b. Withdrawal of North Vietnamese troops from South Vietnam, Laos, and Cambodia.

c. Restoration of the integrity of the demilitarized zone,

d. Insure control of the Government of South Vietnam over all of South Vietnam.

e. Settlement of the conflict of the basis of the 1954 and 1962 Geneva agreements on Vietnam and Laos, respectively.

f. Provide for effective inspection and verification.

2. Conditions for deescalating the war

a. No U.S. government agreement to accept a small number of unrelated Communist de-escalatory measures to create the appearance of progress.

b. Assured security of allied forces.

c. Retention of essential intelligence operations to assure the means of any military arrangements agreed upon.

d. Right of the Government of South Vietnam to move freely throughout its own country.

e. No limitation on the size of the South Vietnamese armed forces.

3. Conditions for a cease-fire

a. Require operational definitions on terms of a cease-fire with respect to constraints and prerogatives of the parties involved

b. No restrictions on the Government of South Vietnam.

c. Provide for patrolling and reconnaissance activities.

4. Conditions for a withdrawal of forces

a. Establish verification procedures and no reliance on assurances

b. Recognize that the Government of South Vietnam is not. yet strong enough to cope with the present political and military threat

c. North Vietnam should "not take advantage" (as stated in the San Antonio formula of 29 September 1967) of a bombing halt and try to improve its position.

d. Establish the normal infiltration rate at the time of the San Antonio at about 7,000 men per month.

SOURCE: JCSM-415-68 (TS) 2 Jul 68

VII. FURTHER POLICY REVIEW & NEW PLANS

The administration's guidelines for implementing its new policies in Southeast Asia were well established by mid-1968. Despite another Communist offensive in May, the partial bombing halt remained in effect with U.S. air strikes on North Vietnam restricted to targets below the 19th parallel. The Allies sought to increase military pressure on the Communists in South Vietnam and Laos while speeding actions to improve the South Vietnamese armed forces. In Paris, American and North Vietnamese negotiators were debating a variety of issues. including Hanoi's insistence that all bombing of its territory had to stop before a peace agreement could be reached.

Deputy Defense Secretary Paul Nitze
Source: National Archives

Review of the War in Saigon and Honolulu

To examine the impact of new policies on the military situation, Secretary Clifford, General Wheeler and other high officials met in Saigon with General Abrams, (who had succeeded General Westmoreland as MACV on 3 July 1968), Ambassador Bunker and their staffs. They reviewed thoroughly all aspects of the war. [1]

With the bombing limited, Vietnamese infiltration south increased dramatically.
Source: People's Army of Vietnam

MACV briefers declared that the major Communist objectives in the Tet offensive of February had been to generate a popular uprising, overthrow the Thieu regime, force the collapse of the Army and isolate Saigon from the United States. Since Tet, the enemy had tried to undermine the U.S. will to continue the war. But his effort had been costly as he continued to suffer attrition and disastrous military failures. Nevertheless, in the North the Hanoi government remained undefeated and was now rebuilding its economy in the areas where bombing had ceased. Receiving more imports and enjoying the shortest LOCs since the war began. the government's revitalized military posture would allow it to launch

Air War – Vietnam

another offensive. MACV estimated that Communist recruitment in the South would average about 3,500 men per month for the next six months. and infiltration from the North 10,000 per month. These gains, weighed against losses, would permit a buildup to 234,000 men by 1 September, a figure close to pre-Tet strength at the end of 1967.

A summary of air operations revealed that by early July, about 63 percent of allied tactical air operations were flown in Vietnam and 37 percent in Laos and southern North. Vietnam. MACV hoped to increase tactical air sorties soon by about 10 percent. About 75 percent of the B-52 effort was expended in the South Vietnam and 25 percent in Laos and North Vietnam. The B-52 strikes "greatly motivated" the .South Vietnamese troops. [2]

The results of the air operations, Secretary Clifford was informed, were gratifying to MACV commanders. Flexibility in shifting the striking power of aircraft prevented a major Communist offensive in May. When Saigon was threatened in June, tactical and B-52 sorties were directed to the III Corps area around the capital. Since the B-52s could "shift firepower rapidly and would continue to have a major, influence on ground battles," there was no need to plan for movements of U.S. maneuver battalions between corps areas to counter enemy threats. The single manager system for controlling air power (inaugurated in I Corps on 8 March 1968) also contributed to flexibility by making it easier to divert tactical air power to where it was needed most.

Air interdiction of vehicular traffic was focused on Laos where 85 percent of enemy trucks moved at night. During the past year, about 72,000 trucks had been sighted in that country and in route package I of North Vietnam. Despite the current rainy season, truck sightings in Laos averaged 25 to 150 per day. Air attacks in Laos and in route package 1 in the period 1 July 1967 through June 1968 cost the enemy an estimated 34,000 tons of supplies; an amount equal 600,000 rounds of 122-mm rockets. The breakdown was as follows

	Number	Tons destroyed
Trucks Destroyed	8,782	17,564
Trucks Damaged	3,138	1,569
Secondary fires and explosions	59,662	14,916
Total		34,049

In conducting the air strikes on vehicular traffic, the Air Force employed a combination of FACs, gunships, flare ships and B-57s, A-ls and sensors. During an average night in April 1968 for example, the airborne command and control center (ABCCC) directed about 84 attack sorties and hit about 49 trucks. About 10 percent of all trucks sighted were destroyed and 15 percent were damaged.

USAF aircraft operating over North Vietnam, MACV reported, had encountered improved enemy air defenses in late 1967 and early 1968. The heavy antiaircraft

Towards A Bombing Halt 1968

fire had reached the point where FAC and other propeller-driven aircraft could no longer risk flying in many areas, or could do so only at higher altitudes. which reduce the ability of pilots to find and destroy trucks at night. Since 3 April. when air operations were limited to below the 19th parallel the number of aircraft receiving fire had doubled. By the end of 1969 it was expected the enemy would have twice as many antiaircraft weapons as at present. The environment had become too dangerous for UC-130s and only FAC 0-2s could be used for marking targets in southwest corner of the Tally-Ho area (above the DMZ). As an alternative, F-4s were being used on flare and reconnaissance missions and other F-4s and F-105s were flying bombing missions with MSQ-77 radar. [3]

The VNAF, according to MACV briefers, was performing well. It was averaging 85 strike sorties per day and its bombing accuracy was comparable to that of USAF units. The quality of personnel was high. and there were sufficient volunteers. The VNAF possessed 348 aircraft of all types, including five A-1 squadrons one A-5 (*Note from Defense Lion Publications. The original text says A-5 but this is a misprint; the aircraft were F-5s*) tactical fighter squadron which had begun operations in June 1967. Current plans called for converting three A-1 squadrons to A-37 aircraft.

Reconnaissance pictures showed truck convoys streaming supplies south. Eventually, these supply dumps would fuel the North Vietnamese armored onslaughts of 1972 and 1975. Source: National Archives.

In a discussion of current North Vietnamese logistic strength, the MACV briefers portrayed a dark picture of allied problems arising from the President's decision of 1 April which restricted bombing to route package I. Compared with 1967, the monthly rate of enemy imports had more than doubled. Tens of thousands of 55-gallon drums, far in excess of military needs, were being distributed along roadsides leading to the DMZ. They would enable the enemy to undertake mechanized operations in and around the zone. The repair of bridges eliminated the trans-shipment of supplies totaling 226,000 tons, an amount equal to 75,300 truckloads. The partial bombing halt also permitted 100,000 of 300,000 LOC maintenance personnel southward. All road, rail, and waterway lines to Thanh Hoa and transit time for supplies from the China border to Thanh Hoa was reduced drastically. Eight major airfields reopened in the new sanctuary area with one of them, Bai Thong, probably serving as a forward staging base. MiGs now new daily as far south as Vinh, and MiG training had increased fivefold. All this enhanced North Vietnam's capability to engage U.S. aircraft and allied ground forces in the South.

Concerning South Vietnamese military strength, Secretary Clifford was advised that on 30 June 1968 there were officially 717,000 men under arms but as a result of vigorous recruiting the true total was 765,000. By the end of fiscal year 1969 there would be 801,000 on the rolls. Saigon's performance in the first half of 1968, including the Tet offensive, was also better than in a comparable period in 1967 in terms of enemy killed and weapons captured. But, "soft spots" remained. Because OSD was still withholding funds, there would be no new equipment (except M-16 rifles) for 84,000 additional recruits until June 1969. Thus, only about 10,000 of them, principally those joining the VNAF, VNN, or in administrative activities, would be able to perform their primary missions adequately. Desertion in South Vietnamese combat units remained a problem. Records for the first five months of 1968 showed the following rise: January, 6,700; February, 8,400; March, 7,700; April, 11,500; and May, 12,900. In conversations with Mr. Clifford, South Vietnam's Vice President Cao Ky attributed the 30 percent desertion rate largely to inadequate pay. He proposed, as a solution, cutting back on bombing and channeling the savings to the Saigon government for use in increasing troop salaries

The outlook on pacification was termed favorable with "a good chance" that 70 percent of the population would be "relatively secure" by the end of the year. South Vietnam's social, political, and economic problems were also discussed. [4]

After leaving Saigon, the Defense Secretary flew to Honolulu to attend a conference between Presidents Johnson and Thieu. In a joint communiqué issued on 20 July, the two Presidents took note of the unabated military activity of the North Vietnamese, the "greatly stepped up infiltration of men and modern equipment" into the South, signs of a renewed offensive and the "negative position" of Hanoi's negotiators in Paris. President Johnson promised that steps would be taken to improve "the fighting power" of the South Vietnamese and that the United States would not impose a coalition government on Saigon. [5] Later, in Washington, Secretary Clifford warned of a new possible enemy offensive in late July, August, or early September. However, he stated there would be no significant change in allied tactics or strategy, and that "spoiling" operations would continue. [6]

Free from the threat of continued bombing, the North Vietnamese were able to repair their wrecked transport system.
Source: National Archives

Towards A Bombing Halt 1968

In a separate report to the President, General Wheeler expressed confidence that if the enemy renewed an offensive, American and allied forces, plus tactical air, B-52s, and artillery could counter it. He said that General Abrams had neither asked for additional forces, and that information obtained from captured enemy personnel, documents. and ralliers, indicated a lower quality of enemy forces. [7]

Deployment Adjustments in the Remainder of 1968

Because General Abrams said he had sufficient troops, U.S. deployments leveled off by midyear. In fact, U.S. military strength in South Vietnam dropped by nearly 1,500 spaces from 1 July to the end of the year, at which time the total stood at about 536,000 (including 59,024 Air Force), although authorized strength was still 549,500. There were, of course, changes in units and personnel, but all increases were compensated by trade-offs to maintain the lowest level of manpower consistent with military needs. [8]

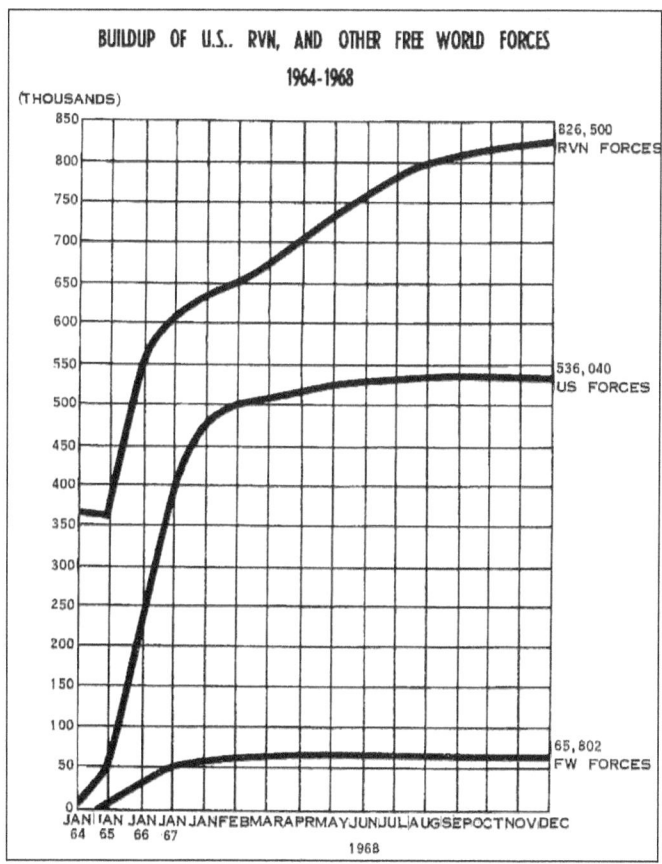

Subsequently, OSD approved two principal changes for the Air Force. The first action, on 27 November, authorized the exchange of new AC-119 gunships for AC-47 gunships (being transferred to the VNAF) or for in-country fighter aircraft. The higher cost of the AC-119s and their tremendous firepower, OSD believed,

Air War – Vietnam

made such a trade-off feasible. [9] The other change, approved on 9 December, allowed one of the three C-130 squadrons, sent to the Pacific to bolster operations during the Tet offensive, to remain in the theater.

The AC-119 Gunships were much more effective than the older AC-47s.

Source: U.S. Air Force

Meanwhile, on 15 August, OSD had approved for Thailand, a new U.S. military personnel ceiling of 47,778, 327 more than on 21 February. Incremental increases in the manpower authorization remained small. largely because of pressure from the American Embassy in Bangkok which feared that an excessive American presence would jeopardize American-Thai working relationships. As a result, the Air Force in previous months had encountered considerable difficulty in deploying more units to Thailand. In the spring of 1968, JCS asked OSD to approve 3,690 more spaces for units to enhance combat operations from that country, but as only 1,594 "offsetting" ones could be found, the manpower ceiling (at that time, 47,451) would have to be increased by 2,096 personnel. [10] OSD however, refused to approve the increase.

Subsequently, in a new effort to find an acceptable formula, General Wheeler on 6 August proposed a "trade-off" of 2,378 badly needed U.S. spaces for 2,162 other spaces that could be saved largely by transferring certain U.S. functions to Thai nationals or to South Vietnam. He also observed that. in contrast with the fears of the American Embassy in Bangkok, American-Thai military relations were satisfactory and that Thai officials displayed no lack of cooperation. [11]

Deputy Defense Secretary Nitze approved most of General Wheeler' proposal on 21 August, but insisted on not going beyond the current U.S. strength figure of

47,778 in Thailand. In accordance with these decisions, 10 more B-52s and 851 personnel were earmarked in September; eight AC-130 Gunship IIs and 414 personnel, and three EC-121 Igloo White aircraft (minus personnel) in October 1968; and 46 propeller aircraft and 916 personnel in February 1969. To offset these augmentations. the Air Force planned to employ 1,474 local nationals in place of U.S. personnel, cancel plans to transfer two DC-130 "Combat Angel" aircraft and 76 supporting personnel to Thailand and take other manpower-saving measures. The American Ambassador subsequently obtained the approval of the Thai government to these changes. [12]

1968 saw the General Dynamics F-111A undergoing its combat evaluations, operating from Thai bases. These showed the aircraft had the potential to be a very effective strike aircraft but also revealed a number of severe problems with it. Source: U.S. Air Force.

In November OSD also approved an Air Force request six more KC-135 tankers (but no personnel) and to withdraw five F-111A Combat Lancer aircraft to the United States. This F-111A unit, consisting initially of six aircraft and 385 personnel was sent to Thailand in mid-March 1968 for combat testing. OSD further approved the deployment to Taiwan of six more KC-135s with 255 personnel to bolster SAC's refueling capability. By the end of 1968, the number of U.S. personnel in Thailand again had risen slightly to 48,301 of which 35,846 were Air Force. [13]

Air War – Vietnam

Additional Planning to Build Up the RVN Forces

The end of the American manpower buildup in Southeast Asia promised to give new impetus to strengthening RVN military forces although General Wheeler foresaw major problems that presumably could retard the effort. Among these were the long lead times required to train crews and technicians for the VNAF and VNN; the adverse impact a larger VNAF and VNN would have on efforts to build up simultaneously the support elements of the South Vietnamese Army; and finally,. the need for the Defense Department to absorb a cut of $3 billion of the $6 billion reduction in federal expenditures ordered by Congress and the administration for fiscal year 1969 [14]

Nevertheless, Secretary Clifford on 7 August informed secretaries, the JCS, and other officials that improving the capability and performance of the armed forces of South Vietnam was "a matter of highest priority." He designated Richard C. Steadman, Deputy Assistant Secretary for East Asia and Pacific in OSD and Adm. William D. Houser of the Strategic Division, JCS as the principal OSD and JCS representatives to accelerate efforts to improve Saigon's forces. The Air Staff's chief representative, appointed by Secretary Brown, was Brig. Gen. Harold V. Larson, Director of Military Assistance in Headquarters USAF. The Air Staff, the Seventh Air Force Advisory Group. the MACV, and PACOM soon pooled their efforts to launch the program. [15]

The redirection of the war effort presented the planners with a number of immediate problems. They quickly determined that using personnel on active duty to train new recruits would degrade the VNAF's combat capability, that neither the Air Force nor the Army possessed adequate facilities to train a large influx of VNAF trainees and. that the construction of a new training center (e. g. on an island offshore from South Vietnam), would be very costly. A basic interservice problem also needed resolution: whether the Air or the Army would have primary responsibility for VNAF helicopter training. Of 3,789 additional VNAF personnel needed for the Phase I training program, 2,336 would be assigned to UH-1 helo units. The Air Staff was also concerned lest the U.S. Army obtain full control of the helicopter training program (and thus seek more funds to expand its training facilities) and make helicopter units organic to Vietnamese Army rather than to the VNAF. [16]

The UH-1 was the iconic helicopter of the Vietnam War and the operation of the type by the ARVN was of more than operational significance. The problem was getting enough Vietnamese ground crews and maintenance personnel.
Source: National Archives.

Towards A Bombing Halt 1968

On 29 August and 2 October, the JCS submitted preliminary and final Phase I plans to assure the increased fighting strength of the South Vietnamese. Both plans assumed continued American participation at the current level. Phase I envisaged 801,000 personnel by the end of fiscal year 1969 and contained only a modest increase of 3,789 in the VNAF which would raise its strength to 20,987. Two H-34 helo squadrons would be converted to UH-1 squadrons (requiring the diversion of 17 helicopters from the Army), and there would be more personnel for the UH-1 wing, aircraft maintenance base supply and civil engineering. Conversion of three A-1 squadrons to A-37s and one C-47 squadron to AC-47s would continue. The South Vietnamese Army force structure was expected to be completed by the third quarter of fiscal year 1970 and the VNAF by the second quarter of fiscal year 1971.[17] Phase I also called for the activation of one UH-1 helo wing and eight UH-1 squadrons (four converted from H-34 squadrons, and four new UH-1 squadrons).

Over a five-year period, the cost of Phase I was placed at around $8.029 billion of which $1.147 billion was allocated for fiscal year 1969. The balance, or $6.881 billion consisted of unprogrammed costs of equipment which could not be absorbed without reducing other Vietnamese force modernization programs.[18]

Supplying the ARVN with UH-1s became a major priority. Source: National Archives.

Deputy Defense Secretary Nitze approved the final Phase I plan on 23 October and asked for a list of requirements for unprogrammed equipment anticipated for fiscal years 1969 and 1970. He conceded that supplying the VNAF would adversely affect the readiness of some units outside of Southeast Asia.[19] Secretary Brown subsequently asked for $13.1 million in fiscal year supplemental funds and $82.4 million in fiscal year 1970 to purchase UH-1 helicopters. Following an Air Staff review, he affirmed the necessity for the army to divert some of its UH-1s to the VNAF and this assure a larger number of helicopters in fiscal year 1970.

To take advantage of Saigon's "mobilization momentum, General Abrams on 4 October urged raising the Phase I RVNAF ceiling from 801,215 to 850,000 with 39,000 of the spaces for Regional Forces. The remainder of the 9,785 spaces, consisting of 1,500 VNAF, 1,700 VNN and 6,585 Army, Republic of Vietnam (ARVN) would be used to allow more on-the-job training and longer leadtime training for Phase II. PACOM endorsed the change as did the Air Staff which observed that 1,500 VNAF spaces were slated primarily for pilots and technicians, some of whom required 22 months of schooling. The JCS concurred on 23

Air War – Vietnam

October. adopted the 850.000 figure as a revision of Phase I, and asked for more funds to support the larger force. [21] Mr. Nitze approved on 1 November. [22]

Post-Hostilities Planning

Closely related to a faster South Vietnamese buildup was U.S. planning for the end of hostilities (T-Day) and the beginning of force withdrawals from South Vietnam (R-Day).

On 25 July 1968 Mr. Nitze asked the service secretaries and the JCS, in cooperation with OSD, to submit troop redeployment proposals to meet each of three alternate U.S. and allied post force structures (designated plans A, B, and C). Plans A and B called for the retention in South Vietnam of 30,000 U.S. and allied troops (a 13,425-man Military Assistance Advisory Group (MAAG) and 16,575 residual personnel) six and 12 months, respectively after all other allied forces were withdrawn from the country. Plan C envisaged 149,030 troops (a 13,425-man MAAG.. and a two-division corps with 135,'605 supporting personnel) 12 months after all allied forces were withdrawn. [23]

In submitting their redeployment proposals the JCS said that the levels provided in plans A and B were inadequate. Air Force and Navy air units, they felt, should be retained in South Vietnam until the VNAF completed its expansion. Communication requirements alone would absorb about 6,500 U.S. personnel, leaving 10,000 spaces for combat and combat support. This would provide room to incorporate other allied units. The substantial manpower in Plan C on the other hand, would leave insufficient troops in the United States, to meet contingencies outside of Southeast Asia if current plans to cut overall American military force levels were carried out. [24]

Much more needed. Source: National Archives.

Under Secretary Hoopes amplified Air Force needs A, B and C but offered an alternative plan D. Submitted to OSD on 2 October as Air Force Operations Plan 12-68, it would stretch out the redeployment of U.S. forces from South Vietnam over an 18-month period, and supporting forces 36 months; enlarge the USAF

Towards A Bombing Halt 1968

posture in the Pacific area to support the VNAF and resume hostilities if necessary; and demonstrate American resolve to help Asian allies. [25]

Mr. Nitze accepted the JCS-prepared redeployment data on a speedy withdrawal from South Vietnam, possibly in accordance with provisions of the Manila Communique of 24 October 1966. He agreed, that a 135,000-man residual force in South Vietnam might be too large, and he saw no need to change current estimates of future U.S. force strength (i.e: the fiscal year 1971 "baseline" force structure in the five-year defense plan). He envisaged returning to a June 1964 post-hostilities defense posture in PACOM. [26]

A group of VNAF pilots on their A-1. The second and fourth pilots from the left were killed in action. Source: National Archives.

However.. the JCS believed that the administration should clarify the meaning of the Manila Communique. The communiqué stated in part: "The people of South Vietnam will ask their allies to remove their forces and evacuate their installations as the military and subversive forces of North Vietnam are withdrawn, infiltration ceases. and the level of violence thus subsides. Those forces will be withdrawn as soon as possible and not later than six months after the above conditions have been fulfilled."

Six months would be insufficient to permit an orderly withdrawal and to dispose of military assets. There was a need, furthermore, to clarify the status of a MAAG and the extent U.S. combat support forces should back an unbalanced South Vietnamese force structure pending its complete modernization. [27]

Post-hostilities planning gained new urgency after the President, on 1 November, ordered a complete halt to the bombing of North Vietnam. On 13 December, the JCS again sent OSD three alternative U.S. force structures to aid the RVN forces after the war's conclusion.

Alternate	MAAG Troops	Support Troops	Other Troops	Total
I	14,313	24,697	none	39,010
II	14,313	24,697	32,303	71,313
III	14,313	24,697	131,519	170,529

Alternate III would comprise a balanced, two-division corps with supporting elements. The Air Force portion for the first force structure would include only headquarters personnel and five advisory teams; for the second, a total of 10,861 personnel; and for the third 25,676 personnel as part of the two-division corps. The second and third force structures would include numerous USAF fighter, reconnaissance, airlift, training, and other units. 28 The JCS, with Air Staff concurrence, also submitted plans to OSD for disposing of the U.S. communication system, much of it Air Force, in South Vietnam. [29]

To facilitate work on post-hostilities arrangements. Deputy Defense Secretary Nitze on 18 December asked the services to maintain quarterly reports of T-Day planning, with emphasis on schedules for U.S. troop redeployments from Southeast Asia and plans for force adjustments on a worldwide basis. [30]

At year's end the Air Staff and other services felt the still needed to clarify the meaning of the Manila Communique of 24 October 1966 regarding troop redeployments from South Vietnam, the status of a MAAG, and what U.S. and allied forces should retain in-country to compensate for RVN military deficiencies combat and technical capability. [31]

VIII. THE COMPLETE BOMBING HALT

Notwithstanding JCS concern about the administration's policy in Southeast Asia, there were indications in early October that the Paris peace talks were leading to a complete bombing halt as a quid pro quo for more fruitful American negotiations with the North Vietnamese. A "break" in the Paris discussions occurred on the 9th and by the 13th Hanoi agreed, in exchange for a halt to all attacks on its territory, to admit the Saigon government to the conference table, and to begin substantive negotiations promptly. It also agreed not to shell indiscriminately the major cities of South Vietnam, nor to violate the DMZ in a manner that jeopardized allied troops. The JCS agreed "under these circumstances" that a bombing halt was acceptable. The understandings were virtually consummated when President Thieu announced that he would not send a delegation to the Paris talks (where representatives of the National Liberation Front also would be present). Nevertheless, the administration decided to proceed without the South Vietnamese. After many weeks of debate on seating arrangements. the South Vietnamese joined the peace talks in early 1969. [1]

The negotiating table in Paris. Source: National Archives

Meanwhile, President Johnson and his military leaders were reviewing the implications of a bombing halt. On 23 October he met with General Momyer, former Seventh Air Force Commander who had become commander of TAC on 1 August, and on the 29th with General Abrams who flew to Washington for the conference. The President was reassured that, under the conditions agreed upon, a complete bombing halt would not endanger American troops. [2]

Air War – Vietnam

Accordingly. President Johnson on 31 October announced an end to the air naval. and artillery bombardment of all North Vietnam and its territorial waters (12-mile limit). He indicated that his decision resulted from an "essential understanding" with Hanoi on de-escalating the war and moving seriously toward peace. He further said.

> The Joint Chiefs of Staff, all military men, have assured me, and General Abrams very firmly asserted to me on Tuesday, that in their military judgment this action should not result in any increase in American casualties. A regular session of the Paris talks is going to take place on next Wednesday, Nov 6, at which the representatives of the Government of South Vietnam are free to participate. [3]

Representatives of the National Liberation Front would also be present, although their attendance "in no way involves recognition." On the basis of the understanding. the President said he expected prompt, productive, serious, and intensive negotiations in Paris in an atmosphere conducive to progress. He pointed to "hopeful events" in South Vietnam, where the government had steadily grown stronger and armed forces had improved. [4]

General Creighton Abrams.
Source: U.S. Army

Secretary Clifford publicly confirmed, the same day, that he had strongly recommended" the bombing halt and that the JCS considered the bombing halt to be "a perfectly acceptable risk." The Saigon government conversely, declared its unhappiness over this "unilateral" U.S. decision. [5]

On 4 November, Secretary Brown commented further on the President's move. He said that, even though bombing had stopped, aerial reconnaissance of North Vietnam would continue and General Abrams would respond to any move threatening American troops. He reported that about 40,000 enemy troops had pulled back from the battle area (in I Corps), thus improving the military situation. The "evolution" of negotiations indicated this was the right time for productive talks with the other side. [6]

Why had the Communists agreed to withdraw certain troops and begin substantive negotiations? Lt. Gen. Lewis D. Walt. Assistant Commandant, Marine Corps credited allied military victories, including those during the February Tet offensive, in which many of the enemy's best troops were wiped out. General Westmoreland, now Army Chief of Staff said that American fighting men had

Towards A Bombing Halt 1968

"raised the cost of aggression to the point where now the enemy apparently wants to negotiate. thus bringing peace one step closer. Brig. Gen. George J. Keegan. Jr., intelligence chief of Seventh Air Force believed that the allied summer air campaign in southern North Vietnam and in Laos from 14 July through 31 October had collapsed the enemy's August- September offensive and forced him to withdraw substantial forces to neighboring sanctuaries. [7]

Enemy Response and Revised Military Operations

Bridge being repaired by North Vietnamese engineers. Source: People's Liberation Army of Vietnam

As expected, the North Vietnamese took immediate advantage of the respite from air and naval attacks to improve their military posture. By mid-November, Air Force and Navy reconnaissance revealed that the movement of trucks down from the 19th to the 17th parallel had increased fourfold. The Communists were repairing roads and bridges, improving airfields, and strengthening anti-aircraft defenses. [8]

Allied forces, under close Washington guidance, adjusted to the new military situation. Outside of I Corps, U.S. troops continued to search out and maintain pressure on the enemy in South Vietnam. Although air operations over the North were limited to reconnaissance, the Americans stepped up tactical and B-52 air strikes in the Barrel Roll and Steel Tiger sectors of Laos. [9]

The withdrawal of about 40,000 enemy troops from the I Corps area enabled General Abrams to announce, on 10 November, that he was transferring his 19,000-man 1st Air Cavalry Division from near the DMZ in I Corps to provinces bordering Cambodia in III Corps where Communist forces threatened to launch a new offensive. Under the code name Liberty Canyon, the troop movement was carried out over a three-week period. principally by transports of the Air Force's 834th Air Division. There was further concern about communist shelling of cities in November and December, in violation of the Paris understandings and about the rapid enemy logistic buildup in the North since the bombing halt. [10]

To forestall a possible enemy thrust, the JCS sought OSD authority for General Abrams to probe into the DMZ to test enemy strength, to conduct a surprise 48-hour air and naval attack against targets in southern North Vietnam up to the 19th parallel or to pursue the enemy into Cambodia with ground forces, tactical aircraft and B-52s. Secretary Clifford requested more information concerning the military effect of such operations in the light of current objectives in Southeast Asia, but took no action on the proposals. [11]

As part of the program designed to protect American and allied forces from surprise attacks, the administration continued over Hanoi's objections, limited reconnaissance of enemy activities in North Vietnam up to the 19th parallel. Following the complete bombing halt, the Air Force flew the first of such missions on 4 November. On the 7th, PACOM approved a MACV plan calling for 20 sorties per day (15 Air Force, five Navy) to observe 66 targets or target areas and this rate continued until the 15th when the JCS limited the flights to 12 per day. Only daytime observation was allowed, weather reconnaissance was charged against the authorized flying rate, drones could not be used and there were other restrictions. [12]

PACOM protested against the low sortie rate, citing the need for at least 90 to 175 sorties per week to fulfill minimum reconnaissance needs. On 20 November, General McConnell sought JCS support in seeking the concurrence of Secretary Clifford for 25 reconnaissance sorties per day, the use of drones, and unlimited weather missions. He pointed to the massive resupply effort under way by the North Vietnamese in the absence of bombing harassment. But in the face of almost certain rejection by the Defense Secretary, service chiefs did not endorse General McConnell's proposal. [13]

Meanwhile, North Vietnamese antiaircraft gunners had gone into action against the reconnaissance flights on 7 November. The Seventh Air Force fighters began to escort the reconnaissance aircraft. On the 13th, the first reconnaissance aircraft was damaged by ground fire, and on the 23d the first RF-4C was downed since the bombing halt. From 4 November through 9 December, 317 reconnaissance missions were flown south of the 19th parallel, of which 96 drew fire. In the same period, four aircraft were lost and four were damaged. [14]

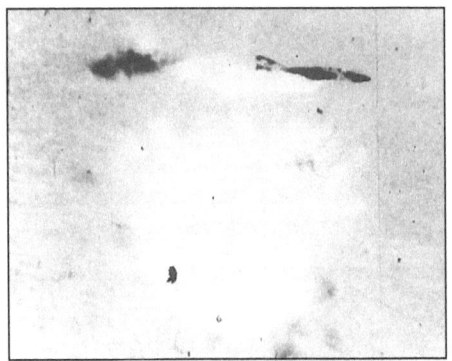

An RF-4C brought down by an SA-2. Source: U.S. Air Force

The stepped-up air action in Laos following the com- took the form of a specially devised air program. Nicknamed Commando Hunt and officially begun on 15 November 1968, it was designed to destroy as many supplies as possible moving South, tie down enemy manpower and further test the effectiveness of the Igloo White sensor system. Directed by the Air Force's Task Force Alpha unit at Nakhon Phanom AB, Thailand. Commando Hunt air strikes concentrated on traffic control and transshipment points. Troop encampments, fleeting targets, and enemy defenses.

Commando Hunt operations encompassed 1,700 square miles of the Barrel Roll and Steel Tiger areas of Laos. Compared with the air effort in October, they shattered records in the total number of tactical attack and B-52 sorties flown in November and December. [15]

	USAF	USN	USMC	Total Tactical Attack	SAC B-52
October	4,681	10	73	4,764	273
November	9,676	2,312	833	12,821	661
December	10,125	3,672	1,344	15,141	687

Two reports, one summarizing the summer air campaign through 31 October and the other from 1 November, dramatized impact of air power in Laos in the last half of 1968. Prepared under the direction of General Keegan, the Seventh Air Force intelligence chief, they showed how, after 1 November, heavy air attacks successfully blocked considerable enemy truck traffic. About 75 percent of the logistic "throughput" was impeded while the remaining 25 percent pushed through on repaired roads and bypasses. About 48 trucks entered Laos every 24 hours through the Nape, the Mugia, and Ban Karai passes, but only eight trucks per day reached South Vietnam. These provided roughly half of the daily minimum logistic of the Communists in northern South Vietnam. [16]

The reports further demonstrated the extent a heavy air campaign, carried out on a 24 hour basis, could successfully interdict key control points, and how new munitions could increase bombing effectiveness. The air strikes forced the Communists to rely more on Sihanoukville and Cambodian LOCs for most of their supplies and munitions in the III and IV Corps areas of South Vietnam. Even if the volume of supplies passing through Laos in November doubled in subsequent months, the Seventh Air Force believed that it would still be insufficient for stockpiling. [17]

Decision to Lower the B-52 Sortie Rate

Although the complete bombing halt prompted OSD to authorize more tactical air sorties in South Vietnam and Laos, it desired, conversely, to reduce the rate of 1,800 B-52 sorties per month authorized by Deputy Defense Secretary Nitze on 22 June.

The Air Staff and the Joint Staff had studied B-52 operations throughout the summer and autumn of 1968. On 18 November, at the request of General Abrams, the JCS urged OSD to maintain the rate of 1, 800 sorties per month through mid-1969. In a reply on the 26th, Mr. Nitze, stressing financial and logistical problems, the cyclical nature of the war, and difficulties in detecting priority targets, said he was considering a variable rate of 1,400 to 1,800 B-52 sorties per month or a monthly average of 1,600 beginning 1 January 1969. This would give General Abrams 19,200 sorties per year and allow him to use the 1,800 sorties in periods of intense combat. The new rate would save about $180 million in fiscal year 1970. He asked for comments on his proposal. [18]

"A mobile reserve force which could singularly influence the outcome of a battle." The Gray Lady at work. Source: U.S. Air Force.

The service chiefs dissented, arguing that the complete bombing halt of North Vietnam since 1 November in itself justified heavier B-52 attacks. They cited the unanimous views of ground commanders who considered the bombers a highly mobile reserve force which could singularly influence the outcome of a battle and obviated the need to transfer large troop contingents from one area to another. General Abrams personally believed that centrally controlled B-52s were so effective that there was "no possible substitute within the conventional arsenal." He equated the present 1,800 per month sortie rate to the "punching power of several ground divisions." Reports of prisoners of war and Communist "ralliers" further attested to the bomber's effectiveness. and the excess of targets over B-52 sortie availability slowed the need to maintain the rate. [19]

Unpersuaded by these arguments, Mr. Nitze on 9 December informed the service chiefs that financial and logistical considerations nevertheless would dictate shortly the 1,400 to 1,800 monthly sortie rate which he had proposed. OSD would assure adequate production of munitions to sustain a rate of 1,800 sorties if necessary. [20]

Towards A Bombing Halt 1968

Both the Air Staff and Joint Staff planned to contest Mr. Nitze's decision, but with a new administration prepared to assume power in Washington on 20 January 1969, action to do so was momentarily deferred. [21]

	\multicolumn{7}{c	}{U.S. ATTACK SORTIES IN SOUTHEAST ASIA (1965 – 1968)}
	USAF	USN	USMC	Total Tactical	SAC	Grand Total
\multicolumn{7}{	c	}{SOUTH VIETNAM}
1965	36,299	18,825	10,798	65,922	1,538	67,460
1966	70,646	21,610	32,430	124,686	4,364	129,050
1967	116,560	443	52,825	169,828	6,609	176,437
1968	134,890	5,427	64,933	205,250	16,505	221,755
Total	358,395	46,305	160,986	565,686	29,016	594,702
\multicolumn{7}{	c	}{NORTH VIETNAM}
1965	11,599	13,783	26	25,408	0	25,408
1966	44,482	32,954	3,695	81,131	223	81,354
1967	54,316	42,587	8,672	105,575	1,364	106,939
1968	41,057	40,848	10,326	92,233	686	92,919
Total	151,454	130,172	22,719	304,347	2,273	306,620
\multicolumn{7}{	c	}{LAOS}
1965	6,235	3,259	363	9,857	24	9,881
1966	31,834	9,044	7,591	48,469	647	49,116
1967	35,328	6,565	2,558	44,451	1713	46,164
1968	58,908	13,022	3,344	75,274	3,377	78,651
Total	132,305	31,890	13,856	178,051	5,761	183,812

SOURCE: USAF Mgt Summary, SEA(S), 24 Jan 69, pp 25 and 37

Air War – Vietnam

The Phase II Plan for RVN Forces

The complete bombing halt also gave further impetus to building up RVN forces. The Phase I plan for modernizing and improving them had been approved by Deputy Defense Secretary Nitze on 23 October, but it became obsolete with the bombing halt on 1 November. General Abrams on his return from Washington, proposed moving faster toward Phase II objectives. On 9 November, he sent the JCS a revised plan to raise South Vietnam's manpower ceiling to 877,000, and the JCS sent it with modifications to OSD on the 15th. 22

Covering a six-year period (fiscal years 1969 through 1974), the Phase II plan was designed to create a self-sufficient military force capable of meeting an insurgency threat after American and North Vietnamese troops withdrew from South Vietnam. It provided for 855,594 personnel (versus 850,000 approved by Mr. Nitze on 1 November) by the end of that period. distributed among the following.

Air Force	32,587
Marine Corps	9,304
Navy	26,100
Regular Army	343,831
Regional Forces	245,632
Popular Forces	178,140
Total	855,594

Since the number of educationally qualified South Vietnamese was limited, the JCS believed that the forces recommended were the maximum possible. The VNAF personnel roster would rise about 50 percent, from 20,987 to 32,587, and its fighter, helicopter, gunship, liaison, reconnaissance, and training strength would increase from 20 to 40 squadrons. There would, however, be fewer fighter and helicopter squadrons than envisaged in the 23 May 1968 plan. It would take six years to complete whereas the force structure of the Army would be the reached by the end of fiscal year 1971. and the Navy's by the end of fiscal year 1973. Equipment and support costs of the

VNAF pilots and ground crews under instruction. The availability of skilled manpower was a major problem
Source: U.S. Air Force.

Phase II plan for the regular RVNAF (i.e. excluding the Regional and Popular Forces) were estimated at $3.139 billion. of which $1.4 billion would be allocated to the expansion of the VNAF.

PHASE II PLAN FOR WAR IMPROVEMENT AND MODERNIZATION Fiscal Years 1969-1974						
	End FY69	End FY70	End FY71	End FY72	End FY73	End FY74
Personnel	20,987	25,239	28,520	30,977	32,587	32,587
Squadrons						
Fighter (A-1s, F-5s, A-37s)	6	6	6	6	8	9
Helicopter (H-34s)	3	1	1	1	1	1
Helicopter (UH-1s)	2	5	11	12	12	12
Helicopter (CH-47s)	0	0	0	1	1	1
Gunship (AC-47s)	1	1	1	1	2	2
Training (U-17s)	1	0	0	0	0	0
Training (U-41s)	0	1	1	1	1	1
Transport (C-47s, C-119s, C-123s)	2	2	2	2	3	5
Liaison (O-1s, U-17s, O-2s)	4	4	5	6	7	7
Reconnaissance (C-47s, RF-5s, U-6s)	1	1	1	1	1	1
Special Air Mission (C-47s, U-17s, UH-1s)	0	1	1	1	1	1
SOURCE: JCSM-678-68 (S), with atch, Phase II Plan for RVNAF Improvement and Modernization, vol I, 15 Nov 68.						

In conjunction with the latest planning for the RVN the JCS cautioned that any withdrawal of U.S. or other free world forces should comply strictly with the conditions of the Manila Communique of 24 October 1966. The removal of troops should take place only after the enemy ceased his operations or pulled back his units in the field and the level of violence subsided. Otherwise the South Vietnamese would not. be able to cope with the remaining enemy forces. Even if early Communist troop withdrawals began U.S. manpower would be needed to offset South Vietnamese deficiencies in combat, logistics. and communications; to transfer installations and dispose of U.S. property; and to advise and support a MAAG. The American military presence would diminish as Saigon's forces attained combat self-sufficiency.[23]

Rebuilding the SA-2 sites was an immediate North Vietnamese priority.
Source: U.S. Air Force.

ESTIMATED UNPROGRAMMED COSTS OF PHASE II REGULAR RVN FORCE PLAN Fiscal Years 1969-1974 (in thousands of dollars)							
Service	FY69	FY70	FY71	FY72	FY73	FY74	Total
Air Force	22,070	112,428	386,432	433,103	429,707	17,786	1,401,526
Navy	30,053	155,274	19,835	13,740	8,162	8,100	235,164
Marine Corps	308	145	7	6	5	1	472
Regular Army	10,708	187,099	264,938	260,122	389,477	389,428	1,501,772
Total	63,139	454,946	671,212	706,971	827,351	415,315	3,138,935
SOURCE: JCSM-678-68 (S), with atch, Phase II Plan for RVNAF Improvement and Modernization, vol I, 15 Nov 68.							

Towards A Bombing Halt 1968

After extensive OSD deliberation. Mr. Nitze on 18 November (Defense Lion Publication notes that there is a penciled notation of unknown reliability correcting this date to 18 December) approved, with some exceptions, the JCS recommendations of 15 November, but he deferred a decision on General Abrams' plan of 9 November to accelerate Phase II until more details could be provided. He warned that the time available to implement Phase II might be short and enjoined the MACV commander and the service secretaries to plan for accelerated training, unit activations, and equipment deliveries. From the JCS he desired a concept of "essential conditions" for ceasing hostilities and a postwar RVNAF plan, Phase III. but only to meet an insurgency threat from the Viet Cong. [24]

On 26 December General Abrams sent to the JCS his plan for accelerating the RVN force buildup. It called for a South Vietnamese armed force of 877,090 personnel by the end of fiscal year 1969, and 877,895 by the end of fiscal year 1971, with all units activated by June 1972 instead of June 1974. By June 1971 manpower would be distributed in the following proportion: [25]

Air Force	32,587
Marine Corps	9,304
Navy	30,805
Regular Army	374,132
Regional Forces	252,927
Popular Forces	178,140
Total	877,895

To assure a more rapid VNAF buildup, the plan provided for the diversion of 60 UH-1 helicopters earmarked for the U.S. Army in Vietnam to the VNAF by June 1969 and conversion of four squadrons of older H-34s to the newer UH-ls. The Army had opposed transfer of its helicopters to the VNAF, and the Air Staff, generally supporting the Phase II plan, expected Army resistance to continue. At year's end the JCS were reviewing the accelerated plan. [26]

While ARVN strength slowly built up, desertions remained a serious problem.
Source: National Archives.

Not yet resolved near the end of 1968 was the alarming desertion rate of South Vietnamese troops which threatened to undermine all planning. Ground combat personnel were abandoning their units at an annual net rate (i.e. less those who returned) of 35 percent of their strength. The following net desertion rate (per 1,000) prevailed in October 1968: Regular Army - 17.2; Regional Forces - 19.2; Popular Forces - 10.2; Navy - 1.3; Marine Corps - 72.1; and Air Force with the lowest rate of all - 0.8. Both General Abrams and Ambassador Bunker were gravely concerned with this problem, discussing it frequently with President Thieu throughout the remainder of the year. [27]

IX. SUMMARY

In retrospect, 1968 was a watershed in U.S. military planning and operations in Southeast Asia. At the beginning of the year, the authorized manpower ceiling of 525,000 for South Vietnam and 45,724 for Thailand still allowed for more deployments. Many officials were optimistic, believing that if the allies maintained military pressure on the North Vietnamese and the Viet Cong and pursued pacification with vigor a negotiated settlement would soon be possible.

The Communist Tet offensive of February 1968, however, shattered the feeling of confidence and changed the administration's overall conduct of the war. To be sure, American commanders in Vietnam believed that the enemy's offensive and his effort to seize the Marine base at Khe Sanh were colossal failures. He suffered enormous casualties, more than 10,000 at Khe Sanh alone, largely from B-52 strikes, and the allies soon routed him out of the urban and rural areas he had overrun temporarily. However, General Westmoreland's request for 206,000 more troops, including air support to capitalize on the enemy's setbacks, shocked many important administration, congressional, and public leaders who believed that the foe had been badly underestimated. Faced with growing financial and other domestic difficulties, the administration was unwilling to increase substantially its commitment in Southeast Asia or risk a wider war by relaxing long-enforced restrictions on combat. It therefore decided to reduce America's involvement and increase the role of South Vietnam in the conflict, and to make a greater effort to disengage through negotiations.

In a first step toward this policy the President, on 1 March 1968, despite strong JCS objections, halted the bombing of North Vietnam above the 20th (and then the 19th) parallel to encourage Hanoi to enter into peace talks. These began in Paris in May. Meanwhile, he limited the increase in U.S. air, naval and ground deployments to South Vietnam to counter the Tet offensive, restricted the rise of the authorized manpower ceiling in that country to 549,500 and ordered a speedup in training the South Vietnamese forces. With less area to bomb, air strikes on enemy territory below the 19th parallel increased.

The initial months of peace talks proved unsuccessful. However, after reaching certain "understandings" with Hanoi's leaders, the President on 1 November ordered a complete bombing halt of the North in exchange for a promise by Hanoi to withdraw some of its forces from I Corps and engage in more substantive discussions in Paris. Simultaneously. the President approved a massive air interdiction program, Commando Hunt, against the infiltration routes of southern Laos. Administration officials made additional plans to hasten the buildup of South Vietnam's air naval, and ground units.

By the end of the year the administration appeared to be making progress in arresting further expansion of U.S. involvement in the war and in moving toward more productive negotiations. U.S. troop strength in South Vietnam stood at about 536,000 (including 59,024 Air Force), well below the authorized ceiling. In Thailand, manpower had leveled off at 48,301 (including 35,846 Air Force). The

Air War – Vietnam

latest plan to assure a self-sufficient South Vietnamese force as soon as possible called for 877,000 men with all military units activated by June 1972. With the bombing halt of North Vietnam, the representatives of Hanoi and the National Liberation. Front seemed ready for more substantive talks with the U.S. and South Vietnamese governments although Saigon momentarily refused to participate.

The bombing cessation of the North and year-end studies the number of expensive B-52 sorties belied the fact that the administration still relied heavily on air power to achieve its goals. In December 1968, U.S. tactical and B-52 attack sorties in South Vietnam remained at a high level and in Laos they had tripled over the average monthly totals of midyear. The attack sorties in both countries exceeded considerably the number flown in previous years. To sustain this effort the United States operated more tactical combat aircraft in theater than at the beginning of the year (1,099 versus 992). Of a total 2,641 U.S. combat and non-combat aircraft and 3,431 in the war theater at the end of 1968, the Air Force 1,177 (including 48 helicopters) in South Vietnam and 595 in Thailand (including 34 B-52's and 36 helicopters). There were a total of 106 B-52s in Thailand, Okinawa and Guam versus only 51 a year previously (OSD Southeast Asia Deployment Program 6. 4 April 1968, and through change 44, 14 March 1969.) Assessing the air effort since the Tet offensive, the Seventh Air Force believed that the allied summer campaign had forced Hanoi in October to withdraw some troops and agree to more serious discussions, and that Commando Hunt operations after mid-November had curbed drastically the enemy's logistic "throughput" from Laos into South Vietnam.

How soon America could attain its objectives by an expeditious buildup in South Vietnam's forces, continuous heavy air operations in South Vietnam and southern Laos pacification, and resolute negotiations in Paris, remained to be seen. The only certainty was that this policy would come under exhaustive review by a new administration under President-elect Richard M. Nixon on 20 January 1969.

APPENDIX I

US MILITARY AND AIRCRAFT STRENGTH IN SOUTHEAST ASIA
(SOUTHEAST ASIA DEPLOYMENT PROGRAM 6, (4 April 68)

U.S. Military Strength In South Vietnam						
	Jan 68	Mar 68	Jun 68	Sep 68	Dec 68	Jun 69
USAF	56,400	57,300	59,900	60,700	60,500	61,500
USN& CG	32,800	35,100	37,200	37,500	37,400	37,300
USMC	78,200	87,100	87,700	82,100	82,000	82,200
US Army	326,900	344,500	353,400	369,100	369,400	368,400
TOTAL	494,300	524,000	538,200	549,400	549,300	549,400
U.S. Military Strength In Thailand						
USAF	32,000	33,500	35,100	35,100	35,100	34,900
USN-USMC & CG	800	800	500	500	500	500
US Army	10,900	12,500	12,700	12,200	12,200	12,200
TOTAL	43,700	46,800	48,300	47,800	47,800	47,600
U.S. Offshore Navy						
	35,200	46,800	48,300	47,800	47,800	47,600
U.S. Fighter and Attack Aircraft In Southeast Asia (by Service)						
USAF	643	686	758	776	776	770
USN	184	202	206	196	198	212
USMC	150	171	191	191	183	183
US Army	0	0	0	0	0	0
TOTAL	977	1,059	1,155	1,163	1,157	1,165
B-52s						
Thailand	20	20	25	25	25	25
Guam	36	***	***	***	***	***
*** Not included						

Air War – Vietnam

	Allied Fighter and Attack Aircraft In Southeast Asia					
	Jan 68	Mar 68	Jun 68	Sep 68	Dec 68	Jun 69
VNAF	70	72	54	36	36	36
RAAF	8	***	***	***	***	***
	USAF Fighter and Attack Aircraft In Southeast Asia (By Type)					
A-1	40	61	61	61	61	61
A-26	12	12	12	12	12	12
AC-47	34 AC-47s erroneously included under non-attack aircraft					
B-57	23	24	24	24	24	18
F-100	218	198	252	270	270	270
F-102	24	30	30	30	30	30
F-104	Phased out in 1967					
F-105	104	108	108	90	90	72
F-4	191	216	234	252	252	288
F-5	Phased out in 1967					
T-28	6	12	12	12	12	12
A-37	25	25	25	25	25	25
TOTAL	643	686	758	776	776	788
	U.S. Helicopters In South East Asia					
USAF	12	15	15	15	15	15
USN	0	0	0	0	0	0
USMC	334	304	304	304	292	292
USA	2,614	2,771	2,835	2,899	3,098	3,410
TOTAL	3,023	3,172	3,234	3,300	3,487	3,799

USAF Helicopters In Southeast Asia (By Type)						
	Jan 68	Mar 68	Jun 68	Sep 68	Dec 68	Jun 69
UH-1	12	15	15	15	15	15
HH-43	30	32	32	32	32	32
HH-53	5	6	6	8	10	10
HH-3	15	22	20	18	18	18
CH-3	13	22	22	22	22	22
TOTAL	75	97	95	97	97	97
US Fixed-Wing Non-Attack Aircraft In Southeast Asia (By Services)						
USAF	809	840	845	867	886	965
USN	52	53	45	42	40	44
USMC	47	46	58	58	82	68
US Army	586	545	564	588	590	612
TOTAL	1,494	1,484	1,512	1,555	1,598	1,689

Air War – Vietnam

USAF Fixed-Wing Non-Attack Aircraft In Southeast Asia (By Type)						
RB-57	3	4	4	4	4	4
EB-66	24	41	41	41	41	41
RF-4	74	76	76	76	76	76
RF-101	17	16	16	16	16	16
AC/C-47	52	46	46	46	46	46
EC-47	43	47	47	47	47	47
EC-121	29	27	27	27	27	27
C-123	79	87	91	91	91	91
HC/C-130	16	17	17	17	17	17
KC-135	39	30	30	30	30	30
C-7A	81	96	96	96	96	96
O-1	173	174	158	150	135	61
O-2	150	147	147	147	147	147
OV-10	0	0	19	47	73	96
U-10	32	32	32	32	32	32
HU-16	Not included					
TOTAL	812	840	847	867	886	965

SOURCE: Memo (S) Dep SECDEF to Secys of Mil Depts et al subj SEA Deployment Prog 6, 4 Apr 68; USAF Mgt Summary,S EA, 26 Jan 68 p 25

APPENDIX II

Equipment Introduced Into Southeast Asia 1968-1969

Calendar Year 1968

Munitions
- LAU-62/A Flare Launcher
- SUU-42/A Dispenser
- SUU-41 Dispenser
- Chemical Weapon BLU-52
- FMU-26A/B Fuze
- Fuze. FMU-56/B
- Long Duration Target Marker LUU-1/B
- CBU- 34A Dispenser and Mine
- CBU-28A Dragontooth Mine
- FMU-57B Proximity Fuze

Reconnaissance
- Printer- Enlarger (EN-99A)
- Photographic Printing. Processing. and Interpretation Facility (ES-73A)
- KA-79 Camera
- KA-80 Panoramic Camera
- M-731 Strike Film Viewer

Electronic Countermeasures
- QRC-312-1/ALT-15 Mod Kit
- QRC-128 Communications Jammer
- AN/ALT-28 ECM Jammer
- QRC-359/ALT-16 Mod Kit
- QRC-335A Seed Sesame
- ALQ-71 ECM Pod
- QRC-337A/ALQ-71 Mod Kits
- ALR-31 (SEE SAM)
- QRC-353-A. Chaff
- QRC-248A IFF Interrogator

Navigation
- ARN-92 Loran D

Aircraft And Missiles
- Tropic Moon I and II
- AIM-4D Pilot Training Missile
- AGM-78A-Standard ARM Missile
- C-130 Gunship II
- F-111A Aircraft
- AIM-7E-2 Sparrow Missile
- OV-10 Aircraft
- AGM-45A Shrike Missile

Air War – Vietnam

AGM-12E Stand-off Cluster Missile
F-4E Aircraft
Black Spot Aircraft

Improved Attack Capability
Laser Guided Bomb
Electro-Optical Guided Bomb
Pave Arrow - Laser Target 'Designator (LTD) and Seeker System
Infrared Guided Bomb

Communications, Command And Control
TPS-50 Radars
Map Overlays for Mobile GCA Units
AN/GPA-129 Video Mapping Group
College Eye Modification C-121
AN/PRC-72 Radio

Personal Life Support
URT-33 Personal Locator Beacon Marker. Signal
SRU-22/P Improved Body Armor
Radio Set URC-64

Operational Support
Truck, Fork-Lift A/S 32H-15
Low Altitude Parachute Extraction System and Platforms
Palletized Mail Systems
Cargo Buffer Stop
Fast Fix Cement
C-130 Ramp Kit
Aircraft Arresting Barrier (BAK-13)
Mobile Electronic Weighing System (A/M 37-U2)
Combat Trap
Cargo Airdrop Release Gate
Hydraulic Flow Comparator

Calendar Year 1969

Munitions
Hard Structural Munition - BLU-31/B
Anti-Vehicle Land Mine (CBU-33)
Anti- Materiel Bomblet (CBU -54B)
Downward Ejection Bomb (CBU-38A)

Reconnaissance
Photo Interpretation Equipment (AR-109A)
F-4D/APX-81 Interrogator
Step and Repeat Printer (FH-701A)
Infra-Data Link (Compass Sight)
Mobile Color Film Processing Facility (EN-75)

Compass Count (AN/AVD-2 Laser)

Electronic Countermeasures
QRC-354-Receiver
QRC-373 Miniaturized Noise Jammer

Navigation
Rotating TACAN Antenna YNl-106
Tactical Instrument Landing System (AN/ARN-97 and AN/TRN-27)
Lightweight TACAN (AN/TRN-26)

Aircraft And Missiles
AGM-45A with Tracking Flare
AGM-78B Standard Arm Missile Hunter I System
AN/ASQ-96 System
Tropic Moon III

Improved Attack Capability
F-4C Laser Bombing System
AN/AVB-1 Lightweight Precision Bombing System

Communications, Command And Control
MSQ-77 Modifications
K-300 All Automatic "Satellite Picture" Recorder
Portable Cloud Height Measuring Device
Rapidly Deployable Antenna Mast

Air War – Vietnam

Towards A Bombing Halt 1968

APPENDIX III

SHEDLIGHT STATUS REPORT, RDT&E FUNDS

SHED LIGHT STATUS REPORT, RDT&E FUNDS DATE: 29 January 1969
(FUNDS IN THOUSANDS OF DOLLARS)

PROGRAM ELEMENT	PROJ	TASK	S/L NR.	SHED LIGHT TITLE	AFRDC OPR	SPEC ATTN	PRI	FY 66/67 FUNDED	FY 68 FUNDED	FY 69 FUNDED	FY 69 ADD'T REQ	FY 70 ADD'T REQ	FY 70 BUDGET
61102F	5635	5	93	RESONANT REGION RADAR	RDDG	S	1-7	430	500	128	0	0	0
64708F	1559	224	93	RESONANT REGION RADAR	RDDH	S	1-7	0	225	0	0	0	0
61102F	7235	--	15A	ISOCON IMAGE AMP DEV	RDDH	S	370	282	120	156	0	--	200
61102F	8601	6	94	AIR/B RECCE MAGNETOMETRY	RDDG	S	370	250	50	0	0	0	0
62101F	7621	7	95	ENVIRON FACTORS LLLTV	RDDH	S	1-7	518	533	0	0	--	--
62403F	5227	3	20	IR NIGHT SENSOR	RDDH	S	370	675	100	0	0	150	--
62403F	4056	6	103	FLIR FABRICATION TECH	RDDG	S	370	230	0	88	100	--	350
62403F	5042	19	101	POLY FREQ SIDE LOOK RADAR	RDDG	S	1-6	2690	750	250	--	--	325
62710F	2563	--	61	TARGET ILLUMINATION PYRO	RDDA	PS	370	452	175	150	0	0	500
63101F	7990	A	96	S/L EFFECTIVENESS MODEL	RDDH	S	1-7	0	0	0	0	0	0
63203F	666A	5B	82	DOPPLER INERTIAL LORAN	RDDG	S	1-7	1417	700	1900	0	0	1950
63208F	665A	C05	23	HIGH ALTITUDE IR SENSOR	RDDH	S	370	550	231	0	0	0	0
63208F	665A	7	35A	FOPEN 1A	RDRM	S	370	900	1897	300	0	0	1000
63208F	665A	C01	100	TAC NEAR REAL TIME RECCE	RDRM	S	1-7	350	780	1000	0	0	100
63208F	665A	C03	137	HIGH RES LOW ALT IR SENSOR	RDRM	S	1-7	0	0	550	0	0	0
63208F	665A	--	53	A/B IR RASTER DISP ABIGD	RDRM	S	370	97	0	0	0	0	0
63215F	698DF	A	15B	ISOCON IMAGE AMPLIFN TEST	RDDH	S	1-7	0	0	20	0	0	0
63215F	698DF	1A	13	ADVANCED LLLTV	RDDE	PS	1-6	1270	178	280	0	0	500
63215F	698DF	1B	18	ADVANCED FLIR	RDDE	PS	1-6	800	329	580	0	0	600
63215F	698DF	1C	31	LASER TARGET RECOGN SYS	RDDE	S	1-6	500	1160	526	0	0	300
63215F	698DF	2A	105	MULTI SENSOR WPN DLVY SYS	RDDE	S	370	0	50	100	0	--	600
63215F	698DF	2B	138	KINEMETIC BOMBING SYS	RDDE	S	370	0	0	625	0	0	0
63215F	698DR	2C	133	CLOSE AIR SUPPORT SYS	RDDE	S	1-7	0	0	600	0	0	200
63302F	679A	6	106	LASER OP GUID INTEG LOGIC	RDDA	PS	1-7	1227	1171	1000	0	0	904

Air War – Vietnam

SHEDLIGHT STATUS REPORT, RDT&E FUNDS

DATE: 29 January 1969

(FUNDS IN THOUSANDS OF DOLLARS)

PROGRAM ELEMENT	PROJ	TASK	S/L NR	SHED LIGHT TITLE	AFPDC OPR	SPEC ATTN	PRI	FY 66/67 FUNDED	FY 68 FUNDED	FY 69 FUNDED	FY 69 ADD'T REQ	FY 70 ADD'T REQ	FY 70 BUDGET
63709F	681A	2	86	LORAN INERTIAL BOMB DEMO	RDD	S	1-7	0	0	0	450	0	0
63716F	670A	P3	114	TARGET MARKING MUNITIONS	RDDA	S	370	0	187	100	---	200	200
63716F	670A	P12	115	BATTLEFIELD ILLUMINATION	RDDA	S	370	0	225	300	0	200	200
64212F	1709		12	LASER RANG AIDED VIS F-4C	RDQRT	PS	1-7	6400	1200	181	0	0	0
64212F	1766		3A	O-2/OV-10 NIGHT AVIONICS SYS	RDDH	PS	1-7	73	0	2250	2250	---	3500
64212F	2701		126	PAVE GAT	RDDA/RDDHS	PS	1-6	0	0	0	---	0	0
64212F	2702		127	NIGHT RECOVERY SYSTEM	RDDH	S	1-6	0	1300	1500	---	0	0
64212F	2707		139	GUNSHIP INERTIAL TGT SYS	RDDH	S	1-6	0	3525	280	0	0	0
4004	4004		119	GUNSHIP DIG FIRE CONTROL	RDDH	S	1-6	0	296	130	0	0	0
64212F[1]	4366		128	C-130 TESTBED AIRCRAFT	RDDH	S	1-6	596	63	250	0	0	180
64212F	5220		10	F-105 T-STICK II	RDQRT	PS	1-7	6699[2]	100	2670	0	0	0
64212F	6033		140	RED FLAME II	RDDH	PS	1-6	15402[2]	3186	352	0	12800	3320
64212F	6038		122	SEC VDICON TUBE	RDDH	PS	1-6	0	148	450	0	0	300
64212F	6041		123	HAVE AUGER	RDDE	PS	1-6	0	750	0	0	550	750
64212F	6041		121	BLACK CROW	RDDH	PS	1-6	0	50	0	0	0	0
64212F	6041		124	PAVE CROW	RDDD	PS	1-6	0	110[3]	0	0	0	0
64212F	XXXX		141	LASER TGT DES IN CLS AIR SUP	RDDH	PS	1-6	0	0	1500	750	2500	0
64212F	XXXX		113	QUIET AIRCRAFT	RDDH	S	1-7	0	0	0	---	---	0
64212F	XXXX		134	RESERVE RANGING	RDDH	S	1-7	0	0	450	0	0	550
64703F	---		80C	AN/TPQ-27	RDDD	S	370	600	0	700	0	0	1500
64708F	1559	198	130	PAVE SPOT	RDDH	S	1-6	---	45	339	0	0	0
64708F	1559	178	110B	MILLIMETER WAVE RADIOMETRY	RDDH	S	1-6	0	449	0	0	0	0
64708F	1559	222	125	PAVE BOX	RDDH	S	1-6	0	0	157	0	0	0
64708F	1559	224**	93	RESONANT REGION RADAR	RDDG	S	1-7	430	225	0	0	0	0
64708F	1559	209***	12	LASER RANG AIDED VIS F-4C	RDQRT	PS	1-7	6400	24	0	0	0	0
64708F	1619	01	73	LASER TGT DES PAVE ARROW	RDDH		1-6	5750	2509	367	1000*	0	1000
64708F	1619	04*	116	LASER TARGET DES PAVE LIGHT	RDDH		1-6	543	565	150	0	0	0
64708F	1619	05*	102	LASER TGT DES PAVE KNIFE	RDDH	S	1-6	---	1991	650	0	0	0

* FY 67/68 were funded from 1559 Tasks 146/157
** FY 66/67 funds under 61102F 5635 Task 5
*** FY 67 funding 64212F 1709

1 Previously 64708 1559 142 funds of 68 and prior are for original Gunship II
2 15.402M funded from other sources
3 60K of this was funded from 1559

Notes For Part One

SHEDLIGHT STATUS REPORT, RDT&E FUNDS

DATE: 29 January 1969

(FUNDS IN THOUSANDS OF DOLLARS)

PROGRAM ELEMENT	PROJ	TASK	S/L NR.	SHED LIGHT TITLE	AFRDC OPR	SPEC ATTN	PRI	FY 66/67 FUNDED	FY 68 FUNDED	FY 69 FUNDED	FY 69 ADD'T REQ	FY 70 ADD'T REQ	FY 70 BUDGET
64708F	6517	--	69A	COIN RANGING & HOMING	RDDD	PS	370	2775	0	250	0	0	350
64708F	2077	--	1A	BIAS	RDDH	PS	1-6	3252	448	0	0	0	0
64708F	4053	--	6	BLACK SPOT	RDDH	PS	1-6	11841	630	0	0	0	0
64709F	--	--	80B	LT WT PRECISION BOMBING SYS	RDDD	S	1-6	300	0	460	0	0	0
64710F	1593	--	33	LASER RECCE ADVANCED DEV	RDRM	PS	1-7	1050	800	2600	600	0	1800
64710F	4010	--	132	IR RECON SET	RDRM	PS	1-6	0	1600	0	0	0	0
64712F	1254	--	99	FORWARD FIRING TGT MK	RDDA	S	1-7	0	0	260	610	--	0
64712F	7053	--	129	GUNS FOR GUNSHIP	RDDA	S	1-7	0	0	0	--	--	0
64712F	8751	--	135	HOMING OPTICAL GUIDANCE SYS	RDDA	S	1-7	0	0	600	0	0	800
XXXX	XXXX	--	136	COMBAT RAM	RDRE	S	1-7	--	QRC 402 PRODUCTION FUNDED				

Air War – Vietnam

NOTES FOR PART ONE

Unless otherwise noted, all primary sources cited (letters, memos, JCS papers) are located in-Headquarters USAF Directorate of Plans File RL (61), (62), (63), or (64) 38-9, depending upon the year of the source.

Chapter I

1. Bureau Int & Rsch, Dept of State, Summary of Principal Events in History of Vietnam, 10 Jan 62 (hereinafter referred to as RFE-14) in AFCHO; R. W. Lindholm, ed, Vietnam, The First Five Years (Michigan State, 1959), p 4; A. B. Cole, ed, Conflict in Indo-China and International Repercussions, A Documentary History 1945-1955 (New York 1956), p 195; Southeast Asia in Perspective (New York, 1956), p 170.

2. RFE-14, 10 Jan 62.

3. Cole, pp 195 and, 255; British Information Service, Vietnam (London, 1961), p 18.

4. See note above; King, p 171; Robert Scigliano, South Vietnam, Nation Under Stress (Boston, 1963), pp 162-63.

5. Cole, pp 195, 255; King, p 171; Scigliano, pp 162-63; British Info Svc, Vietnam, pp 14-15; House Hearings before Subcmte on Far East and Pacific of Cmte on Foreign Affairs, 27 Jul - 15 Aug 59, Current Situation in the Far East, p 35.

6. Capt Mack D. Secord, "The Viet Nam Air Force" Air University Review, Nov - Dec 63, p 60; Journal of Mutual Security (JMS), prep by Asst for Mutual Security, May 57, pp 154 - 55, p 126, and Mar 59, p 149

7. Cole, pp 251-56; RFE-14, 10 Jan 62; British Info svc, Vietnam, p 15; Bernard F. Fall, The Two Vietnams, A Political and Military Analysis (New York, 1963), p 219.

8. JMS, Nov 60, p 35; British Info Svc, Vietnam, p 14.

9. Scigliano, pp 138, 163-64; Secord, p 60; JMS May 57, pp 154 – 55; Mar 58, p 126, and Mar 59, p 149.

10. Dept of State, Vietnam,: Guidelines for Policy and Operations, Feb 63, p 19; RAND RM-4140 PR, Jul 64, The Role of North Vietnam in the Southern Insurgency pp 31-36.

11. Ltr, CINCPAC to JCS, 27 Mar 61, subjt SVN Internal Security Situation, 1960, Dept of State Press Release 287, 4 May 61.

12. Memo, Dep D/Plans for Policy to C/S USAF, 3 Nov 61, subj: Background Paper on CINCPAC Oplan 32-59; Hist, CINCPAC, 1961, Pt 1, pp 168-69.

13. Ltr, CINCPAC to JCS, 18 Jan 61, subj: Increased Force Levels for RVN Armed Forces; memo, P. M. Nitze, Asst SOD/ISA to Chmn JCS, 25 Jan 61, subj: U.S. Support for Addit Mil Forces in VN and Thai; memo, SOD to Secy State, 30 Jan 61, no subj; rpt VN: Dept of State Guidelines for Policy and Ops, Apr 62; JCS 1992/917, 9 Feb 61; Hist, CINCPAC, 1961, pt 1, pp 169-70.

14. Memo, M/G D. A. Burchinal, D/Plans to C/S USAF, 20 Jun 61, subj: Increase in GVN Forces.

15. Ltr, CINCPAC to JCS, 27 Mar 61, subj: SVN Internal Security Sit, 1960; JCS 2343/21/1, 8 Nov 61; Hist, CINCPAC, 1961, Pt 1, pp 169-70.

16. Memo, Col A. N. Williams, Dep D/Plans for Policy to C/S USAF, 9 May 61, subj: Prog of Action for VN; Hist, CINCPAC, 1961, Pt 1, p 172; rpt, Prog of Action to Prevent Communist Domination of SVN, 6 May 61; NSA memo 52, 11 May 61.

17. See note above; Dept of State rsch memo, RFE-14, 10 Jan 62

18. Memo, SAF to SOD, 9 May 61, subj: Prog of Action for VN, in OSAF 1257-61; memo, Burchinal to C/S USAF, 17 May, 61, same subj.

19. JCS 2343/1, 7 Jul 61; JCS 2343/67, 13 Jan 62; rpt, Jt Action Prog Proposed by the VN-US Spec Financial Gps to Pres Diem and Kennedy, Jul 61; ltr, CINCPAC to JCS, 9 Jul 51, subj: Increase in Force Level for RVN; memo; Burchinal to C/S USAF, 20 Jul 61, subj: Increase in GVN Forces; ltr, Actg SOD to Secy State, 3 Jul 61, no subj; JCS 2343/5, 26 Jul 61; NSA memo 65, 11Aug 61.

20. NSA memo 65, 11 Aug 61; memo, SOD to Chmn JCS, 5.Feb 62, subj, Increase of all GVN Forces; JCS 2343/67, 13 Jan 62; memo, Dep SOD to SA, et.al, 18 Aug 61, subj: Joint Prog of Action by GVN, in OSAF 1257-61.

21. JCS 2343/25, 9 Oct 61; memo, Dep SOD to Chmn JCS, 5 Oct 61, subj: Concept For Use of SEATO Forces in VN; memo SOD to SA, et al., 21 Oct 61, subj: SEA; New York Times, 12 Oct 61; Public Papers of the President of the United States. John F. Kennedy 1961 (GPO, 1962), pp 680-81; Dept of Defense Annual Report for Fiscal Year 1963, p 192; New York Times, 2 Jul 63.

22. Msg 38287, CINCPAC to JCS, 24 Oct 61; ltr, Gen Maxwell D. Taylor, Mil Rep Pres, to Pres, 3 Nov 61, no subj, with atchd rpt on Gen Taylor's Mission to SVN and anx A through I.

23. Draft memo, SoD to Pres, 6 Nov 61, no subj; JCS 2343/36, 9 Nov 61.

24. JCS 2343/40, 13 Nov 61; memo, M/G J. W. Carpenter, III, Dep D/Plans, to C/S USAF, 13 Nov 61, subj: SVN; memo, Burchinal to C/S USAF, 20 Nov 61.

25. NSA memo 111, 22 Nov 61; Dept of State Bul, 4 Dec 51, pp 920-21.

Notes For Part One

26. NSA memo 111, 22 Nov 61; rpt, Vietnam: Dept of State Guidelines for Policy and Ops, Apr 62; Dept of State Press Release 8, 4 Jan 62; Dept of State Pub 7308, 8 Dec 61, A threat to the Peace: North Vietnam's Efforts to conquer South Vietnam, pts 1 and 2

27. Memo, C/S USAF to JCS, 5 Dec 61, subj: SVN; JCS 2343/70, 13 Jan 62; memo, SOD to Pres, no subj, 27 Jan 62.

Chapter II

1. Memo, B/G G. S. Brown, Mil Asst-to SOD to SA et al., subj SVN; memo, SOD to JCS, 27 Nov 61, subj: First Phase of the VN Prog; memo, SOD to Pres, 22 Dec 6, subj: Mil Cmd in SVN: memo L/Col S.B. Berry, Jr., Mil Asst to SOD to SA, et al., 4 Dec 61; Dept of State Bul 4 Dec 61, pp 920-21,

2. Memo, Dep SOD to Chmn JCS, et al., 4 Jan 62, subj: Pub Affairs Security Aspects of Ops in VN.

3. JCS, 2343/46, 22 Nov 61.

4. Memo, Asst SOD/ISA to SOD, 1 Dec 61, subj: VN Comd Arrangements; ltr, Secy State to SOD, 18 Dec 61, no subj.

5. Ibid: ltr, SOD to Secy State, 7 Dec 61, no subj; memo, SoD to Pres, 22 Dec 61, no subj; Baltimore Sun, 9 Feb 62.

6. Memo, SoD to Pres, 22 Dec 61, no subj; JCS 2343/62, 8 Feb 62; ltr Pres to all Ambassadors, 29 May 61; msg 2393, JCS to CINCPAC, 28 Nov 61; msg 01600, CINCPAC to CNO et al., 8 Feb 62, in Hist Study, Contemporary Historical Evaluation Of Counter-Insurgency Operations (hereinafter cited as Project CHECO SEA Rpt), prep by Hq PACAF- (May 1964), pt IV-A, docs 1 and 8; DOD Press Release 204-62, 8 Feb 62.

7. Memo, Carpenter to C/S USAF, 17 Sep 63,. subj: Proposed Change in Service Manning in C/S Position, USMAC/V; msg 28784, PACAF to C/S USAF, 8 Dec 61; msg PACAF to C/S USAF, 13 Dec 61; msg 43936, PACAF to C/S USAF, 18 Feb 62; msg 63603, Hq USAF to PACAF, 21 Feb 62.

8. Memo, Carpenter to C/S USAF, 17 Sep 63; rpt, C/S Visit to SVN, 24 Apr 62; memo, Col F. R. Pancake, Asst Dep D/Plan to C/S USAF, 5 Jan 63; subj Review of Mil Sit in SVN; msg 43936, PACAF to C/S USAF, 18 Feb 62.

9. Msg 020145, CINCPAC to JCS, 2 Nov 61; msg 072013, CINCPAC to Ch USMAAG/V, 7 Nov 61; msg 252015A, CINCPAC to ch MAAG/V, 25 Nov 61; msg 16790, PACAF to C/S USAF, 29 Nov 61.

10. Hist, 2d ADVON, 15 Nov 61-8 Oct 62, pp xviii, 17; JCS 2343/191, 4 Feb 63.

11. Memo, SOD to SA et al., 21 Oct 61, subj: SEA; memo, Brown to SA et al., 27 Nov 61, subj: SVN; Charles H. Hildreth USAF Counter-insurgency doctrine and capabilities, 1961 – 1962 (AFCHO, 1964) pp 12 – 14 in AFCHO, Hist, PACAF, Jul - Dec 61, I, pt 2, p 23)

12. Msg 25979, JCS to CINCPAC, 6 Dec 61; msg 44865, JCS to CINCPAC, 26 Dec 61; msg 41464, CINCPAC to PACAF, 20 Dec 61; memo, L/Col J.B. Owens, Combined Plans Div, D/Plans, to C/S USAF, 13 Feb 62, subj: Actions Concerning SVN; Hist, CINCPAC, 1961, Pt 1, pp 187-88; Rcrd, SOD Honolulu Conference (hereinafter referred to as Hono Conf), 16 Dec 61, in AFCHO.

13. Hist, PACAF, Jul - Dec 61, I, pt 2, pp 19-20.

14. Ibid.; Hist, 13th AF, Jul- Dec 61, pp 62-70.

15. Hist, 13th AF, Jul - Dec 61, p 85; Hist, PACAF Jul-Dec 62, I, Pt 2, pp 27-28; Rcrd, SOD Hono Conf, 19 Feb 62.

16. Rcrd, SOD Hono Conf, 15 Jan 62 and 19 Feb 627 Hist, PACAF, Jul – Dec 61, I, pt 2, p 28; Hist, 13th AF, Jul-Dec 61, p 86.

17. Memo, Col R. P. Baldwin, Air Def Dev, D/Ops to Engr Div, D/Ops, 28 Nov 61, subj: Talking Paper, TAC Control Capability; Hist, CINCPAC, 1961, pt 1, pp 33, 175; Hist, 2d ADVON, 15 Nov 61 - 8 Oct 62, p 10.

18. Rcrd, SOD Hono Conf, 16 Dec 61 and 15 Jan 62; Hist, 2d ADVON, 15 Nov 61 - 8 Oct, 62, pp 91-93.

19. Rcrd, SOD Hono Conf, 16 Jan 62; Hist, CINCPAC, 1962, p 164; Hist, D/Ops, Jan - Jun 62, pp 57-58; Hist, 2d ADVON, 15 Nov 61-8 Oct 62, p 95 and app D, item 21;

20 "Tactical Air Control in the VNAF," Air University Review, Sep-Oct 63, pp 75-81, prep by AU Rev staff.

21. JCS 2343/68, 9 Jan 62; memo, Dep SOD to SA, et al., 13 Jan 62; Rcrd, SOD Hono Conf, 15 Jan 62; Hist, D/Telecom, Jul - Dec 62, pp 108-09; Hist, AFLC, 1 Jul 62 - 30 Jun 63, pt, II, pp 53-57.

Chapter III

1. Rcrd, SOD Hono Conf, 15 Dec 61, item 4.

2. Ibid., 15 Jan 62, item 6.

3. Memo, Carpenter to C/S USAF, 22 Mar 62, subj: Buildup in SVN 6 Dec 61; memo, C/S USAF to SOD, 4 Apr 62, subj: Estab of Quick Reaction Forces in SVN; memo, Chmn JCS to SOD, 17 Nov 62, subj: Viet Cong Attacks on Strat Hamlets; rpt C/S USAF Visit to SVN, 16 – 21April 62, prep by D/Plans, 24 Apr 62 (hereinafter cited as LeMay Rpt); Hist, CINCPAC 1962, p 179.

Notes For Part One

4. See note above; draft memo by C/S Army, 11 Mar 62, subj: Estab of Quick Reaction Forces in SVN; msg 85592, Hq USAF to PACAF, 5 May 62; msg 67265, Hq USAF to PACAF, 18 Jul 62; Hildreth, Counterinsurgency, 1961 – 1962, pp 24 - 25.

5. Memo, SAF to SOD, 4 Dec 61, no subj, in OSAF 1257-61; memo, Dep SOD to Chmn JCS et al., 4 Jan 62 subj: Pub Affairs and Security Aspects of Ops in VN; Rcrd, SOD Hono Conf, 19 Feb 62, p 33.

6. Memo, Owens to C/S USAF, 13 Feb 62, subj: Actions Concerning SVN; Rcrd, SOD Hono Conf, 15 Jan 62, pp 3-1 to 3-5; JCS 2343/96, 19 Mar 62; Hist, PACAF, Jul-Dec 61, pt 2, pp 23-28.

7. Hist, 2d ADVON, 15 Nov 61-8 Oct 62, pp 127-30.

8. Ibid, p 133; memo, M/G R.F/ Worden, Dep D/Plans, to C/S USAF, 23 Jan 64, subj: JCS Briefing by Gen Anthis.

9. Rcrd, SOD Hono conf, 19 Feb 62, pp 15 - 16; LeMay rpt, 24 April 62; L/Col E.T. Cragg, Asst Dep D/War Plans, D/Plans to C/S USAF 2 Mar 62, subj: Actions Concerning SVN; memo for rcrd by M. C. Loughlin, Cold War Div, D/Plans, 12 Mar 62, no subj; msg 38634, PACAF to C/S/ USAF, 4 Apr 62; New York Times, 19 Feb 62; msg 67718 Hq USAF to PACAF 8 Mar 62, in project CHECO SEA Rpt, pt V-A doc no 60 and Pt III pp 26-27.

10. Hist, 2d ADVON, 15 Nov 61-8 Oct 62, p 133; New York Times, 27 Mar 62; Baltimore Sun 12 Jul 62.

11. Memo, M/G W.W. Momyer, D/Ops Rqts to DCS/O, 3 Apr 62, subj: Suspected Communist Night Activity in SVN; Hist, CINCPAC, 1962, pp 188-89; Hist, 13th AF, pp 75, 81-82; LeMay Rpt, 24 Apr 62; JMS Mar 62, p 171; Washington Post, 3 Apr 62.

12. Rcrd, SOD Hono Conf, 19 Feb 62; memo, M/G S.W. Agee, D/Ops under SAF, 20 Mar 62, subj: Ranch Hand; Hist CINCPAC 1962 pp 187-88; Hist of Airlift in SVN, Dec. 61 – 0ct. 62, prep by 6492d Combat Cargo Gp (P), 17 Dec 62, pp 15, 18 – 19, in AFCHO Hildreth, USAF Special Air Warfare Doctrines and Capabilities, 1963 (AFCHO, 1964) p 32, in AFCHO.

13. Msg 25786, PACAF to Hq USAF, 13 Mar 62; Hist, 13th AF, 1962, p 63.

14. Rcrd, SOD Hono Conf, 16 Dec 61, item 2; memo, R/Adm J. H. Wellings, Vice D/Joint Staff to Asst to SOD Spec Ops, 3 Feb 62, subj: Bien Hoa Ops of 21 Jan 62; memo for rcrd by Owens, 7 Feb 62, subj: Bien Hoa Cambodian Village Incident.

15. See note above; ltr, U. Alexis Johnson, Dep Under Secy of State to W. P. Bundy, Asst SOD/ISA, 24 Jan 62, no subj; memo, Chmn JCS to SOD, 30 Jan 62, no subj; msg 36258, PACAF to C/S USAF, 12 Feb 62.

16. Msg 36257, PACAF to C/S USAF, 12 Feb 63; msg 36121, PACAF to C/S USAF, 12 Feb 61; msg 36122, PACAF to C/S USAF, 13 Feb 62; Rcrd, SOD Hono Conf, 19 Feb 52, item 5.

17. Msg 14539, PACAF to USAF, 16 Mar 62; Hist, 2d ADVON, 15 Nov 61 - 8 Oct 62, pp 162 - 64; msg 1173, Dept of State to AmEmb Saigon, 4 Apr 62; New York Times, 16 Mar 62; New York Herald Tribune. 25 Mar 62.

18. LeMay Rpt, 24 Apr 62.

Chapter IV

1. Rcrd, SOD Hono Conf, 11 May and 23 Jul 62; rpt, Visit to SEA by SOD, 8 - 11 May 62, Ch III, pp 1- 3; New York Times 12 May 62; Philadelphia Inquirer, 22 May 62; New York News, 2 Jun 62; Baltimore Sun, 12 Jul 62; rpt, 0SI to IG Hq USAF, Jun 63, subj: Viet Cong, in OSAF 290 - 63.

2. Rcrd, SOD Hono Conf, 23 Jul and 8 Oct 62, memo, V. H, Krulak, Off of Spec Asst for COIN and Spec Activities, OSD to C/S Army et al. 29 Nov 62, subj: Three-Year Prog for U.S. Mil Pers and Materiel Sup for SVN; JCS 2343/119, 4 Feb 63.

3. Rcrd, SOD Hono Conf, 23 Jul 62, pp 7-1 to 7-2.

4. Memo, SOD to Chmn JCS, 23 Aug 62, subj: Three-Year Prog for U.S. Mil Personnel and Materiel Sup for SVN.

5. JCS 2343/191, 4 Feb 63; Rcrd, Discussions on VN at PAC0M Hq, Dec 17-18, 1962 pp 5-9; Hist, 2d ADVON, 15 Nov 61 - 8 Oct 62, app D, item 18.

6. Rcrd, Discussions on VN at PACOM Hq, Dec 17-18, 1962, pp 5-9; msg 11889, PACAF to C/S USAF, 28 Dec 62.

7. JCS 2343/191, 4 Feb 63; memo, B/G G.C. Kelleher, Asst C/S J-3 to Senior Advisors in I, II, III, and IV Corps, 21 Feb 63, subj: NCP.

8. Msg 52507, PACAF to C/S USAF, 1 Dec 63; msg 060737, AmEmb Saigon to DOD, et al., 6 Jul 63; Hist, 2d ADVON, 15 Nov 61 - 8 Oct 62, app D, item 18.

9. Msg 060837, AmEmb Saigon to DOD, et al., 6 Jul 63; msg 32186, PACAF to C/S USAF, 10 Jul 63; msg 39276, PACAF to C/S USAF 30 Aug 63.

10. Msg 69799, Hq USAF to PACAF, 15 Mar 62; msg 16838, PACAF to Hq USAF, 17 Mar 62; JCS 2343/128, 16 Jul 62; Hist, 2d ADVON, 15 Nov 61 - 8 Oct 62, pp 153 - 155; Hist, AFLC, 1 Jul 62 - 30 Jun 63, p 26; Rcrd, S0D Hono Conf, 8 Oct 62, p 5, JCS J-3 ops 200-62-,r 20-Nov 62; JCS 2343/175, 4 Dec 62; Hist, D/Ops, Jan-Jun 63, pp 64-65; memo, Dep SOD to Chmn JCS, 31 Dec 62.

Notes For Part One

11. Memos, Carpenter, D/Plans, to C/S USAF, 7 Feb 63 and 28 Feb 63, subj: Air Aug, SVN; Army Staff memo 36-63 to D/Jt Staff, 27 Feb 63; JCS 2343/202, 28 Feb 63; Hist, CINCPAC, 1963, p 213; Hist, D/Ops, Jul-Dec 63, Sec V, p 3; Hist, D/Aerospace Progs, Jan-Jun 63, p 25; Hildreth, Special Air Warfare, 1963, pp 30-32.

12. Talking Paper for Chmn JCS for SOD Mtg, 18 Feb 63; memo, Worden to C/S USAF, 30 Nov. 63, subj: Mil Sit in RVN; JCS 2343/21, 25 Mar 63 Rcrd, SOD Hono Conf, 6 May 63; Hist, D/Ops, Jan-Jun 63, pp.54-55; Hist, 13th AF, Jan - Jun 63, pp 72-73; msg 58388 PACAF to C/S USAF, 21 Mar 63.

13. Memo, C/S USAF to JCS, 1 Apr 63, subj: PCS Tsfr of USAF Forces.

14. Ltr, DAF to PACAF, 17 Jun 63, subj: Activation of the 1st Air Commando Sq (C) and Certain Other USAF Unit Actions, in AFCHO; Hist, TAC, Jan - Jun 63, p 71, Hist, D/Ops, Jul - Dec 63, Sec V, p 3; Hist, D/Aerospace Progs, Jan-Jun 63, p 25.

15. Hist, 2d ADVON, 15 Nov 61- 8 Oct 62, pp 153-55.

16. Msg 8518, PACAF to C/S USAF, 20 Sep 62; Ltr, LeMay to O'Donnell, 1 Sep 52, no subj; Proj CHECO SEA &t, pt V, pp 51-55; Hist, 2d ADVON 15 Nov 61 - 8 Oct 62, pp 146-50; excerpt, SAF testimony before House Cmte on Armed Services, 21 Feb 63, in SAFOI

17. Hist, 2d ADVON, 15 Nov 61 – 8 Oct 62, pp 146, 150-51; msg 55773 PACAF to C/S USAF, 11 Sep 62; msg 1975, PACAF to C/S USAF, 15 Sep 62.

18. Hist, D/Ops, Jan - Jun 62, pp 47 – 48; JCS 2343/135, 1 Aug 62; memo Carpenter to C/S USAF, 3 Aug 62, subj: Decca Navig Sys for SVN; memo, SAF to U. Alexis Johnson, Dep Under Secy State for Pol Affairs, 28 Jan 63, no subj, in OSAF 290-63, ltr, SAF to Sen John Stennis, Chmn Subcmte on Prepared Invest Cmte on Armed Services, 16 Jul 64, no subj, in OSAF 101-64; msg 73737, Hq USAF to AFLC et al., 10 Aug 62; Hist, D/Maint-Engr, Jul-Dec 62, p 53; Hist, AFLC, 1 Jul 62-30 Jun 63, pt II, pp 45 - 52.

19. Hist, 2d ADVON, 15 Nov 61 - 8 Oct 62, p 133; Talking Papers on msg Dept of State to AmEmb Saigon (Harriman to Nolting), Mar 63; see app 1, 2, 3, and 4.

20. Hist, 2d ADVON, 15 Nov 61- 8 Oct.62, p 133, DoD Press Release 16-23, 5 Jan 63; see app 5 and 6.

21. Hist, TAC, Jan - Jun 63, pp 585-87; Talking Papers on msg, Dept of State to AmEmb Saigon, Mar 63; msg COMUSMAC/V to JCS et al Jan 63; see app 2.

22. See note above.

Air War – Vietnam

23. Memo, JCS 2343/221, 25-Mar-63; memo, Worden to C/S USAF, 30 Nov 63, subj Mil Sit in RVN; Talking Paper for Chmn JCS for SOD Mtg 18 Feb 62; Rcrd, SOD Hono Conf, 6 May 63, pp 1-a-1 to 1-2-6; Hist D/Ops Jan – Jun 63, pp 54 – 66; Hist, 13th AF Jan – Jun 63, pp 72-73; msg 58388, PACAF to C/S USAF, 21 March 63

24. Msg C/S USAF to PACAF, 19 Apr 63; msg, PACAF to C/S USAF, 4 Apr 63; Rcrd, SOD Hono Conf, 6 May 63, pp 1-a-5 to 1-a-6; msg 7467, PACAF to C/S USAF, 8 May 63; Hist, TAC, Jan – Jun 63 pp 432, 594 – 95; memo, T. L. Hughes, Bur Int & Rsch, Dept of State to Secy State 22 Oct 63, subj: statistics on war effort show unfavorable Trend, Proj CHECO SEA Rpt, Pt VI pp 54 – 55.

25. Msg 82504, C/S USAF to PACAF, TAC 27 Jul, 63, doc 60 in Proj CHECO SEA rpt, Oct 61- Dec 63, VoI V-A

26. Senate Hearings before subcmte on DOD Appropriations, 88th Cong 2d Sess, DoD Appropriations for 1965, pt 1, pp 14 – 15; Rcrd, SOD Hono conf, 20 Nov 63;. msg 020459, COMUSAC/V to JCS et al., 8 Nov 63 and other weekly COMUSMAC/V "Headway" rpts Nov – Dec 63; msg 180120 CINCPAC to COMUSMAC/V, 1 Feb 64; ltr, SAF to Sen. John Stennis, 16 Ju164; New York Times. 10 and 23 Dec 63; Washington Post 14 Nov 63.

27. Ltr, SAF to Rep Carl Vinson, Chm Cmte on Armed Services, 13 May 64, no subj, in OSAF 101-64; memo for Rcrd by L/Col W.T. Calligan, Dep Ch, Cong Invest Div, SAFLL, 24 June 64, subj: Hearings by Senate Preparedness Invest Subcmte, Senate Cmte on Armed Services, in OSAF 101-64; see app 1, 2, 3, 4, and 5.

28. Msg 132015, COMUSMAC/V to JCS, et al., 13 Jun 63; rpt. AF Study Gp on VN, prep by OSAF, May 64, pt-III; 290724, COMUSMAC/V to JCS et al., 29 Nov 63; memo for rcrd L/Col J.L. Crego, Off D/Plans, 16 Jan 64, no subj; Talking Paper on USAF/U.S. Army a/c losses and damages in VN, 17 Jan 64, in OSAF 101-64; see app 7.

29. Hist, D/Ops, Jan - Jun 64, pp 35 – 40; Hist, 2d AD, Ch I, Jan - Jun 64 pp 45-51; memo, T.D. McKiernan, Asst Dep D/Plans for Policy, D/Plans, to AFCHO, 3 May 65, subj: Draft Study of AFCHO Hist Study in AFCHO.

Chapter V

1. DOD Press Release 16-63, 5 Jan 6l; Washington Star, 6 Jan 63, Washington Post. 6 Jan 63; JCS 2343/l9l, 4 Feb 63; Proj CHECO Rpt pp 89-96.

2. Intvw, author with Col W.V. McBride, Ch Sp Warfare Div, DCS/P&O, 9 Jan 64; JCS 2343/191, 4 Feb 63.

3. Briefing Paper, 11 Jan 63, subj: Air Force Briefing for Gen Wheeler and Others, in Sp Warfare Div, DCS/P&P; Hildreth, Counter-insurgency, 1961 - 1962. pp 25-36.

Notes For Part One

4. JCS 2343/191, 4 Feb 63.

5. Rpt, 11 Feb 63, subj: Air Staff Observations During Trip to SVN, prep by L/G D. A. Burchinal, DCS/P&P.

6. Intvw, author with McBride, 9 Jan 64 and 4 Aug 64; Hist, D/Plans, Jan-Jun 63, p 43.

7. Msg, Dept of State to AmEmb Saigon, 22 Mar 63; memo, L/Col A. T. Sampson, Sp Warfare Div, D/Plans, to C/S USAF, 25 Mar 63, subj: State Msg from Nolting, From Mr. Harriman, in Cold War Div, D/Plans.

8. Memo, Pancake to Carpenter, 14 May 63, subj: Value of Interdiction Sorties in SVN.

9. Ibid.; Rcrd, SOD Hono Conf, 6 May 63.

10. Memo, Pancake to Carpenter, 14 May 63; Rcrd, SOD Hono Conf, 6 May 63.

11. Msg 37957, PACAF to C/S USAF, 9 Jul 62, msg. 68086, Hq USAF to PACAF, 20 Jul 62; memo, Col L. H. Richmond, Dep D/Plans for War Plans to Cold War Div, D/Plans, 21 Nov 62, subj: Rules of Engagement for Air Ops in SVN; msg 16838, PACAF to Hq USAF, 17 Mar 62; msg 38569, PACAF to C/S USAF, 4 Jul 62; JCS 2343/128, 15 Jul 62.

12. Msg 59249, PACAF to USAF, 3 Mar 62; msg 66331, Hq USAF to PACAF 5 Mar 62; msg 14539, PACAF to USAF, 16 Mar 62; LeMay report 24 Apr 62; DOD Press Release 16-23, 5 Jan 63; see app 6.

13. Memo, L/G G. P. Disosway, DCS/O and Momyer, D/Ops Rqts, to C/S USAF, 22 Dec 52, subj: Trip Rpt to SVN; (hereinafter cited as Disosway Rpt), in OSAF 11-62; Proj CHECO, SEA Rpt, Pt V, pp 93-94; Intvw with McBride, 4 Aug 64.

14. Proj CHECO SEA Rpt, pt V, pp 17-18; pt 1, pp 60-61.

15. Memo, Col C. C. Wooten, Ch Spec Advsy Gp, Off Asst C/S Intel to Asst C/S Intel, 7 Nov 62, subj: The Role of Intel in COIN Ops.

16. Memo, Richmond to D/Plans, 21 Nov 52, subj: Rules of Engagement; memo, Worden to C/S USAF, 30 Nov 63, subj: Mil Sit in RVN; Proj CHECO SEA Rpt, pt V, pp 6-10 and pt II, pp 67-91 msg 170525, COMUSMAC/V to JCS 924, 17 Jan 64.

17. Hist, D/Telecon, Jul - Dec 62, p 108; Hist, D/Plans, Jan - Jun 63, pp 212 - 43; Hist, D/Maint-Engr, Jan-Jun 63, p 71 Proj CHECO Rpt; pt VI. pp 46 - 49; memo, Worden to C/S USAF, 23 Jan 64, subj: JCS Briefing by Anthis.

18. Proj CHECO SEA Rpt, pt VI, p 13 - 17.

19. Ibid., pt IV, pp 38 – 39, app 6.

Air War – Vietnam

20. Msg 2204, JCS to CINCPAC, 23 Apr 62.
21. Msg, CINCPAC to JCS, 26 Apr 62, msg, CINCPAC to Actg Chmn JCS, 22 Apr 62.
22. Msg 56208, PACAF to C/S USAF, 24 Jul 62; msg 19949, PACAF to C/S USAF, 11 Aug 62, msg 36161, PACAF Lo C/S USAF, 25 Aug 56; Disosnay Rpt, 22 Dec 62.
23. Hist, CINCPAC, 1963, p 133; msg 19949, CINCPAC to JCS, 11 Aug 62; msg 88023, C/S USAF to PACAF, 4 Oct 62; Hist, D/Plans; Jan - Jun 63, pp 43, 180; Proj CHECO SEA Rpt, pt IV, pp 41 - 42.
24. JCS 2428/240-1, 7 Nov 63, memo, Pres to Chmn JCS, 2 Dec 63, no subj; memo, SOD to SA, 6 May 64, no subj; Hist, D/Plans, Jul - Dec 63, p 233, msg 83999, C/S USAF to PACAF, 11 Jan 64.
25. Memo, Carpenter to C/S USAF, Apr 64, subj: Svc Resp for Manning Posn of C/S USMAC/V; Chart, dtd 1961 – 64 [on Manpower Auth in SVN], in Off of D/M&O, DCS/P&P; see app 8.

CHAPTER VI

1. Hist, CINCPAC, 1961, pt 1, pp 162, 183.
2. Memo, SOD to Secys of Mil Depts, et al., 5 Sep 61, subj: Exper comd for sub-limited war, in OSAF; Rcrd, SOD Hono conf, 16 Jan 62, item 15.
3. Msg 63306, Hq USAF to PACAF, 21 Feb 62 memo, Burchinal to C/S USAF, 8 Jun 62, subj: SOD/JCS Weekly Intel/Ops Briefing, SVN (Project Headway); msg 67265, Hq USAF to PACAF, 18 Jul 62; JCS 2343/131, 18 Jul62; msg 65801, Hq USAF to PACAF, TAC, AFSC, 23 Jul; JCS 2343/129, 28 Jul 62; memo, JCS to SOD, no subj.
4. Msg 78216, Hq USAF to PACAF, 28 Aug 62; .JCS 2343/190, 12 Feb 63, JCS Team Trip to SVN, 4 Mar 63; memo, Brown, to Dep SOD, 16 Jan 63, subj: Equip Testing in SVN, in OSAF 290-63; Hist, CINCPAC 1962, pp 168-69.
5. Msg 83087, C/S USAF to PACAF, 15 Sep 62; msg 83574 C/S USAF to PACAF, 18 Sep 62; memo, C/S USAF to JCS, 28 Sep 62, subj: Estab of Army Test Unit, VN; msg 89547, C/S USAF to TAC, 10 Oct 62.
6. Msg 46129,PACAF to C/S USAF, 12 Oct 62; Study, Sit in SVN, 17 Dec 62; msg 94785, Hq USAF to PACAF, 7 Nov 62; msg 94853, Hq, USAF to TAC, 7 Nov 62; memo, Worden to C/S USAF, 17 Dec 62, subj: Sit in SVN; JCS 2343/203, 4 Mar 63; msg 66967, C/S USAF to PACAF, 9 Jan 63; msg 97337, C/S USAF to PACAF, et al., 23 Nov 62.
7. Memo for Rcrd by L/Col H. M. Chapman, combined plans Div, D/Plans, 19 Nov 62, subj: Army Test Unit; JCS 2343/190, 12 Feb 63.
8. JCS 2343/203, 4 Mar 63

Notes For Part One

9. Memo, C/S USAF to JCS, 21 Aug 63, subj: Test Plan Air Assault Task Force; memo, C/S USAF to JCS, 23 Oct 63, subj: Serv Test Prog in SVN. memo, Pancake to C/S USAF, 5 Jan 63, subj: Review of Mil Sit in SVN; Burchinal Rpt 11 Feb 63.

10. JCS 2343/203, 4 Mar 63; memo, Chmn JCS to SOD, 11 Apr 63; subj: R&D Comd Relations; memo, SOD to JCS, 23 Apr 63, same subj.

11. Memo Carpenter to C/S USAF, 6 Feb 64, subj: Mtg with DDR&E and D/O&MP,OSD; Hist, CINCPAC, 1963, pp 223-24.

12. See note above.

13. Burchinal Rpt, 11 Feb 63; Study, Sit in SVN, 17 Dec 62.

14. Memo, Col A.S. Pouliot, Off D/Ops to C/S USAF, 21 Oct 63, subj: Results of the Employment of OV-1 Mohawk, in support of COIN ops.

15. Burchinal Rpt, 11 Feb 63; House Hearings before Subcmte on Appropriations for 1964, pt 2 pp485, 494-95. Senate Hearings before Armed Helicopters 88th Cong, 1 Sess, Military Procurement Authorization FY 1964 pp 314 – 15

16. Ltr, JOEG/V-ARPA Field Unit to CINCPAC, 25 July 63, subj: JOEG/V's Eval of Armed Helicopters; lst Ind, Hq USMAC/V, 15 Aug 63; 2d Ind CINCPAC to JCS, 25 Sep 63, same subj; JCS 2343/270-1, 27 Nov 63.

17. Ltr, JOEG/V-ARPA Field Unit to C/S Army thru CINCPAC, 7 Dec 63, subj: Employment of CU-2B Caribou, in Support of COIN Ops.

18. Final Rpt, Opl Test and Eval, YC-123H in RVN, 1 Jun 63, prep by Hq 2d AD, in Hist, 13th AF, Jul - Dec. 63, Vol III memo, JOEG/V-ARPA Field Unit thru COMUSMAC/V to C/S USAF 26 Aug 63, subj: 2d AD Test and Eval of YC-123H in RVN.

19. Final Rpt, Opl Test and Eval, U-10 in RVN, prep by 2AD, 1 Jun 63, in Hist, 13th AF, Jul - Dec 63, III; memo, JOEG/V-ARPA Field Unit to C/S USAF, 3 Sep 63, subj: Eval of Test Results of Opl Test and Eval of U-10.

20. Memo, JOEG/V-ARPA Field Unit to C/S USAF, thru COMUSMAC/V and CINCPAC, 11 Oct 63, subj: 2d AD Test and Eval of TAPS in VN.

21. Final St, Tac Anlys of T-28B A/C in RVN, 30 Apr, 63, in Hist, 13th AF, Jul - Dec 63, II, 13 Hist, 13th AF, Jul-Dec 63, pp 73-74; msg 77337, C/S USAF to PACAF, et al., 23 Nov 62.

22. Final Rpt, Tac Anlys of C-123B A/C in RVN, 15 Apr 63 and Tac Anlys of TF-102 A/C in RVN, 30 Jun 63, both in Hist, 13th AF, Jul - Dec 63, II; memo, Col R. B. Shick, Mil-Asst to SAF, 14 Oct 63, subj: TAC Anlys of C-123B A/C in RVN, in OSAF 290-63.

23. Msg 96883, C/S USAF to PACAF 18 Sep 63
24. Memo, Burchinal to CINCPACAF, 26 Dec 63, no subj
25. Memo, Dep SoD to Pres, 21 Nov 61, subj: Defol Ops in VN, 21 Nov 61; Rcrd, SOD Hono Conf, 16 Dec 61 and 15 Jan 62
26. Memo, Brown to SA et al., 29 Nov 61, subj: SVN; memo, P.F. Hilbert, Dep for Reqts Rev, Off of Under SAF, to Bundy, 12 Dec 61, no subj, in OSAF 1257-61 Rcrd, SOD Hono Conf, 16 Dec 61; memo, McKiernar, to AFCHO, 3 May 65, in AFCHO.
27. Rcrd, SOD Hono Conf, 15 Jan 62, p 13, 59-60; Hist PACAF, Jul-Dec 61, I, pt 2, p 28; Hist, 13th AF, Jul- Dec 61, p 86.
28. Rcrd, SOD Hono Conf, 15 Jan 62 and 19 Feb 62; Baltimore Sun, 25 Jan 62.
29. Memo, Col W. J. Meng, Exec Vice C/S to SAF 15 May 62, subj: Status of Defol Proj SVN; ltr, CINCPAC to JCS, 24 Jul 62, no subj.
30. Ltr, Ch CHECO Team to J. W. Angell, Ch AFCHO, 3 Jun 63, subj: Herbicide Defol, in AFCHO; Hist, D/Ops, Jul - Dec 62, pp 43-44; NSA memo 178, 9 Aug 62, subj: Destruction of Mangrove Swamps in SVN; Rcrd, Discussions on VN at PACOM Hq, Dec 17- 18, 1962, pp 41-43, JCS 2343/214, 21 Mar 62.
31. Memo, Chmn JCS to SOD, 28 Sep 62, subj: Rev and Opl Eval of Defol; memo, Chmn JCS to SOD; 9 Nov 62, subj: Defol Proj in SVN.
32. Memo, Dep SOD to Chmn JCS, 13 Oct 62, subj: Herbicide Proj; memo, Bundy to SOD, 27 Nov 62, Subj: Defol/Herbicide Prog in SVN.
33. Rcrd, SOD Hono Conf, 23 Jul 62, LeMay Rpt, 24 Apr 62; memo, Chmn JCS to SOD, 28 Jul 62, subj: Chemical Crop Destruction, SVN; SOD to Pres, 8 Aug 62, same subj; memo, McBride to C/S USAF, 11 Sep 62, same subj; memo, Secy State to SOD, 28 Aug 62, subj: VN proj for Crop Destruction.
34. Rcrd, SOD Hono Conf, 8 Oct 62; Hist, CINCPAC, 1962, pp 185-87; Hist, CINCPAC, 1963, p 227; Hist, D/Ops, 1 Jul - 31 Dec 62, p 44; ltr, CINCPAC to JCS, 22 Mar 63, subj: Rpt Concerning Psych Aspects of Use of Defol in RVN; ltr, CHECO Team to Angell, 3 Jun 63.
35. Memo, Chmn JCS to SOD, 17 Apr 63, subj: Defol and Crop Destruction in SVN; Hist, D/Plans, Jan - Jun 63, p 238.
36. Hist, CINCPAC, 1963, pp 227-30; Hist, 13th AF, Jul - Dec 63, pp 68-69.

CHAPTER VII

1. Memo L/Col N. F. Lambertson, Off D/Ops, to Dep D/plans, 20 Jul 61, subj: increase in GVN Border Patrol and insurgency suppression

Notes For Part One

      capabilities; memo, Burchinal to C/S USAF, 20 Jun 61, subj: Increase in GVN Forces.

2. Memo for Rcrd by Off, COMUSMAC/V, 29 Sep 61, subj: Mtg at Independence Palace, Saigon; JCS 2343/27, 19 Oct 61.

3. JCS 2343/29, 25 Oct 51; memo, Burchinal to C/S USAF, 26 Oct 61, subj: SEA; memo, C/S USAF to JCS, 27 Oct 61, subj: SEA.

4. Memo for Rcrd by Off, COMUSMAC/V, 29 Sep 61; Rcrd, SOD Hono Conf, 16 Dec 61, item 7; Proj CHECO SEA Rpt, pt IV, pp 33-38.

5. Journal of Military Assistance (JMA), prep by Asst for Mutual Security, Sep 61, p 153; Dec 61, p 144; Jun 62, p 179 memo, Jt Staff to Cfmn JCS, 14 Nov 61, subj: SVN; memo, Brown to SA et al., 29 Nov 61, subj: SVN; Rcrd, SOD Hono Conf, 15 Jan 62, Hist, CINCPAC 1962, p 190; Hist, 13th AF, 1962, pp 103 – 04, Hist, 2d ADV0N, 15 Nov 61 - 8 Oct 62, app D, item 21; Air Force and Space Digest, Sep 64, p 103.

6. Rcrd, SOD Hono Conf, 21 Mar 62 and 23 Jul 62; Hist, 13th AF, 1962, pp 103-04; Hist, AF Study Gp on VN, May 1964, pt III.

7. LeMay Rpt, 24 Apr 62; Hist, CINCPAC, 1962, p I90.

8. Msg 3612, PACAF to C/S USAF, 18 Jan 63; Ltr, CINCPAC to JCS, 15 Jan 63, subj: Comprehensive Plan for SVN; JCS 2343/203, 4 Mar 63; msg 3668r, PACAF to C/S USAF, 2d ADVON, 17th AF, 15 Aug 63 Disosway Rpt, 22 Dec 62; Talking Paper for Chmn JCS for SOD Mtg, 25 Mar 63; no subj, 25 Mar 63; Hist, CINCPAC, 1963, pp 204-05.

9. Study, Sit in SVN, 17 Dec 62; msg 22419, PACAF to C/S USAF, 22 Feb 63; msg 33847, CINCPAC to Hq USAF, 15 Apr 63; memo, Dep SOD to Chmn JCS, 31 Dec 62, subj: Farmgate Aug, in OSAF 11-62; msg 160406, CINCPAC to JCS, 16 Apr 63, in Proj CHECO SEA Rpt, Pt VI-B, doc 25; Hildreth, Special Air Warfare, 1963, pp 34-35.

10. Hildreth, Special Air Warfare, 1963, pp 34-35; Rcrd, SOD Hono Conf, 6 May 63; app 6; JMA p 179.

11. JMA, Jun 63, p 193; Sep 63, p 187; Hildreth, Special Air Warfare, 1963 pp 34-35, Proj CHECO SEA Rpt, Pt VI, p 28,

12. JMA, Sep 6, p l63.

13. Memo, L/G J. A. Dabney, Actg Asst SOD/ISA to Chmn JCS, 19 Aug 61, no subj; JCS 2343/22, 7 Oct 61; msg 33743, Dep of State-DoD to AmEmb Saigon, 19 Oct 61; JCS 2343/186, 15 Jan 63.

14. Rcrd, SOD Hono Conf, 23 Jul, 61, pp 1-5; JCS 2343/186, 15 Aug 63; Disosway Rpt, 22 Dec 62; Proj CHECO SEA Rpt, Pt III, pp 20-21.

15. JCS 2343/186 15 Jan 63; ltr, SAF to SOD, 16 Mar 63, no subj.

16. Memo, B/G A. N. Williams, Dep O/Plans for Policy to Cold War Div, D/Plans, 15 Feb 63, subj: Geneva Agreements and Jets; msg 92474, C/S USAF to PACAF, 11 Apr 63.

17. Memo, SOD to Chmn JCS, 17 May 63, subj: Jet A/C for SVN; memo, SOD to SAF, 27 May 63, subj: A/C and Pilots for VNAF; Rcrd, SOD Hono Conf, 6 May 63; Proj CHEC0 SEA Rpt, pt III, p 22.

18. Msg 25890, PACAF to C/S USAF, 8 Jan 64; proj CHECO SEA Rpt. pt III, p 22.

CHAPTER VIII

1. Msg 062345, COMUSMAC/V to JCS, et al., 6 Dec 62; msg COMUSMAC/V to JCS, et al., 10 Jan 61; New York Times, 18 Oct 64.

2. Christian Science Monitor, 20 Sep 62; New York Herald Tribune, 20 Sep 62; Washington Star, 8 Oct 62; New York Times, 13 Dec. 62; Baltimore Sun; 21 Jan and 5 Feb 63; Washington Post 2 Feb 63; Transcripts of SOD Press Confs, 23 and 30 Jan 63, in SAFOI; Senate Hearings before Cmte on Appropriations, 87th Cong, Foreign Assistance and Related Agencies Appropriations for 1963, p 771.

3. Memo, Worden to Cold War Div, D/Plans, 17 Dec 62, subj: Sit in SVN: An Appraisal; memo, Pancake to C/S USAF, 5 Jan 63, subj: Review of Mil Sit in SVN; SAF Press Statements, 8 Jan, 23 Feb 63, in SAFOI; Rcrd, SOD Hono-Conf, 6 May 63, pp 1-a-8 to 1-a-9; Study, VC Infiltration, prep by Hq USMAC/V, 31 Oct 64.

4. Memo, Worden to Cold War Div, D/Plans, 12 Dec 62, subj: Sit in SVN.

5. Memo, Pancake to C/S USAF, 5 Jan 63, subj: Review of Mil Sit in SVN.

6. Ltr CINCPAC to Asst SOD/PA, 26 Nov 62, subj: News Correspondents View Concerning SVN War and Govt.

7. Rpt by Sen Mike Mansfield, et al., to Senate Cmte on Foreign Relations, 88th Cong, lst Sess, Vietnam and Southeast Asia, 1963, pp 6 - 8.

8. JCS 2343/221, 29 Mar 63; msg 56517, CINCPAC to JCS, 30 Apr 63; JCS 2343/241-1, 9 May 63; memo, Chmn JCS to SOD, 28 May 63; JCS 2343/248, 18 May 63, Rcrd, SOD Hono conf, 6 May 63, pp 1-1to 1-4; Hist, CINCPAC, 1963, pp 240-41.

9. Msg 19655, PACAF to C/S USAF, 5 Apr 63; Rcrd, SOD Hono Conf, 5 May 63, pp 4-a-1 to 4-a-3; Proj CHECO SEA Rpt, pt III, pp 45 - 49, and pt III-A, doc 49.

10. Rcrd, SOD Hono Conf, 6 May 63, pp 1-1 to 1-4; New York Times, 9 & 10 May 63; New York Times, 12 June 26 Aug, 10 Sept and 17 Sep 63; Washington Post, 31 Oct 63.

11. Memo, Worden to C/S USAF, 27 Aug 63, subj: Background Paper on Pol Sit in SVN; New York Times, 3 Sep 63; Washington Post, 17 Sep

Notes For Part One

63; msg 38307, PACAF to C/S USAF; 29 Aug 63; House Hearings before subcmte on Appropriations, 88th Cong, 2d Sess, DOD Appropriations for 1965, pt 4, p 11.

12. Memo, Chnm JCS to Pres, 9 Sep 63, subj: draft ltr, SOD to Secy state, 4 Nov 63; Senate Hearings before Subcmte on DoD appropriations, 88th Cong, 2d Sess, DOD Appropriations for 1965, pt 1, pp 14 - 15; New York Times, 3 –Oct 63.

13. New York Times. 3 Nov 63; Washington Post, 11 Nov 63; JMS, Dec 63, p 177.

14. Rcrd SOD Hono Conf, 6 May 63; memo, Chmn JCS to SOD, 20 Aug 63, subj: Summary Rpt on 8th SOD Conf, 7 May 63; msg 4992, PACAF C/S USAF, 3 Aug 63; Washington Post, 15, 16 Nov and.3 Dec 63; New York Times, 15 and Dec 63.

15. NSA memo 273, 26 Nov 63; memo, SOD to SA et al., 6 Dec 63, subj: NSA Memo 273, 26 Nov 63.

16. Memo, Pres to Chmn JCS, 2 Dec 63; msg 50231, CINCPAC to JCS, 26 Jan 64; New York Times, 21, 22 Dec 63 and 2 Jan 64.

17. Memo, C/S USAF to PACAF, I Dec 63; msg 85559, C/S USAF to PACAF, 17 Jan 64; memo, C/S USAF to JCS; 22 Jan 64, subj: Increase in Aerial Recon Capability in SEA; memo, JCS to SOD, 22 Jan 64, subj: VN and SEA; House Hearings before Subcmte on Appropriations, 88th Cong, 2d Sess, DOD Appropriations for 1965, Pt 4, p 12.

Air War – Vietnam

Notes For Part Two

NOTES TO PART TWO

Unless otherwise noted, all primary sources cited (letters, memos, JCS papers) are located in Headquarters USAF Directorate of Plans File RL (64) and (65) 38-9, depending on the year of the source.

Chapter I

1. Jacob Van Staaveren, Plans and Policies In South Vietnam, 1961- 1963, (forms Part One of this volume), VIII (TS); N.Y. Times, 12 Jan 64.
2. Hist, 2d AD, Jan - Jun 64, pp 24-25 (S); msg 45203, PACAF to C/S USAF, 4 Mar 64 (S); N.Y. Times, 12 Jan 64.
3. Memo, M/Gen. J.W. Carpenter III, D/Plans, DCS/P&O to C/S USAF, 25 Jan 64, subj: Pacification Plan for Long An Prov, RVN, (S); msg 50231, CINCPAC to JCS, 26 Jan 64 (TS).
4. Memos, Chmn JCS to SOD, 22 Jan 64, subj: VN and SEA (TS); C/S USAF to JCS, 22 Jan 64 (TS).
5. Ibid; MAC/V Comd Hist, 1964, p 47 (TS).
6. Current History, Mar 64, p 192; N.Y. Times, 30 and 31 Jan 64; Wash Post, 1 Feb 64.
7. Hist, 2d AD, Jan - Jun 64, I, pp 21-25 (S); Statement by Secretary McNamara, 17 Feb 64, in House Hearings before Subcmte on Appropriations, 88th Cong, 2d Sess, DOD 1965 Appropriations, IV, pp 12 - 13; N.Y. Times, 2 and 19 Feb 64.
8. NASM 280, 14 Feb 64 (S); memo for rcrd by M.V Forrestal, subj: SVN, 20 Feb 64 (S); msg 3-3-46, Hq USAF to PACAF, 2 Mar 64 (TS); Wash Post, 25 Feb 64.
9. Memo, Chmn JCS to Dir Jt Staff, 5 Feb 64, subj: Revitalized SVN Campaign (S).
10. JCS 2343/317-2, 13 Feb 64 (TS); memo, Wm V. McBride, Chief, Spec Warfare Div to Dep D/Plans, DCS/P&O, 27 Feb 64, subj: VN and SEA (S); Army Staff memo 62-64, 27 Feb 64 (TS).
11. Hist, 2d AD, Jan - Jun 64, I, pp 25 - 29 (S).
12. Hist, 2d AD, Jan - Jun 64, I, pp 29 - 33 (S).
13. Memo, Carpenter to C/S USAF, 25 Jan 64 (S); memo, SAF to SOD, 4 Feb 64, subj: Pacification Plan for Long An Prov (S); JCS 2343/317-2, 13 Feb 64 (TS); msg 93264, C/S USAF to PACAF, 15 Feb 64 (S).

Air War – Vietnam

14. Msg 71, PACAF to C/S USAF, 1 Feb 64 (S); Hist, 2d AD, Jan - Jun 64, I, pp 34 - 36 (S); msgs 37391 and 37408, PACAF to C/S USAF, 14 May 64 (S); msg 19580, PACAF to USAF, 10 Jun 64 (S).

15. Memo, C/S USAF to JCS, 21 Feb 64, no subj (TS); msg 3-3-46, Hq USAF to PACAF, 2 Mar 64 (TS).

16. Memo, SOD to Chmn JCS, 21 Feb 64, subj: SVN (S); N.Y Times, 19 Feb 64.

17. JCS 2343/326-6, 1 Mar 64; JCSM-168-64, 2 Mar 64 (TS); JCSM-174-64, 2 Mar 64 (TS).

18. JCSM-174-64, 2 Mar 64 (TS).

19. JCS 2343/236-6, 1 Mar 64 (TS); memo, Carpenter to C/S USAF, 25 Jun 64, subj: Mtg with PACOM Planners (S); Hist, D/Plans, Jul-Dec 64, pp 46-57 (TS); Hist, CINCPAC, 1964, p 54 (TS).

20. N.Y. Times, 20 - 28 Feb 64.

21. Msg 54337, PACAF to C/S USAF, 10 Mar 64 (TS); Unsigned memo of Conversation between SOD and P.M. of SVN, 13 Mar 64 (TS).

22. Msg 54337, PACAF to C/S USAF, 10 Mar 64 (TS); msg 58129, PACAF to C/S USAF, 12 Mar 64 (S).

23. NSAM 288, 17 Mar 64 (TS); JCSM-245-64, 20 Mar 64 (TS); JCS 2343/347-1, 20 Mar 64 (TS): Hist, CINCPAC, pp 49-51 (TS); JCSM-222-64, 14 Mar 64 (TS); DOD Pamphlet for Armed Forces Info and Educ, 15 Apr 63, Vol 3, No 20, subj: U.S. Policy in VN (U); N.Y. Times, 14 and 20 - 21Mar 64 and 21 Oct - 4 Nov 64; Balt Sun 9, 10 and 17 Dec 64; msg 05822, 2d AD to C/S USAF, 29 Oct 64 (TS);,Talking Paper for the JCS for SOD-JCS Mtg, 2 Nov 64, 30 Oct 64, subj: Probl of Cambodia Border Incident (TS).

24. Hist, CINCPAC, 1964, pp 49 - 51 (TS); JCSM-541-64, 24 Jun 64 (TS).

CHAPTER II

1. N.Y. Times, 17 - 20 Apr 64.

2. JCSM-298-64, 14 Apr 64 (TS); Hist, D/Plans, Jul-Dec 64, pp 58 – 59 (TS).

3. Memo for Rcrd, W.P. Bundy, Chief OSD/ISA, 29 Apr 64, subj: Discussion of Poss Extended Action • • • in VN (TS); JCS 2343/360-1, 22 Apr 64 (TS); N.Y. Times, 13 - 24 Apr 64.

4. JCS 2343/360-1, 22 Apr 64 (TS).

5. Memo, Wheeler to SOD and JCS, 22 Apr 64(subj: Trip Rpt, VN (TS); Checo SEA Rpt, Jul – Dec 64, pp 92 - 93 (TS).

6. Msgs 37391 and 37408, PACAF to C/S USAF, 14 May 64 (TS).

Notes For Part Two

7. Asst SOD/ISA News Release 389-64, 14 May 64; N.Y. Times, 15 and 19 May 64.

8. JCSM-429-64, 19 May 64 (S); JCSM-468-64, 28 May 64 (S); JCSM-470-64, 30 May 64 (S); memo, SOD to Chmn JCS, 23 May 64, no subj (U); msg 19580, PACAF to Hq USAF, 10 Jun 64 (S).

9. Memos, Carpenter to C/S USAF, 3 Mar and 5 Apr 64, subj: VN and SEA (S); memo, JCSM-288-64, 8 Apr 64 (S); Hist, CINCPAC, 1964, pp 306 - 08 and Chart IV (S); Hist, 2d AD, Jan-Jun 64, I, pp 84-92 (S); Hist Rpt, D/Policy, Hq PACAF, Apr 64, p 2, in Hist, PACAF, Jan-Jun 64, I, pt 2 (TS).

10. Ibid.

11. Memo, J.A. Mendenhall, Office of SEA Affairs, State Dept to McGeorge Bundy, Spec Asst to the Pres et al, 27 May 64, no subj (S).

12. Talking Paper for JCS Mtg on 27 Jul 64, 27 Jul 64, subj: Actions Relevant to SVN, with atch draft memo to Pres (TS); N.Y. Times, 20 - 22 May 64; msg 58614, CJCS to CINCPAC, 19 Apr 64 (TS).

13. JCSM-469-64, 30 May 64 (TS).

14. CSAFM-459-64, 28 May 64 (TS).

15. CSAFM-459-64, 28 May 64 (TS); memo, M/G. R. F. Worden, Dep D/Plans, DCS/P&D to C/S USAF, 29 May 64 with atch papers (TS); memo, Chmn JCS to SOD, 4 Jun 64, subj: Obj and C/A-SEA (TS); CM-1454-64, 5 Jun 64 (TS); CM-1450-64, 2 Jun 64 (TS); memo, SOD to Chmn JCS, 10 Jun 64 (U); JCS 2343/423 11 Jul 64 (TS); Hist, D/Plans, Jan- Jun 64, p 20 (TS).

16. Memo, L/Col J.B. Owens, Off of Dep Dir of Plans for Policy, DCS/P&O, 10 Jul 64, subj: Outline Plans for Air Strikes Against NVN, with atch memo to Pres (TS).

17. Memo, B/G R.A. Yudkin, Dep D/of Plans for Policy, DCS/P&O, 26 Jun 64, subj: SOD Hono Conf, 1 and 2 Jun 64 (TS); N.Y. Times, 1 and 2 Jun 64.

18. Chmn JCS to J Staff,4 Jun 64, subj: Required Action Resulting Hono Mtg, 1-2 Jun 64 (TS); JCS 2343/411, 4 Jun 64 (TS).

19. Talking Paper for JCS for SOD-JCS Mtg on 20 Apr 64, 18 Mar 64, subj: RCMD 11, NSAM 288(TS); JCSM-541-64, 24 Jun 64(TS).

20. JCS 2343/426, 26 Jul 64 (TS); JCSM-639-64, 27 Jul 64 (TS).

21. Hist, CINCPAC, 1964, pp 51-52 (TS).

22. Memo, Carpenter to C/S USAF, 16 Jul 64, subj: SVN (S); msg 73043, C/S USAF to PACAF, 2 Jul 64 (S); Hist, CINCPAC,1964, p 2 (TS); N.Y. Times, 21, 24, and 29 Jun 64; Wash Post, 21 Jun 64.

23. Talking Paper for JCS Mtg on 27 Jul 64,27 Jul 64(TS); JCSM-665-64,4 Aug 64 (TS); memo, SOD to Chmn JCS, 7 Aug 64(S).

24. JCS 2343/426,26 Jul 64 (TS); JCS 2343/431, 2 Aug 64(TS); N.Y. Times, 7 and 20 Jul 64, Balt Sun, 7 and 20 Jul 64.

25. Ibid; Hist, D/Plans, Jul-Dec 64, p 320 (TS).

26. Talking Paper for JCS Mtg on 27 Jul 64, 27 Jul 64 (TS); memo, SOD to Chmn JCS, 7 Aug 64, no subj (S); N.Y. Times, 28 Jul 64; Wash Star, 27 Jul 64, MAC/V Comd Hist,1964, p 15 (S)

CHAPTER III

1. Hist, CINCPAC, 1964, pp 366-72 (TS); Hist, D/Plans, Jul-Dec 64, p 52 (TS); DOD News Releases 570-64 and 571-64, 4 Aug 64, and 575-64, 5 Aug 64; Intrvw, CBS-TV with SOD, 5 Aug 64; Wash Post, 6 Aug 64, Times, 8 Aug 64.

2. DOD Press Release 575-64, 5 Aug 64; N.Y. Times, 6 Aug 64; Hist, CINCPAC, 1964, p 372 (TS); Hist, D/Ops, Jul - Dec 64 p 55 (S), JCSM-718-64, 19 Aug 64 (S); memo, SOD to Chmn JCS, 31 Aug 64 (TS).

3. N.Y. Times, 11 Aug 64; Senate Rpt, Background Information Relating to Southeast Asia and Vietnam, prep by Cmte on Fgn Relations, 89th Cong, 1st sess (Revised, 16 June 1965), pp 124-28.

4. Hist, CINCPAC, 1964, p 373 (TS); N.Y. Times, 6 thru 12 Aug 64.

5. Msg 50148, PACAF to C/S USAF, 9 Aug 64 (TS); ltr, Yudkin to PACAF, 28 Sep 64, subj: Trip Rpt, Jt Fact-Finding Team Visit to RVN (S).

6. Memo, Carpenter, Asst DCS/POO for JCS, to C/S USAF, 17 Aug 64, subj: Next C/A (TS); memo, Carpenter to C/S USAF, 12 Aug 64, subj: Recm C/A in SEA (TS); JCSM-746-64, 26 Aug 64 (TS).

7. Memo by W.P. Bundy, Asst Secy State for Far Eastern Affairs, 11 Aug 64, subj: 2d Draft on Next C/S in SEA (S).

8. Ibid: Hist, CINCPAC, 1964, pp 438 - 41 (TS); msg, Saigon to Secy State, 6 Sep 64 (TS); memo, B/G. P.D. Wynne, Jr, Acting Asst C/S Intel to SAFOS, 8 Sep 64 (S), subj: SNIE 53-64, in OSAF 101-64; N.Y. Times, 16 Aug thru 13 Sep 64.

9. JCSM-779-64, 24 Aug 64 (TS).

10. JCSM-746-64, 26 Aug 64 (TS); House Hearings before Subcmte on Appropriations for 1966, pt 3, p 915 (U); MAC/V Comd Hist, 1964, p 68 (TS).

11. Memo, Chmn JCS to SOD, 9 Sep 64, subj: C/A for SVN (TS); Hist, D/Plans, Jul-Dec 64, pp 58-59 (TS).

Notes For Part Two

12. JCS 2343/450, 31 Aug 64 (TS); memo, Chmn JCS to SOD, 9 Sep 64 (TS); Hist, D/Plans, Jul-Dec 64, pp 50 - 51, 58 - 59, and 319 (TS); N.Y. Times, 25 Aug 64; Balt Sun, 25 Aug 64; Chicago Tribune, 1 Sep 64.

13. JCS 2343/457, 9 Sep 64 (TS); Balt Sun 1 Sep 64.

14. NSAM 314, 10 Sep 64 (TS).

15. Hist, CINCPAC, 1964, pp 373-77 and 385-86 (TS); Times, 19, 20, and 21 Sep 64.

16. Hist, CINCPAC, 1964, p 55 (TS); Hist, D/Plans, Jul-Dec 64, pp 56-58 (TS); msg 6555, C/S USAF to PACAF, 1 Apr 64 (TS).

17. Memos, C/S USAF to JCS, 2 and 9 Oct 64 (TS); JCS 2343/477, 8 Oct 64 (TS); Hist, CINCPAC, 1964, pp 50-52 (TS).

18. JCS 2343/439, 12 Aug 64 (TS); JCSN:-835-64, 30 Sep 64 (TS); Talking Paper for the Chmn JCS for Mtg with Amb Taylor on 30 Nov 64, 29 Nov 64, subj: Proposed Discussion Items (TS); Study, 31 Oct 64, subj: VC Infiltration (S), prep by J-2 Div, Hq MAC/V.

19. JCSM-835-64, 30 Sep 64 (TS); msg 57320, COMMAC/V to JCS, 23 Oct 64 (TS).

20. Talking Paper for Chmn JCS Mtg with Amb Taylor on 30 Nov 64 (TS); JCSM-893-64, 2 Oct 64 (TS); msg 57320, COMMAC/V to JCS, 23 Oct 64(TS); N.Y. Times, 13 - 15 and 27 Sep 64.

21. CSAFM-J-24-64, 12 Oct 64 (TS); JCSM-893-64, 21 Oct 64(TS).

22. JCSM-902-64, 22 Oct 64 (TS); memo, SOD to Chmn JCS, 29 Oct 64, no subj (TS).

23. Msgs 50226 and 50227, PACAF to C/S USAF,20 Oct 64(TS); Hist, D/Plans, Jul-Dec 64, p 54 (S).

24. Wash Post, 27 Oct 64; Chicago Tribune, 31 Oct 64, N.Y. Times, 1 Nov 64 and 13 Jun 65.

CHAPTER IV

1. Memo, Gen J.P. McConnell, Vice C/S USAF to SAFOS, 16 Nov 64, subj: Bien Hoa Attack (S); Hist, CINCPAC, 1964, pp 381-82 (S); Phila Inquirer, 3 Nov 64.

2. Hist, D/Plans, Jul - Dec 64, pp 58 - 59 (TS).

3. JCSM-933-64, 4 Nov 64 (TS); Hist, D/Plans, Jul-Dec 64, pp 58-59 (TS); Hist, CINCPAC, 1964, pp 381 - 83 (TS); file, The Bien Hoa Incident, 8 Jan 65 (TS); N.Y. Times, 2 and 3 Nov 64.

4. Ibid.

Air War – Vietnam

5. JCSM-933-64, 4 Nov 64 (TS); memo, Chmn JCS to Dir Jt Staff, 2 Nov 64, subj: C/A in SEA (TS); Hist, D/Plans, Jul - Dec 64, pp 59 - 60(TS).

6. Memo, SOD to Chmn JCS, 13 Nov 64, subj: Recm U.S. C/A in Retaliation to VC Attack on Bien Hoa (TS).

7. Memo, McConnell to SAF, 16 Nov 64, subj: AB Defense (S); memo, M/G. W.K. Martin, Asst DCS/Pers, to SAF, 11 Dec 64, subj: Chronological Summary of Hist Background to VC Mortar Attack on Bien Hoa, 1 Nov 64(S), both in OSAF 101-64; memo, L/G.D.A. Burchinal, Dir Jt Staff to Chmn JCS et al, 1 Sep 64, subj: Scty of AB in SVN (S).

8. Ibid.

9. Memo, McConnell to SAF, 16 Nov 64 (S).

10. Ibid; memo, L/Gen. K.K. Compton, Insp Gen Hq USAF to L/Gen. W.H. Blanchard, DCS/P&O, 4 Dec 64, subj: Airfield Scty in VN (S); Talking Paper for JCS Mtg on 11 Dec 64, 10 Dec 64, subj: Scty Forces in RVN (TS).

11. Talking Paper for JCS Mtg on 11 Dec 64, dated 10 Dec 64 (TS)

12. JCS 2343/501, 9 Dec 64 (TS).

13. Hist, D/Plans Jul - Dec 64, p 61 (TS); Rpt, NSC Working Gp on VN, 13 Nov 64 (TS).

14. JCSM-955-64, 14 Nov 64 (TS); AF Planners Memo 143-64, 18 Nov 64 (TS).

15. Ibid.

16. JCSM-967-64, 18 Nov 64 (TS).

17. JCSM-982-64, 23 Nov 64 (TS); Hist, D/Plans, Jul - Dec 64, pp 61 - 64 (TS).

18. Hist, D/Plans, Jul - Dec 64, pp 61 - 64 (TS).

19. Ibid; Wash. Post, 2 Dec 64; Balt Sun 4 Dec 64.

20. Hist, D/Plans, Jul - Dec 64, pp 61 - 66 (TS); memo, Carpenter to C/S USAF, 29 Dec 64, subj: Ops in Laos (TS).

21. Hist, D/Plans, Jul - Dec 64, pp 61 - 66 (TS).

22. Talking Paper for Chmn JCS on an Item to be Discussed at JCS Mtg, 8 Feb 65, subj: C/A in SEA (TS); N.Y. Times, 12 and 28 Dec 64.

23. JCSM-1047-64, 17 Dec 64 (S); memo, SOD to Chmn JCS, 13 Jan 65, subj: Increase in RVNAF (S).

24. Ibid.

25. &L_ Times, 11, 16, and 21 - 24 Dec 64; Balt Sun, 15 Dec 64.

Notes For Part Two

26. N.Y. Times, 23 - 31 Dec 64; Wash Post, 22 - 24 Dec 64.
27. N.Y. Times, 24 - 31 Dec 64; JCSM-1076-64, 28 Dec 64 (TS); JCSM-70-65, 29 Jan 65 (TS); ltr, Gen. H. Harris, Comdr PACAF to McConnell, 3 Jan 65, no subj (TS).
28. AFCHO Intrv with LeMay, 27 Jan 65 (S).

CHAPTER V

1. Van Staaveren, USAF Plans and Policies in SVN 1961 - 1963. p 104 (in this volume, Part I, Appendix 6 & 7), Hist,2d AD, Jan - Jun, I, pp 3 - 4 (S)
2. Hist, 2d AD, Jan - Jun, I, pp 3 - 5 (S)
3. Ibid; pp 45 - 51(S); Hist, D/Ops, Jan - Jun 64, pp 35-38(S).
4. Memos for rcrd by L/Col W.T. Galligan, Dep Chief, Cong Invest Div., Off of L&L (on Hearings before House Cmte on Armed Services and Senate Preparedness Invest Subcmte), no subj, 21 May - 24 Jun 64,(S), in OSAF 101-64. N.Y. Times, 14 and 21 May 64; Balt Sun, 13 and 27 May 64.
5. Charles H. Hildreth, USAF Special Air Warfare Doctrines and Capabilities, 1963 (AFCHO,1964), pp 50 - 54 (S), JCSM-211-64, 12 Mar 63 (TS); Hist, D/Ops, Jan - Jun 64, pp 35 - 36 (S); Hist, Aerospace Progs, Jan-Jun 64, p 37 (S); Hist, 2d AD, Jan-Jun 64, I, pp 61-62 (S); Life Magazine,4 May 64.
6. Msg 41318, JCS to CINCPAC, 29 Feb 64 (TS); msg 51492, PACAF to C/S USAF, 7 Mar 64 (TS); memo, SOD to Chmn JCS, 20 Mar 64 (TS); msg 54337, PACAF to C/S USAF, 10 Mar 64 (TS); Hist, D/Ops, Jan-Jun 64, pp 37-39 (S); Rpt of USAF Study Gp on VN, May 64 (S),in OSAF.
7. JCSM-350-64, 29 Apr 64 (TS); Hist, 2d AD, Jan - Jun 64, I, pp 51-56 (S); Hist, TAC, Jan-Jun 64, pp 508-09 (S).
8. Ibid.
9. Ibid; Hist, 2d AD, Jul - Dec 64, Vol I, p 69 (S); Hist. Aerospace Progs, Jan-Jun 64, pp 25-26 (S).
10. Hist, Aerospace Progs, Jan - Jun 64, pp 25-26 (S), and Jul - Dec 64, p 33 (S); Hist, 2d AD, Jul - Dec 64, Vol I, p 132 (S).
11. Ibid.
12. Hist, 2d AD, Jan - Jun 64, I, pp 61 – 62 (S).
13. Ibid; JCS 2343/328, 28 Feb 64 (TS); msg 3-3-46, Hq USAF to PACAF, 2 Mar 64 (TS); JCSM-169-64, 2 Mar 64 (TS); JCSM-193-64, 5 Mar 64 (TS); Hist Rpt, D/Policy, Hq PACAF, Feb 64, p 3 (TS).
14. Msg 51492, PACAF to C/S USAF, 7 Mar 64(TS).

Air War – Vietnam

15. Memo, Carpenter to C/S USAF, 16 May 64, subj: SOD Trip to SVN (S); msgs 37391 and 37408, PACAF to C/S USAF, 14 Mar 64 (TS); Hist, CINCPAC, 1964, pp 361 - 64 (TS); msg 13507, C/S USAF to PACAF, 27 Sep 64 (S); memos, SOD to Chmn JCS, 20 and 30 Mar 64 (S); JCS 2343/351-4, 25 Mar 64 (TS).

16. Memos, SOD to Chmn JCS, 20 and 30 Mar 64 (S); JCS 2343/351-4, 25 Mar 64 (TS); Hist, CINCPAC, 1964, pp 361- 64 (TS); Hist, D/Ops, Jan-Jun 64, pp 39 - 40 (S).

17. Ibid.

18. Hist, 2d AD, Jul - Dec 64, I, pp 62 - 70 (S); Hist, PACAF, Jan - Jun 64, I, pt 2, ch 5, p 93 (TS)

19. Ibid

20. Ibid

21. Msg 27228, CINCPAC to C/S USAF, 6 Oct 64 (TS); Hist, CINCPAC, 1964, p 372 (TS); Hist, D/Ops, Jul - Dec 64, p 85 (S).

22. JCSM-665-64, 4 A64 (S); memo, SOD to Chmn JCS, 7 Aug 64, same subj (S); Hist, D Plans, Jan-Jun 64, p 322 (S}; Hist, D/Plans, Jan - Jun 64 (s); Hist, SAWC (TAC), Jul-Dec 64, pp 72-73 (S; Checo SEA Rpt, Jul - Dec 64, p 59 (TS).

23. JCS 2343/459, 4 Sep 64 (TS); Hist, D/Ops, Jul - Dec 64, p 82 (S); Hist, 2d AD, Jul - Dec 64, I, pp 65 and 68 (S).

24. Memo for rcrd by Maj. C. D. Thompson, Dir of Ops, DCS/P&O, 25 Aug 64, subj: SAR Forces (C); Hist, D/Ops, Jul - Dec 64, p 34 (S); Hist, 2d AD, Jan - Jun 64, ch 1pp 105 - 110, Jul - Dec 64, I, pp 135-36, and Jul - Dec 64, II, p 22 (S)

25. Hist, D/Ops, Jul - Dec 64, pp 45 - 46 (S).

26. Ibid; Hist, 2d AD, Jan - Jun 64, ch 1, pp 63 - 64, and Jul - Dec 64, I, pp 129-33 (S); JCS 2343/451, 28 Aug 64 (S); JCSM-785-64, 15 Sep 64 (S); msg 14632, JCS to CINCPAC, 28 Sep 64 (S).

27. JCSM-785-64, 15 Sep 64, subj: Retention of the 19th TASS (S); Hist, 2d AD, Jan - Jun 64, ch 1, p 64 (S); JCS 2343/451, 28 Aug 64 (S); Checo SEA Rpt, Jul-Dec 64, pp 60-62 (TS).

28. Ibid; memo, Yudkin to M/G. A. C. Agan Jr., D/Plans, DCS/P&O, 14 Dec 64, subj: Liaison A/C-RVN (s); memo, Yudkin to SAW Div, DCS/P&O, 2 Oct 64, subj: Addit L/19 A/C for RVN (S); JCS 2343/459, 4 Sep 64 (TS); Hist, D/Ops, Jul-Dec 64, p 82 (S); Hist, 2d AD, Jul-Dec 64, I, p 62 (S).

29. Hist, 2d AD, Jul-Dec 64, II, p 116; ltr, Hist Div, PACAF to AFCHO, 23 Sep 65, subj: Signif Events, PACAF, FY-65 (S).

Notes For Part Two

30. Checo SEA Rpt, Jul - Dec 64, pp 135 and 138 (TS).
31. Ibid; Hist, 2d AD, Jul-Dec 64, I, pp 85-97 (S).

CHAPTER VI

1. Van Staaveren, USAF Plans and Policies in SVN 1961 - 1963. p 104 (in this volume, Part I, Appendix 6 & 7); Journal of Military Assistance (JMA), prep by Eval Div. Asst for Mutual Scty, DCS/S&L, Dec 63, pp 178 - 79 (S); memo, Wheeler to SOD and JCS, 22 Apr 64, subj: Trip Rpt (TS).
2. Rpt of AF Study Gp on VN, 64 (S); Hist, CINCPAC, 1964, pp 318 and 432 (TS); Chronology of 2d AD, Jan - Jun 64, prep by 2d AD Hist Div, p 6 (U).
3. Msg 54337, PACAF to C/S USAF, 10 Mar 64 (TS); memo, SOD to SA et al, 17 Mar 64, subj: Imp of SVN Prog (S); Hist, Rpt, D/Policy, Hq PACAF, May 64, p 2 (S), in Hist, PACAF, Jan - Jun 64, I, pt 2.
4. Ibid.
5. Memo, M/G. J. K. Hester, Asst Vice C/S to SAF, 29 Jun 64, in OSAF 101- 64 (S); memo, L/C J.C. Price, D/Ops to C/S USAF, 11 Dec 64, subj: Supp for SVN (TS); msgs 37391 and 37408, PACAF to C/S USAF, 14 May 64 (TS); msg 2d AD to USAF, 13 May 64 (TS); Hist, CINCPAC, 1964, pp 318 - 21 (TS); Hist for Asst for Mutual Scty, Jan-Jun 64, PP 49 -50 (S).
6. Msg 14632, JCS to CINCPAC, 28 Sep 62 (S); Hist of Mutual Scty, Jul - Dec 64, p 41 (S).
7. Hist, CINCPAC, 1964, pp 318 - 21 (TS); JCSM-630-64, 24 Jul 64 (S); JCSM-875-64, 15 Oct 64 (S).
8. Memo, SOD to Chmn JCS, 6 Nov 64, subj: VNAF Ftr Sqs (S).
9. Msg 60009, C/S USAF to PACAF, 21 Dec 64 (S); msg 20676, PACAF to C/S USAF, 12 Dec 64 (TS).
10. Hist, CINCPAC, 1964, pp 361 - 64 (TS).
11. Ibid; msg 135-7, C/S USAF to PACAF, 27 Sep 64 (S).
12. JCS 2343/436-1, 25 Aug 64 (TS), msg 91689, C/S USAF to PACAF, 20 Nov 64 (S)
13. Msg 77348, C/S USAF to PACAF, 1 Oct 64 (S); Checo SEA Rpt, Jul - Dec 64, pp 42 - 43 (TS).
14. Checo SEA Rpt, Jul - Dec 64, pp 42 - 43 (TS).
15. Hist, 2d AD, Jan - Jun 64, ch I, pp 59 - 61 (S).

16. Hist, 2d AD, Jul-Dec 64, II, 116 (S); JMS, Dec 64, p 182 (S); USAF Mgt Summary, 3 Mar 65 (S).

17. Van Staaveren, USAF Plans and Policies in SVN, 1961-1963, pp 46 – 48 (in this volume, Part I, p 49 – 51) (TS).

18. Ibid; Talking Paper for Chmn JCS for His Mtg with CINCPAC on 8 Sep 64 (S).

19. JCSM-514-64, 12 Jun 64 (S); CM-1427-64, 15 Jun 64, subj: Deputy Comdr MAC/V (S); msg MAC 3077, Westmoreland to Taylor, 18 Jun 64 (S); Talking Paper for Chmn JCS for Mtg with CINCPAC on 8 Sep 64 (S).

20. Talking Paper for Chmn JCS for Mtg with CINCPAC on a Sep 64 (S).

21. Hist, 13th AF, 1964, pp 35-40 (S).

22. Ibid; CSAFM-742-64, 28 Aug 64 (S); CSAFM-754-64, 2 Sep 64 (S).

23. Ltr, Yudkin to PACAF, 28 Sep 64, subj: Trip Rpt, Joint Fact-Finding Team Visit to RVN (S).

24. Msg 21566, CINCPACAF to C/S USAF, 2 Oct 64 (S); msg 61575, C/S USAF to PACAF, 28 Dec 64 (S); msg 33725, CINCPACAF to C/S USAF, 20 Dec 64 (S); Hist, D/Plans, Jul-Dec 64, p 265 (S).

25. Talking Paper for Chmn JCS for his Mtg with CINCPAC on Senior AF Repr, 3 Sep 64; Gen Off Br, DCS/P, 19 Oct 64.

26. Van Staaveren, USAF Plans Policies in SVN, 1961- 1963, p 18, 44-45 (in this volume, Part I, p 14, 47– 49) (TS).

27. Ibid.

28. Msg 37391 and 37408, PACAF to C/S USAF, 14 May 64 (TS).

29. Memos for rcrd by L/Col W.T. Galligan, Off of L&L, 21 May – 24 Jun 64 (S); Rpt of AF Study Gp on VN, May 64 (S); ltr, SAF to Chmn; Subcmte on Preparedness Invest Cmte on Armed Svcs, U.S. Senate, 16 Jul 64, no subj (S); Life Magazine, 4 May 64.

30. JCS 2343/380,20 May 64(S).

31. JCS 2343/380,20 May 64(S); Talking Paper for Chmn JCS for Discussion with SOD on 2 Nov 64, subj: USAF Activities in SVN (TS).

32. Msg 11003, PACAF to C/S USAF, 6 Jun 64 (S); Hist Rpt, D/Plans, Hq PACAF, Jun 64, pp 2 - 3,(S), in Hist, PACAF, Jan - Jun 64, I pt 2.

33. Hist, D/Plans, Jul-Dec 64, p 327 (S); msg 14213, PACAF to C/S USAF, 28 Sep 64(S); Hist, 2d AD, Jul - Dec 64, II, pp 42 - 44 (S).

34. MAC/V Comd Hist, 1964, pp 80 - 82 (TS).

Notes For Part Two

CHAPTER VII

1. Dept of State Bulletin, 13 Aug 62, p 259.

2. JMA, Sep 62, pp 144 and 150 (S); msg 87185, C/S USAF to PACAF, 24 Jan 64 (S); Hist, 2d AD, Jan - Jun 64, ch 1, pp 120 - 25 (S); Hist, PACAF, Jan - Jun 64, I, pt 2, p 149 (S).

3. Hist, 2d AD, Jan - Jun 64, ch 1, pp 120-25 (S); Hist, TAC, Jul - Dec 64, p 598 (S); Hist, SAWC, Jan - Jun 64, pp 46 - 47 (S); msg 34496, PACAF to C/S USAF, 26 Feb 64 (S).

4. Hist, TAC, Jul - Dec 64, p 599 (S); Hist, PACAF, Vol I, pt 2, p 151 (TS); Hist, CINCPAC, 1964, pp 261-66 (TS); JCS 2344/81,25 May 64 (TS).

5. Van Staaveren, USAF Plans and Policies in SVN, 1961 – 63, pp 18 - 19 (in this volume, Part I, pp 29 - 30) (TS); Hist, CINCPAC, 1964, pp 269-73 (TS); Hist, PACAF, Jan-Jun 64, I, pt 2, pp 65 - 73 (TS); Hist, 2d AD, Jan - Jun 64, ch 1, pp 116 - 20, and ch 2, pp 61 - 64 (S).

6. N.Y. Times, 22 May 64.

7. Hist, CINCPAC, 1964, pp 261 - 66 (TS); Hist, TAC, Jul - Dec 64, p 599 (S); Hist, 2d AD, Jan - Jun 64, ch. l, p 125 (S); Hist, PACAF Jan - Jun 64, Vol I, pt 2, p 151(S); JCS 2344/81, 25 May 64 (TS); DJSM-882-64-25 May 64 (TS); memo for C/S USAF, 29 May 64, same subj (TS); Checo SEA Rpt, Jul - Dec 64, p 176 (TS).

8. Hist, PACAF, Vol I, pt 2, ch 6 (TS); Hist, TAC, Jul-Dec 64, pp 598 – 600 (S); Hist, SAWC, Jul - Dec 64, pp 61 - 63 (S); Hist, D/Ops, Jul – Dec 64, p 39 (S).

9. Checo SEA Rpt, Jul - Dec 64, pp 149 - 66 (TS).

10. Hist,2d AD, Jan - Jun 64, ch 2, pp 61- 64 (S); Checo SEA Rpt, Jul - Dec 64, pp 149 - 66 (TS); JMA, Sep 64, p 163; memo, Col W.P. Anderson, Chief, Spec Warfare Div, DCS/P&O to AFCHO, 6 Dec 65, subj: Draft of AFCHO Hist Study, in AFCHO.

11. JCS 2344/100, 17 Oct 64 (TS); Hist, PACAF, I, pt 2, ch 5, pp 83 - 85 (TS); Checo SEA Rpt, Jul - Dec 64, p 4(TS); Hist, CINCPAC,1964, pp 262 - 263 (TS); Hist, D/Ops, Jul - Dec 64, pp 42 - 43 (S).

12. CSAFM 498 - 64, 15 Jun 64 (TS); JCSM 595-64, 10 Jul 64 (TS); JCSM-645-64, 29 Jul 64 (TS); memo, SOD to Chmn JCS, 1 Aug 64, no subj.

13. Hist, 13th AF, Vol I, p .35 (S).

14. JCSM-870-64, 13 Oct 64 (TS); memo, L/Col. R. L. Kolman, Off of Dir of Plans, to C/S USAF 26 Oct 64, subj: Ops in Laos (TS); JCSM-889-64, 20 Oct 64 (TS); memo, M/G A.C. Agan, D/Plans, DCS/P&O to Carpenter, 9 Nov 64, subj: T-28 Strikes in Laos (TS); memo, Carpenter to C/S USAF, 12 Nov 64, subj: Air Action Laos Corridor (TS).

15. JCSM-870-64, 13 Oct 64 (TS); JCSM-889-64, 20 Oct 64 (TS).
16. CSAFM K-72-64, 23 Nov 64, subj: Mil Action in Laos (TS); Hist, D/Plans, Jul - Dec 64, p 330 (S); Hist, CINCPAC,1964, pp 21 - 22 (S).
17. Memo, SOD to Chmn JCS, 21 Oct 64, no subj (S).
18. CSAFM-K-72-64, 23 Nov 64, subj: Mil Action in Laos (TS); memo, J.T. McNaughton, OSD/ISA to Chmn JCS, 26 Nov 64, no subj (S); JCSM-997-64, 28 Nov 64 (TS); L/Col E.S. Minnich, Off of Dir of Plans, DCS/P&O to Asst for Jt and NSC Matters, 3 Dec 64 (TS), subj: Infil Through Laos (TS); memo, Carpenter to C/S USAF 22 Nov 64, Mil Action in Laos, with atch Background Paper (TS).
19. JCSM-1041-64, 11 Dec 64 (TS); memo, Carpenter to C/S USAF, 29 Dec 64, subj: Ops in Laos,(TS).
20. Memo, Carpenter to C/S USAF, 29 Dec 64, subj: Background Paper on Ops in Laos (TS); Checo SEA Rpt, Jul-Dec 64, pp 201-17; Hist. D/Ops, Jul-Dec 64, p 56 (S).
21. Ibid; Hist, CINCPAC,1964, p 272 (TS).

Notes For Part Three

NOTES FOR PART THREE

Chapter 1

1. Hist of 2d AD, Jan - Jun 65, Vol II, p 4 (S); JCS 2343/634, 21Jul 65 (TS).

2. Statement by SOD Robert S. McNamara before subcmte on appropriations and the Cmte on Armed Services, 89th Cong, lst Sess, DoD appropriations, 1966, Pt 1, p17; N.Y. Times, 27 Jan 65.

3. Hist of 2d AD, Jan - Jun 65, Vol II p 7 – 8 (S); Jacob van Staaveren, USAF Plans and Policies in South Vietnam and Laos, 1964, (hereinafter referred to as van Staaveren); pp 43 and 47 (TS) in AFCHO; N.Y. Times, 20 – 31 Dec 64, 9 - 13 Jan, and 27-31 Jan 65.

4, Hist of 2d AD, Jan - Jun 65, Vol II, p 1 - 2 and 5 - 9 (S); HQ MAC/V Comd Hist, 1965, pp 46 and 269 (S); CINCPAC Comd Hist, 1965, An B, p 22 (S); Van Staaveren, p 95 (TS)

5. Msgs 1455 and 1466, CINCPACAF to C/S USAF, 10 Jan 65 (S); Hist of 2d AD, Jan - Jun 65, Vol II, p 1, 6-9 (S).

6. JCS 2343/488 and JCS 2343/488-1, 14 Nov 64 (TS), Hist of D/Plans, Jul – Dec 64, p 58 (TS); memo, Col J.S. Berger, Tac Div, D/Ops, DCS/P&O, to C/S USAF, 9 Jan 65, subj: Security of U.S. forces, RVN (S); unsigned memo, 6 Jan 65, subj: List of Actions Pending in SEA (TS).

7. Memo, Berger to C/S USAF, 9 Jan 65 (TS); Unsigned memo, 6 Jan 65 (TS).

8. Memo, Maj Gen J.W. Carpenter, III, Asst DCS/P&O for JCS to C/S USAF, 5 Jan 65, subj: Mtg of NSC principals, 5 Jan 65, with atch Background Paper and Transition Phase (TS); unsigned memo, 6 Jan 65; JCSM-7-65, 7 Jan 65 (TS); USAF Mgt, summary, 13 May 65, p 15 (S).

9. JCSM-28-65, 15 Jan 65 (TS); JCSM-70-65, 29 Jan 65 (TS); memo, P. Solbert, Dep Asst OSD (ISA) to Chm JCS, 29 Jan 65, subj: Ops in Laos (TS)

10. DJSM-1938, 10 Dec 64 (S); memo, Carpenter to D/Jt Staff, 9 Jan 65, subj: Air Force comments on establishing an International Force in SVN (TS); Hist of D/Plans, Jul - Dec 64, p 54 (S).

11. Msgs 1465 and 1466, CINCPACAF to C/S USAF, 10 Jan 65, (S); memo. Col W.P Anderson, Asst Chief, Spec Warfare Div, D/Plans,. DCS/P&O to C/S USAF, 27 Jan 65, subj: Emergency Use of Jet Aircraft (TS) msg JCS to CINCPAC, 27 Jan 65 (TS); Hist of D/Ops, Jan - Jun 65, pp 87-88 (S); Hist of 2d AD, Jan - Jun 65, Vol II, p 6 - 9 (S); Wash Evening Post 7 Feb 65.

Air War – Vietnam

12. Hist of 2d AD, Jan - Jun 65, Vol II, p 23-24 (S); Wash Post, 2 and 8 Feb 65; Wash Evening Star, 7 Feb 65, OSD/PA News Release 86-65, 11 Feb 65.

13, OSD/PA news release 77-55, 1 Feb 65; State Dept Bul, 22 Feb 65, p 239.

14. Memo, Berger to C/S USAF, 4 Mar 65, subj: B/R and R/T (TS); Hist of PACAF, 1 Jul 64 - 30 Jun 65, Vol I, pt 1, p 19 (S); OSD/PA news release 77-65, 8 Feb 65; N.Y. Times. Balt Sun, and Wash Post. 8 and 9 Feb 65.

15. Hist of PACAF, 1 Jul 64 - 30 Jun 65, Vol I, Pt 2, p 95 (TS).

16. Hist of 2d AD, Jan - Jun 65, Vol II, p 25 - 26 (S); memo, Berger to C/S USAF, 4 Mar 65 (TS); N.Y. Times. 14 Feb 65; statement by McNamara before House Subcmte Hearings of the Cmte on Appropriations, 89th congr 2d Sess, supplemental defense appropriations for 1965, pp 33 and 37.

17. Memo, Maj Gen S.J. McKee, D/Plans, DCS/P&O to C/S USAF, 7 Mar 65, subj: Use of B-52s in SVN (TS); N.Y. Times, 14 Feb 65; Hist of SAC, Jan – Jun 65, Vol II, p 239 (TS).

18. JCSM-100-65, 11 Feb 65 (TS); memo, Lt Col E.S. Minnich, Combined Plans Div, DCS/P&O to C/S USAF, 10 Feb 65, subj, C/A in SEA (TS); JCSM-149-65, 4 Mar 65 (TS).

19. CM-424, 11 Feb 65 (TS); memo, Carpenter to C/S USAF, 1 Mar 65, subj: Air Strikes Against NVN (TS); memo, Carpenter to C/S USAF, 1 Mar 65, subj: Piecemeal Planning, SEA (TS); memo, Col. J.H. Germeraad, Asst Dep D/Plans for War Plans, DCS/P&O to C/S USAF, 1 Mar 65, subj: Scty Situation in SVN (TS); JCSM-149-65 (TS).

20. Hist of 2d AD, Jan - Jun 65, Vol II, p 22-28 (S).

21. Hist of 2d AD, Jan - Jun 65, Vol II, p 33 - 43 (S); CINCPAC Comd Hist, 1965, Vol II, pp 114 - 15, 420 (TS); N.Y. Times 25 and, 26 Feb 65.

22. Hist of 2d AD, Jan - Jun 65, Vol II, p 33-36 (S); N.Y. 16 to 26 Feb 65.

23, CSAFM-B-54-65 and CMC-11-65, 11 Feb 65 (TS); memo McKee to C/S USAF, 13 Feb 65, subj: Scty of U.S. Instl in SVN (TS); JCSM-110-65, 16 Feb 65 (TS); JCSM-121-65, 20 Feb 65 (TS).

24. Hist of 2d AD, Jan - Jun 65, Vol II, p 41 (S); N.Y. Times, 25 and 26 Feb 65, and 1 and 11 Mar 65; JCSM-130-55, 24 Feb 65 (TS); memo, Berger to C/S USAF, 24 Feb 65, subj: Improved Scty and Readiness Measures in RVN (TS).

25. Hist of PACAF, 1 Jul 64 - 30 Jun 65, Vol I, Pt 2, pp 101-104 (TS); memo, McKee to C/S USAF, 4 Mar 65, subj: Use of B-52s in SVN(TS); Hist of PACAF, 1 Jul 64 - 30 Jun 65, vol 1, Pt 2, p 104 (TS); N.Y. Times, 3, 8 and 11 Mar 65; Wash Post. 3 Mar 65.

Notes For Part Three

26. Hist of PACAF, 1 Jul 64 - 30 Jun 65, Vol I, pt 2, p 104 (TS); memo, McKee to C/S USAF, 7 Mar 65, subj: Use of B-52s in SVN (TS) memo, Berger to C/S USAF, 21 Jun 65, subj: Arc Light I (TS).

27. Hist of D/Ops, Jul - Dec 65, p 48; N.Y. Times, 15 Mar 65; Wash Post, 15 Mar 65.

Chapter II

1. DJSM-138-64, to Dec 64 (TS); Unsigned memo, 6 Jan 65, subj: List of Actions Pending in SEA (TS); memo, Carpenter to C/S USAF, 9 Jan 65, subj: AF Comments on Establishment of an International Force (TS); memo, Carpenter to C/S USAF, 16 Jan 65, subj: International Military Asst to VN (C); JCS 2343/515, 25 Jan 65 (TS).

2. Memo, Minnich to C/S USAF, 29 Jan 65, subj: Situation in SEA, with atch Talking Paper on Deployment of Forces, SEA (TS).

3. Army Staff memo 23-65, 10 Feb 65 (TS).

4. Memo, Minnich to C/S USAF, 10 Feb 65, subj: C/A, SEA (TS); memo for rcrd, by Col W.V. Mcbride, 16 Feb 65, subj: Ops Plan 32-65 (TS); memo, Lt Gen W.H. Blanchard, DCS/P&O to SAF, 18 Feb 65, subj: U.S. Posture in The Far East (TS); memo, Col J.T. Scepansky, Dep Dir of Pers planning, DCS/P to SAF, 17 Feb 65, subj: Reserve Force Rqmts (TS).

5. Memo, SAF to SOD,19 Feb 65, subj: Review of Oplan 32-64 and Oplan 39-65 (TS); JCSM-176-65, 11Mar 65 (TS); memo, Gen Earle G. Wheeler, Chmn JCS to CNO et al., 17 Feb 65, subj: Prelim Action Toward Increased Readiness (C); memo, Germeraad to C/S USAF, 19 Feb 65, subj: Review of Log and Admin Capabilities of CINCPAC Oplan 32-64 and 39-65 (TS).

6. CM-481-65, 11 Mar 65 (TS).

7. Memo, Germeraad to C/S USAF, 12 Mar 65, subj: Contingency Planning for SEA/Westpac (TS); CSAFM-J-67-65, 12 Mar 65 (TS).

8. Rprt, Gen H.K. Johnson, CSA to SOD et al., 14 Mar 65, subj: VN trip from 5 - 12 Mar 65, Tab B (TS).

9. Two memos, Germeraad to C/S USAF, 8 Mar 65, subj: Scty Situation in SVN (TS).

10. Ibid; N.Y. Times, 5 - 13 Mar 65.

11. Rprt, Gen Johnson to SOD et al., 14 Mar 65 (TS).

12. Memo, Carl T. Rowan, Dir of USIA to Pres, 16 Mar 65, no subj (TS).

13. JCSM-197-65, 17 Mar 65 (TS).

Air War – Vietnam

14. Memo, McKee to C/S USAF, 18 Mar 65, subj: briefing on views of Ambassador Taylor, CINCPAC, COMUS/V On Certain Proposals Made by Chmn JCS (TS).

15. Memo, Germeraad to C/S USAF, 17 Mar 65, subj: Ground Forces to SEA (TS); memo, McKee to C/S USAF, 17Mar 65, subj: Deployments to SEA (TS); CMCM-27-65, 18 Mar 65; CMCM-28-65, 19 Mar 65 (TS).

16. CSAFM-J-78-65,17 Mar 65 (TS); memo, Germeraad to C/S USAF, 25 Mar 65, Subj: To Consider Implications of Withdrawal of Subj Paper from: JCS agenda by Dir of Jt staff (TS).

17. JCSM-204-65, 20 Mar 65 (TS)

18. CSAFM-J-94-65, 24 Mar 65 (TS); Talking Paper for the Chmn JCS for use at mtg of JCS on 5 Apr 65, 4 Apr 65 (TS).

19. JCSM-216-65, 25 Mar 65 (TS); memo, SOD to Chmn JCS, 31 Mar 65, no subj (S); CSAM-163-65t, 29 Mar 65 (TS).

20. Msg 10437, PACAF to C/S USAF, 28 Mar 65 (TS); memo Carpenter to C/S USAF, 30 Mar 65, subj: Concept for Phased Deployments to SEA (TS); memo, Carpenter to C/S USAF, 30 Mar 61, subj: Deployment of a U.S. Army Div to the Central Highlands (TS).

21. JCSM-238-65, 2 Apr 65 (TS).

22. Memo, SOD to Chmn JCS, 14 May 65, subj: Policies and Proceedings for More Eff Prosecution of the War (TS).

23. Memo, Carpenter to C/S USAF, 28 Mar 65, subj: Recent actions by JCS and Recm to Higher Auth Concerning SEA (TS).

24. Hist of 2d AD, Jan - Jun 65, Vol II, p 45-46 (S).

25. Hist of 2d AD, Jan - Jun 65, Vol II, p 46-47 (S); Hist of SAC, Jan – Jun 65, Vol II, p 253 (TS).

26. Hist of PACAF, 1 Jul 64 - 30 Jun 65, Vol I, pt 1, p 22 (S); Wash Post 25 Mar 65; N.Y. Times. 29 Mar and 1 Apr 65.

27. CSAFM-J-77-65, 17 Mar 65 (TS); JCSM-202-65, 20 Mar 65 (TS); msg 98352, PACAF to C/S USAF, 27 Mar 65 (TS); Hist of SAC, Jan - Jun 65, Vol II, p 231 (TS).

28. Memo, Germeraad to C/S USAF, 25 Mar 65; JCSM-218-65, 24 Mar 65 (S); CSAFM-J-86-65, 27 Mar 65 (S); JCSM-215-65, 25 Mar 65 (TS); JCSM-221-6, 27 Mar 65 (TS); JCS 2343/551-1, 14 Apr 65 (TS).

29. Hist of PACAF, 1 Jul 64 - 30 Jun 65, Vol I, pt 1, pp 71-79 (S); USAF mgt Summary, 7 Jun 65, p 15 (S).

Notes For Part Three

Chapter III

1. Memo, Carpenter to C/S USAF, 28 Mar 65, subj: Recent Actions by JCS and Recm to Higher Auth concerning SEA (TS); HQ MAC/V comd hist, 1965, p 46 (C).
2. NSAM 328, 6 Apr 65 (TS); Talking Paper for Chmn JCS for use at mtg of JCS on 5 Apr subsequent to mtg with SOD, 4 Apr 65 (TS).
3. N.Y. Times, 3 Apr 65.
4. N.Y. Times, 8 Apr 65.
5. Memo, SOD to Chmn, JCS, 1 Apr 65, subj: Immed Deployment of F-4C sq to Thai (TS); memo, SOD to Chmn JCS, 6 Apr 65, subj: Manpower increases to RVN (S); memo, Maj W.P. Paluch, Tac Div, D/Ops to C/S USAF, 4 Apr 65,. Subj: Rqmt for USAF tac ftrs to Thai (TS); JCS 2343/559-1, 7 Apr 65 (TS); CINCPAC Comd Hist, 1965, Vol II, pp 279 and 285 (TS).
6. Memo, Germeraad to C/S USAF, 12 Apr 65, subj: Recn Action for SVN (TS); CSAFM-R-44-65, 12 Apr 65 (TS).
7. Memo, McGeorge Bundy, Spec Asst to the Pres for NSC Affairs to Secy State, et al., 9 Apr 65, subj: NSAM 330 (S).
8. N.Y. Times. 14 Apr and 4 May 65.
9. JCSM-265-45, 8 Apr 65 (TS); JCSM-288-65, 17 Apr 65 (TS); SOD to Chmn JCS, 5 Apr 65, no subj (S); memo, Germeraad to C/S USAF, 27 Apr 65, subj: deployment of forces to SVN (TS); HQ MAC/V Comd Hist, l965, p 270 (C); CINCPAC Comd Hist, 1965, Vol II, p 280 (TS).
10. Hist of 2d AD, Jan - Jun 65, Vol II, p 49 - 55 (S); msgs 54872, 54886, 54999, CINCPACAF to C/S USAF, 23 Apr 65 (TS).
11. Ibid; memo, SOD to Pres, 21 Apr 65, no subj (TS).
12. Ibid.
13. Memo, SOD to Pres, 21 Apr 65 (TS); JCSM 2343/564-7, 25 Apr 65 (TS); JCSM 231-65, 30 Apr 65 (TS); Hist of D/Plans, Jan- Jun 65, pp 111 – ll2 (TS).
14. JCS 2343/543-2, 30 Apr 65 (TS).
15. JCSM-376-65, 19 May 65 (TS); memo, McKee to C/S USAF, 16 May 65, subj: Contingency Planning for SEA/WESTPAC (JCS 2339/182) (TS).
16. Memos, Germeraad to C/S USAF, 8 and 10 Jun 65, subj: U.S. Allied Troop Deployments to SVN (TS); CINCPAC Comd Hist, Vol II, p 291 (TS), N.Y. Times, 23 and 27 May, and 8, 17, and 18 Jun 65.
17. Memos, Germeraad to C/S USAF, 8 and 10 Jun 65 (TS).

18. Memo, Germeraad to C/S USAF, 10 Jun 65 (TS).
19. JCSM-457-65, 11Jun 65 (TS); JCSM-482, 17 Jul 65: (TS); memo Germeraad to C/S USAF, 24 Jul 65, subj: Addn Deployments to SVN (TS); msg 81901 C/S USAF to CINCPACAF, 20 Jul 65 (TS).
20. Balt Sun, 12 Jun 65.
21. JCSM-482-65, 17 Jun 65 (TS); msg 81901, C/S USAF to PACAF, 20 Jun 65 (TS); memo, Germeraad to C/S USAF, 24 Jun 65 (TS).
22. N.Y. Times. 10 and 17 Jun 65; Wash Post, 9 Jun 65.
23. Memo, Germeraad to C/S USAF, 28 Jun 65, subj: Further Deployments to SVN (TS); memo, Lt Col A.W. Braswell, Combined Plans Div, D/Plans, DCS/P&O to C/S USAF, 10 Jul 65, subj: Reexamination of Concepts for SVN (TS); memo, SOD to SA, 28 Jun 65, subj: Helo Companies (S).
24. Ibid.
25. CSAFM-105-65, 30 Jun 65 (TS).
26. JCSM-515-65, 2 Jul 65 (TS).
27. HQ MAC/V Comd Hist, 1965, pp 162 and 166 (C).
28. Hist of 2d AD, Jan - Jun 65, Vol II, p 45-52 (S); msg 514872, CINCPACAF to C/S USAF, 23 Apr 65 (TS); JCSM-530-65, 3 Jul 65 (TS); Hist of SAC Jan – Jul 65, Vol II, pp 247-48 (TS); CINCPAC Comd Hist, 1965, Vol II, p 420 (TS).
29. Hist of 2d AD, Jan - Jun 65, Vol II, p 56-88 (S); DJSM-744-65, 26 Jun 65; unsigned rprt, 22 Jul 65, subj: Item 12 of SOD questions and Amb Taylor's Answers (TS); rprt, Col H.N. Brown, Chief, SAW Div, D/Ops. to AFCHO, 21 Jan 66, subj: Chronology on Eff of Air Power in RVN (U); N.Y. Times. 13 May 65.
30. Memo, Maj Gen R.H. Curtin, D/Civil Engring, DCS/P&R to SAF, 22 Jun 65, Subj: AF Capability to Construct Expedient runways (S); Hist of Aerospace Progs, Jan - Jun 65, p 31 (S); Hist of 2d AD, Jan - Jun 65, Vol II, p 49 - 88 (S); JCS 2343/559-34, 30 Nov 65 (TS).
31. Hist of 2d AD, Jan - Jun 65, Vol II, p 176-77 (S); Fact Sheet on Bien Hoa, 18 May 65 (S), in OSAF 132-65; N.Y. Times, 17 May 65.
32. CINCPAC Comd Hist, 1965, Vol II, pp 420 - 21 (TS).
33. Hist of SAC, Jan - Jun 65, Vol II, pp 247 – 64 (TS).
34. Ibid; Hist of 2d AD, Jan - Jun 65, Vol II, p 59 and 86-87 (TS); memo, Carpenter to C/S USAF, 27 Apr 65, subj: Utilization of Arc Light B-52 forces (TS); Berger to C/S USAF, 21 Jun 65, subj: Arc Light I (TS); Carpenter to C/S USAF, 8 Jul 65, subj: Use of Strat Forces in SEA (TS); CSAFM-F-45-65, 12 Jul 65 (TS); Statement by McNamara on 4 Aug

Notes For Part Three

65 before Senate Subcmte Hearings on appropriations, 89th Cong, 1st Sess, DoD appropriations for 1966, pt 2, pp 793-94; N.Y. Times. 19 May and 17 – 20 Jun 65, address by Gen McConnell before The Dallas (Tex) Council on World Affairs, 16 Sep 65, in SAFOI.

35. Hist of 2d AD, Jan - Jun 65, Vol II, p 84 - 85 (S); msg 4426, PACAF to C/S USAF, 7 Jul 65 (TS).

36. Hist of D/Ops, Jan - Jun 65, p 88 (S); CM-534-65, 6 Apr 65 (TS); JCSM-498-65; 2 Jul 65 (TS); CINCPAC Comd Hist, 1965, Vol II, p 423 (S).

37. Memo, Carpenter to C/S USAF, 5 Apr 65, subj: F-104 Sq (TS); CM-534-65, 6 Apr 65 (TS); CSAFM-R-40-65 12 Apr 65 (TS)

38. N.Y. Times 13 - 20 May 65.

39. Ibid;.JCSM-300-65, 22 Apr 65 (TS); D/Ops briefing for SAF, 1 Oct 65, no subj (TS); N.Y. Times, 5 Apr and 11 Jul 65.

40. JCSM-415-65, 27 May 65 (TS), JCSM-442-65, 7 Jun 65 (TS); .JCSM-498-65, 26 Jul 65 (TS); memo. Col F.L. Kaufman, Asst Dep Dir of Plans for War Plans, DCS/P&O to C/S USAF, 28 Aug 65, subj: Air Ops against NVN (TS); JCS 2343/546 -6, 7 Jul 65 (TS); CSAFM-F-40-65, 7 Jul 65 (TS).

41. Memo, Kaufmann to C/S USAF, 28 Aug 65 (S); N.Y. Times, 11 Jul 65; N.Y. Herald-Tribune. 12 Jul 65.

42. JCSM-514-65, 1 Jul 65, subj: Response to Significant Incidents in VN (TS); memo, SOD to Chmn JCS, 7 Jul 65, same subj (TS); memo, McKee to C/S USAF, 12 Jul 65, same subj (TS); memo, Col B.W, Lucia, Asst Dep D/Plans for war plans, DCS/P&O to Asst for Jt and NSC matters, 9 Jul 65, subj: Blockade and aerial mining study (TS); Secret Supplement to the Staff Digest, 3 Sep 65 (S).

43. CINCPAC Comd Hist, 1965, Vol II, p 415 (S); memo, HQ USAF to Dir for Ops, Jt staff, 6 Apr 65, subj: AF statement of Nonconcurrence in JCS 2344/11-1 (TS); CSAFM-R-47-65, 14 Apr 65 (TS).

44. CINCPAC Comd Hist, 1965, Vol II, pp 402 - 03 (TS); JCS 2344/113, 4 Apr 65 (TS); HQ MAC/V Comd Hist 1965, p 209 (TS); Hist of PACAF, 1 Jul 64 - 30 Jun 65, Vol I, Pt 2, pp 82-83 (TS).

45. Hist of PACAF, 1 Jul 64 - 30 Jun 65, Vol I, Pt 2, pp 82-83 (TS); CINCPAC Comd Hist, 1965, Vol II, p 402-03 (TS); JCS 2343/559-34, 30 Nov 65 (TS).

46. CINCPAC Comd Hist, 1965, Vol II, p 402 - 06 (TS).

47. Ibid; HQ MAC/V briefing, 28 Nov 65 (TS).

48. CINCPAC Comd Hist, 1965, Vol II, p 313 (TS).

Air War – Vietnam

49. Msgs 64258, and 64539, C/S USAF to PACAF, 15 Apr 65 (TS); memo Maj. S.J. Smith, Combined Plans Div, D/Plans, DCS/P&O to C/S USAF, 27 Apr 65 (TS); JCS 2448/3, 26 May 65 (TS); JCSM-319-45, 28 Apr 65 (TS).

50. Ibid; HQ MAC/V Comd Hist, 1965, p 50 (S); AF News Service release 8-20-65-587; AF and Space Digest, Sep 65, p 120.

Chapter IV

1. CINCPAC Comd Hist, 1965, Vol II, pp 460 - 61 (S); Wash Post, 2 Jul 65, Balt Sun. 7 Jul 65.

2. Msg telecon 360, 1971305, 2d AD to PACAF, 16 Jul 65 (TS); memo, McKee to C/S USAF, 22 Jul 65, subj: COMUSMAC/V submission (TS); memo, McKee to C/S USAF, 19 Jul 65, subj: Issues Raised by SOD as a result of VN trip (TS); HQ MAC/V Comd Hist, 1965, p 191 (TS); Ofc of Pers, D/State.

3. Memo, McKee to C/S USAF, 19 Jul 65, Issues Raised by SOD as a result of VN trip (TS)

4. JCSM-551-65, 19 Jul 65 (TS).

5. Hist of D/Plans, 1 Jul - 31 Dec 65, p 176 (TS); msg 90300, C/S USAF to CINCPACAF, 31 Jul 65 (TS).

6. Memo, SOD to Pres, 1Sep 65, no subj (TS); memo, Maj Gen R.N. Smith, D/Plans to DCS/P&O, 9 Sep 65, subj: Deployment of units (TS); Wash Post 29 Jul 65; N.Y. Times. 29 Jul 65.

7. N.Y. Times. 2 Aug 65; Balt Sun. 5 Aug 65; memo, McKee to C/S USAF, 5 Aug 65, subj: U.S. Mil Posture (TS); CSAFM-N-17-65, 6 Aug 65.

8. Memo, SAF to SOD, 13 Aug 65 (S), in OSAF 132-65; memo, Asst SOD (S&L) to Secys of Mil Depts et al, 30 Aug 65, subj; VN Supporting Exped Task Force (U); unsigned memo, 2 Aug 65 (U); memo, Col E.W. Lenfest, Asst Dep D/Plans for war plans, DCS/P&O to AFCHO, subj: draft of Hist study (TS); Hist of TAC, Jul - Dec 65, pp 36-37 (S).

9. CSAFM-F-61-65, 21 Jul 65 (TS); JCSM-652-65, 27 Aug 6:5 (TS); Hist of O/Plans, Jul - Dec 65, pp 170-71 (TS); memo, SOD to Chmn JCS, 13 Sep 65, subj: Concept for VN (TS); memo, Germeraad to C/S USAF, 25 Jan 66, subj: Deployments to SVN (TS).

10. Memos, SOD to Pres, 1 and 22 Sep 6!, subj: Additional Forces to SVN (TS); Hist of D/Plans, Jul - Dec 65, p 176 (TS); memo, Gen Smith to DCS/P&O, 9 Sep 65, subj: Deployment of Units (TS); JCSM-643-65, 23 Aug 65 (TS); Memo, Germeraad to C/S USAF, 25 Jan 66 (TS).

11. Msg 96668, C/S USAF to CINCPACAF, 13 Sep 65 (TS).

12. Hist of D/Ops, Jul - Dec 65, p 39 (S).

Notes For Part Three

13. Msg 96668, C/S USAF to CINCPACAF, 13 Sep 65 (TS).; Hq MAC/V Comd Hist, 1965 p 44 (TS).

14. Memo, Germeraad to C/S USAF, 14 Oct 65, subj: Phase I (TS); CINCPAC Comd Hist, 1965, Vol II, p 307 (TS); Hist of D/Plans, Jul - Dec 65, pp 172-73 and 178-79 (TS); Hist of D/Ops, Jul - Dec 65, p 50 (S).

15. Ibid; HQ MAC/V Comd Hist, 1965, p 44 (TS).

16. JCSM-811-65, 10 Nov 65 (TS); JCSM-814-65, 10 Nov 65 (TS); CINCPAC Comd Hist, 1965, vol II, pp 307- 08 (TS).

17. Hq MAC/V Comd Hist, 1965, pp 165-85 and 269 (S); memo, Germeraad to C/S USAF, 3 Aug 65, subj: Airfield Const (C), Apps 6, 7 and 11; memo, Lenfest to AFCHO, 13 Sep 65 (TS).

18. Ltr, COMUSMAC/V to CG 173rd Airborne Brig, 7 Jul 65, subj: Minimizing Combat Casualties (U), 1st Ind, McConnell to CINCPACAF, 8 Sep 65 (U); msg 90300, C/S USAF to PACAF, 31 Jul 65 (TS).

19. HQ MAC/V Comd Hist, 1965, pp 186, 195 and 481 (TS); Gen Smith's briefing. Bk for SEA, Vol II, Tab B, Dec 1965 (TS); JCS 2343/559-34 30 Nov 65 (TS).

20. Memo, Col H.N. Brown, Chief SAW Div, D/Ops to AFCHO, 21 Jan 66, with atch rprt, Chronology on Eff of Air Power in RVN (U), in AFCHO.

21. Msg 99742, C/S USAF to CINCPACAF, 24 Nov 65 (TS); Hq MAC/V briefing on Log, 28 Nov 65 (TS); msg 95074, CINCPACAF to C/S USAF, 1 Dec 65 (TS).

22. Hist of D/Aerospace Progs, Jul - Dec 65, p 30 (S); Gen Smith's briefing-Bk on SEA Vol II, Tabs B and D, Dec 65 (TS); Hist of D/Ops, Jul - Dec 65, pp 25-26 (S).

23. HQ USAF Daily Staff Digest No 17, 25 Jan 66 (C); memo, Berger to C/S USAF, 18 Aug 65, subj: Utilization of Arc Light Forces (TS); JCSM-642-65, 21 Aug 65 (TS); memo, SOD to Chmn JCS, 29 Sep 65 (TS).

24. Gen Smith's briefing Bk for SEA, Vol II, Tab B, Dec 65 (TS).

25. Hq MAC/V Comd Hist, 1965, pp 191-92 (TS); Hist of D/Ops, Jul- Dec 65, pp 98 - 99 (S); D/Ops Briefing for SAF, 1 Oct 65, no subj (TS); N.Y. Times, 27 Aug 65; Wash Post, 31 Aug 65.

26. Hist of SAC, Jan - Jun 65, Vol II, pp 264-77 (TS); memo George W Ball, Under Secy of State to SOD, 31 Jul, 65, no subj (S); Hist of D/Ops, Jul - Dec 65, p 95 (S).

27. JCSM-600-65, 3 Aug 65 (S).

Air War – Vietnam

28. Memo, Germeraad to C/S USAF, 3 Aug 65, subj: Airfield const (S); JCSM-638-65, 23 Aug 65 (TS).

29. Memo, SAF to SOD, 5 Oct 65, subj: Air Force Capability to Const Exped Runways (S).

30. CINCPAC Comd Hist, 1965, An B, pp 110 - 11 (S); Statement by Capt Foster Lalor, Asst Chief for Const, Bureau of Yards and Docks, U.S. Navy, on 11 Jan 66, before House Subcmtes of the Cmte on Appropriations, 89th Cong, 2d Sess, supplemental defense appropriations for 1966, pp 182 – 83 (U).

31. Memo, McKee to C/S USAF, 13 Sep 65, subj: Scty of U.S. bases in SVN (TS); Memo, Lt Col J.L. Milton to C/S USAF, 29 Oct 65, same subj; HQ MAC/V Comd Hist, 1965, pp 481 (U), N.Y. Times, 2 Jul and 25 Aug 65.

32. Memo, McKee to C/S USAF, 13 Sep 65 (TS); CSAFM-C-22-65, 13 Sep 65 (TS).

33. Memo, Lt Col J.F. Milton, SAW Div, D/Ops, DCS/P&O to C/S USAF, 29 Oct 65; Hist of D/Ops, Jul - Dec 65, pp 27 - 29 (S); memo, Lenfest to AFCHO, 13 Sep 65 (TS).

34. USAF Mgt summary, 7 Jan 66, pp 28 - 32 (S); CINCPAC Cmnd Hist, 1965, Vol II, pp 377 - 8 (TS); Hist of D/Ops, Jul - Dec 65, p 50 (S).

35. Ltr, Gen W.H. Blanchard, Vice C/S to Deputies, Dirs, and Chiefs of comparable Off, 13 Aug 65, subj: USAF Attrition (U); Hist of D/Ops, Jul – Dec 65, p89 (S); HQ MAC/V Comd Hist, 1965, p 205 (S); Wash Post. 18 Oct 65; CINCPAC Comd Hist, 1965, Vol II, pp 377-83 (TS).

36. JCSM-670-65, 2 Sep 65 (TS); JCSM-686-65, 11Sep 65.(TS); demo, Kaufman to C/S USAF, 28 Aug 65, subj: Air Ops Against NVN (TS); Talking Paper for Chmn JCS on Item to be Discussed at JCS Mtg on 10 Sep 65, subj:. Air Strikes against NVN (TS); msg 90300, C/S USAF to PACAF, 31 Jul 65 (TS); memo, McKee to C/S USAF, 3 Aug 65, subj: Blockade and Aerial Mining Study (TS).

37. Memo, Dep SOD to Chmn JCS, 18 Aug 65, subj: Blockade and Aerial Mining Study (TS); memos, SOD to Chmn JCS and Dir CIA, 15 Sep 65, subj: Air Strikes on NVN (TS); ltr, W.P. Bundy, Asst Secy of State for Far Eastern Affairs to OSD/ISA, 26 Sep 65 (S).

38. Hist of D/Ops, Jul - Dec 65, p 51 (S).

39. Memo, McKee to C/S USAF, 7 Nov 65, subj: Air Ops Against NVN POL system (TS); JCSM-810-65, 10 Nov 65 (TS), JCSM-811-65, 10 Nov 65 (TS); CINCPAC Cmd Hist, 1965, Vol II, p 353 (TS).

40. HQ MAC/V Cmd His, 1965, pp 210 - 12 (TS) CINCPAC Comd Hist, Vol II, 1965, p 415 (TS). Gen Smith's Briefing Bk Vol II, Tab K, Dec 65 (TS).

Notes For Part Three

41. HQ MAC/V Comd Hist, 1965, pp 210 – 11 (TS).
42. CINCPAC Comd Hist, 1965, Vol II, pp 410 – 11 and 430 (TS),
43. Ibid. p. 315 Project CHECO SEA Rprt, USAF Ops From Thailand, 10 Aug 66, pp 13 - 16 (S).

Chapter V

1. HQ MAC/V Comd Hist, 1965, pp 269 and 283 (C); Balt Sun, 17 Oct 65, N.Y. Times, 23 Oct 65.
2. HQ MAC/V Comd Hist, 1965, p 44 (TS).
3. HQ MAC/V briefing, 28 Nov 65 (TS).
4. HQ MAC/V Comd Hist, 1965, pp 196-97 (TS).
5. Hq MAC/V Briefing, 28 Nov 65 (TS), memo, Westmoreland to SOD, Nov 65, Subj: Special Needs to Support Tiger Hound (TS); Gen Smith's Briefing Bk for SEA, Vol II, Dec 65 (TS); Hq MAC/V Comd Hist, 1965, pp 197-98 and 213 (TS).
6. HQ MAC/V Briefing, 28 Nov 65 (TS).
7. Extracts of statements by McNamara on the outlook in South Vietnam, 1 Jan 65 through 1 Jan 66 in hearings before House Subcmtes of the Cmte on appropriations, 89th Cong, 2d Sess, Supplemental Defense Appropriations for 1966, pp 85-86.
8. Ibid.
9. Memo, Germeraad to C/S USAF, 23 Dec 65, subj: Additional Jet Airfield Rqts, SEA (TS); Hist of D/Ops, Jul - Dec 65, p 104 (S).
10. Ibid.
11. CSAFM-30-65, 6 Dec 65 (TS); Special Rprt, SVN Action List. 10 Dec 65 (TS).
12. HQ MAC/V Comd Hist, 1965, p 45 and 259 (TS).
13. Memo, N.S. Paul, Acting SAF to Dep SOD, 31 Dec 65, subj: Reprogrammed Phase Rqmts for CY 1966 (TS); Memo, Col F.J. Coleman, Asst Dep D./Plans For War Plans, to Asst for Jt and NSC matters, DCS/P&O, 6 Jan 66, Subj: Capabilities to meet CINCPAC/COMUSMAC/V rqts (TS); CINCPAC Comd Hist, 1965, Vol II, pp 308-310 (TS); memo, Macdonald to C/S USAF, 25 Feb 66, subj: A deployment schedule for SEA and other Pacific areas (TS).
14. CINCPAC Comd Hist, 1965, Vol II, p 423 (S).
15. Memo, Brown to AFCHO, 21 Jan 65 (U); HQ MAC/V Comd Hist, 1965, pp 197- 99 (TS); HQ USAF Daily Staff Digest No 17, 25 Jan 66 (C).

16. Ltr, L. Unger, Deputy Asst Secy of State for Far Eastern affairs to A. Friedman, Deputy ASD/ISA, 15 Dec 65, subj: Additional B-52 bases (TS).

17. Hq MAC/V Comd Hist, 1965, pp 201 - 05 (S); CINCPAC's Comd Hist, 1965; Vol II, p 423 (S); Balt Sun, 2 Dec 65; N.Y. Times 10 Dec 65; App 6 (C).

18. Gen. Smith's Briefing Bk for SEA, Vol II, Tab H (TS).

19. CINCPAC Comd Hist, 1965, Vol II, pp 377 - 83 (TS); App 8 (S).

20. Memo, Lt Gen J.T. Carroll, Dir of DIA to SOD, 21 Jan 65, subj: An Appraisal of the Bombing of SVN (S).

21. Hq MAC/V Comd Hist, 1965, pp 197, 213-217 (TS); GEN Smith's Briefing Bk on SEA, Vol II, Dec 65, Tab A (TS); N.Y. News 14 Dec 65; CINCPAC Comd Hist, 1965, Vol II, pp 411, 430 (TS).

22. HQ MAC/V Comd Hist, 1965, p 261 (S); Gen Smith's Briefing Bk on SEA, Vol II, Dec 65 (TS); JCSM-907-65, 27 Dec 65 (TS).

23. JCSM-16-66, 8 Jan 66 (TS); N.Y. Times. 25 - 31 Dec 65.

24. Hq MAC/V Comd Hist, 1965, p 269 (S); CINCPAC Comd Hist, 1965, An B, App G (S); Gen Smith's Briefing Bk on SEA, Vol II, Dec 65 (TS); Hist of Aerospace Progs, Jul - Dec 65, pp 29 - 30 (S); Apps 1, 3 (S).

25. HQ MAC/V Hist, 1965, p 283 (C); CINCPAC Comd Hist, 1965, Vol II, pp 591-92 (S).

26. Gen Smith's Briefing Bk on SEA, Vol II, Dec 65 (TS);; memo, Col E.F. Macdonald, Asst Chief, Combined Plans Div, DCS/P&O, to C/S USAF 29 Jan 66, subj: Consequences of shifting to an Enclave Strategy in SVN (TS).

27. Statement by McNamara on 26 Jan 66 before House hearings before Subcmtes of the Cmte on Appropriations, 89th Cong, 2d Sess, Supplemental Defense Appropriations for 1956, pp 32-33 (U).

NOTES FOR PART FOUR

Chapter I

1. Hist (TS), CINCPAC, 1965, vol II, pp 326 and 328; Project CHECO SEA Rprt (TS), 15 Dec 66, subj: Comd and Control, 1965, pp 1-7; memo (TS), Lt Col B. F. Echols, Exec, Dir/Plans to AFCHO, 27 Nov 6?, subj: Review of Draft Hist Study, "The Air Campaign Against NVN."

2. Hist (TS), CINCPAC, 1965, vol II, pp 326 and 328; Testimony of Gen J. P. McConnell, CSAF on 9 May 66 before Senate Preparedness Investigating Subcmte of Cmte on Armed Services, 89th Cong, 2d Sess (U) g-10 May 66, USAF Tactical Air Ops and Readiness, pp 25-26.

3. Rpt (TS), An Eval of the Effects of the Air Campaign Against NVN and Laos, prepared by Jt Staff, Nov 66, in Dir/Plans; Talking paper for the JCS for the State-JCS Mtg on 1 Apr 66 (TS), Undated, subj: Discussions with Mr. Bundy on Far Eastern Matters, in Dir/plans; Hist (TS), CINCPAC, 1965, vol II, pp 339-41; memo (TS), Col D. G. Gravenstine, Chief Ops Review Gp, Dir/Ops to AFCHO, 22 Nov 6?, subj: Draft of AFCHO Hist Study.

4. Memo (TS), Col J. C. Berger, Asst Dir for Jt Matters, Dir/Ops to CSAF, 10 Aug 66; Background Paper on Division of R/T Area (TS), Mar 66, both in Dir/Plans; Excerpts from Gen Moore's Presentation to the JCS (TS), 13 Jul 66, in OSAF; Project CHECO SEA Rprts (TS), 15 Dec 66, subj: Comd and Control, 1965, pp 1-9; and 1 Mar 67, subj: Control of Air Strikes in SEA, pp 95-97; memo (TS), Echols to AFCHO, 27 Nov 67.

5. Van Staaveren (TS), 1965, pp 7L-74; N.Y. Times, l Feb 66.

6. Memo (TS), Col J.H. Germeraad, Asst Dep Dir of Plans for War plans, Dir/Plans to CSAF, 10 Jan 66, subj: Strat for SEA; Background paper on Pertinent Testimony by SECDEF and JCS given on 20 Jan 66 (TS), 20 Jan 66, both in Dir/Plans.

7. JCSM-16-66 (TS), 8 Jan 66.

8. Memo (TS), Lt Gen J.T. Carroll,, Dir DIA to SECDEF, 2L Jan 66, subj: An Appraisal of the Bombing of NVN, in Dir/Plans; JCSM-41-66 (TS), 18 Jan 66.

9. JCSM-56-66 (TS), 25 Jan 66.

10. JCS 2343/751 (TS), 13 Jan 66; SM-82-66 (TS), 22 Jan 66.

11. Memo (TS), SECDEF to Chmn JCS, S Jan 66, no subj: in Dir/Plans; CM-1135-66 (TS), 22 Jan 66.

12. Testimony of Secy McNamara on 26 Jan 66 before House Subcmte on Appns, 89th Cong, 2d Sess (U), Supplemental Def Appns for 1966, p 31.

13. Ibid., p 32; background briefing by U.S. officials (U), 31 Jan 66, in SAFOI.

14. Memo (TS), SECDEF to Pres, 24 Jan 66, subj: The Mil Outlook in SVN, in Dir/Plans; Hist (TS), CINCPAC, 1966, vol II, p 605.

15. Wash Post,11 Feb 66; N.Y. Times, 1 Feb 66.

16. Intvw (U), McConnell with Hearst Panel, 2I Mar 66, in SAFOI; Hist (TS), CINCPAC, 1966, vol II, p 49I; Rprt (TS), Dir/Ops, 20 Apr 66, subj: SEA Counter-Air Alternatives, p A -28, in AFCHO.

17. Memo (TS), Col D. G. Cooper, Ofc Dep Dir of Plans for War Plans, Dir/Plans to CSAF, 12 Feb 66, subj: The Employment of Air Power in the War in NVN; Briefing of JCS R/t Study Gp Rprt (TS), 6 Apr 66, subj: Air Ops Against NVN, App A; Rprt (TS), An Eval of Effect of the Air Campaign Against NVN and Laos, all in Dir/Plans; Hist (TS), CINCPAC, 1966, vol II, pp 493-44; Jacob Van Staaveren, USAF Deployment Planning for SEA (AFCHO, 1966) (TS), pp 1-2 and 26 (hereinafter cited as Van Staaveren, 1966).

18. CM-I147-66 (TS), 1 Feb 66.

19. Hist (TS), CINCPAC, 1966, vol II, pp 510-11; Van Staaveren (TS), 1966, chi II.

20. Memo (U), Lt Gen H. T. Wheless, Asst Vice CSAF to Deeps, Dies, Chiefs of Comparable Offices, 1? Feb 66, subj: Analysis of Air Power, in Dir/Plans; Van Staaveren, 1966, pp 10-15.

21. Memo (S), Lt Gen R. R. Compton, DCS/P&O to DCS/P&R, 21 Feb 66, subj: Organization in SEA, in Dir/Plans.

22. Memo (TS), Maj Gen S. J. McKee, Asst DCS/Plans and Ops for JCS to CSAF, 18 Feb 66, subj: Air Ops Against NVN; JCSM-113-66 (TS), 19 Feb 66, both in Dir/Plans.

23. Testimony of Secy McNamara on 25.Jan 66 before House Subcmte on Appns, 89th Cong, 2d Sess (U), Supplementary Def Appns for 1966, pp 33 and 39; memo (TS), Cooper to CSAF,1 2 Feb 66 Subj: The employment of Air Power in the War in VN; memo (TS), McKee to SECDEF, 24 Mar 66, subj: Air Ops against NVN, both in Dir/Plans; N.Y. Times, 5 Feb 66.

Chapter II

I. Jacob Van Staaveren, USAF Plans and Operations in Southeast Asia (AFCHO, 1965) (TS), p 50 (hereinafter cited as Van Staaveren 1965) Van Staaveren, 1966, pp 4 and 19.

Notes For Part Four

2. Rprt (S), SEA Air Ops, Mar 66, pp 2-3, prepared by Dir/Tac Eval, Hqs PACAF (hereinafter cited as PACAF rprt); JCS R/T Study Gp Rprt (TS), 6 Apr 67, App A; ltr (TS), CINCPAC to JCS, 18 Sep, subj: An Eval of CY 66-67 Force Rqmts; rprt (TS), Eval of Effects of the Air Campaign Against NVN and Laos, Nov 66, all in Dir/Plans; JCSM-I53-66 (TS), 10 Mar 66.

3. Memo (TS), McKee to Gen. W. H. Blanchard, Vice CSAF, 23 Mar 66, subj: Air Ops Against Aflds in NVN, in Dir/Ops; Hist (TS) MACV, 1966, p 431; Hist (TS), CINCPAC, 1966, Vol II, p 494.

4. Memo (TS), McKee to CSAF, 25 Mar 66, subj: Acft Losses Over NVN, watch Talking Paper, in Dir/plans; intvw (U), McConnell with Hearst Panel, 21 Mar 66 in SAFOI; rprt (TS),Dir/plans,20 Apr 66, p A-34; N.Y. Journal American, 20 Mar 66.

5. Hist (S), Dir/Ops, Jul-Dec 66, p 10; Hq USAF Ops Analysis Initial progress Rprt (S), Mar 66, subj: Analysis of Effectiveness of Interdiction in SEA, in AFCHO.

6. Hq USAF Ops Analysis Second Progress Rprt (S), May 66, subj: Analysis of Effectiveness of Air Interdiction in SEA, chi V in AFCHO.

7. Summary of Action by JCS (TS), 2b Mar 66, subj: Air Ops Against NVN, in Dir/Plans; Hist (TS), CINCPAC, 1966, vol II, p 497.

8. CSAFM-W-66 (TS), 20 Jan 66; CSAFM-P-23-66 and CMCM-33-66 (TS), 18 Apr 66; Talking Paper on Air Interdiction NVN/Laos (TS), 6 Jul 66; rprt (TS), An Eval of the Effects of the Air Campaign Against NVN and Laos, Nov 66, all in Dir/Plans; Hist (TS), CINCPAC, 1966, vol II, p 497; Hist (TS), MACV 1966, p 431.

9. CSAFM-W-66 (TS), 20 Jun 66; rprt (TS), An Eval of the Effects of the Air Campaign Against NVN and Laos, Nov 66, PACAF rprt (S), SEA Air Ops, Apr 66, pp 3-8, all in Dir/Plans.

10. DAF Order No 559N (U), 26 Mar 66, in AFCHO; Hist (TS), CINCPAC, 1966, vol II, p 468; tel to Ofc of Asst for Gen Officer Matters, DCS/P (U), 15 Aug 67.

11. PACAF rprt (S), SEA Air Ops, Apr 66, p 388, in Dir/Ops; Seventh AF Chronology, 1 Jul 65-30 Jun 66 (S), p 48; Hq USAF Ops Analysis Second Progress Rprt (S), May 66, pp 39-44, both in AFCHO; project CHECO SEA Rprts (TS), 15 Jul 67, subj: R/T, Jul 6b-Dec 66, p b0, and 21 Jul 67, subj: Expansion of USAF Ops in SEA, f966, pp 100-03; Hist (TS), CINCPAC, 1966, vol II, p 575.

12. Seventh AF Chronology, I Jul 65-30 Jun 66, p 11; PACAF rprt (S), SEA Air Ops, Apr 66, pp 3-8.

13. Background Paper on the Division of the R/T Area (TS), Mar 66; Talking Paper on the Division of the R/T Area (TS), Mar 66, both in Dir/Plans; Hist (TS), CINCPAC, 1966, vol II, pp 494-95.

14. Memo (TS), McKee to CSAF, 16 Apr 66, subj: Priority of Air Effort in SEA; memo (TS), SECDEF to Chmn JCS, 14 Apr 66, no subj: ltr (TS),CINCPAC to JCS, 18 Sep 66, subj: Eval of CY 66-67 Force Rqmts w/atch MACV Rprt (TS), 5 Sep 66; CM-1354-66 (TS), 20 Apr 66; Background Paper on R/T Areas (TS), Mar 66, all in Dir/Plans; Hist (TS), CINCPAC, 1966, vol II, pp 494-97; memo (TS), Gravenstine to AFCHO, 22 Nov 67.

15. JCS 2343/805-1 (TS), 14 Apr 66.

16. CSAFM-P-30-66 (TS), 20 Apr 66; memo (TS), Maj Gen L. D. Clay, Dep Dir of Plans to CSAF, 26 Jul 66, subj: U.S. Strat for SEA and S.W. Pacific; JCS 2343/805-1 (TS), 14 Apr 66; JCS 23431805-5, 22 Jlu'L66, all in Dir/Plans.

17. JCS R/T Study Gp Rprt (TS), 6 Apr 66, subj: Air ops Against NVN; memo (TS), McKee to CSAF, 13 Apr 66, subj: R/T Stuay Gp Rprt, Air Ops Against NVN; memo (TS), Gravenstine to AFCHO, 22 Nov 66.

18. CSAFM-P-22-66 (TS), 13 Apr 66; memo (TS), McKee to CSAF, 13 Apr 66; JCSM-238-66 (TS), 14 Apr 66, all in Dir/Plans.

19. Transcript (U), Secy Brown's remarks on "Meet the Press, " 22 May 66, in SAFOI.

20. Memo (S), Berger to CSAF, 15 Sep 66, subj: 7th AF Ops in RP II, ltr, and IV; PACAF rprt (S), SEA Air Ops, May 66, pp 1-8, both in Dir/Plans.

21. PACAF rprt (S), SEA Air Ops, May 66, pp L-8; Seventh AF Chronology, 1 Jul 65 to 30 Jun 66, p 52; ltr (TS), CINCPAC to JCS, 18 Sep 66; Project CHECO SEA Rprts (TS), I Sep 66, subj: Night Interdiction in SEA, pp 33 - 37, and 25 May 67, subj: Interdiction in SEA (1965-1966), pp 39-69.

22. Testimony of McConnell on 9 May 66 before Senate Preparedness Investigating Subcmte (TS), pp 16-17 (AFCHO's classified copy); PACAF rprt (S), SEA Air Ops, May 66, pp 1-8 and 22; CINCPACFLT Analysis Staff Study 9-66 (TS), 12 Jul 66, subj: Combat Effectiveness of the SA-2 through Mid-1966, both in Dir/Plans.

23. Memo (S), Maj Gen R. N. Smith, Dir of Plans to DCS/P&O, 3 May 66, subj: Capabilities for Aerial Blockade; msg 87716 (TS), CSAF to SAC, PACAF, TAC, USAFE, 6 May 66, both in Dir/Plans.

24. Msg 95413 (TS), CINCPACAF to CSAF, 24 May 66, in Dir/Plans.

25. Hist (S), Dir/Ops, Jul-Dec 66, p 126; PACAF rprt (S), SEA Air Ops, Jun 66, pp 6-9; Seventh AF Chronology, 1 Jul 65-30 Jun 66, (S), p 52; ltr

Notes For Part Four

(TS), CINCPAC to JCS, 18 Sep 66; Project CHECO SEA Rprt (S), 9 Aug 67, subj: Combat Skyspot, pp 6 and 19; Project CHECO SEA Rprt (TS), 1 Sep 66, subj: Night Interdiction in SEA, pp 33-37.

26. PACAF rprt (S), SEA Air Ops, Jun 66, pp 6-9; project CHECO SEA Rprt (TS), I Sep 66, subj: Night Interdiction in SEA, pp 33-37.

27. Project CHECO SEA Rprt (TS), 25 May 67, subj: Interdiction in SEA, 1965-1966, pp 60-61.

Chapter III

1. Memo (TS), R. Helms, Acting Dir CIA to Dep SECDEF, 27 Dec 65, subj: Probable Reaction to U.S. Bombing of POL, Targets in NVN, in Dir/Plans.

2. Memo (TS), McKee to SECDEF, 24 Mar 66, subj: Air Ops Against NVN; memo (S), C. R. Vance, Dep SECDEF to Chmn JCS, 25 Apr 66, same subj; memo (TS), W.W. Rostow, Spec Asst to pres to Secys State and Def, 6 May 66, no subj, all in Dir/Plans; study (TS), 27 Oct 66, subj: Effectiveness of Air Strikes Against NVN, prepared by Sys Analysis Div, Dept of Navy, in OSAF.

3. Memo (TS), Smith to CSAF, 16 Jun 66, subj: NVN Strike prog, in Dir/Plans; Hist (TS), CINCPAC, 1966, vol II, p 498.

4. Ibid.; Testimony of McConnell on g May 66 before Senate preparedness Investigating Subcmte of the Cmte on Armed Services (U), p 27.

5. Project CHECO SEA Rprt (TS), t5 Jut 6?, subj: R/T, Jul 65-Dec 66, p 59; N. Y. News, 24 Jun 66; Wash Post, 30 Jun 66, N.Y. Times, 1 Jul 66.

6. Hist (TS), CINCPAC, 1966, vol II, pp 499-500; Hist (TS), MACV 1966, p 431; Wash Post, 26 Jun 66; Balt Sun, 27 Jun 66.

7. Project CHECO SEA Rprt (TS), 15 Jul 62, subj: R/T, Jut 65-Dec 66, p 64; Hist (TS), CINCPAC, 1966, vol II, pp 499-500; Van Staaveren, 1966, p 42; N. Y. Times, 1 Jul 66.

8. Wash Post, 30 Jun 66.

9. N. Y. Times, I Jul 66; Van Staaveren, 1966, p 42.

10. Ltr (TS), CINCPAC to JCS, 4 Aug 66, subj: CINCPAC Briefing for SECDEF, 8 JUL 66; memo (TS), A. Enthoven, Asst SECDEF for Sys Analysis to Secys of Mil Depts et al, 12 Jul 66, subj: CINCPAC July 8, 1966 Briefing, both in Dir/Plans. Hist (TS), CINCPAC, 1966, vol II, pp 510-11.

11. Ltr (TS), CINCPAC to JCS, 4 Aug 66; memo (TS), Enthoven to Secys of Mil Depts et al, 12 Jul 66.

12. Van Staaveren, 1966, pp 42-53.

Air War – Vietnam

13. PACAF rprt (S), SEA Air Ops, Jul 66, pp 4-7; Rpt (TS), An Eval of the Effect of the Air Campaign Against NVN and Laos, Nov 66; ltr (TS), CINCPAC to JCS, 4 Aug 66.

14. Hist (S), Dir/Ops, Jul-Dec 66, pp 13 and 20-22.

15. Memo (TS), Berger to CSAF, 15 Sep 66; Excerpts from Gen Moorer's Presentation to the JCS (TS), 13 Jul 66; PACAF rprt (S), SEA Air Ops, Jul 66, pp 4-7; memo (TS), Gravenstine to AFCHO, 22 Nov 67.

16. Talking Paper for JCS for Their Mtg with Adm Sharp at the JCS Mtg of 23 Sep 66 (TS), 22 Sep 66, in Dir/Plans; PACAF rprt (S), SEA AirOps, Aug 66, pp 1-2; Hist (TS), CINCPAC, 1966, vol II, pp 500-02.

17. Memo (TS), M/Gen J. E. Thomas, Asst CS/I to SAF, 14 Oct 66, subj: PACAF Rprt on the NVN POL Situation, in Dir/plans.

18. PACAF Rprts (S), SEA Air Ops, Jul 66, pp 4-b, Aug 66, pp 1-3; Sep 66, pp 4 and 8; and Oct 66, pp 10-ll, all in Ops Review Gp, Dir/Ops.

19. Talking Paper for JCS for Their Mtg with Adm Sharp on 23 Sep 66 (TS), 22 Sep 66; PACAF rprts (S), SEA Air Ops, Jul 66, pp 4-5 and 20; Aug 66, p 22; Sep 66, p 23; and Oct 66, p 23.

20. PACAF rprt (S), SEA Air Ops, Jul 66, pp 4-5 and 20; N. Y. Times, 8 Jul 66 and 9 Aug 66; Wash Star, 8 Aug 66; Balt Sun, 22 Sept 66

21. Project CHECO SEA Rprt (TS), I Sep 66, subj: Night Interdiction in SEA, pp 37-38; ltr (TS), CINCPAC to JCS, 18 Sep 66; Hist (TS), MACV, 1966, p 434; NJ. Times, 31 Jul 66.

22. Project CHECO SEA Rpts (TS), 9 Sep 66, subj: Night Interdiction in SEA, pp 37-38; 21 Nov 66, subj: Operation Tally-Ho, pp vi and 1-12; 15 Feb 67, subj: Air Ops in the DMZ Area, pp 35-42; and 15 May 67, subj: Air Interdiction in SEA, pp 6I and 64; briefing (TS), by Brig Gen C. M. Talbott, Dep Dir Tac Air Control Center, 7th AF for SECDEF et al (Saigon), 10 Oct 66, Doc No 13 in Project CHECO SEA Rprt, 11 Feb 67 pt II; PACAF rprt (S), SEA Air Ops, Jul 66, pp ?-8; Wash Star, 1 Aug 66.

23. Memo (TS), Rear Adm F. J. Bloui, Dir Fast East Region, OSD to Dir of Jt Staff, 1 Jun 66, subj: Air Ops in the DMZ; msg (TS), JCS to CINCPAC, 20 Jun 66, both in Dir/plans; Hist (TS), MACV, 1966, pp 24-25.

24. PACAF rprt (S), SEA Air Ops, Aug 66, p 6; JCSM-603-66 (TS), t? Sep 66; N.Y. Times, 31 Jul 66.

25. Memo (S), McConnell to Dep SECDEF, 25 Aug 66, no subj, in Dir/plans; Hist (S), Dir/Ops, Jul 66, p 255; project CHECO SEA Rprt (TS), 21 Nov 66, subj: Operation Tally-Ho, pp 24-25.

Notes For Part Four

26. PACAF rprt (S), SEA Air Ops, Oct 66, p 2; Project CHECO SEA Rprt (TS), 15 Feb 6?, subj: Air Ops in the DMZ area, pp 22, 26-28, 37, and 41.

27. Project CHECO SEA Rprt (TS), 25 May 67, subj: Air Interdiction in SEA, 1965-1966, pp 64-65.

28. Memo for record (S), by Lt Col L. F. Duggan, Exec Asst Ofc, Dir Jt Staff, 13 Oct 66, no subj; memo (TS), undated, subj: JCS Assessment of the Threat, both in Dir/Plans; Briefing (TS), by Brig Gen Talbott, 10 Oct 66; Project CHECO SEA Rprt (TS), 15 Feb 67, subj: Air Ops in the DMZ area, 1966, pp 24-25 and 51; PACAF rprt (S), SEA Air Ops, pp 1-7 and 17.

29. Memo (TS), Holloway to SAF, 19 Oct 66, subj: Results of Air Effort Upon Movement Through NVN/SVN DMZ During Aug 66, in Dir/Plans.

30. Project CHECO SEA Rprt (TS), 25 May 6?, subj: Air Interdiction in SEA, 1965-1966, p 68; Doc 96 in Project CHECO SEA Rprt, 15 Feb 67, Pt II.

Chapter IV

t. Hist (S), Dir/Ops, Jul-Dec 66, pp 20-23.

2. Memo (S), Col F.W. Vetter, Mil Asst to SAF to Vice CSAF, 3 Aug 66, subj: Significance of Watercraft Destroyed in NVN, in Dir/Plans.

3. Ibid.

4. Hist (S), Dir/Ops, Jul-Dec 66, pp 23-24; memo (TS), Gravenstine to AFCHO, 22 Nov 66.

5. Memo (TS), SECDEF to SAF, SN, 2 Sep 66, subj: Night Ops in SEA, in OSAF.

6. Ibid.

7. Memo (S), SN to SECDEF, 28 Sep 66, subj: Study Results: Night Ops in NVN, in OSAF.

8. Memo (S), SAF to SECDEF, 10 Nov 66, no subj; study (TS), 2? Oct 66, subj: Effectiveness of Air Strikes Against NVN.

9. Memo (TS), SN to SECDEF, 3 Nov 66, subj: Study of Effectiveness of Air Strikes Against NVN w/atch study (TS), 27 Oct 67, subj: Effectiveness of Air Strikes, both in OSAF; memo (TS), Gravenstine to AFCHO, 22 Nov 67.

10. Memo (TS), SAF to SECDEF, 10 Nov 66.

11. Memo (S), SAF to SECDEF, 19 Jul 66, subj: A/C Attrition in SEA, in Dir/Plans.

Air War – Vietnam

12. Ibid.

13. Memo (S), SAF to SECDEF, 24 Aug 66, subj: Questions Resulting from Briefing on Night Ops in SEA; memo (TS), McConnell to Dep SECDEF, 25 Aug 66, subj: JCS 2343/894-1, 25 Aug 66, both in OSAF.

14. Memo (S), Clay to CSAF, 25 Aug 66, subj: SEA Tac Ftr Attrition and A/C Proc Prog; memo (S), Holloway to Chmn JCS, 29 Aug 66, subj: SEA Tac Ftr Attrition and A/C Procur, both in Dir/Plans.

15. N. Y. Times, 23 Sep 66.

16. Briefing Rprt of Factors Affecting A/C Losses in SEA (S), 26 Sep 66, prepared by Col. H.W. Hise, Chrmn, JCS A/C Losses Study Gp; ICS A/C Losses Study Gp Rprt (TS), Nov 66, subj: Factors Affecting Combat Air Ops and A/C Losses in SEA, both in Dir/Plans.

17. Msg 20135 (S), CINCPACAF to CSAF, 20 Oct 66, in OSAF; CINCPACFLT Analysis Staff Study 9-66 (TS), 12 Jul 66, subj: Combat Effectiveness of the SA-2 Through Mid-1966; Briefing Rprt of Factors Affecting A/C Losses in SEA (S), 26 Sep 66, both in Dir/Plans; Hist (S), Dir/Ops, Jul-Dec 66, pp 272-74.

18. Msg 20135 (S), CINCPACAF to CS,AF, 20 Oct 66; Briefing Rprt of Factors Affecting A/C Losses in SEA (S), 26 Sep 66.

19. Memos (S), Clay to CSAF, 23 and 27 Sep and 3 Oct 66, same subjs: Factors Affecting A/C Losses in SEA, in Dir/Plans; JCSM-651-66, 10 Oct 66.

20. Memo (VI, 22 Oct 66, subj: Secy Brown's Questions Concerning the Hise Rprt, in OSAF; Talking Paper for Chmn JCS on an Analysis of Air Ops in NVN to be discussed with SECDEF on 12 Nov 66 (TS), Il Nov 66, subj: Analysis of Air Ops in NVN, both in Dir/Plans; JCS 2343/956-f (TS), 15 Nov 66.

21. Memo (S), SECDEF to Chmn JCS, 17 Sep 66, subj: SEA Utilization of A/C, in OSAF; transcript (U), SECDEF News Briefing, 22 Sep 66, in SAFOL

22. Memo (TS), Chief, PAC Div, Jt Staff to J-3, 17 Sep 66, subj: Utilization of A/C in SEA; in OSAF; JCSM-646-66 (TS), 6 Oct 66.

23. JCSM-645-66 (TS), 6 Oct 66; JCSM-646-66, 6 Oct 66.

Chapter V

1. Van Staaveren, 1966, chi V.

2. CM-1906-66 (TS), 8 Nov 66; memo (TS), Gravenstine to AFCHO, 22 Nov 67.

Notes For Part Four

3. Memo (TS), SAF to SECDEF, 10 Nov 66, no subj, w/atch Interim Reply on Air staff Action Items Resulting from SECDEF Trip to SEA, 10-14 Oct 66, in OSAF.

4. PACAF rprt (S), SEA Air Ops, Nov 66, pp l-4; rprt (TS), An Eval of the Effects of the Air Campaign on NVN and Laos, Nov 66, both in Dir/Plans; Van Staaveren, 1966, pp 63-66.

5. PACAF rprts (S), SEA Air Ops, Nov 66, pp 1-9; Dec 66, pp 1-8, both in Dir/Plans.

6. Ibid.; Project CHECO SEA Rprt (TS), 15 Jul 67, subj: R/t, Jul 65-Dec 66, pp 98-99; Hist (TS), CINCPAC, 1966, vol II, pp 504-05 and 512; Balt Sun 18 Dec 66; N.Y. Times, 16 Dec 66.

7. Balt Sun, 14 Dec 66; N.Y. Times, lb Dec 66; Wash Post 15 and 16 Dec 66.

8. Project CHECO SEA Rprt (TS), to Jul 67, subj: R/T, Jul 65-Dec 66, pp 99-100.

9. Ibid.; N.Y. Times, 27 Dec 66.

10. Project CHECO SEA Rprt (TS), 2b May 6?, subj: Air Interdiction in SEA, 1965-1966, p 68; PACAF rprt (S), SEA Air Ops, Nov 66, pp l-9; Dec 66, pp l-8.

11. Ibid.; app I and 2; N.Y. Times, 26, 27 Dec 66, and 3 Jan 67.

12. CASFM-D-25-66 (TS), 23 Nov 66; memo (TS), Brig Gen E.A. McDonald, Dep Dir of Plans for War Plans to Dir/Plans, 16 Dec 66, subj: Combat Beaver, both in Dir/Plans; Hist (S), Dir/Ops, Jul-Dec 66, pp 2-3 and 254.

13. Memo (TS), McDonald to Dir/Plans, 23 Nov 66; Hist (S), Dir/Ops, Jul-Dec 66, pp 2-3; Project CHECO SEA Rprt (TS), 15 Jul 67, subj: R/T, Jul 65-Dec 66, pp 94-95.

14. Project CHECO SEA Rprt (TS), 21 Jul 66, subj: Expansion of USAF Ops in SEA, 1966, p 111; PACAF rprts (S), SEA Air Ops, Nov 66, p 22; and Dec 66, p 25.

15. PACAF Chronology, Jul 65-Jun 66 (S), in AFCHO; PACAF rprts (S), SEA Air Ops, Nov 66, pp 1-9; Dec 66, pp l-8; project CHECO SEA Rprt (TS), 15 Jul 67, subj: R/T, Jul 65-Dec 66, p 118; USAF Mgt Summary (S), 6 Jan 67, p 70; Hist (TS), CINCPAC, 1966, vol II, pp 522-28; app 10 and 11.

16. Ltr (TS), CINCPAC to JCS, 22 Nov 66, subj: SA-Threat Conf Rpt, in Dir/Plans; Hist (TS), CINCPAC, 1966, vol II, pp 516-19.

17. Ltr (TS), CINCPAC to JCS, 22 Nov 66; JCS 23431977 (TS), 16 Dec 66.

18. Memo (TS), Col E. T. Burnett, Dep Chief, Tac Div, Dir/Ops to Asst Dir of Plans for Jt and NSC Matters, 28 Nov 66, subj: Major Recommendations of the SA-2 Conf, in Dir/Plans; JCS 23431977 (TS), 16 Dec 66; Hist (TS), CINCPAC, 1966, vol II, p 519.

19. Van Staaveren, 1966, pp 71-74.

20. Address (U), Gen McConnell before Jt Activities Briefing, Hq USAF, 23 Nov 66, in SAFOI; Testimony of McConnell on 9 May 66 before Senate Investigating Preparedness Subcmte (U), p 29; Van Staaveren, 1966, pp 71-74.

21. Address (U), Gen McConnell before the Houston, Texas Forum, 29 Nov 66, in SAFOI.

22. Project CHECO SEA Rprts (TS), 1 Mar 67, subj: Control of Air Strikes in SEA, pp 81-99; and 23 Oct 67, subj: The War in VN, pp 44-45; memo (TS), SAF to SECDEF, 3 Jun 67, subj: Possible Course of Action in SEA; memo (TS), SAF to SECDEF, 9 Jun 67, no subj, both in Dir/Plans; memo (TS), Echols to AFCHO, 27 Nov 67.

23. Hist (TS), CINCPAC, 1966, vol II, pp 510-12 and 606-07.

24. Address (U), Secy Brown before Aviation/Space Writers Assoc Mtg, Wash D. C., 8 Dec 66, in SAFOI; Balt Sun, 9 Dec 66; rprt (U), Selected Statements on VN by DOD and other Admin Officials, I Jan-30 Jun 67, p 33, in SAFOI.

25. Testimony of Secy McNamara on 20 Feb 67 before House Subcmtes of the Cmte on Appns, 90th Cong, 1st Sess, Supplemental Def Appns for 19-67. p 21; Van Staaveren, 1966 pp 48-50.

Notes For Part Five

NOTES TO PART FIVE

Chapter I

1. Wash Post, 11 Jan 67; Balt Sun, 11 Jan 67; N. Y. Times, 12 Jan 67; W. W. Rostow, Sp Asst to the Pres, to SECDEF, 30 Jan 67, subj: Gen Taylor's Rprt.

2. Stmt by SECDEF McNamara on 25 Jan 67 in Hearings before the Senate Subcmte of the Cmte on Appns, 90th Cong, 1st Sess, DOD Appns for FY 1968, pt 1, pp 18-24; N. Y. Times, 18 Jun and 26 Oct 66; 12 Sep 66; Stmt by McNamara on 23 Jan 67 in Hearings before Senate Cmte on Armed Services and Subcmte on DOD of the Cmte on Appns, 90th Cong, 1st Sess, Supplemental Mil Procur and Const Auths, FY 1967, pp 12-20.

3. Memo (TS) Brig Gen E. A. McDonald, Dep Dir of Plans for War Plans to Dir/Plans, 3 Jan 67, subj: Prog Eval SEA; Rprt (U), Selected Stmts on VN by DOD and other Admin Officials, 1 Jan-30 Jun 67, prep by Ofc of Research and Anal, OSAF, 14 Jul 67, p 68; Jacob Van Staaveren, USAF Deployment Planning for Southeast Asia, 1966 (TS) (AFCHO, Jun 67), pp 70-74; and USAF Plans and Operations: The Air Campaign Against North Vietnam 1966 (TS) (AFCHO, Jan 68), pp 67-71.

4. Van Staaveren, USAF Deployment Planning for Southeast Asia, 1966, pp 70-74; The Air Campaign Against North Vietnam, 1966, pp 67-71.

5. Stmt by McNamara in Hearings before House Subcmte of Cmte on Appns, 90th Cong, 1st Sess, Supplemental Def Appns for 1967, p 21; Stmt by Secy Brown on 2 Feb 67 before Senate Subcmte of the Cmte on Appns, 90th Cong, 1st Sess, p 816; Van Staaveren, The Air Campaign Against North Vietnam, 1966, pp 13 and 69-70.

6. Van Staaveren, USAF Deployment Planning for Southeast Asia, 1966 p 63.

7. Stmt by Secy McNamara on 20 Feb 67 in Hearings before House Subcmte of Cmte on Appns, 90th Cong, 1st Sess, Supplemental DOD Appns for FY 1967, p 18; USAF Mgt Summary, 13 Jan 67, p 70; Hist (TS), CINCPAC, 1967, Vol II, pp 567-68.

8. Memo (S) Asst SECDEF (SA) to Secys of Mil Depts et al, 13 Feb 67, subj: SEA Deployment Prog 4 through Change 11.

9. USAF Mgt Summary (S), 3 Feb 67, p 73.

10. Ibid.

11. Hist (S), Seventh AF. Jan-Jun 67, p 34; Senate Hearings before Cmte on Armed Services and Subcmte on DOD of Cmte on Appns, Supple-

Air War – Vietnam

	mental Mil and Procur Auth, FY 1967, 90th Cong, 1st Sess. 23, 24 and 25 Jan 67, pp 55-57; USAF Mgt Summary (S), 13 Jan 67, p 68.
12.	Ltr (S), C. E. Hutchins, Jr., DCS/P&O Hq PACOM to SECDEF et al, 2 Jun 67, subj: Prog 4 Strength Accounting; DJSM-331-67, 15 Mar 67; memo (S), Dep SECDEF to JCS, 31 Mar 67, subj: Prog 4 Pers Strengths in SVN: Hist (S) Seventh Air Force, Jan-66-Jun 67, P 70.
13.	Hist (S), Dir/Ops. Jan-Jun 67, pp 157-59; memo (S) SECDEF to Secys of Mil Depts et al, 13 Feb 67: AF News Service, 5 May 67.
14.	Hist (TS). Dir /Plans, Jan-Jun 67, pp 391-98; memo (S) SECDEF to Secys of Mil Depts, et al, 13 Feb 67.
15.	Memo (TS), M/Gen L. D. Clay. Jr., Dir/Plans to Dir Doctrine Concepts and Objectives, 22 Dec 66, subj: B-52 Sortie Rate; memo (S) Col B. M. Shotts, Strat Div, Dir/Ops to OSAF, 14 Oct 67, subj: Increased B-52 Capability in SEA; msg JCS 7903 (TS), JCS to CINCPAC. CINCSAC, 4 Mar 67; memo (TS), N. E. Paul, Under SAF to SECDEF, 17 Mar 67, subj: B-52 Capability at U-Tapao; Proj CHECO SEA Rprt (TS), 22 Mar 67, subj: Arc Light, Jan -Jun 67, ch II; Van Staaveren. USAF Deployment Planning for Southeast 1966, pp 66-67.

Chapter II

1.	JCS 2339/253 (TS), 19 Feb 67; memo (TS). Lt Col S.G. Smith. Strat Plans Br, War Planning Div to CSAF. 20 Feb 67, subj: Mil Action Prog in SEA.
2.	JCS 2339/253 (TS), 19 Feb 67; CSAFM-A-48-67 (TS), 20 Feb 67; memo (TS), CSAF to Chmn JCS. 27 Feb 67, subj: Mil Action Prog for SEA.
3.	Memo (TS), Smith to CSAF. 20 Feb 67; memo (TS), Col H. W. Lauterbach. Asst Dep Dir of Plans for War Plans to Asst Dir of Plans for Jt and NSC Matters. 10 Apr 67, subj: Deployments and Mil Action SEA. with atch Background Paper on JCS 2339/255-1; JCS 2339/253 (TS). 19 Feb 67.
4.	Memo (TS), Clay to OSAF, 24 Mar 67, subj: Force Rqmts SEA; memo (TS), Mayland to CSAF, 30 Mar 67, subj: MACV FY 1968 Force Rqmts with atch Background Paper on MACV FY 1968 Rqts; ltr (TS), COMUSMACV to CINCPAC. 5 Apr 67, subj: FY 68 Force Rqmts w/atch rprt on MACV Force Rqts, FY 1968; memo (TS). Lauterbach to Asst Dir of Plans for Jt and NSC Matters, 10 Apr 67, subj: Deployments and Mil Action.
5.	CM-2192-67, 22 Mar 67; DJSM-374-67 (TS), 25 Mar 67; memo (TS), Lauterbach to Asst Dir of Plans for Jt and NSC Matters, 10 Apr 67; memo (S), Dep SECDEF to JCS. 21 Mar 67, subj: Prog 4 Personnel Strength for SVN.

Notes For Part Five

6. CM-2192-67 (TS). 22 Mar 67; DJSM-374-67(TS). 10 Apr 67; Hist (TS). CINCPAC. 1967, Vol II. p 522.

7. Memo (TS). Lauterbach to Asst Dir of Plans for Jt and NSC Matters. 10 Apr 67; memos (TS). Brig Gen R. G. Dupont. Asst Dir of Plans for Jt and NSC Matters to CSAF. 13 and 16 Apr 67. subj: Deployments and Mil Action; memo (TS). Mayland to CSAF. 9 Apr 67. subj: CINCPAC Force Rqmts.

8. CASFM-57-67 (TS). 14 Apr 67; Dupont to CSAF. 13 Apr 67, subj: Deployments and Mil Action. w/atch Talking Paper on Deployments and Mil Action; CM-2233-67 (TS). 13 Apr 67; JCS 2472/21-1 (TS). 14 Apr 67.

9. Memo (TS). Dupont to CSAF. 16 Apr 67. subj: Deployments and Mil Actions JCSM-218-67 (TS). 20 Apr 67.

10. JCSM-218-67 (TS). 20 Apr 67; CM-2255-67 (S). 20 Apr 67.

11. Memo (TS). Clay to CSAF. 12 May 67. subj: Alternative C/A. w/atch Background Paper on Alternative C/A; CM-2381-67 (TS). 29 May 67.

12. CSAFM-M-64-67 (TS). 21 Apr 67; memo (TS). Clay to CSAF. 16 May 67. subj: Ops Against NVN.

13. Memo (TS). Clay to CSAF. 16 May 67; JCSM-286-67 (TS). 20 May 67.

14. Memo (TS). Clay to CSAF. 12 May 67. subj: Alternate C/ A (JCS 2472/561). w/atch Talking and Background Paper; CM-2377-67 (TS). 24 May 67; memo (TS). Col F. W. Vetter Jr., Mil Asst to SAF To Maj Gen G. B. Simler. Dep Dir of Ops. 30 Jun 67. no subj.

15. Memo (TS). Clay to CSAF. 12 May 67. subj: Alternate C/A. w/atch Talking Paper on Alternative CIA.

16. CM-2377-67 (TS), 24 May 67.

17. CM-2353-67 (TS). 20 May 67. w/atch Draft Memo to the Pres. Dtd 19 May 67; Hist (S). Dir/Plans Jan-Jun 67, pp 165-66; CM-2381-67 (TS). 29 May 67.

18. Memo (TS). McDonald to CSAF. 30 May 67. subj: Draft Memo for the Pres on Future Actions in VN.

19. Hist (TS). Dir/Plans Jan-Jun 67. pp 165-66; JCSM-307-67. 1 Jun 67.

20. Hist (TS). Dir/Plans. Jan-Jun 67. pp 212-14; CM-2255-67 (TS). 20 Apr 67; JCSM-288-67. 20 May 67; JCS 2101/538. 13 Apr 67.

21. Memo (TS). SECDEF to Chmn JCS. SAF. SN. Dir CIA. 20 May 67. no subj.

22. Memo (TS). McDonald to CSAF. 26 May 67. subj: Air Ops Against NVN; JCSM-312-67 (TS). 2 Jun 67.

Air War – Vietnam

23. Memo (TS). SAF to SECDEF. 3 Jun 67. subj: Possible CIA in SEA; memo (TS), SAF to Asst SECDEF (SA). 5 Jun 67, no subj. in OSAF; memo (TS). SAF to SECDEF. 9 Jun 67, subj: Possible CIA in SEA; Hist (S). Dir/Ops. Jan-Jun 67. pp 15-16.

24. Memo (TS). SN to SECDEF. 2 Jun 67, subj: Alternative Bombing Prog.

25. Memo (TS). SAF to •SECDEF. 9 Jun 67. no subj: memo (TS). SAF to SECDEF. 13 Jun 67. subj: SEA Analyses; JCSM-341-67 (TS), 16 Jun 67; memo (U). Col B. S. Gunderson. OSAF to Asst Secy of State for East Asia Affairs. 9 Jun 67. no subj. in OSAF.

Chapter III

1. N.Y. Times. 7-12 Jul 67; Chicago Daily News. 5 Jul 67; Balt Sun. 7 Jul 67; Hist (TS). CINCPAC. Vol II. p 531.

2. Rprt (TS). Briefing for SECDEF in Saigon. 7-8 Jul 67, watch briefing by Amb Bunker. Gens Westmoreland and Momyer. Adm Sharp. et al.

3. Msg 221235 (S). COMUSMACV to CINCPAC and Chmn JCS. 22 Sep 67; memo (S). Col P.M. Spencer. Asst Dep Dir of Plans for Policy to SAF. 1 Oct 67. subj: SACSA Briefing to SECDEF on RVNAF Capabilities w I atch Background Paper on VNAF Improvement Prog.

4. Wash Post. 12 Jul 67; Phil Inquirer. 13 Jul 67; N.Y. Times. 13, 14 Jul and 4Aug 67; memo (S). SECDEF to Asst SECDEF (SA). 7 Sep 67. subj: New Manpower and Force Structure Authorization.

5. Hist (TS). Dir/Plans. Jul-Dec 67. pp 385 and 388-89; memo (TS). King to CSAF. 12 Jul 67. subj: U.S. Force Structure w I atch Background Papers on U.S. Force Structure and on Force Packages; CM-2499-67 (TS). 12 Jul 67; CM-2506-67 (TS). 13 Jul 67; JCS 24721115 (TS). 18 Jul 67; memo (S). SECDEF to Chmn JCS. 13 Jun 67, subj: Increased Use of SVN Civilians for U.S. Troop Support; Hist (TS). CINCPAC. 1967, p 531.

6. Memo (TS). King to CSAF, 11 Sep 67, subj: U.S. Force Objectives VN. w/atch Background Paper on U.S. Force Deployments. VN: JCSM-416-67 (TS). 20 Jul 67.

7. Memo (TS). Brig Gen R. D. Rheinbold. Dep Dir of Plans. to CSAF. 18 Jul 67, subj: U.S. Force Deployments. VN. w I atch Talking Paper on JCS 24721115. watch Background Paper on U.S. Force Structure in VN; memo (TS). King to CSAF. 11 Sep 67. subj: U.S. Force Deployments watch Background Paper on U.S. Force Deployments. VN; JCSM-416-67 (TS). 20 Jul 67; Hist (S). Dir/Plans. Jul-Dec 67. p 385; Jt Staff Rprt (TS), 29 Jan 68, subj: 1967 Year-end Review of VN. App B. Memo (S). Clay to other Air Staff Offices (S). 7 Aug 67; memo (S). SECDEF to Chmn JCS. 10 Aug 67. subj: Marine Tac Air Support in SVN; memo (S). Dep SECDEF to SAF. 4 Oct 67. subj: F-4D Ready Sq.

Notes For Part Five

8. Hist (S), Dir/Plans, Jul-Dec 67, p 386; memo (S), King to CSAF, 11Sep 67; msg (S), 83726, CSAF to CINCPACAF, 11 Aug 67; memo (S), SECDEF to Chmn JCS, 10 Aug 67, subj: FY 1968 Force Rqmts in SVN (Prog 5).

9. Hist (S), Dir/Plans, Jul-Dec 67, pp 386, 389-90; memo (S), Maj Gen R. H. Ellis, Dir/Plans to CSAF, 11 Sep 67, subj: U.S. Force Deployments VN; memo (S), Rear Adm J. V. Cobb, Dep Dir, Jt Staff to SECDEF, 25 Jul 67, no subj: JCSM-505-67 (TS), 15 Sep 67; CSAFM-B-41-67 (TS), 14 Aug 67; CM-2493-67 (S), 10 Jul 67; JCS 2101/539-2 (TS), 14 Aug 67; memo (TS), Ellis to CSAF, 27 Aug 67, subj: Worldwide U.S. Mil Posture as Related to General Purpose Forces.

10. Memo (S), SECDEF to Secys of Mil Depts and Chmn JCS, 5 Oct 67, subj: FY 68 U.S. Forces Deployments, VN; Hist (TS), CINCPAC, 1967, Vol II, pp 537-38; Hist (S), Dir/Plans, Jul - Dec 67, p 386.

11. Memo (S), R. Murray, Ofc Asst SECDEF (SA) to SECDEF, 9 Aug 67, subj: Addn AF Pers for Thai. in OSAF; memo (S), Dep S ECDEF to SAF, 15 Aug 67, subj: AF Pers Rqmts in Thai; Summary of USAF Orgn and Manning in Thai, 24 Aug-18 Sep 67; memo (S), SAF to Dep SECDEF, 1 Sep 67, subj: AF Pers Rqmts in Thai, in OSAF; memo (S), SAF to Dep SECDEF, 3 Oct 67, same subj; memo (S), SECDEF to SAF, Chmn JCS, 13 Oct 67, subj: U.S. Troop Strength in Thai; Hist (S), Dir I Plans, Jul - Dec 67, p 381.

12. Hq MACV Rprt (S), Review and Analysis System for RVNAF Prog, 16 Sep 67; memo (S), Spencer to CSAF, 21 Sep 67, subj: Increase in RVNAF Force Structure; JCSM-530-67 (S), 28 Sep 67; msg 221235 (TS), COMUSMACV to CINCPAC, 22 Sep 67; memo (S), SECDEF to Chmn JCS, 7 Oct 67, subj: Increase in FY 68 RVN Force Level; Hist (S), Dir/Plans, Jul - Dec 67, pp 167-68; Hist (TS), CINCPAC, 1967, Vol II, p 559.

13. Memo (TS), Maj R. A. Owens to Asst Dir of Plans for Jt and NSC Matters, 14 Aug 67, subj: To reduce Impact on U.S. Air Ops when only four CVAs are in WESTPAC; memo (TS), Owens to CSAF, 16 Aug 67, subj: Replacement of *Forrestal* Air Ops Capability; memo (TS), Asst SECDEF (SA) to Secys of Mil Depts et al, 16 Oct 67, subj: SEA Deployment Prog 5 through Change 4; JCSM-468-67 (TS), 24 Aug 67; memo (C), SECDEF to Chmn JCS, 13 Sep 67, subj: Replacement of Forrestal Air Ops Capability; N. Y. Times, 2 Aug 67; Hist (TS), CINCPAC, 1967, Vol II, p 986.

14. Hist (TS), CINCPAC, 1967, Vol II, pp 501-03; memo (S), SECDEF to Chmn JCS, 16 Aug 67, subj: Expansion of Dye Marker Obstacle Sub-system; Herman S. Wolk, USAF Plans and Policies: Logistics and Base Construction in Southeast Asia, 1967 (AFCHO, Oct 68), ch IV; Proj CHECO SEA Rprt, 31 Jul subj: IGLOO WHITE (Initial Phase), prepared by Dir/Tac Eval. CHECO Div, Hq PACAF, to pages 42 – 46

15. Hist (S), Dir/Aerospace Frogs, Jul-Dec 67, pp 45 and 67; memo (S), Col R. G. Echols to Asst Dir for Plans for Jt and NSC Matters, 1 Aug 67, subj: Airborne AECM Support; memo (S) SECDEF to SAF, Chmn JCS, 23 Oct 67, subj: Deployment Adj Req; memo (S), Asst SECDEF (SA) to Secys of Mil Depts et al 24 Nov 67; subj: SEA Deployment Prog 5 through Change 6; Hist (TS Dir/Plans, Jul-Dec 67, p 383; USAF Mgt Summary (S), 5 Jan 68, p 25.

Chapter IV

1. Balt Sun, 13 Aug, 19 Aug, 1 Sep 67; N.Y. Times, 4 Sep, 5 Sep 67.

2. Memo (TS), Shotts to Asst Dir of Plans for Jt and NSC Matters, 13 Sep 67, subj: Concept for an Optimum Air Campaign Against NVN; memo (TS), Col R. L. Gleason, Asst Chief, Sp Warfare Div, Dir/ Plans to Asst Dir of Plans for Jt and NSC Matters, 27 Sep 67, subj: Increased Pressures on NVN; JCSM-555-67 (TS), 17 Oct 67; Talking Paper for the JCS for the SECDEF-JCS Mtg, 23 Sep 67, subj: Increased Pressures on NVN.

3. JCS 2472/158-2 (TS), 29 Sep 67; msg 98610 (TS), COMUSMACV to CINCPAC, 16 Sep 67; Proj CHECO SEA Rprt (S), 5 Jan 68, subj: Op Neutralize; Hist (S), CINCPAC, 1967, Vol II, pp 619-20; Jt Staff Rprt (TS), 29 Jan 68, subj: 1967 Year-end Review of VN, App B, CSAFM-L-20-67 (TS), 19 Jul 67.

4. Memo (TS), CMC to Chmn JCS, 22 Sep 67, w/atch Memo for Pres; CMCM-31-67 (TS), 24 Sep 67.

5. Memo (TS), Shotts to CSAF, 28 Sep 67, subj: Sit in the DMZ w/atch Talking Paper on Sit in the DMZ; CM-2668-67 (TS), 28 Sep 67.

6. Memo (TS), Shotts to CSAF, 28 Sep 67; memo (TS), Col L. W. Bray, Jr., Asst Dir of Plans for Jt and NSC Matters to CSAF, 30 Sep 67, subj: Sit in the DMZ; JCS 2472/158-2 (TS), 29 Sep 67.

7. Talking Paper for Chmn JCS on an Item to be Discussed at the JCS Mtg of 25 Oct 67 (TS), 24 Oct 67.

8. N. Y. Times, 22 and 30 Sep 67.

9. Memo (TS), Col K. L. Collings, Ofc of Dep Dir of Plans for War Plans to CSAF, 5 Dec 67, subj: Reduced Levels of Activity in VN, w/atch Background Paper on Dev of J-5 P 2159-1, same subj: memos (TS), Lauterbach to Asst Dir of Plans for Jt and NSC Matters, 20 and 26 Dec 67, subj: Reduced Levels of Activity in VN; memo (S), W. W. Rostow, Sp Asst to Pres, to SECDEF, 30 Jan 67, subj: Gen Taylor's Rprt; JCSM-107-67, 21 Feb 67; memo (S), CSAF to SAF, Under SAF, 4 Jan 68, subj: Graduated Deescalation, w/atch Bradford Proposal and Study, "Graduated Deescalation," prepared by Concepts and Objectives Div, Dir/Doctrine, Concepts and Objectives, Dir/P&O; N.Y. Times, 23 Feb 67, Wash Post, 7 Feb 68.

Notes For Part Five

10. Memo (TS), Collings to CSAF, 5 Dec 67, subj: Reduced Levels of Activity in VN; memos (TS), Lauterbach to Asst Dir of Plans for Jt and NSC Matters, 20 and 26 Dec 67.

11. Memo (TS), Ellis to CSAF, 10 Oct 67, subj: Review and Summary of JCS Positions on C/A in VN, w/atch Talking Paper, same subj.

12. JCS 2339/266-1 (TS), 13 Dec 67; CSAFM-H-20-67 (TS), 15 Dec 67; memo (TS), Col J. A. Larson, Dep Asst for NSC Matters, Dir I Plans to Dir I Plans and Policy, J-5, 14 Dec 67, no subj: CM-2803-67 (TS), 5 Dec 67; JCSM-698-67 (TS), 16 Dec 67; Clay to CSAF. 21 Feb 67, subj: Settlement of the Conflict in VN; JCSM-107-67 (TS), 21 Feb 67.

13. Memo (TS), CMC Point Paper, 11 Oct 67; memo (TS), Ellis to CSAF, 17 Oct 67; subj: Planning Guide for a Study to Formulate Altn Mil CIA for SEA, w/atch Background Paper on Planning Guidance; memo (TS), King to CSAF, 12 Oct 67, subj: C/A in VN (J-5 TP 113-67); SM-709-67 (TS), 18 Oct 67. Hist (S), Dir/Plans, Jul-Dec 67, pp 178-79 and 424; memo (TS), Ellis to CSAF, 12 Oct 67; Talking Paper for JCS at their Mtg of 13 Oct 67, subj: J-5 TP ll3-67; CSAFM-R-23-67 (TS), 13 Oct 67; MC Flimsy 88-67 (TS), 13 Oct 67, subj: J-5 TP 113-67.

14. JCSM-555-66 (TS), 17 Oct 67; JCS 2472/173-1 (TS), 29 Nov 67, w/atch Draft Plan for an Air Campaign Against NVN; Hist (TS), Dir /Plans, Jul-Dec 67, pp 382-83.

15. Memo (TS), Ellis to CSAF, 14 Nov 67, subj: Increased Pressures in VN (CM-2754-67); DJSM-1381-67 (TS), 13 Nov 67; memo (TS), Ellis to CSAF, 21 Nov 67, subj: Policy for the Conduct of Ops over the Next Four Months.

16. Memo (TS), Shotts to Asst Dir of Plans for Jt and NSC Matters, 13 Sep 67, subj: Concept for an Optimum Air Campaign Against NVN; Talking Paper for the JCS Mtg with Adm Sharp on 16 Oct 67 (TP-42-67), 16 Oct 67; Hist of Dir/Ops (S), Jul-Dec 67, p 35. DJSM-1265-67 (S), 14 Oct 67; MR (U) by Col W. H. Holt, Staff Off, Dir /Ops, 19 Oct 67, subj: Estbl of a Jt Planning Gp; Memo (TS), Maj R. A. Owens, Tac Div, Dir/Ops to CSAF, 16 Oct 67, subj: CINCPAC Mtg with JCS, w/atch Background Paper; DJSM-1312-67 (TS), 24 Oct 67.

17. Memo (TS), Col Gail Stubbs, Dep Dir of Policy, Dir/Plans to Dep Asst Dir for Jt Matters, 24 Nov 67, subj: Draft Plan for Air Campaign Against NVN; memo (TS), Col R. W. Beck, Asst Dep Dir of Plans for War Plans to Dep Asst Dir for Jt Matters, 22 Nov 67, subj: Comments on Draft Plan For Air Campaign Against NVN; JCS 2472/173-1 (TS), 29 Nov 67; memo (TS), Ellis to CSAF, 30 Nov 67, subj: Plan for an Air Campaign Against NVN. CSAFM-H-1-67 (TS), 1 Dec 67; JCS memo (TS), 4 Dec 67, no subj; memo (TS), War Planning Div to Maj Gen R.N. Smith, Asst DCS/ P&O, 18 Jan 67, subj: Aeriel Interdiction and Resupply Denial in NVN.

Air War – Vietnam

18. Memo (TS). Ellis to CSAF. 21 Nov 67. subj: Policy for the Conduct of Ops in SEA over the Next Four Months; CM-2752-67 (TS). 10 Nov 67; JCSM-633-67 (TS). 27 Nov 67; JCS 2572/222 (TS). 1 Feb 68.

19. Memo (TS). Ellis to CSAF. 21 Nov 67; CSAFM-D-23-67 (TS). 22 Nov 67.

20. Hist (S). Dir/Plans. Jul-Dec 67. pp 206-07; JCS 2572/222 (TS). 1 Feb 68; memo (TS). Of c. War Planning Div to Gen Smith. 18 Jan 68. subj: Aerial Interdiction and Resupply Denial in NVN.

Chapter V

1. Hist (TS). CINCPAC. 1967, Vol II. pp 911-28.

2. Hist (S). Dir/Plans. Jul-Dec 67. pp 206-07; memo (TS). Spencer to CSAF. 13 Nov 67. subj: Priority Progs in SVN; DJSM-1381-67 (TS), 13 Nov 67.

3. Hist (S). Dir/Plans. Jul-Dec 67, pp 206-07; DJSM-1381-67 (TS). 13 Nov 67.

4. DJSM-1381-67 (TS). 13 Nov 67.

5. Wash Post. 17 Nov 67; Phila Inquirer. 17 Nov 67; Balt Sun. 16 and 20 Nov 67; Weekly Compilation of Pres Docs. 20 Nov 6 pp 1. 562-65.

6. SM-778-67 (S). 18 Nov 67; msg (TS). CSAF to CINCPAC. 28 Nov 67.

7. Msg (TS). CSAF to CINCPAC. 28 Nov 67; Weekly Compilation of Pres Docs (U), 20 Nov 67. p 1. 583; Wash Post. 22 Nov 67; Msg (S). CSAF to CINCPACAF. 22 Nov 67.

8. Msg (TS). CSAF to CINCPAC. 28 Nov 67.

9. Hist (TS). Dir/Plans. Jul-Dec 67. p 207; Journal of Mil Asst (S). Dec 67. p 185; Jt Staff Rprt (TS). 29 Jan 68. subj: 1967 Year-end Review of VN. pp 7-29; memo (TS). Gleason to Asst Dir of Plans for Jt and NSC Matters. 5 Dec 67. subj: RVNAF FY 68 and FY 69 Forces.

10. Hist (TS). Dir I Plans. Jul-Dec 67. p 384; memo (S). SECDEF to Chmn JCS. 23 Oct 67. subj: Deployment Adj Rqmt; CM-2668-67 (TS). 28 Sep 67; DJSM-1378 (TS). 9 Nov 67.

11. Hist (S). Dir /Ops. Jul-Dec 67, pp 181-82; Chicago Trib. 13 Dec 67.

12. Memo (TS), CSAF to SAF, 4 Oct 67, subj: Increase Arc Light Ops; CSAFM-R-10-6 (TS), 4 Oct 67; CSAFM-R-14-67 (TS), 6 Oct 67; memo (TS), SAF to SECDEF, 6 Oct 67, subj: Increased Arc Light B-52 Sorties; CSAFM-R-20-67 (TS), 12 Oct 67; JCSM-554-67 (TS), 14 Oct 67, memo (TS), Shotts to CSAF, 14 Oct 67, subj: Increased B-52 Capability in SEA; MR (TS), by Lt Col Billy J. Moore, War Planning Div, 7 Dec 67; CSAFM-H-24-67 (TS), 15 Dec 67; memo (TS), SAF to SECDEF, 15 Dec 67, subj: Increased B-52 Sortie Rates; Hist (S), Dir/Ops, Jul-Dec 67,

Notes For Part Five

p 386; memo (S), Asst SECDEF (SA) to Secys of Mil Depts et al, 29 Dec 67, subj: Deployment Prog 5 through Change 8: Proj CHECO SEA Rprt (S), 29 Nov 68, subj: The Air War in VN, Jul-Dec 67, pp 48-49.

13. Hist of Dir/Ops (S), Jul-Dec 67, p 333; Hist (TS), CINCPAC, 1967, Vol II, pp 504 and 921; memos (S), Asst SECDEF (SA) to Secys of Mil Depts, 29 Dec 67, subj: Deployment Prog 5 through Change 8; memo (S), Asst SECDEF to Secys of Mil Depts et al, 15 Feb 67, subj: Deployment Prog 5 through Change 13; Stmt by Secy McNamara on 16 Feb 68 before House Subcmte of the Cmte on Appns, 90th Cong, 2d Sess, DOD Approps for 1969, pt l, pp 79 and 196-73; N. !• Times, 20-26 Dec 67.

14. Rprts (S), Trends, Indicators, and Analysis, prepared by Ops Review Gp, Dir/Ops, Hq USAF; Dec 67, p 3-36, and Jan 68,' pp 2-l to 2-8, and 3-36; MACV Comd Hist (TS), 1967, Vol I. pp 36 and 108-ll.

15. Hist (TS), Dir/Ops, Jul-Dec 67, pp 19 and 30-33; memo (TS), Col F. W. Vetter Jr., Mil Asst to SAF to Maj Gen G. B. Simler, Dep Dir /Ops, 30 Jun 67, no subj: memo (TS), Vetter to Vice CSAF, 14 Aug 67, no subj: Eval of the Effectiveness of the Air War in NVN; memo (S), W. W. Rostow, Sp Asst to Pres, to SECDEF, Dir CIA et al, 25 Oct 67, subj: VN Data and Prog Indicators; Daily Staff Digest, Hq USAF (C), 13 Nov 67; Vetter to Dr. H. Brown, 30 Nov 67, no subj; memo (C), Col C. G. Whitely, Chief Aero- space Doctrine Div to Maj Gen R. A. Yudkin, Dir of Doctrine Concepts and Objectives, 22 Dec 67, subj: Trends, Indicators and Analysis; memo (S), G. A. Carver Jr., CIA to Rostow. 30 Dec 67, subj: Task Force Progress Rprt.

16. Stmt by Secy McNamara on 16 Feb 68 before the House Subcmte of the Cmte on Appns, 90th Cong, 2d Sess, DOD Appns for 1969, pt l, p 173.

17. Memo (S), Asst SECDEF to Secys of Mil Depts et al, 29 Dec 67, subj: SEA Deployment Prog 5 through Change 8; memo(S):" Asst SECDEF (SA) to Secys of Mil Depts et al, 15 Feb 67, subj: SEA Deployment Prog 5 through Change 13; USAF Mgt Summary (S), 26 Jun 67, p 25 and 5 Jan 68, p 25; Jt Staff Rprt (TS), 1967 Year-end Review of VN, 29 Jan 68, ch 3; Stmt by Secy McNamara on 22 Jan 68 before Senate Armed Services Cmte on FY 1969-73 Def Prog and 1969 Def Budget, p 105 (draft copy).

18. Memos (S), Asst SECDEF (SA) to Secys of Mil Depts et al, 29 Dec 67 and 15 Feb 68; Jt Staff Rprt (TS), 29 Jan 68, subj: 1967 Year-end Review of VN, ch 6.

Air War – Vietnam

Notes For Part Six

NOTES TO PART SIX

Chapter I.

1. Public Papers of the Presidents of the United States, Lyndon B. Johnson (1968-69) (GPO, 1970), I, p-2

2. Jacob Van Staaveren, The Air Force in Vietnam: The Search for Military Alternatives, 1967 (TS) (Off/AF Hist, 1967), pp 54-59 (Hereafter-cited as AF in VN, 1967) Forms Part Four of Airwar: Vietnam Plans and Operations Volume One published by Defense Lion Publications.

3. ibid pp 61-63; Jt Staff Study (TS), chap 2, subj: 1967 Year-End Review of VN, 29 Jan 69.

4. CM-2922-68 (TS), 19 Jan 68, w/atch MR (TS), by Maj Gen Dupuy, subj: Conversation with Amb Robert Komer and Maj Gen George I. Forsyth, CORDS, MACV 17 Jan 1968.

5. Memo (S) CSAF to Pres, subj: National Strat in VN, 1Feb 68; N.Y. Times 30 Jan 68.

6. Memos (TS), CSAF to SAF and Under SAF. subj: Graduated De-escalation with atch Bradford Proposal and Study, same subj, 4 Jan 68; (TS), Maj Gen Richard A. Yudkin, Dir/Doctrine, Concepts, and DCS/P&O to Dir/Ops. Et al., same subj, 7 Dec 68.

7. Memo (TS), Maj Gen Richard H. Ellis, Dir/Plans to CSAF, subj: Reduced Levels of Activity in VN, 4 Jan 68; JCS 2472/205 (TS), 12 Feb 68; JCS 2472/205-3 (TS), 12 Feb 68.

8. Memo (TS) CSAF to CINCPAC, 4 Jan 68; memo (TS), Maj John• J. Nolan, PAC SEA Br, Dep Dir/Plans for Plcy to Asst Dir/Plans for Jt and NSC Matters, subj: Sea Cabin, 24 Jan 68; memo (TS) Ellis to CSAF, same subj; 25 Jan 68; JCSM-62-68 (TS), 31 Jan 68; N.Y. Times, 30 Sep 67 and 2 and 3 Jan 68; •Wash Post, 2 and 3 Jan 68.

9. Christian Science Monitor, 5 Jan 68, N.Y. Times, 8 and 18 Jan 68; CM-2927-68 (S), 20 jan 68; stmt by Clifford on 25 Jan 68 before Senate Armed Services Cmte, 90th cong, 2d Sess, Nomination of Clark M Clifford as SECDEF, p 9.

10. All strength figures in this section are based on: Van Staaveren, (TS) AF in VN, 1967, pp 59-65, 78-79; USAF Mgt Summary, SEA (S), 5 Jan 68, p 65.

Chapter II.

1. Charles H. Hildreth, Bernard C. Nalty, and Anne B. Mascolino, The Air Force. Response to the Pueblo Crisis, (TS) (Off/ AF Hist, 1968), p 2; Hist (S) Dir/Ops, Jan - Jun 68, p 152.

Air War – Vietnam

2. Hist (TS), SAC, Jan-Jun 1968, vol I, p 129.

3. Stmt of the Pres, 26 Jan 68, Weekly Compilation of Presidential Documents, vol 4, no 4, p 135.

4. Adm U.S. Grant Sharp, CINCPAC and Gen William C. Westmoreland, COMUSMACV, Rprt on VN (as of 30 Jun 68) (U) {GPO, 1969}, pp 162 – 163 and 170 - 74; CM-2973-68 (TS), 13 Feb 68; CM-3003-68 (TS), 12 Feb 68; Hist (S), MACV, 1968, vol II, p 923; stmt by Gen Wheeler on 68 before House Subcmte of Cmte on Approps, 90th Cong, 2d Sess, subj DoD Approps for FY1969, pt 1, pp 45-46.

5. Hist (S), Dir/Ops. Jan - Jun 1968, p 152; Hist (S). SAC, Jan - Jun 1968, vol 1, p 125; Proj CHECO SEA Rprt (TS), subj: Khe Sanh (Op Niagara). 22 Jan - 31 Mar 68, 13 Sep 68, pp 4 - 11, p 37 hereinafter cited as CHECO Rprt Khe Sanh; Hist (S), MACV, 1968, vol I, pp 24 - 26.

6. CM-2944-68 (TS), 3 Feb 68; CSAFM-B-12-68 (TS), 7 Feb 68; Weekly Compilation of Pres docs (U). 18 Feb 68; Wash Post, 10 and 17 Feb 68; N.Y. Times 5 Feb 68.

7. CM-2944-68 (TS), 3 Feb 68.

8. Sharp and Westmoreland (U), Rprt on VN, p 164; CM-2944-68 (TS), 3 Feb 68; CM-2988-68 (TS); JCS 2472/277 (S), 17 Apr 68 Wash Post 14 Feb 68.

9. Memo (S), Ellis to CSAF,. subj: Increased Arc Light Sorties, 19 Mar 68 memo (S), Lt Col Warren W. Halstead, Comd Planning Div to Asst Dir/Plans for Jt and NSC Matters, subj: Increased Air Mun in Support of Arc Light Sorties, 24 Feb 68; Hist (S), SAC, Jan-Jun 1968, vol 1 pp 127-130; Balt Sun. 8 and 15 Feb 68.

10. Memo (S), Col Jerry F. Hogue, Dep Chief, Tac Div, Dir/Ops to Asst Dir/Plans for Jt and NSC Matters, subj: Use of Thai-based Acft Against Targets in SVN, 15 Feb 68; CHECO Rprt Khe Sanh (TS), p 7; Hist (S), CINCPAC, 1968, vol III, p 54.

11. Hist rprt (S), 7/13 AF 1969, subj: Hist of Task Force Alpha, 1 Oct 67 - 30 Apr 68, pp 56-57; memo (TS), Lt Gen Alfred D. Starbird (USA), Dir, Def Comm Planning Gp (DCPG) to SECDEF. subj: Briefing in DCPG Program, 15 Feb 68; Proj CHECO SEA rprts (S), subj: Igloo White (Initial Phase), 31 Jul 68, pp 1- 39; and subj: Khe Sanh, 13 Sep 68, p 29; Hist (S), MACV, 1968, vol I, pp 423 -24; Sharp and Westmoreland (U), Rprt on VN pp 171-72.

12. Hist (S), Dir/Dev Ofc. DCS/R&D, Jul - Dec 1968, pp 165 - 68; N.Y. Times 17 Feb 68; CHECO Rprt, Khe Sanh (TS), p 89

13. JCS 1478/125-2 (TS), 4 Apr 68; CSAFM-D-ll-68 (TS), and CSAFM-D-12-68; JCSM-237-68 (TS), 19 Apr 68; memo (C), Dep SECDEF to

Notes For Part Six

Chmn JCS subj: Ops Control of III MAF Av Assets, 15 May 68; Sharp and Westmoreland (U) Rprt on VN, p 173.

14. CHECO Rprt, Khe Sanh (TS). pp 90-91, 112-14; Moyers S. Shore, II. The Battle for Khe Sanh (Wash, DC: Hq USMC, 1969), pp 95-102 (hereinafter cited as Shore, Khe Sanh).

15. Shore, Khe Sanh, pp 101-103; Sharp and Westmoreland, Rprt on VN, p 163, stmts by Gen McConnell on 6 May 68 before Senate Subcmte on Approps, subj DOD Approps for FY 1969, 90th Cong. 2d Sess, pt 1, p 89 and on 16 Apr 69 before Senate Cmte on Armed Services. subj: Auth for Mil Procur, Research, and Dev, FY70, and Res Strength, 91st Cong, 1st Sess, pt 1, p 940; Hist (S), MACV, 1968, vol I, p 425.

16 CHECO Rprt, Khe Sanh (TS), p 92; Shore, Khe Sanh, pp 101-102.

17. CHECO Rprt, Khe Sanh, pp 73-74; Sharp and-Westmoreland (U), Rprt on VN, p 172; Hist (S), MACV, 1968, vol I, p 426; N.Y. Times, 7 Mar 68

18. Proj CHECO SEA Rprts (S), subj: Air Response to the Tet Offensive, 30 Jan – 29 Feb, 68, 12 Aug 68, pp 2 - 3 (hereinafter cited as CHECO Rprt. Tet (S)); CHECO Rprt Khe Sanh (TS), pp 5-6; CM-2927-68 (TS), 20 Jan 68; Sharp and Westmoreland (U). Rprt on VN, p 158; Phila Bull, 11 Jan 68;   Chicago Trib; 18 Jan 68; Balt Sun, 21 Jan 68.

19. CHECO Rprt, Tet (S), p 55 CM-2973-68 (TS) 13 Feb 68; N.Y. Times 26 Jan 68.

20. CHECO Rprt, Tet (S), pp 3-4; CHECO Rprt, Khe Sanh (TS), pp 24-25; CHECO Rprt (S), subj: The Defense of Saigon, 14 Dec 68, p 4 (hereinafter cited as CHECO Rprt, Saigon (S)); Hist (S). MACV, 1968, vol I p 378; Sharp and Westmoreland (U) Rprt on VN, pp 158-159; Senate rprt (U) Background Information Relative to Southeast Asia and Vietnam (Rev ed, Mar 68), prep by US Senate Cmte on Foreign Relations, 90th Cong, 2d Sess, p 56; N.Y. Times, 1Feb 68.

21. CHECO Rprt, Tet (S), pp 12-21; Rprt (S) Tac Eval, Hq PACAF, SEA Air Ops, Feb 1968, p 13; Sharp and Westmoreland (U) Rprt on VN, p 158; rcrds (S) Hq USAF Comd Post, May 1968; USAF Mgt Summary, SEA (S), 8 Mar 68, p i; Senate rprt (U), Background Info Relating SEA and VN, p 56

22. Sharp and Westmoreland (U), Rprt on VN, p 44; Hist (S), MACV, 68 vol II p 426 - 29.

23. CHECO Rprt, Tet (S) pp 10, 55 – 60.

24. Senate rprt (U), Background Info Relating to SEA and VN, p 67; Hist (TS), CINCPAC, 1968, vol III, p 122; N.Y. Times, 2 and 5 Feb 68; Balt Sun, 5 and 7 Feb 68.

Air War – Vietnam

25. Compilation of Pres Docs (U), 5 Feb 68, pp 198 - 203; Senate Info Relating to SEA and VN, p 67; Def Intel Digest (S) Jun 1969, p.17; Sharp and Westmoreland (U), Rprt on VN, p 161.

26. Senate rprt (U) Background Info Relating to SEA and VN, p 57; N.Y. Times 2 and 5 Feb 68; Balt Sun, 5, 7, and 15 Feb 68; Hist (TS), CINCPAC, 1968, vol III, p 122; memo (TS), Ellis to CSAF, subj: Air Campaign Against NVN, 1 Feb 68; CSAFM-A-34-68 (TS), 31 Jan 68.

27. Ibid.

28. CM-3003-68 (TS), 12 Feb 68.

Chapter III

1. CM-2944-68 (TS), 3 Feb 68; MR (C), Thomas D. Morris, Asst. SECDEF (I&L), subj: Action in Response to MACV Msg on Addn Help required, 3 Feb 68; memo (S), Ellis to CSAF, subj: Addn MACV Rqmt, 8 Feb 68; Wash Post, 3 Feb 68.

2. Hist (S) Dir/Ops, Jan-Jun 1968, p 141; memo (S). Russell Murray, SECDEF to Secys of Mil Depts, et. al., subj: Deployment Program 5, Through Change 13, 15 Feb 68.

3. Memo (S) Glenn V. Gibson, Dep Asst SECDEF (I&L) to SECDEF, no subj, 9 Feb 68.

4. Memo (TS), Ellis to CSAF, subj: Addn MACV Rqmts; memo (S), 8 Feb 68; Gibson to SECDEF, 9 Feb 68; stmt by Secy McNamara on 14 Feb 68before Senate Subcmte on Approps, 90th Cong, 2d Sess, subj: DOD Approps for FY 69, pt 1, p 43; Hist (S), Dir of Aerospace Progs, Jun 1968, p 46.

5. Memo (TS), Ellis to CSAF, 8 Feb 68; memo (S), Gibson to SECDEF, Feb 68; Hist (S). MACV, 1968, vol I, p 428.

6. JCSM-91-68 (TS), 12 Feb 68; CM-2973-68 (TS), 13 Feb 68.

7. JCSM-91-68 (TS), 12 Feb 68; Sharp and Westmoreland (U), Rprt on VN p 161.

8. CM-2944-68 (TS), subj: Emerg Reinforcement of COMUSMACV, 3 Feb 68; Sharp and Westmoreland (U), Rprt on VN, pp 172-73; Senate rprt (U) Background Info Relating   SEA and VN, p 51.

9. Memo (TS), Col Bryan M. Shotts, Asst for Jt Matters, Dir/Ops to CSAF subj: Energ Reinforcement of COMUSMACV, 11 Feb 68; JCS 2472/226-2 (TS), 11 Feb 68; JCSM-91-68 (TS). 12 Feb 68.

10. Hist (TS), Dir/Plans, Jan – Jun 1968 pp 71-72; CM-3003-68 (TS), 12 Feb 68.

11. CM-3003-68 (TS), 12 Feb 68.

Notes For Part Six

12. Hist (S) Dir/Ops, Jan-Jun 1968, p 142; msgs 24226 and 24418 (S), JCS to CSAF, CINCPAC, et. al. 13 Feb 68; JCS 2472/231 (TS), 13 Feb 68; memo (S) Actg Asst SECDEF (S) to Secys of Mil Depts, et.al., subj: SEA Deployment Prog 5, Through Change 13, 15 Feb 68; Sharp and Westmoreland (U) rprt on VN, p 165; N.Y. Times, 14 Feb 68; Wash Post, 12 and 14 Feb, Balt Sun,'24 Feb 68.

13. JCSM-96-68 (TS), 13 Feb 68; JCSM-99-68 (TS) 14 Feb 68.

14. JCSM-99-68 (TS). 14 Feb 68.

15. Rprt (TS) Chmn on Sit in VN and MACV Force Rqmts, 28 Feb 68 (JCS 2472/237, 28 Feb 68); N.Y. Times, 7 Feb 70; Townsend Hoopes, The Limits of Intervention (An Inside Account of How the Johnson Policy of Escalation in Vietnam Was Reversed) (New York: David McKay Co., Inc 1969) p 161.

16. Hist (TS) Dir/Plans, Jan-Jun 1968, pp 78-79; Hoopes, Limits of Intervention, chaps VII and VIII; N.Y. Times, 7 Feb 70; Wash Post 7 Feb 70.

17. Memo (S), Under SAF to SECDEF, subj: Possible Addn Forces, SEA 28 Feb 68.

18. CM-3082-68 (TS), 3 Mar 68.

19. MEMO (TS), Dep SECDEF to Pres, no subj, 4 Mar 68.

Chapter IV

1. The three strategies, or campaigns, which are outlined in this chapter can be found in the Memo (S), SAF to Dep SECDEF, subj: SEA Alternatives with Atch Views of Air Staff and Rand-Ops Analysis SEA Gp 4 Mar 68.

2. Ibid

3. Ibid

4. Memo (U) CSAF to SAF. subj: Concepts of Ops for SEA with increased Emphasis on Air Ops, 3 Apr 68, with atch DAF Air Staff Summary Sheet, 2 Apr 68, same subj; Balt Sun, 17 Apr 68; CSAFM-C-68 (TS), 18 Mar 68.

5. MR (S), Brig Gen Charles W: Lenfest, Dep Dir/Plan : for War Plans, DCS/P&O, subj SEA Alternative Strategies, 9 Apr 68; memo (S); Under SAF to SECDEF subj SEA Alternative Strategies, 12 Apr 68; N.Y. Times, 2 Apr 68.

6. Memo (S), Under SAF to Chmn JCS, subj: Alternate Air Strategies, 10 May 68; memo (S), Under SAF to SECDEF, subj: VN. 17 May 68.

7. Hoopes, Limits of Intervention, p 177; memo (TS), Col Donald H. King, Asst Dep Dir/Plans and Plcy to CSAF, subj: Strategies with atch

Background and Talking Papers on Special Cmte's Recommendations Concerning In-Depth Study of VN Policy, 17 Mar 68.

8. Hoopes, Limits of Intervention, pp 180 - 81.

9. Memo (S), Dep SECDEF to Chmn JCS, subj: SEA Deployments, 14 Mar 68; Hist (TS), Dir/Plans, Jan-Jun 1968, pp 78-79.

10. Memo (S), SAF to SECDEF, subj: Deployments to SEA, 15 Mar 68; CSAFM-C-63-68 (TS), 18 Mar 68; memo (TS). King to CSAF - subj: MACV Force Rqmts, 22 Mar 68; N.Y. Times, 12 Apr 68.

11. Memo (S), SAF to SECDEF, 15 Mar 68; JCSM-159-68 (TS), 14 Mar 68; Clifford "A Vietnam Reappraisal," Foreign Affairs (Jul, 1969), p 610.

12. CSAF to 7th AF. AF Advsy Gp, 15 Apr 68; CM-3128-68 (S) CM-3131-68 (TS), 21 Mar 68; memo (S), Dep SECDEF to Chmn JCS subj: Increase in FY 1968 RVNAF Force Levels, 4 Apr 68.

13. Hist (S), Dir/Plans, Jan-Jun 1968, pp 83 - 84; memo (S), CSAF to SAF 27 Feb 68; msg (S), 1977 State-Def to Amemb Seoul and to CINCPAC, 26 Feb 68; memo (S), CSAF to SAF, subj: ROKAF Air Def Capability and possible ROKAF Participation in the VN Conflict, 27 Feb 68; msg (S) CSAF to CINCPACAF, 11 Mar 68.

14. Memo (S), SAF to SECDEF, subj: Deployment to Nam Phong AB, 4 Mar 68, CSAFM-C-63-68 (TS); 18 Mar 68; memo (TS), Col Charles B. Hodges, Asst Ch for Jt Matters, War Planning Div to Asst Dir/Plans for Jt and NSC Matters, with atch Background Paper on Dev of Nam Phong AB, 25 Mar 68; memo (TS), Col Charles W. Abbott, Asst Dep Dir/Plans for War Plans to Asst Dir/Plans for Jt and NSC Matters, subj: Nam Phong AB, 20 Mar 68.

15. Memo (TS) Hodges to Asst Dir/Plans for Jt and NSC Matters, 25 Mar (S), Dep SECDEF to SAF, subj: Dev of Nam Phong, 23 Mar 68.

16. Memo (TS), King to CSAF, subj: Strategies, with atch Background and Talking papers on Army Study on SVN/SEA Alternatives, 17 Mar 68.

17. Memo (TS),: King to CSAF, subj: Strategies, with atch Background and Talking papers on ISA's Independent Memo to the Pres on Alternate Strategies in SEA, 17 Mar 68.

18. Memo (TS), King to CSAF, subj: Strategies, with atch Background Paper on Special Cmte's Recommendation Concerning In-Depth Study of VN Policy, 17 Mar 68.

19. Ibid.

20. Ibid.

Notes For Part Six

Chapter V

1. Testimony of General McConnell before the House Cmte on Appns, 90th Congress, 2nd Sess, DOD 1969 Appns, Pt I, 28 Feb 68, pp 801-802, 808.
2. USAF Tactical Panel Mtg #67-30(S), subj: Tropic Moon III, 22 Nov 67.
3. Interim Rprt (S), USAF Ops Analys/RAND SEA Study Gp, 22 Feb 68.
4. Ibid.
5. Summary Rprt (TS), The U.S. Interdiction Effort in North Vietnam & Laos; Some Assessments of Results and Opportunities, prep by USAF Ops Analys/RAND SEA Study Gp, 1 Jul 68.
6. Rprt (S), Engineering Dev in a Crisis, prep by USAF Ops Analys/ RAND SEA Study Gp, 14 Jul 68.
7. Ibid.; Summary Rprt (TS), The U.S. Interdiction in North Vietnam & Laos: Some Assessments of Results and Opportunities, prep by USAF Ops Analys/RAND SEA Study GP, 1 Jul 68
8. Rprt (S), Engineering Development in a Crisis prep by USAF Ops Analys/RAND SEA Study Gp, 14 Jul 68.
9. Memo (S), Col Brian S. Gunderson, SAFOS, to Asst SAF (R&D) and Asst Vice CSAF, subj: AFGOA/RAND Rprt, Engineering Development in a Crisis, 29 Aug 68.
10. Ltr (S), Gen John P. McConnell, CSAF to Dirs of Air Staff, subj: USAF Ops Analyst/RAND Studies and Report, 16 Sep 68.
11. Ltr (S), Lt Gen Seth J. McKee Asst VCSAF to Asst SAF (R&D), subj: AFGOA I RAND Rprt, Engineering Development in a Crisis, 9 Oct 68
12. Atch (S), subj: Air Staff Comments on Spec Progs, to Ltr (S), Lt Gen Seth J. McKee, Asst VCSAF to Asst SAF (R&D), subj: AFGOA/ RAND Rprt, Engineering Development in a Crisis, 9 Oct 68.
13. Rprt (S), Speeding Decision on Which Projects Deserve Special Treatment, 15 Jan 69, by Hq USAF Ad Hoc Study Gp.
14. Air Staff White Paper (S), subj: Air Force Development/ Procurement Actions in Response to SEA Problems, 29 Jan 69.
15. Ibid.
16. Ibid.
17. Ibid.
18. Memo (S). Dr. Alexander H. Flax. Asst SAF (R&D) to SAF. subj: USAF Ops Analys/RAND Rprt. Engineering Development in a Crisis. 28 Jan 69.
19. Ibid.

20. Ibid.
21. Ibid.
22. DCS/ R&D Staff Study (U). subj: SEAOR Mtg. 15 Nov 67; Final Rprt (S). Credible Comet. prep by Hq USAF. 1 Mar 68.
23. Leonard Sullivan. Jr .• Deputy DDR&E (SEA Matters). "Ten Lessons from Southeast Asia." Journal of Defense Research (S). Spring 1969.
24. DCS/R&D Staff Study (U). subj: SEAOR Mtg. 15 Nov 67; Final Rprt (S). Credible Comet. prep by Hq USAF. 1 Mar 68.
25. Atch (U) to Ltr (U). Gen James Ferguson. Comdr. AFSC to Hq USAF. subj: Southeast Asia Operational Requirements. 4 Dec 67; Agreements Reached During General Officers' SEAOR Review. 15-16 Nov 67.
26. Ltr (U). Ferguson to Hq USAF. subj: Southeast Asia Operational Requirements (SEAOR's). 4 Dec 67.
27. Ibid.
28. Ibid.
29. Ltr (U). Lt Gen Joseph R. Holzapple. DCS/R&D to Dir of Opl Rqmts & Dev Plans. et al. subj: Review of SEAOR's. 12 Dec 67.
30. Ibid.
31. Briefing (U). Dir of Opl Rqmts & Dev Plans to DCS/R&D subj: Recommendations of SEAOR Review Board. Mar 68.
32. Ibid. Intvw. Herman S. Wolk with Lt Col Orville A. Reed. Jr .• Dir of Opl Rqmts & Dev Plans. 7 Jul 69.
33. Intvw with Lt Col Reed. 9 Jul 69.
34. The MacDonald Rprt (S). Planning for the Effective Use of Advanced Technological Resources in. Southeast Asia. 7 Jan 69.
35. Final Rprt (S). Credible Comet. prep by Hq USAF. 1 Mar 68.
36. Hq USAF Southeast Asia Management Summary. 31 Jan 69.
37. WSEG Rprt #16 (S), The North Vietnamese Air Defense Environment: Air-to-Air Encounters in Southeast Asia, Apr 68; Msg (S), Momyer to Ryan. subj: North Vietnamese GCI Advantage. 29 Jan 68.
38. Ibid. Electronic Warfare Panel Mtg #68-15. subj: Have Dart, 3 Apr 68.
39. Msg (S). Momyer to Ryan, subj: North Vietnamese GCI Advantage. 29 Jan 68.
40. Ltr (S). CSAF to AFSC, AFLC, TAC & ADC. subj: GCI Capability. 31 Jan 68.

Notes For Part Six

41. AFSC Summary Rprt. vol I (S). subj: Have Dart Task Force Study. 13 Mar 68.

42. Ibid.; Electronic Warfare Panel Mtg #68-15 (S). subj: Have Dart. 3 Apr 68.

43. Ibid.

44. Electronic Warfare Panel Mtg #68-15 (S). subj: Have Dart. 3 Apr 68.

45. Final Rprt (S). Credible Comet. prep by Hq USAF. 1 Mar 68, Chaps v. IX. x.

46. Ibid., Chap IX.

47. Final Rprt (S). Credible Comet. prep by Hq USAF. 1 Mar 68, Chap X.

48. Rprt (TS)•.Night Song. Updated by JCS. 25 Apr 68.

49. Ibid.

50. Ibid.

51. Testimony of General Holzapple before House Subcmte on Appns. 90th Cong. 2nd Sess. DOD 1969 Appns. 5 Mar 68. Part II. p 177.

52. Final Rprt. Combat Lancer. vol I (S). prep by TAC Tactical Fighter Weapons Center. Nellis AFB. May 69.

53. RAND Corp Doc D-17861-PR. subj: An Analysis of F-IllA Radar Bombing Capability in Combat (Combat Lancer). 1 Oct 68.

54. Ibid.

55. Final Rprt. Combat Lancer. vol I (S). prep by TAC Tactical Fighter Weapons Center. Nellis AFB. May 69.

56. Memo (S). Vice Adm. Nels C. Johnson. Dir/Jnt Staff to CINCPAC. subj: New Bombing Capabilities. 17 Sep 68.

57. Ibid.; Air Force Planner's Memo #398-68 (S). subj: New Bombing Capabilities (DJSM-1063-68). 11 Sep 68.

58. Memo (S). Dep SECDEF Nitze to SAF. subj: F-105 T-Stick II Modification. 28 Sep 68.

59. Ltr (S). CINCPAC to JCS. subj: PACOM Significant R&D Problem Areas. 4 Mar 68.

60. Encl (S) to Memo (S). SAF to Dep SECDEF. subj: SEA Alternative Strategies. 4 Mar 68.

61. Atch (S) to Memo (TS). Dir/Plans to CSAF. subj: Maj Gen Blood. USAF Dir of Ops. Hq 7AF will Brief on COMMANOO HUNT. 7 Sep 68; Dir/Plans Talking Paper on Air Staff Position Re Commando Hunt; Encl

Air War – Vietnam

(S) to Memo (S). SAF to Dep SECDEF. subj: SEA Alternative Strategies. 4 Mar 68.

62. Rprt (S). Southeast Asia Air Ops. prep by Hq PACAF. Nov 68; Briefing (S). by Maj Gen David C. Jones. Vice Comdr. 7 AF. subj: Northeast Monsoon Interdiction Campaign. Jun 69.

63. JCSM-669-68 (S). CJCS to SECDEF. subj: IGLOO WHITE/DUEL BLADE/DUFFEL BAG Eval Comm Rprt. 7 Nov 68; Rprt (S). Southeast Asia Air Ops prep by Hq PACAF. Nov 68.

64. Memo (S). Sullivan to DDR&E. Dir/ARPA. Dep Dirs/DDR&E and Members of the Senior PROVOST Steering Gp. subj: Vietnam Trip Rprt #12. (Jun 2-17. 1969). 18 Jun 69.

65. Briefing (S), by Maj Gen Jones, subj: Northeast Monsoon Inter- diction Campaign, Jun 69.

66. Atch (S) to Memo (S), Dir/Plans to CSAF, subj: Disestablishment of JTF-728, 20 Jun 68.

67. Ibid., Memo for Rcrd (S), by Lt Col C. H. Johnson, Dir/Plans, subj: IGLOO WHITE DUEL BLADE Status, 8 Jul 68.

68. Ibid.

69. Atch (S) to Memo (TS), Dir/Ops to CSAF, subj: A Briefing by Adm. RusselL USN (Ret) on the DCPG Study, 15 Oct 68.

70. JCSM-669-68, Memo (S), CJCS to SECDEF, subj: IGLOO WHITE/DUEL BLADE/DUFFELL BAG Evaluation Comm Rprt, 7 Nov 68.

71. Memo (S), Foster to Dir/DCPG, subj: Guidance on Future Course of DCPG, 2 Dec 68.

72. Tactical Panel Mtg #67- 30 (S), subj: Tropic Moon III, 22 Nov 67.

73. Hist (S), Dir/Dev, DCS/R&D. Jul-Dec 68, p 166.

74. Encl (S) to Ltr (S), Lt Gen Marvin L. McNickle, DCS/R&D to Gen Ryan, Vice CSAF, subj: Rprt on Testing in SEA, 1 Apr 69.

75. Ibid.; Hist (S), Dir/Dev, DCS/R&D, Jul-Dec 68, pp 157-158.

76. Ibid.

77. Encl (S) to Ltr (S) McNickle to Ryan, subj: Rprt on Testing in Southeast Asia, 1 Apr 69.

78. Ibid.; Hist (TS), MACV 1968, vol II. p 754.

79. Staff Study (S), subj: Tropic Moon III, Dec 68, by USAF SEA Projs Div; Tactical Panel Mtg #67-30 (S), subj: Tropic Moon III, 22 Nov 67.

Notes For Part Six

80. CSAF Decision Paper (S), subj: Tropic Moon III, attached to Memo (U), Col Roger D. Coleson, Dir /Secretariat to Office of AF History, 15 Jul 68.

81. Tactical Panel Mtg #67-30 (S), subj: Tropic Moon III. 22 Nov 67; Atch (S), subj: Air Staff Response to Recommendations Contained in Trip Rprt of Gen F. K. Everest to Ltr (S), Maj Gen Henry B. Kucheman. Jr., Dir/Dev. DCS/R&D to Asst Dir. OT&E. DDR&E, subj: Comments on Trip Rprt, Gen Frank K. Everest, Jr., 13 Mar-3Apr 69; Hist (S), Dir/Dev, Jul-Dec 68.

82. Testimony of Dr. Brown before House Cmte on Appns, 90th Cong, 2nd Sess, DOD 1969 Appns, Pt 1

Chapter VI

1. Clifford, "A Vietnam Reappraisal," pp 610-13.

2. Public Paper of the President, Lyndon. B. Johnson 1968 – 1969, I, (Washington, 1970), p 470.

3. Ibid., 470 – 473, N.Y. Times, 1Apr 68; Clifford (U), "A Vietnam Reappraisal" p 614.

4. Van Staaveren (TS), AF in VN, 1967, pp 18, 22, 23, and 28; CINCPAC Qtr Rprt (S), subj: Measurement of Progress in SEA, 30 Sep 68, p 75; Hearings before Senate Preparedness Investigating Subcmte, 16 Aug 67, pt 2 pp 132 – 133, and on 22 Aug 67, pt 3, pp 234 - 35; Hist (TS), CINCPAC, 1968, vol III, p 124.

5. Ltr. (TS), CINCPAC to JCS, subj: Est Effects of Restrictions on Cessation of Air Ops Against NVN, 12 Aug 68, with atch CINCPAC study, 15 Apr 68.

6. Rprts (S), Trends, Indicators, and Analyses, prepared by Ops Review Gp, Dir/Ops, Hq USAF, May 1968, pp 1-1, and Jun 1968, pp 1-1; memo (S), Col Brian S. Gunderson, Exec Asst to SAF, to CSAF, no subj, 9 May 1968; with atch memo (TS), SAF to SECDEF, no subj, 3 May 68.

7. Memo (TS), Alfred B. Fitt, Asst SECDEF (Manpower and Res Affairs) to SECDEF, 2 Apr 68; DJSM-380-68 (TS), 2 Apr 68.

8. Memo (TS), SECDEF to Secys of Mil Depts and Chmn JCS, subj: Res Callup 4 Apr 68; CSAM-173-68 (TS). 3 Apr 68.

9. Memo (TS), 6 Apr 68; CM-3187-68 (TS), 6 Apr 68.

10. Ibid., JCSM-221-68 (TS). 10 Apr 68.

11. Memo (S), SAF to SECDEF, subj: "Res Recall and FAC-ALO Tng, 6 Apr. 1968; memo (S), SAF to SECDEF, subj: Callup of AFRES Units, 8 Apr, 1968.

12. N.Y. Times, 12 Apr 1968; Phila Inquirer, 12 Apr 1968.

13. Memo (S), Dep SECDEF to Secys of Mil Depts, et. al., subj: SEA Deployment Prog 6, 4 Apr 1968; Hist (TS), Dir/Plans, Jan - Jun 1968, pp 78 - 81; Hist (TS), CINCPAC, 1968, vol III, p 32; Hist (S), TAC, Jan - Jun 1968, pp 493-99.

14. Memo (S); Dep SECDEF to Chmn JCS, subj: Tac Acft Deployments to SEA, 5 Apr 1968; memo (S), Ellis to CSAF, same subj, 18 Apr 1968.

15. Memo (S), Ellis to CSAF, subj: Tac Acft Deployments to SEA, 18 Apr 1968; JCSM-255-68 (S), 22 Apr 68; Hist (TS), Dir/Plans. Jan – Jun 68, pp 77-78.

16. Memos (S), Dep SECDEF to Cbmn JCS, subj: USAF Posture in Korea, 22 May 68; Dep SECDEF to Chmn JCS, subj: Tac Acft Deployment to SEA, 24 May 68.

17. Memo (S) Dep SECDEF to Chmn JCS, subj: B-52 Sortie Rate, 15 Apr 1968

18. JCSM-257-68 (S), 23 Apr 68.

19. Memo (S) Col Abbott C. Greenleaf, Mil Asst to SECDEF. subj: B-52 Sortie Rate, 29 Apr 68; CSAFM-D-3-68 (S), 3 Apr 68; memo (S), SAF to SECDEF subj: B-52 Retention, 23 Apr 68; memo (S), Dep SECDEF to SAF, subj: B-52 Retention. 15 May 68.

20. Hist (S), Dir/Ops, Jul - Dec 1968, p 360; CM-3354-68 (S), 23 May 68; memo (S) Maj Gen Thomas N. Wilson, Dep Dir/Plans to CSAF. subj: B-52 Sortie Rate, 27 May 68; JCSM-333-68 (S), 29 May 68.

21. CM-3354-68 (S), 23 May 68

22. Memo (S) Dep SECDEF to Chmn JCS, subj: B-52 Sortie Rate, 22 June 68.

23. Memo (S), George F. Bogardus, Asst for Poltl Affairs, Dir/Plans to Asst Dir/Plans for Jt and NSC Matters, subj: AF Above Threshold Change to Prog 6, 28 Jun 68; Hist (S), SAC, Jan - Jun 1968, vol I, pp 132 - 35.

24. Memo (S), Ellis to CSAF, subj: Accelerated Expansion of RVNAF, 11Apr 68; memo (S), Dep SECDEF to Chmn JCS, subj: RVNAF Improvement and Modernization, 16 Apr 68.

25. JCSM-233-68 (S), 15 Apr 68.

26. Memo (S), Col Paul Spencer, Asst Dep Dir/Plans for Plcy to CSAF, subj: US Force Posture and Planning, 17 May 68, with atch Background Paper on VNAF Force Structure; JCSM-324-68 (S). 23 May 68.

27. Hist (S), Dir/Plans, Jan-Jun 1968, p 68; memo (S) Dep SECDEF to Chmn JCS, subj: Accelerated Expansion of RVNAF, 24 May 68; memo (S), Dep SECDEF to Chmn JCS, subj: RVNAF Improvement and Modernization 25 Jun 68.

Notes For Part Six

28. Memo (S), Lt Col Billy J. Moore, Chief War Planning Div to Asst for Jt and NSC Matters, 30 Jun 68; memo (S), Dep SECDEF to 25 Jun 68; Dep SECDEF to Secys of Mil Depts, et. al., subj: Improvement and Modernization, 30 Jun 68.

29. Memo (S), Brig Gen Leslie W. Bray, Jr. Asst Dir/Plans for Jt and NSC Matters to CSAF, subj: Prog to Give Greater Recognition to RVN, l0 Jul 68; JCS 2472/313-1 (S), 11 Jul 68.

30. Memo (TS), Lt Col Robert B. Kenworthy, Intl Affairs Div, Dir/Plans to CSAF, subj: Negotiations on SEA, 15 Apr 68; memo (S), Ellis to CSAF, 18 Apr 68; SM-278-68 (S), 23 Apr 68; memo (TS), Dep SECDEF to Chmn JCS subj: Redeployment of Forces, 26 Apr 68; memo (TS), to CSAF, subj: A Concept for VN During Negotiations, 7 May 68.

31. Memo (TS) Spencer to CSAF, subj: JCS Discussions with SECDEF atch Talking Papers on Bombing of NVN, 3 May 68.

32 Ibid; JCSM-289-68. (TS), 8 May 68

33. Memo (TS), Spencer to CSAF, subj: A Concept for Negotiations, 7 May 68, .with atch Talking Papers on JCS 2472/282-1, 7 May 68; JCSM-290-68 (TS), 8 May 68.

34. Ltr (TS), Chmn JCS to Amb W. Averell Harriman, no subj, 3 May 68.

(Defense Lion Publications notes that two pages in the original are missing.)

21. Memo (S), Ofc, Eastern Regional Div, Dep Dir/Plans for Plcy to Asst Dir/Plans for Jt and NSC Matters, subj: RVNAF Improvement and Modernization, 17 Oct 68; JCSM-633-68 (S). 25 Oct 68.

22. Memo (S), Dep SECDEF to Secys of Mil Depts, et. al., 1Nov 68.

23. Memo (S), Dep SECDEF to Chmn JCS, et. al., subj: T-Day Planning, Hist (TS), Dir/Plans, Jul-Dec 68, pp 366-68.

24. Hist (TS), Dir/Plans, Jul-Dec 1968, pp 366-68; JCSM-531-68 (TS), 3 Sep 68

25. Memo (S), Under SAF to SECDEF, subj: T-Day Planning, 2 Oct 68; Hist (TS), Dir/Plans, Jul-Dec 1968, pp 366-68.

26. Memo (S), Dep SECDEF to Chmn JCS, subj: T-Day Planning, 17 Oct 68; CM-3737-68 (S), 30 Oct 68; Hist (TS), Dir/Plans, Jul – Dec1968 pp 366-68; Senate rprt (U), Background Info Relating to SEA and NVN p 207.

27. Memo (S) Dep SECDEF to Chmn JCS, subj: T-Day Planning, 17 Oct 68; CM-3737-68 (S), 30 Oct 68; Hist (TS), Dir/Plans, Jul – Dec 68 pp 366-68.

28. JCSM-733-68 (TS), 13 Dec 68.

29. Memo (S), Asst Dir/Plans for Jt and NSC Matters to CSAF 13 Nov 68, subj: Comm T-day Planning; JCSM-684-68(S), 19 Nov 68.

30. Memo (S) Dep SECDEF to Mil Depts, et. al, subj: T-Day Planning, 18 Dec 68.

31. Memo (S)., Dir/Plans to CSAF, subj: Briefing on "Withdrawal Planning VNAF Improvement and Modernization, 29 Nov 68; Talking Paper for Chmn JCS for a mtg with SECDEF on 2 Dec 68 (TS). subj: T Day Planning and RVNAF Progress, 2 Dec 68.

Chapter VIII

1. Clifford "A Vietnam Reappraisal " p 60lff; N.Y. Times, 24 Oct 68; Wash Post 24 and 31 Oct 68; Christian Science Monitor, 17 Oct 68.

2. Wash Post, 24 and 31 Oct and 1 Nov 68; Life, 15 Nov 68, pp 84a - 92.

3. Public Papers of the Presidents, Lyndon B. Johnson, 1968-69, II p 1100

4. Ibid.

5. N.Y. Times,1Nov 68; Senate rprt (U), Background Info Relating SEA and VN (5th rev ed, Mar 1969), p 66.

6. Intv (U), Peter Hackes, NBC News with Secy Brown, 4 Nov 68; Selected Stmts on Vietnam by DOD and Other Administration Officials, 1Jul - 31 Dec 68 (prepared by Research and Anlys Div, OSAF), p 119.

7. Phila Bull, 8 and 12 Nov 68 memo (S), Ellis to CSAF, subj: A Briefing by Brig Gen Keegan, DCS/Intel, 7th AF, 19 Dec 68.

8. Phila Inquirer, 5 Nov 68; N.Y. Times, 16 Nov 68.

9. Rprt (S), SEA Air Ops, Nov 68, p 7;. Hist (TS), MACV, 1968, vol I pp 371 – 401, 431 - 33.

10. Wash Post, 8 Nov 68; Chicago Trib, 8 and 10 Nov 68; Balt Sun. 12 Nov 68; N.Y. Times, 14 Nov 68; transcript of SECDEF News Conference, 10 Nov 68, in SAFOI CM-3800-68 (TS), 2 Dec 68; OSAF AF News Svc, Feature No 11-15-68-278F, subj: USA VN Battle Rprt.

11. Hist.(TS), Dir/Plans, Jan-Jun 1968, p 69; JCSM-741-68 (TS), 12 Dec 68; JCSM-742-68 (TS), 13 Dec 68; CM-3800-68 (TS), 2 Dec 68; USAF Mgt Digest, SEA 10 Jan 69, p 6.

12. Memo (S), Shotts to CSAF, subj: Estb of an Adequate NYN Recon Prog, 20 Nov 68; CSAFM-K-33-68 (TS). 20 Nov 68; Balt Sun. 6 Nov 68.

13. Ibid.

14. JCS 2472/400 (TS), 12 Dec 68; rprt (S)n SEA Air Ops, Dec 68, p 7; USAF Mgt Summary, SEA (S), 10 Jan 69, p 28.

Notes For Part Six

28. Memo (S), Lt Col Billy J. Moore, Chief War Planning Div to Asst for Jt and NSC Matters, 30 Jun 68; memo (S), Dep SECDEF to 25 Jun 68; Dep SECDEF to Secys of Mil Depts, et. al., subj: Improvement and Modernization, 30 Jun 68.

29. Memo (S), Brig Gen Leslie W. Bray, Jr. Asst Dir/Plans for Jt and NSC Matters to CSAF, subj: Prog to Give Greater Recognition to RVN, 10 Jul 68; JCS 2472/313-1 (S), 11 Jul 68.

30. Memo (TS), Lt Col Robert B. Kenworthy, Intl Affairs Div, Dir/Plans to CSAF, subj: Negotiations on SEA, 15 Apr 68; memo (S), Ellis to CSAF, 18 Apr 68; SM-278-68 (S), 23 Apr 68; memo (TS), Dep SECDEF to Chmn JCS subj: Redeployment of Forces, 26 Apr 68; memo (TS), to CSAF, subj: A Concept for VN During Negotiations, 7 May 68.

31. Memo (TS) Spencer to CSAF, subj: JCS Discussions with SECDEF atch Talking Papers on Bombing of NVN, 3 May 68.

32. Ibid; JCSM-289-68. (TS), 8 May 68

33. Memo (TS), Spencer to CSAF, subj: A Concept for Negotiations, 7 May 68, .with atch Talking Papers on JCS 2472/282-1, 7 May 68; JCSM-290-68 (TS), 8 May 68.

34. Ltr (TS), Chmn JCS to Amb W. Averell Harriman, no subj, 3 May 68.

(Defense Lion Publications notes that two pages in the original are missing.)

21. Memo (S), Ofc, Eastern Regional Div, Dep Dir/Plans for Plcy to Asst Dir/Plans for Jt and NSC Matters, subj: RVNAF Improvement and Modernization, 17 Oct 68; JCSM-633-68 (S). 25 Oct 68.

22. Memo (S), Dep SECDEF to Secys of Mil Depts, et. al., 1Nov 68.

23. Memo (S), Dep SECDEF to Chmn JCS, et. al., subj: T-Day Planning, Hist (TS), Dir/Plans, Jul-Dec 68, pp 366-68.

24. Hist (TS), Dir/Plans, Jul-Dec 1968, pp 366-68; JCSM-531-68 (TS), 3 Sep 68

25. Memo (S), Under SAF to SECDEF, subj: T-Day Planning, 2 Oct 68; Hist (TS), Dir/Plans, Jul-Dec 1968, pp 366-68.

26. Memo (S), Dep SECDEF to Chmn JCS, subj: T-Day Planning, 17 Oct 68; CM-3737-68 (S), 30 Oct 68; Hist (TS), Dir/Plans, Jul – Dec1968 pp 366-68; Senate rprt (U), Background Info Relating to SEA and NVN p 207.

27. Memo (S) Dep SECDEF to Chmn JCS, subj: T-Day Planning, 17 Oct 68; CM-3737-68 (S), 30 Oct 68; Hist (TS), Dir/Plans, Jul – Dec 68 pp 366-68.

28. JCSM-733-68 (TS), 13 Dec 68.

Air War – Vietnam

29. Memo (S), Asst Dir/Plans for Jt and NSC Matters to CSAF 13 Nov 68, subj: Comm T-day Planning; JCSM-684-68(S), 19 Nov 68.
30. Memo (S) Dep SECDEF to Mil Depts, et. al, subj: T-Day Planning, 18 Dec 68.
31. Memo (S)., Dir/Plans to CSAF, subj: Briefing on "Withdrawal Planning VNAF Improvement and Modernization, 29 Nov 68; Talking Paper for Chmn JCS for a mtg with SECDEF on 2 Dec 68 (TS). subj: T Day Planning and RVNAF Progress, 2 Dec 68.

Chapter VIII

1. Clifford "A Vietnam Reappraisal " p 60lff; N.Y. Times, 24 Oct 68; Wash Post 24 and 31 Oct 68; Christian Science Monitor, 17 Oct 68.
2. Wash Post, 24 and 31 Oct and 1 Nov 68; Life, 15 Nov 68, pp 84a - 92.
3. Public Papers of the Presidents, Lyndon B. Johnson, 1968-69, II p 1100
4. Ibid.
5. N.Y. Times,1Nov 68; Senate rprt (U), Background Info Relating SEA and VN (5th rev ed, Mar 1969), p 66.
6. Intv (U), Peter Hackes, NBC News with Secy Brown, 4 Nov 68; Selected Stmts on Vietnam by DOD and Other Administration Officials, 1Jul - 31 Dec 68 (prepared by Research and Anlys Div, OSAF), p 119.
7. Phila Bull, 8 and 12 Nov 68 memo (S), Ellis to CSAF, subj: A Briefing by Brig Gen Keegan, DCS/Intel, 7th AF, 19 Dec 68.
8. Phila Inquirer, 5 Nov 68; N.Y. Times, 16 Nov 68.
9. Rprt (S), SEA Air Ops, Nov 68, p 7;. Hist (TS), MACV, 1968, vol I pp 371 – 401, 431 - 33.
10. Wash Post, 8 Nov 68; Chicago Trib, 8 and 10 Nov 68; Balt Sun. 12 Nov 68; N.Y. Times, 14 Nov 68; transcript of SECDEF News Conference, 10 Nov 68, in SAFOI CM-3800-68 (TS), 2 Dec 68; OSAF AF News Svc, Feature No 11-15-68-278F, subj: USA VN Battle Rprt.
11. Hist.(TS), Dir/Plans, Jan-Jun 1968, p 69; JCSM-741-68 (TS), 12 Dec 68; JCSM-742-68 (TS), 13 Dec 68; CM-3800-68 (TS), 2 Dec 68; USAF Mgt Digest, SEA 10 Jan 69, p 6.
12. Memo (S), Shotts to CSAF, subj: Estb of an Adequate NYN Recon Prog, 20 Nov 68; CSAFM-K-33-68 (TS). 20 Nov 68; Balt Sun. 6 Nov 68.
13. Ibid.
14. JCS 2472/400 (TS), 12 Dec 68; rprt (S)n SEA Air Ops, Dec 68, p 7; USAF Mgt Summary, SEA (S), 10 Jan 69, p 28.

Notes For Part Six

15. Rprt (S). SEA Air Ops, Nov 1968, pp 7 - 23 and Dec 1968, pp 6 and 16; CSAFM-I-10-68 (S), 9 Sep 68; Hist (S), MACV, 1968, vol I, p 409.

16. Memo (S), Ellis to CSAF, subj: The Interdiction Campaign in Laos, 20 Dec 68; memo (S). J-3 Background Paper 48-68, for Chmn JCS for a mtg with SECDEF on 23 Dec 68, 20 Dec 68; memo (S), Ellis to CSAF, subj: Briefing by Capt Mills. Ofc of Asst DCS/I; on Effect of 1 Nov Bombing Suspension, 12 Dec 68.

17. Ibid.

18. Hist (S), Dir/Ops, Jul-Dec 1968, pp 360 - 61; Lt Gen Glenn W. Martin, DCS/P&O to Asst Dir/Plans for Jt and NSC Matters, subj: Arc Light Follow-on Study, 16 Aug 68; memo (S), Dep SECDEF to Chmn JCS, subj: Arc Light Sortie Rate, 26 Nov 68; memo (S), Ellis to CSAF. subj: Arc Light Sortie Rate with atch Background Paper on B-52 Sortie Rate, 3 Dec 68.

19. Hist (S), Dir/Ops, Jul-Dec 1968, pp 360-62; JCSM-711-68 (S), 4 Dec 68 CM-3805-68 (S), 4 Dec 68.

20. Memo (S). Dep SECDEF to Chmn JCS, 9 Dec 68, subj: Arc Light Sorties.

21. Hist (S) Dir/Ops Jul - Dec 1968, p 363; Hist (S). Dir/Plans. Jul – Dec 1968 pp 200 – 01

22. Hist (S) CPAC, 1968, vol III, pp 79-80; Senate rprt (U), Background Info Relating to SEA and VN, p 207.

23. JCSM-678-68 (S), with atch Phase II Plan for RVNAF Modernization and Improvement, vols I, II and III, 15 Nov 68.

24. Memo (S) Dep SECDEF to Secys of Mil Depts, et. al., no subj, 18 Dec 68

25. Ltr (S), Hq MACV to CINCPAC, JCS, et. al., subj: RVNAF Improvement and Modernization, Phase II, 25 Dec 68; Hist (S), CINCPAC, 1968 vol III, pp 80-81; memo (S), Ellis to CSAF, subj: Phase II RVNAF Force Structure with atch Talking and Background Papers, 2 Jan 69.

26. Memo (S) Ellis to CSAF, subj: Chmn JCS Paper on RVNAF Improvement and Modernization Prog, 20 Dec 68; memo (S), Moore to Asst Dir/Plans for Jt and NSC Matters, subj: RVNAF Phase II, 29 Dec 68; Background Paper for Chmn JCS for a mtg with SECDEF, on 23 Dec 68, subj: Mil Situation in SVN; Hist (S), CINCPAC, 1968, vol Ill, pp 80-81; memo (S) Ellis to CSAF, subj: Phase II RVNAF Force Structure. With Talking and Background Papers, 2 Jan 69.

27. Memo (S) Ellis to CSAF, subj: Chmn JCS Paper on RVNAF Improvement and Modernization Prog, 20 Dec 68.

Plans and Operations 1965

www.ingramcontent.com/pod-product-compliance
Lightning Source LLC
Chambersburg PA
CBHW020631230426
43665CB00008B/123